Springer Series in Optical Sciences Volume 32
Edited by Arthur L. Schawlow

Springer Series in Optical Sciences

1 **Solid-State Laser Engineering**
By. W. Koechner

2 **Table of Laser Lines in Gases and Vapors**
3rd Edition
By R. Beck, W. Englisch, and K. Gürs

3 **Tunable Lasers and Applications**
Editors: A. Mooradian, T. Jaeger, and P. Stokseth

4 **Nonlinear Laser Spectroscopy**
By V. S. Letokhov and V. P. Chebotayev

5 **Optics and Lasers**
An Engineering Physics Approach
By M. Young

6 **Photoelectron Statistics**
With Applications to Spectroscopy and Optical Communication
By B. Saleh

7 **Laser Spectroscopy III**
Editors: J. L. Hall and J. L. Carlsten

8 **Frontiers in Visual Science**
Editors: S. J. Cool and E. J. Smith III

9 **High-Power Lasers and Applications**
2nd Printing
Editors: K.-L. Kompa and H. Walther

10 **Detection of Optical and Infrared Radiation**
2nd Printing
By R. H. Kingston

11 **Matrix Theory of Photoelasticity**
By P. S. Theocaris and E. E. Gdoutos

12 **The Monte Carlo Method in Atmospheric Optics**
By G. I. Marchuk, G. A. Mikhailov, M. A. Nazaraliev, R. A. Darbinian, B. A. Kargin, and B. S. Elepov

13 **Physiological Optics**
By Y. Le Grand and S. G. El Hage

14 **Laser Crystals** Physics and Properties
By A. A. Kaminskii

15 **X-Ray Spectroscopy**
By B. K. Agarwal

16 **Holographic Interferometry**
From the Scope of Deformation Analysis of Opaque Bodies
By W. Schumann and M. Dubas

17 **Nonlinear Optics of Free Atoms and Molecules**
By D. C. Hanna, M. A. Yuratich, D. Cotter

18 **Holography in Medicine and Biology**
Editor: G. von Bally

19 **Color Theory and Its Application in Art and Design**
By G. A. Agoston

20 **Interferometry by Holography**
By Yu. I. Ostrovsky, M. M. Butusov, G. V. Ostrovskaya

21 **Laser Spectroscopy IV**
Editors: H. Walther, K. W. Rothe

22 **Lasers in Photomedicine and Photobiology**
Editors: R. Pratesi and C. A. Sacchi

23 **Vertebrate Photoreceptor Optics**
Editors: J. M. Enoch and F. L. Tobey, Jr.

24 **Optical Fiber Systems and Their Components**
An Introduction
By A. B. Sharma, S. J. Halme, and M. M. Butusov

25 **High Peak Power Nd : Glass Laser Systems**
By D. C. Brown

26 **Lasers and Applications**
Editors: W. O. N. Guimaraes, C. T. Lin, and A. Mooradian

27 **Color Measurement** Theme and Variations
By D. L. MacAdam

28 **Modular Optical Design**
By O. N. Stavroudis

29 **Inverse Problems in Laser Sounding of the Atmosphere**
By V. E. Zuev and I. E. Naats

30 **Laser Spectroscopy V**
Editors: A. R. W. McKellar, T. Oka, and B. P. Stoicheff

31 **Optics in Biomedical Sciences**
Editors: G. von Bally and P. Greguss

32 **Fiber-Optic Rotation Sensors and Related Technologies**
Editors: S. Ezekiel and H. J. Arditty

33 **Integrated Optics: Theory and Technology**
By R.G. Hunsperger

34 **The High Power Iodine Laser**
By G.A. Brederlow, E.E. Fill, and K.J. Witte

35 **Engineering Optics**
By K. Iizuka

36 **Transmission Electron Microscopy**
By L. Reimer

37 **Opto-Acoustic Molecular Spectroscopy**
By V.S. Letokhov and V.P. Zharov

38 **Photon Correlation Techniques**
Editor: E.O. Schulz-DuBois

Fiber-Optic Rotation Sensors

and Related Technologies

Proceedings of the First International Conference
MIT, Cambridge, Mass., USA, November 9–11, 1981

Editors:
S. Ezekiel and H.J. Arditty

With 331 Figures

Springer-Verlag Berlin Heidelberg GmbH 1982

Professor SHAOUL EZEKIEL

Research Laboratory of Electronics, Massachusetts Institute of Technology,
Cambridge, MA 02139 USA

Dr. HERVÉ J. ARDITTY

Thomson-CSF, Laboratoire Central de Recherches, Domaine de Corbeville, B.P. 10,
F-91401 Orsay Cedex, France

DOI 10.1007/978-3-540-39490-7

Library of Congress Cataloging in Publication Data. Main entry under title: Fiber-optic rotation sensors. (Springer series in optical sciences ; v. 32). Bibliography: p. Includes indexes. 1. Optical gyroscopes--Congresses. 2. Fiber optics--Congresses. I. Ezekiel, Shaoul. II. Arditty, H. (Hervé), 1951- . III. Series. TL589.2.06F53 1982 681'.753 82-16971

Originally published by Springer-Verlag Berlin Heidelberg New York in 1982.

MyCopy version of the original edition 1982

2153/3130-543210

www.springer.com/mycopy

Preface

Currently there is considerable interest in the application of optical methods for the measurement of absolute rotation. Active approaches, so-called ring laser gyros, have been under serious development for at least 15 years. More recently, passive approaches using ring resonators or multiturn fiber interferometers have also demonstrated much promise. The only previous conference devoted exclusively to optical rotation sensors, held in 1978 in San Diego, California, was organized by the Society of Photo-optical Instrumentation Engineers(S.P.I.E.). Although the main emphasis at that conference was on ring laser gyros, a number of papers were also included that described the early development of fiber gyroscopes. Since then the field of fiber optic rotation sensors has grown so rapidly that a conference devoted primarily to this subject was needed.

The First International Conference on Fiber-Optic Rotation Sensors was held at the Massachusetts Institute of Technology, Cambridge, Massachusetts, November 9-11, 1981. The purpose of the conference was to bring together the many researchers and interested personnel from universities, industry, and government to discuss and exchange ideas on the many recent developments in fiber-optic rotation sensors and related technologies. The program consisted of tutorial papers as well as invited and contributed papers.

The conference was attended by 180 scientists and engineers from a number of countries including Australia, Canada, England, France, Germany (FRG), Israel, Japan, Scotland, Sweden, and the United States of America. Unfortunately, scientists from Italy and the Peoples' Republic of China who were to present papers at the conference were unable to attend. However, their papers are included in this proceedings.

A number of people have contributed to the success of the conference. We would especially like to thank the members of the program committee, F. Aronowitz (USA), B. Culshaw (England), W.C. Goss (USA), R.W. McAdory (USA), H.J. Shaw (USA), H. Taylor (USA), Y. Ueno (Japan), R. Ulrich (Germany), for their help in organizing the program. We are extremely indebted to the staff of the MIT Research Laboratory of Electronics, particularly Debi Lauricella, Barbara McCarthy, and Virginia Lauricella, for attending to all the problems of running a conference and doing it so cheerfully and so well.

Summer 1982 *S. Ezekiel · H.J. Arditty*

Contents

4. Fiber Optic Rotation Sensor Systems

4.1 Open-Loop Operation

4.2 Closed-Loop Operation

5. Limiting Factors

5.1 Non-Reciprocal Error Sources

5.2 Detection Noise

6. Advanced Concepts

Part 1

Tutorial Review

Fiber-Optic Rotation Sensors. Tutorial Review

S. Ezekiel

Research Laboratory of Electronics, Massachusetts Institute of Technology
Cambridge, MA 02139, USA

H.J. Arditty

Thomson CSF, Laboratoire Central de Recherches, BP 10, Domaine de Corbeville
F-91401 Orsay, France

Why the Interest?

The measurement of rotation is of considerable interest in a number of areas. For example, inertial navigation systems as used in aircraft and spacecraft depend critically on accurate inertial rotation sensors. The allowable errors in rotation sensor performance depend on the particular application. Typical requirements for aircraft navigation lie between 0.01 and 0.001 degree/hour. In terms of earth rotation rate Ω_E = 15 degrees/hour this becomes 10^{-3} to 10^{-4} Ω_E. A number of other applications of rotation sensors exist such as surveying where the accurate determination of azimuth and geodetic latitude is important [1]. In this case performance of 10^{-6} Ω_E or better is needed. Geophysics applications include the determination of astronomical latitude, and the monitoring of polar motion caused by wobble, rotation, precession and wandering effects [1]. A highly precise rotation sensor may be used to measure any changes in the length of the day and to detect torsional oscillations in the earth caused by earthquakes. Finally, ultraprecise sensors may find applications in relativity related experiments such as the determination of the preferred frame, dragging of inertial frames, etc. [2].

Methods of Rotation Sensing

The popular sensor of rotation over the past few decades has been the mechanical gyroscope which depends for its operation on the high angular momentum generated by a spinning wheel or a spinning ball. The angular momentum of spinning nuclei has also been investigated for use as a rotation sensor [3]. The advent of the laser in 1960 rekindled the interest in the use of the Sagnac [4],[5] effect for the sensing of inertial rotation by purely optical means. The so-called ring laser "gyroscope" [6] which has been under active development for almost 2 decades has succeeded in achieving inertial grade performance and has recently been selected for use in the new Boeing 757 and

767 aircrafts and also in other navigation systems. More recently the availability of low loss single mode fibers opened up a very active and very promising area of research in fiberoptic rotation sensors also based on the Sagnac effect [7].

Interest in Optical Rotation Sensors

The advantages of optical "gyroscopes" over mechanical ones include the absence of moving parts, warm up time and g-sensitivity. However, it is the promise of the projected low cost of the optical devices that is driving their development.

In the next few sections we will give a brief and simple derivation of the Sagnac effect in a vacuum and also in a medium; discuss techniques of implementing the Sagnac effect for the measurement of rotation together with the fundamental limits on sensitivity in each case. The basic principles of fiberoptic rotation sensors will then be considered with emphasis on techniques and problem areas.

Sagnac Effect in a Vacuum

All the optical rotation sensors under development are based on the Sagnac effect [5] which generates an optical path difference ΔL that is proportional to a rotation rate Ω. For example, if we have a disc of radius R that is rotating about an axis perpendicular to the plane of the disc with angular velocity Ω, as shown in Fig. 1, the optical path difference, ΔL, experienced by light propagating in opposite directions along the perimeter is given by

$$\Delta L = \frac{4A}{c_0} \Omega \tag{1}$$

where A is the area enclosed by the path, i.e., $A = \pi R^2$ and c_0 is the velocity of light in vacuum.

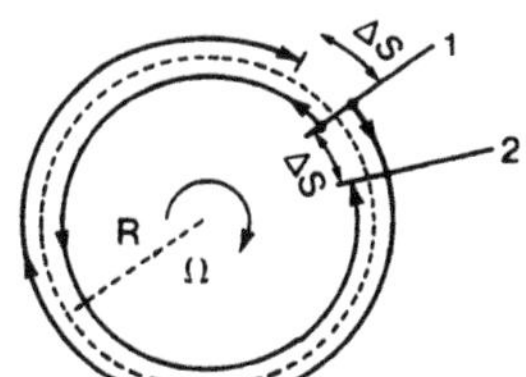

Fig.1 Sagnac effect

The rigorous derivation of this formula is based on the propagation of light in a rotating frame [5], i.e., an accelerating frame of reference, where the general theory of relativity must be used to perform the calculation. However, a simple way of explaining [5] this formula is to consider the rotating disc in Fig. 1. At a given point on the perimeter, designated by 1 in Fig. 1, identical photons are sent in clockwise and counterclockwise directions along the perimeter. If $\Omega = 0$, the photons which travel at the speed of light c_0, will arrive at the starting point 1 after covering an identical distance $2\pi R$ in a time $t = 2\pi R/c_0$. Now in the presence of the angular rotation Ω, the ccw photon will arrive at the starting point on the disc, now located at position 2, after covering a distance L_{ccw} which is shorter than the perimeter $2\pi R$ given by

$$L_{ccw} = 2\pi R - R\,\Omega t_{ccw} = c_{ccw}\, t_{ccw} \quad (2)$$

where $R\Omega$ is the tangential velocity of the ring and t_{ccw} is the time taken to cover the distance L_{ccw}. In addition, L_{ccw} is also given by the product of the velocity of light c_{ccw} in the ccw direction and t_{ccw}. For propagation in a vacuum $c_{ccw} = c_o$. Similarly, the photons propagating in the cw direction experience a larger perimeter L_{cw} given by

$$L_{cw} = 2\pi R + R\Omega\, t_{cw} = c_{cw}\, t_{cw} \quad (3)$$

Using (2) and (3) we can solve for t_{cw} and t_{ccw} as given in Fig. 1 so that the difference Δt between clockwise and counterclockwise propagation times becomes

$$\Delta t = t_{cw} - t_{ccw} \approx \frac{2\pi R \cdot 2R\Omega}{c_o^2} = \frac{4\pi R^2 \Omega}{c_o^2} = \frac{4A}{c_o^2}\,\Omega \quad (4)$$

The path length ΔL travelled by light in a time Δt is therefore given by

$$\Delta L = c_o\, \Delta t = \frac{4A}{c_o}\,\Omega \quad (5)$$

Sagnac Effect in a Medium

In the case of light propagation in a medium [5] of index n, the velocity of propagation must take into consideration the relativistic addition of the velocity of light in the medium, i.e., c_o/n and the tangential velocity of the medium, i.e., $R\Omega$ so that c_{cw} becomes

$$c_{cw} = \frac{\frac{c_o}{n} + R\Omega}{1 + \frac{R\Omega}{nc_o^2}} = \frac{c_o}{n} + R\Omega\,(1 - 1/n^2) + \ldots \quad (6)$$

to first order in v/c. Similarly c_{cw} is given by

$$c_{ccw} = \frac{\frac{c_o}{n} - R\Omega}{1 - \frac{R\Omega}{nc_o^2}} = \frac{c_o}{n} - R\Omega\,(1 - 1/n^2) + \ldots \quad (7)$$

Therefore Δt in a medium becomes

$$\Delta t = t_{cw} - t_{ccw}$$

$$= 2\pi R\; \frac{2R\Omega - (c_{cw} - c_{ccw})}{c_{cw}\, c_{ccw}} \quad (8)$$

Upon substitution for $c_{cw} - c_{ccw}$ from above, we get

$$\Delta t = 2\pi R \frac{2R\Omega - 2R\Omega(1 - 1/n^2)}{c_o^2/n^2}$$

$$= 2\pi R \frac{2R\Omega}{c_o^2} = \frac{4A}{c_o^2}\Omega \tag{9}$$

which is identical to that in a vacuum. If the medium is a fiber wound in a coil of N turns, then Δt becomes

$$\Delta t = \frac{4AN}{c_o^2}\Omega \tag{10}$$

This corresponds to a nonreciprocal phase shift $\Delta\phi$, given by

$$\Delta\phi = 2\pi\frac{\Delta t}{\lambda_o/c_o} = 2\pi\frac{\Delta t}{\lambda/c} = \frac{8\pi AN}{\lambda_o c_o}\Omega \tag{11}$$

where λ, c are the wavelength and velocity of light in the medium, respectively. In terms of path length difference we get

$$\Delta L = \frac{\Delta\phi}{2\pi}\lambda_o = \frac{4AN}{c_o}\Omega \tag{12}$$

For a fiber of length L wound in a coil of diameter D, we have

$$A = \frac{\pi D^2}{4} \quad \text{and} \quad N = \frac{L}{\pi D} \quad \text{so that}$$

$$\Delta L = \frac{4AN}{c_o}\Omega = \frac{LD}{c_o}\Omega \tag{13}$$

or

$$\Delta\phi = \frac{2\pi LD}{\lambda_o c_o}\Omega \tag{14}$$

How Large is the Sagnac Effect?

In order to get a feel for the magnitude of ΔL, let us assume an area $A = 100\ cm^2$ and a rotation rate of $10^{-3}\ \Omega_E$ (i.e., 0.015°/hr or 7×10^{-8} rad/sec). For a single turn fiber enclosing such an area we get $\Delta L \approx 10^{-15}$ cm. This is not a very large effect considering the diameter of a hydrogen atom is about 10^{-8} cm. Clearly a large number of turns N is necessary to increase the magnitude of ΔL.

Methods of Optical Rotation Sensing

For the sake of completeness we show in Fig. 2 the various schemes for the measurement of ΔL. On the extreme right is the multiturn fiber interferometer method [7] mentioned above. On the extreme left is the ring laser approach [6] and in the middle is the passive resonator approach [8]. In both the active and the passive resonator approaches a nonreciprocal path length difference ΔL due to the Sagnac effect becomes a nonreciprocal change in the resonance frequency of the cavity Δf for cw and ccw propagation where

$$\Delta f = \frac{4A}{\lambda_0 P} \Omega \tag{15}$$

where P is the perimeter of the path. In the active resonator (i.e., ring laser) approach the cw and ccw outputs of the laser have a frequency difference Δf which is automatically generated when the laser is subjected to a rotation. In the case of the passive resonator Δf has to be measured by means of lasers external to the cavity [8],[9].

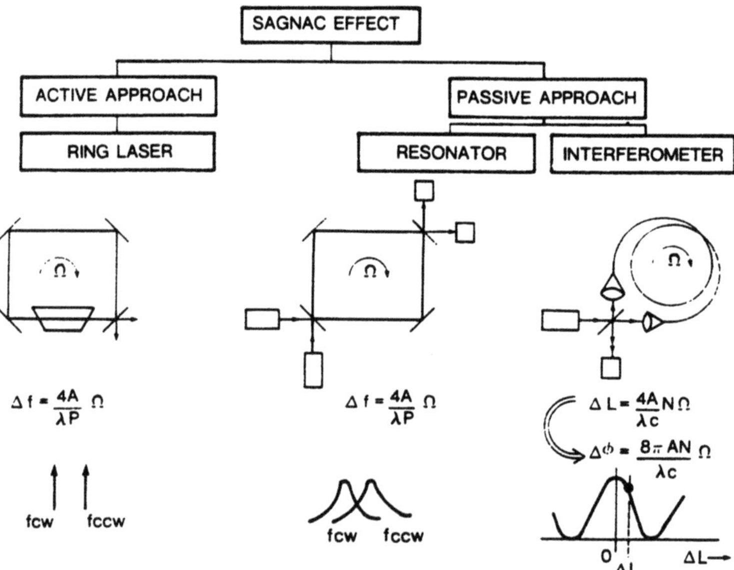

Fig.2 Techniques for measurement of ΔL

Fundamental Limits in Optical Rotation Sensors

In this section we will state without derivation the approximate quantum noise limit for all three cases. For the ring laser, the quantum limit comes from spontaneous emission [10] in the gain medium and gives an uncertainty $\delta\Omega$ in the measurement of Ω given by

$$\delta\Omega \approx \frac{\lambda_o P}{4A} \frac{\Gamma_c}{(n_{ph}\tau)^{1/2}} \tag{16}$$

where Γ_c is the linewidth of the ring laser cavity with no gain; n_{ph} is the number of photons/sec in the laser beam and τ is the averaging time. For the passive resonator case the limit is determined by photon shot noise [8],[9] and is given by

$$\delta\Omega \approx \frac{\lambda_o P}{4A} \frac{\Gamma_c}{(n_{ph}\ \eta_D \tau)^{1/2}} \tag{17}$$

where η_D is the quantum efficiency of the photo detector. As can be seen the passive and active resonator approaches give approximately the same limit. For the multiturn fiber interferometer [11], the photon shot noise limit gives

$$\delta\Omega \approx \frac{c_o}{LD} \frac{\lambda_o/2}{(n_{ph}\ \eta_D \tau)^{1/2}} \tag{18}$$

where n_{ph} is the number of photons/sec leaving the interferometer. All these limits may be compared for A = 100 cm^2; P = 60 cm; λ_o = 6 x 10^{-5} cm; n_{ph} = 3 x 10^{15} photons/sec corresponding to 1 mW; L = 400 m (i.e., N = L/P = 1000); Γ_c = 300 kHz; η_D = 0.3 and τ = 1 sec. Using these parameters we get a $\delta\Omega \approx 4 \times 10^{-4}\ \Omega_E$ or 0.006°/hr for the ring laser case, a $\delta\Omega \approx 7 \times 10^{-4}\ \Omega_E$ or 0.01°/hr for the passive resonator, and a $\delta\Omega \approx 5 \times 10^{-4}\ \Omega_E$ or 0.008°/hr for the fiber interferometer.

Fiberoptic Rotation Sensors

A simple configuration of a multiturn fiberoptic rotation sensor [7] is shown in Fig. 3. Light from a laser or some other suitable light source is divided into 2 beams by a 50-50 beam splitter and then coupled into the two ends of a multiturn single mode fiber coil. The light emerging from the two fiber ends is combined by the beam splitter and detected in a photodetector. In the absence of rotation, the two emerging beams interfere either destructively or constructively depending on the type of beam splitter used. For a 50-50 lossless beam splitter the emerging beams, as

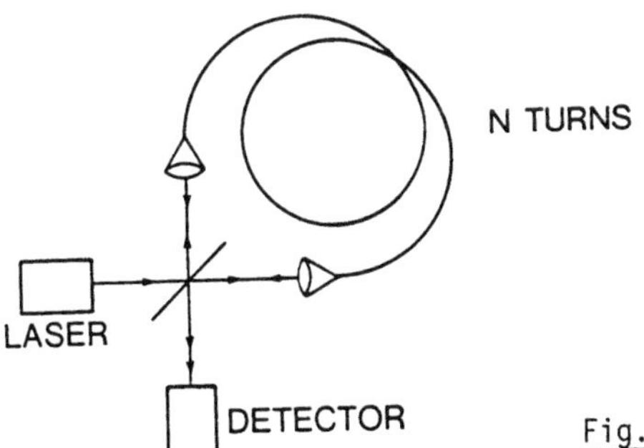

Fig.3 A simple fiberoptic rotation sensor

shown in Fig. 3, interfere destructively. However, if one examines the beams that return back to the light source, one finds them interfering constructively, i.e., at the peak of a fringe. In the presence of a rotation rate Ω, a ΔL will be generated given by

$$\Delta L = L_{cw} - L_{ccw} = \frac{4AN}{c_o}\Omega = \frac{LD}{c_o}\Omega \tag{19}$$

where A, N, L, D have been defined earlier. This ΔL will therefore cause a fringe shift Δz given by

$$\Delta z = \frac{LD}{\lambda_o c_o}\Omega \tag{20}$$

or a phase shift

$$\Delta\phi = \frac{2\pi LD}{\lambda_o c_o}\Omega \tag{21}$$

For $A = \pi D^2/4 = 100$ cm^2 and $N = L/\pi D = 1000$ (i.e., $D \approx 11.3$ cm and $L \approx 355$ meters) and if we assume $\lambda_o = 0.63$ µm, we get a phase shift of 1.3 rad for a rotation rate of 1 rad/sec. Therefore, to detect the full earth rotation, we must measure a phase shift of 9.1×10^{-5} radians and for typical navigation applications ($10^{-3}\ \Omega_E$) the phase shift reduces to about 10^{-7} radians.

For a given size sensor, i.e., a fixed coil diameter D, the sensitivity may be enhanced by increasing the length of the fiber L by winding more turns. Unfortunately one cannot increase L indefinitely because of the finite attenuation in the fiber. Typically, for a fiber attenuation of 1 dB/km, the optimum length is several kilometers.

Photon Shot-Noise Limit in Fiberoptic Gyro

Earlier we gave without proof a comparison of the basic limits to the measurement of rotation using the three techniques. In this section we will derive an approximate formula for the limit in a fiber rotation sensor.

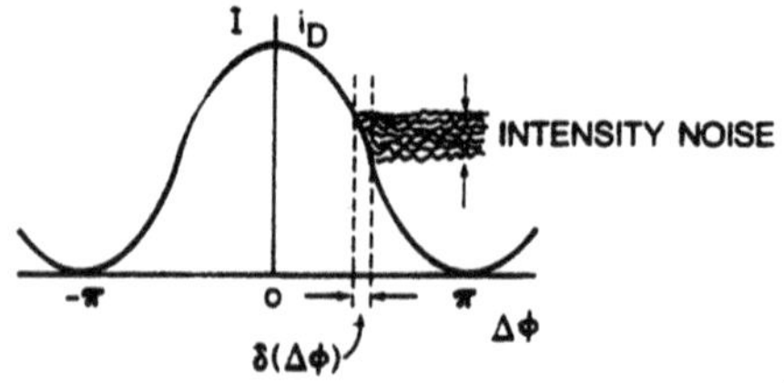

Fig.4 Plot of detector output vs. $\Delta\phi$

Figure 4 shows a plot of intensity I or detector output current i_D versus nonreciprocal phase shift $\Delta\phi$. In this case, the peak intensity, due to constructive interference, is shown centered on $\Delta\phi = 0$ for a zero rotation rate. In the presence of a rotation, $\Delta\phi$ shifts from zero and therefore we get a change in detector current i_D. The greatest change in i_D for a given small change in $\Delta\phi$ clearly occurs at the point on the fringe with the maximum slope, i.e., where $\Delta\phi = \pm\ \pi/2$. Therefore, by applying a fixed nonreciprocal bias of $\pi/2$, we can maintain the operating point where the sensitivity to rotation is a maximum. In this way, an applied rotation causes a $\Delta\phi$ which in turn generates a change in the light intensity at the detector that is proportional to the rotation. A problem arises when the intensity of the light source varies since this cannot be distinguished from a change in intensity due to a rotation. Therefore the uncertainty in the measurement of a given rotation rate, i.e., a given $\Delta\phi$, must be influenced by the intensity noise on the light. Although there are many ways of compensating for intensity variations of the light source, it is not, however, possible to reduce the effect of photon shot noise because it is a random process. Therefore under ideal conditions the uncertainty in the measurement of $\Delta\phi$ is limited only by the photon shot noise[12]. This uncertainty $\delta(\Delta\phi)$ is therefore given by

$$\delta(\Delta\phi) = \frac{\text{photon shot noise}}{\text{fringe slope}} \tag{22}$$

and this is a minimum where the fringe slope is maximum. In other words,

$$\delta(\Delta\phi) \approx \frac{(2e\ i_D B)^{1/2}}{i_D/\pi} \approx \frac{(n_{ph}\ \eta_D \tau)^{1/2}}{n_{ph}\ \eta_D\ \tau/\pi} \tag{23}$$

where e is the electron charge; B is the bandwidth of the detection system; n_{ph} is the number of photons/sec falling on the detector; η_D is the quantum efficiency of the detector and τ is the averaging time = 1/2B. Since

$$\Delta\phi = \frac{2\pi LD}{\lambda_o c_o}\,\Omega$$

the uncertainty in the measurement of Ω, i.e., $\delta\Omega$, becomes

$$\begin{aligned}\delta\Omega &= \frac{\lambda_o c_o}{2\pi LD}\,\delta(\Delta\phi)\\ &= \frac{c_o}{LD}\,\frac{\lambda_o/2}{(n_{ph}\eta_D\tau)^{1/2}} = \frac{c_o}{LD}\,\frac{\lambda_o/2}{(i_D/2\ eB)^{1/2}}\end{aligned} \tag{24}$$

which is the same expression given earlier.

Ideal Performance

The ideal performance of a fiberoptic "gyro" may therefore be described by a random drift that is limited by photon shot noise and a zero bias or drift in the absence of rotation. For a given rotation, the stability of the scale factor, i.e., $2\pi LD/\lambda_o c_o$, which relates Ω to $\Delta\phi$, would be limited by the stability of L, D and λ_o.

Measurement of Nonreciprocal Phase Shift

In order to reach the ideal performance discussed in the previous section, a number of problems must be overcome. In this section we restrict our attention to the measurement of nonreciprocal phase shift $\Delta\phi$ with an uncertainty that is limited only by the photon shot noise.

A simple way of measuring $\Delta\phi$ is illustrated in Fig. 5 where a $\pi/2$ bias is applied so as to operate at the point of maximum slope. In this way an increase in intensity corresponds to a negative $\Delta\phi$ and vice versa. Among the disadvantages of this method is the stability of the bias and the need to compensate for laser intensity fluctuations. A better method might be to employ a differential scheme in which two detectors are placed astride a fringe as shown in the lower diagram of Fig. 5. This scheme has twice the sensitivity of the first, and better discrimination against intensity variations. However, it still suffers from the instability of the operating points and requires high common mode rejection.

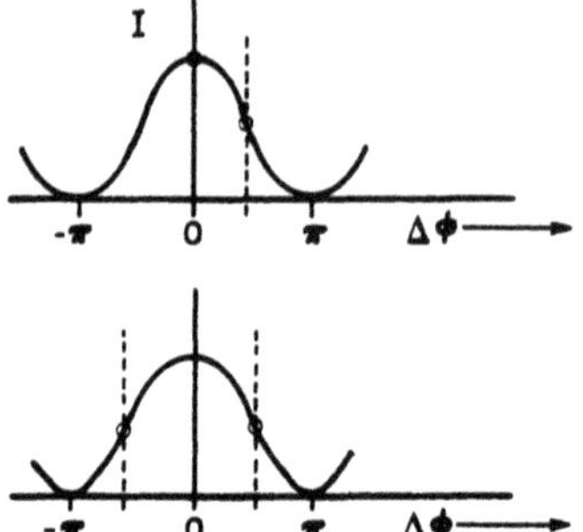

Fig.5 Measurement of $\Delta\phi$ - d.c. method

A much better method is to use an a.c. modulation scheme employing non-reciprocal phase dither [11] as shown in Fig. 6. The requirements for optimum operation are that the amplitude of the phase modulation should be $\pm\,\pi/2$ and the rate of the modulation should be high enough so that the detector noise is dominated by photon shot noise. Figure 7 is a sketch of a typical noise spectrum of a laser showing the large "1/f" noise component at low frequencies. The start of the shot noise limited region depends on the particular light source and can be anywhere from a few kHz to a few hundred kHz. Using such a modulation scheme, the output of the photodetector is demodulated in a phase sensitive demodulator followed by a low pass filter. In this way one obtains a zero output at $\Delta\phi = 0$, a positive voltage for $\Delta\phi < 0$ and a negative voltage for $\Delta\phi > 0$. The main advantages of the modulation method are that the peak of the interference pattern is used as the reference point (i.e., no need for external offset) and that the null point is independent of intensity fluctuations as long as the modulation rate is high enough as mentioned above.

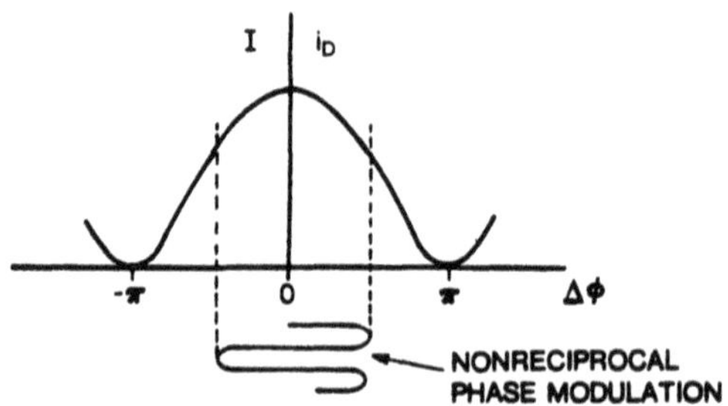

Fig.6 Measurement of $\Delta\phi$ - a.c. modulation method

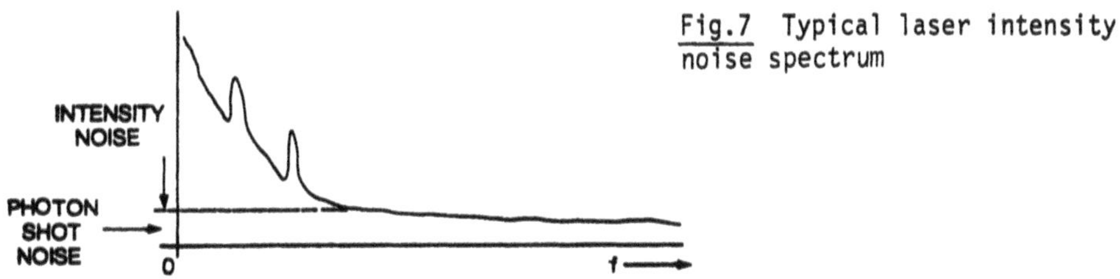

Fig.7 Typical laser intensity noise spectrum

Methods of Nonreciprocal Phase Modulation

In this section we discuss the various possibilities for achieving nonreciprocal phase modulation at high rates. We start with the phase shift ϕ that light experiences in propagating along a single mode fiber of length L and refractive index n given by

$$\phi = \frac{2\pi f n L}{c_0} \tag{25}$$

where f is the frequency of the light in Hz. It should be noted that the magnitude of ϕ does not depend on the direction of propagation so that $\phi_{cw} = \phi_{ccw} = \phi$. This implies that if we vary L or n or f we cannot generate a nonreciprocal phase shift.

Therefore in order to generate a nonreciprocal phase shift we must either make $L_{cw} \neq L_{ccw}$ or $n_{cw} \neq n_{ccw}$ or $f_{cw} \neq f_{ccw}$.

One possibility of making $L_{cw} \neq L_{ccw}$ whether in a medium or in a vacuum is to mechanically dither the interferometer at a high angular rate so that the Sagnac effect itself resulting from this motion can provide necessary nonreciprocal phase shift. This method has in fact been used in early fiber gyros [13] but is clearly not very desirable in general.

In order to generate a nonreciprocal refractive index, i.e., $n_{cw} \neq n_{ccw}$ there are a number of methods that could be used. A simple scheme would be to make the polarization of, say, the cw beam orthogonal to the polarization of ccw beam and then use an E/O phase modulator to generate a polarization dependent refractive index. Such a scheme has been discussed and demonstrated [11] but suffers from the fact that orthogonally polarized beams propagating in the fiber do not experience the same index thus generating a temperature dependent bias.

Another method of making $n_{cw} \neq n_{ccw}$ makes use of the Faraday effect either in the main fiber coil or in a separate length of fiber. By applying a longitudinal magnetic field to the fiber it is possible to cause the index for right circularly polarized light to be different from that for left circularly polarized light. Even though the Verdet constant, i.e., the constant of proportionality between magnetic field strength and index difference, is small, it can be enhanced considerably by using a long length of fiber. Again, this scheme has been demonstrated recently [14].

Yet another nonreciprocal index scheme that has enjoyed much popularity is the time delay modulation method [15]-[18] illustrated in Fig. 8. In this case one takes advantage of the comparatively long time the photons spend in the fiber. A typical set-up would be to place a phase modulator near the beam splitter as shown in Fig.8. The phase modulator can be an electrooptic crystal, a fiber wound around a piezo-electric cylinder (PZT) etc. Now if the phase modulator is driven at a frequency f_m it is then possible to generate a nonreciprocal phase shift given in [18]

$$\phi_{cw} - \phi_{ccw} \approx 2\phi_o \sin\left(2\pi f_m \frac{\tau_D}{2}\right) \tag{26}$$

where ϕ_o is the magnitude of the reciprocal phase shift generated by the phase modulator; τ_D is the delay time in the fiber which is given by nL/c_o.

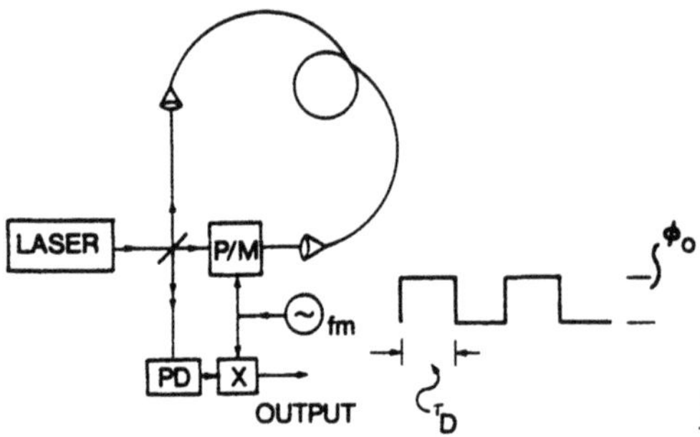

Fig.8 Time delay modulation method

To maximize the nonreciprocal phase shift it is necessary to make the argument of the sin function $\approx \pi/2$, i.e., by choosing $f_m \approx 1/2\tau_D$. If f_m is less than $1/2\tau_D$ then ϕ_o has to be increased to achieve the appropriate amplitude of the nonreciprocal phase shift. For a 1 km long fiber the optimum f_m is about 100 kHz.

Now we turn to the generation of nonreciprocal phase modulation by the frequency method [11][18]. In this scheme f_{cw} is made different from f_{ccw} so that

$$\phi_{cw} - \phi_{ccw} = \frac{2\pi}{c_o} nL \,(f_{cw} - f_{ccw}) \tag{27}$$

A simple way of implementing this method is by employing two acoustooptic (A/O) frequency shifters placed symmetrically on either side of the beam splitter within the interferometer. By driving the A/O with independent oscillators it is possible to generate both nonreciprocal phase modulation as well as fixed nonreciprocal phase shifts. For example, in a 1 km fiber a frequency difference $f_{cw} - f_{ccw}$ of 50 kHz generates a nonreciprocal phase shift of $\pi/2$.

Related frequency methods have also been investigated [19,20].

Open Loop and Closed Loop Operation

The open loop sensor system is shown in Fig. 9 where a nonreciprocal phase modulator NRPM is placed near one fiber end and driven at f_m. The output of the photo detector is then demodulated at f_m in a phase sensitive demodulator. After low pass filtering, the demodulator output is a sinusoidal function of $\Delta\phi$ as illustrated in Fig. 9. For any given $\Delta\phi$, a d.c. voltage output is therefore obtained which is proportional to $\Delta\phi$. The disadvantages of the open loop system include (a) the calibration of the demodulator output since this depends on the gains of the various amplifiers that precede it as well as on the intensity of the light source, and (b) the nonlinear behavior of the demodulator output with $\Delta\phi$.

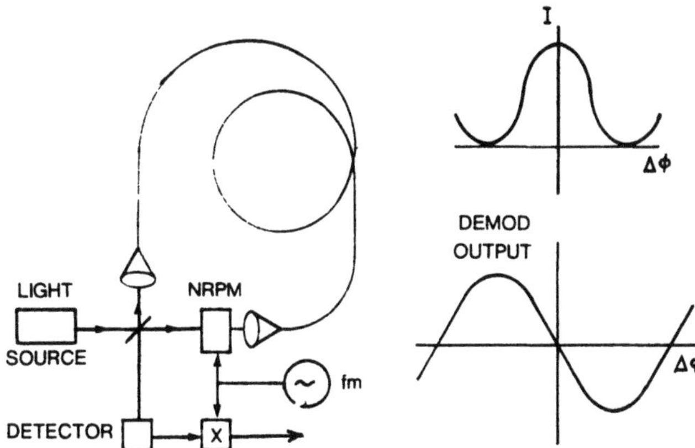

Fig.9 Simplified open loop system

In the closed loop system [11],[18], shown in Fig. 10, the output of the demodulator is passed through a servo amplifier which then drives a nonreciprocal phase transducer (NRPT) placed within the fiber interferometer. In this way, the sensor is always operated at null, i.e., at $\Delta\phi = 0$ by generating a suitable nonreciprocal phase shift in the NRPT that is equal to but opposite in sign to that generated by a rotation Ω. The output of the system is then the output of the NRPT. Therefore the NRPT becomes a critical element.

The advantages of the closed loop system over the open loop system include (a) the output is independent of light source intensity variations since the system is always operated at null. (The modulation frequency must be high enough to reach the photon shot noise.) (b) the output is independent of the gains of individual components in the measurement system as long as a very high open loop gain is maintained, and (c) the output linearity and stability depends only on the NRPT.

The NRPT transducer could, for example, be a Faraday effect device or an acoustooptic frequency shifter. If a Faraday device is used, then the stability depends on the stability of the length of the fiber and the stability of the magnetic field/phase shift transfer function. However, if the NRPT is an acoustooptic crystal then a frequency difference $\Delta f = f_{cw} - f_{ccw}$ is generated to offset a $\Delta\phi = 2\pi LD/\lambda_0 c_0\ \Omega$ caused by a rotation. Therefore we have

$$\Delta\phi = \frac{2\pi}{c_0}\,\Delta f\ nL = \frac{2\pi LD}{\lambda_0 c_0}\,\Omega \tag{28}$$

which implies that

$$\Delta f = \frac{D}{n\lambda_0}\,\Omega \tag{29}$$

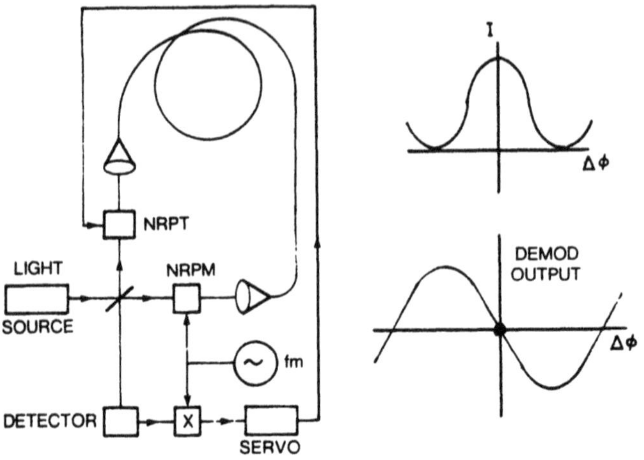

Fig.10 Simplified closed loop system

Eq. (29) indicates that the scale factor stability in this case depends on the coil diameter D and on n and λ_0. If we multiply the numerator and denominator of (29) by $\pi D/4$ we get

$$\Delta f = \frac{\pi D^2/4}{n\lambda_0 \pi D/4}\,\Omega = \frac{4A}{\lambda_0 P}\,\Omega \tag{30}$$

where P is the optical perimeter = $n\pi D$ of the fiber coil. It should be noted that (30) is identical with (15) for either the ring laser or the passive resonator approach.

Reciprocity

In what preceeds, reciprocity has been taken for granted. We will see in the following section why and how it conditions the reliable operation of the Sagnac interferometer.

Although reciprocity has a very general physical sense, we will only consider here its implications with regards to E.M. wave propagation. A two port device is reciprocal if one cannot tell wether one propagates through it in the direct (say left to right) or reverse (say right to left) direction. In other words, a device to which one can apply time reversal (or phase conjugation) without further ado.

At this point one should distinguish between non-reciprocal phenomena and reciprocal phenomena implemented in a way that is not reciprocal.

It turns out that very few phenomena are non-reciprocal in nature. If one excludes :

- non-linear phenomena,
- magnetic phenomena,
- time varying phenomena,
- relativistic phenomena,

all others are reciprocal.

Let us here take time to give some insight on those forms of reciprocity failure.

- Assume an experiment in which a beam propagates in series into a non-linear medium, then in a linear absorber. In the direct direction the non-linear medium sees a full amplitude wave and behaves accordingly. The result is then propagated through the absorber. In the reverse direction the wave is attenuated by the absorber before it reaches the non-linear medium, which then presents totally different characteristics to its propagation : hence non-reciprocity. Linear phenomena do not care. We will see in this conference how the non-linear interaction in the fiber medium can be the source of a measurement error.
- In most magnetic field interactions, the vector representation is 3-dimensional and completely defines the triedron direction. One can then distinguish between the real-world and its mirror image. Reversing the direction of propagation changes the sign of the interaction : hence non-reciprocity. Electric field interactions do not care. As we will see in several papers, magneto-optic interaction is a way of biasing the fiber optic gyroscope.
- In most time varying phenomena, reciprocity is simply not defined because the time at which the direct and reverse experiments are done matters. This non-reciprocity will turn out to be the privileged way of implementing a bias, but also a cause of drift (Schupe).
- The Sagnac effect is the example of a relativistic non-reciprocal phenomenon. Just as for the magnetic field, the rotation completely orient the experiment space and makes it different from its mirror image ; hence the sign reversal of the phase shift associated with the reverse direction of propagation. Another example is worth mentioning: that of stationary flow of matter along the direction of propagation. Again the sign of this flow defines a direction of propagation. Its influence, the Fresnel-Fizeau drag, is non-reciprocal (a drift phenomenon well known in the laser gyroscope).

Now even though all other physical phenomena are reciprocal in nature, one can design an experiment that behaves in a fashion that looks nonreciprocal.

A good example is the half silvered absorbing glass now used to make window panes in warm climates and bad taste sun-glasses everywhere. It certainly looks very non-reciprocal : one can see through it from inside with very little loss and people outside cannot see anything.

None of the bona-fide non-reciprocal phenomena is present. In fact, a properly conducted experiment shows that there is no transmission coefficient difference between the direct (absorbing glass then reflecting layer) and the reverse (reflecting layer then absorbing glass) directions. The difference is that the reflection coefficient depends very much on the side chosen : the signal (transmitted) to noise (reflection) ratio is much larger in the direct direction (poor reflection coefficient of a fairly dark environment) than in the reverse direction (high reflection coefficient of a much brighter environment). And reciprocity does not imply that the reflections should be symmetrical.

Why all this fuss about reciprocity ? We are setting up to detect and measure slow varying interferometric path differences on the order of 10^{-15} m over path lengths of 10^{3} m in a physical medium. This 10^{18} ratio is clearly not realistic in the general case, if only because of the path thermal expansion. The most thermally stable material exibits expansion coefficients in the $10^{-7}/°C$ range. Does a 10^{-11} °C temperature regulation mean anything ? Well our only chance is that thermal expansion, like essentially all the other parasitic effects, is reciprocal while the Sagnac effect is not. The game will be to use reciprocity to identically cancel the large unwanted drifts, to detect and measure reliably the small rotation induced phase change.

The Sagnac interferometer is clearly roughly reciprocal in spirit. Because its two waves propagate in opposite directions, through one and the same light path, reciprocal perturbations of this light path cancel out while the non-reciprocal Sagnac effect is doubled.

Recombining two waves with strictly identical histories, i.e. strictly in phase, one expects to obtain exactly constructive interferences. However, a simple experimentation of the set-up proposed by Sagnac yields very disturbing results : it shows a ring shaped two-wave (raised sine) interference pattern [35] which shifts in and out in phase with applied inertial rotation but also drifts both in time and with temperature, (Fig. 11).

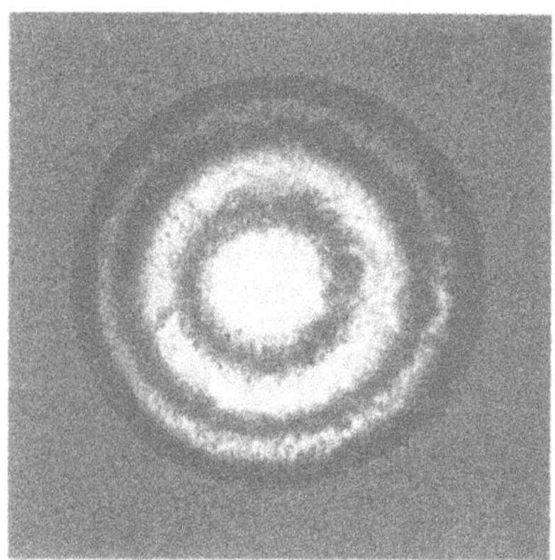

Fig. 11 Sagnac interferometer output example

An intensity cross-section analysis shows shifting regions of constructive and destructive interferences (Fig. 12).

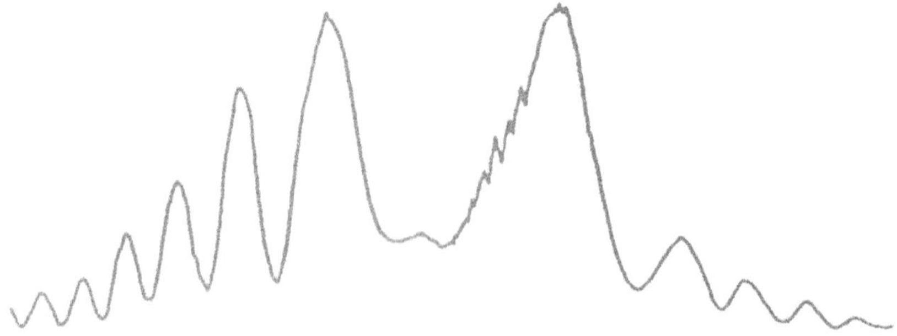

Fig. 12 Intensity cross section through the interference pattern at rest

Let us review the reasons for this seemingly non-reciprocal behaviour.

The first reason is the way the beam-splitter is utilized to separate and recombine the two beams of the Sagnac interferometer.

Consider the classical Sagnac interferometer implementation (Fig. 13) in which only reciprocal phenomena are assumed at this time. The light wave coming from the source hits the beam-splitter where it is separated into two waves which will, from then on, accumulate independent phases. Let us keep track of these respective phases. One of these waves is reflected on the side "a" of the beam-splitter where it experiences a phase shift φ_{ra} It then propagates counter-clock wise though the common light path where it experiences a phase shift φ_1. Upon return on the beam-splitter, one part of that wave is reflected on its side "b" where it experiences a phase shift φ_{rb}. It then proceeds to the detector. The other part of the CCW beam is transmitted through the BS where it experiences a phase shift φ_t, back into the light source.

The second half of the original wave goes through the BS on the way in, experiencing φ_t (the same φ_t as before by reciprocity) then around the loop, experiencing φ_1 (the same φ_1 by reciprocity). Upon return on the beam-splitter, one part transmits through the BS experiencing φ_t (same remark) towards the detector while the other part is reflected on BS side "a", experiencing φ_{ra} (reciprocity again) back to the light source.

Two interference phenomena hence take place :

- One in the detection arm where the two waves have accumulated respectively :

$$\varphi_{ra} + \varphi_1 + \varphi_{rb} \quad \text{and} \quad \varphi_t + \varphi_1 + \varphi_t \tag{31}$$

- and one back into the source arm where the two waves have seen respectively :

$$\varphi_{ra} + \varphi_1 + \varphi_t \quad \text{and} \quad \varphi_t + \varphi_1 + \varphi_{ra} \tag{32}$$

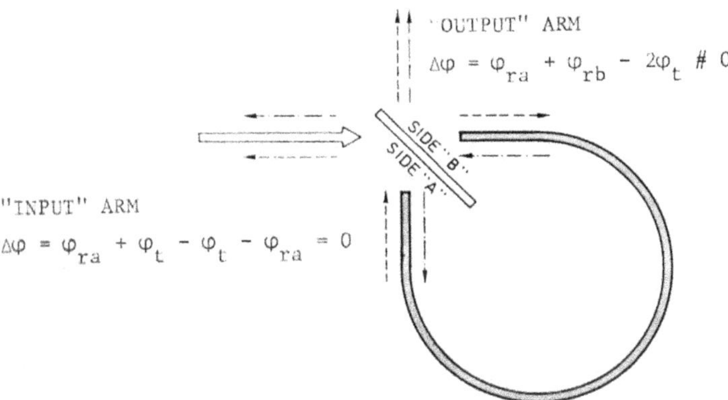

Fig. 13 The two output ports of a Sagnac interferometer

The last two waves have clearly strictly identical phase shifts : their interference is exactly constructive. If one then introduces a non-reciprocal phase shift on φ_1, the power fed back to the light source will be a raised cosine function of that phase shift with strictly no bias.

This holds without any assumption on the type or quality of the beam-splitter, apart from the reciprocity requirement.

On the other hand, the two waves recombining into the detector arm have different phase histories. They differ by :

$$2\varphi_t - \varphi_{ra} + \varphi_{rb} \tag{33}$$

The analysis of the value of this difference is straight-forward :

- if the BS is lossless, a purely theoretical case, one can apply the principle of conversation of energy. What ever is not returned to the source must get to the detector : the interference in the detector arm is the complementary or square sine function of the non-reciprocal phase shift applied. The interference in the detector arm is strictly destructive and the above phase difference is exactely π,
- if, as always is real life, the beam-splitter presents some losses, the energy sent to the detector differs from the ideal square sine function by the loss terms. The interference is no longer strictly destructive, the above phase difference differs from π by a bias which depends on the characteristics of the BS.

Because the BS characteristics depend on time and environment, this non-reciprocity is not acceptable : in order to operate a Sagnac interferometer in a strictly reciprocal fashion, one must not use a detector at its traditional place but some how measure the light returning back to the light source. Among possible technical solutions are the use of a second beam-splitter (for which no special properties are required, not even reciprocity, although operation at close to 3 dB is optimal), the use of a light switch (taking advantage of the finite propagation time to substitute the detector for the source at the appropriate instant) or the utilization of the light source as a detector.

The second reason has to do with degrees of freedom. Light propagation is free space is dealt with using wave equations which have a continuous set of solutions, the number of which is infinite. Although not strictly correct, the same treatement is applied to real life discrete optical systems.

However, when one deals with waveguides, particularly when their transverse dimensions are not very large compared with the wavelength, the wave equation only has discrete solutions, in a finite number.

Each of these few solutions to Maxwell's equations through the waveguide is called a mode and constitutes an independent degree of freedom.

Each of the modes constitutes an independent light path, capable of carrying energy independently of the others and susceptible of experiencing environment related perturbations that differ from what the other modes see. If the CW and CCW lights, although travelling through the same optical system, do so through different modes, reciprocity is clearly not achieved.

One obvious solution is to restrict the entire optical system to operate on one mode only. Although this is in principle possible, in integrated optics for example, it is generally not practical.

Is it possible to utilize a multimode light path in a Sagnac interferometer in a perfectly reciprocal fashion ?

The idea is of course that, since each mode is an independent solution, orthogonal to all others, one mode can be selected for transporting the energy, the others remaining unused. Long multimode systems, however because of perturbations, small in amplitude but acting over long interaction lengths, present unpredictable exchanges of energy between individual modes. Although distributed, these mode couplings behave like ordinary (reciprocal) four-port power dividers. Their first effect is only to deplete the original mode of propagation from some of the energy carried. Because they also work in reverse, however, they bring to this mode light that has travelled elsewhere, hence bearing phase information with a different history, an obvious source of possible non-reciprocity.

Although a more elegant and complete demonstration is quite straighfoward, using for example the matrix mode coupling formalism, let us take an example to illustrate the path to the solution. Fig. 14-a represents an arbitrary four-mode light path which includes eight random perturbations giving rise to local mode coupling. Fig. 14-b illustrates the mechanism described above : the light, entering through mode "one" only, on side "A",

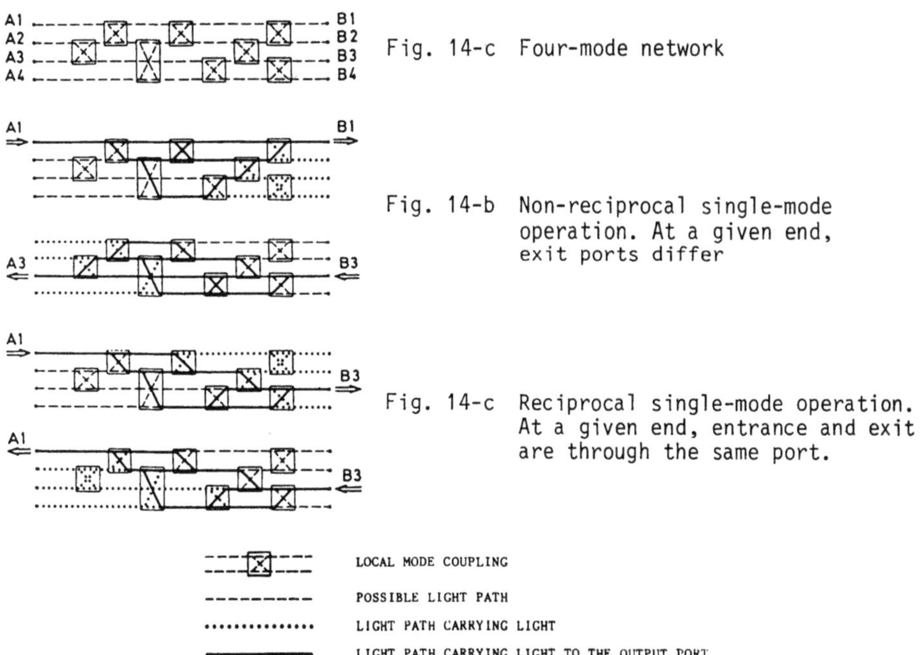

Fig. 14-c Four-mode network

Fig. 14-b Non-reciprocal single-mode operation. At a given end, exit ports differ

Fig. 14-c Reciprocal single-mode operation. At a given end, entrance and exit are through the same port.

Fig. 14 Two-way single-mode operation of a multimode interferometer light path

is divided and recombined at each coupling point to finally excite all four modes of the structure. If one elects to use light exiting through mode "one" on side "B", it is convenient to follow all possible paths for the energy that will ultimately contribute to that output (thick continuous lines). Phase perturbations applied anywhere on one of these paths will influence the phase of the output signal. On the other hand, perturbations applied on paths carrying no light (obvious) or on paths carrying light not contributing to the output (dotted line) have no influence on the output phase. A necessary condition (a more thorough analysis shows that it is also sufficient) to the reciprocal operation is that the thick continuous line maze be the very same for light entering through side "A" and light entering though side "B". This is clearly not achieved if, although in a given direction a same mode is conserved, a different single mode is used in the reverse direction. Fig. 14-c, however, shows how this is always achieved if, at each end, a same mode is utilized for the way-in and the way-out. This holds for any choice of mode, or combinaison of modes, at a given end, independantly of the choice made at the opposite end.

Devices that restrict locally the operation of a light path to a true single mode do exist. They usually consist of a spatial filter (a diffraction limited pin-hole between two lenses) and a polarizer to remove the remaining degeneracy between the E and H fields. A short portion of dielectric single mode waveguide constitutes an excellent mode filter because the ratio between the loss experienced by the guided mode and the loss seen by the (unwanted) non-guided light can be made arbitrarily small at very low energy loss for the guided part.

The combination of the three above conclusions :

- the necessity to use the same arm of the interferometer for input and output,
- the necessity to restrict, at each end of the closed light path, the operation to a same mode for the entry and exit of the light,
- the necessity, once reciprocity is achieved by the above two requirements, to bias and modulate in order to detect and measure the Sagnac phase shift,

leads to an interferometer design somewhat more complicated than the original Sagnac set-up.

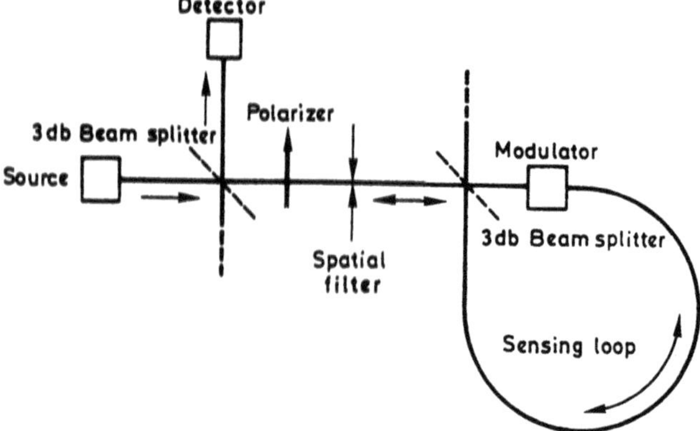

Fig. 15 The minimum reciprocal configuration

This configuration, depicted in Fig. 15, uses two beam-splitters, a single-mode filter situated on the common input-output path between the beam-splitters, and one or several modulators at the end of the closed loop light path. Because it is the simplest implementation that guarantees strict reciprocal operations in non-ideal conditions, we call it the "minimum configuration".

Two comments can be made on this configuration :

- Although no simplification can be made on the "minimum configuration" without destroying its reciprocity, hence rendering the long term measurement of extremely small Sagnac phase shifts impossible, the implementation of simpler set-ups may be justified if the requirements can be limited. This is true of low performance devices which only detect phase shifts commensurate with ordinary mechanical and thermal stability. It can also hold if only AC or at least very short term operation is required.
- The "minimum configuration" is fascinating because it has no requirement for stability what so ever (at least till we get to second order problems). All the constrains for perfection are concentrated in the mode filter, a device that dielectric waveguide allow to construct to an arbitrary rejection ratio. Adjustments are of course needed, if only to assure that a non-zero amount of light returns to the detector to provide an acceptable signal to noise ratio, but perfect reciprocity is acheived no matter how far-off any of the adjustments may be.

Technology

In the past 6 years, a lot of the theoretical problems associated with the fiber optic gyroscope have been uncovered and solved. A few more will be described in this conference : they may well be the only relevant ones left. Today the principal challenge seems to be technology : the problem of constructing at a reasonnable cost a fiber gyroscope of manageable size without degrading too much the theoretically obtainable performances.

A common denominator to all gyros of concern here is the fiber optic loop or coil. Fiber single mode at the wavelength of operation is now widely commercialy available from the telecommunication industry. Although generally of very low losses (less than 8 dB/km in the red, less than 3 dB/km at 0.85 μm and less than 1 dB/km at 1.3 μm), a characteristic desirable both because more energy reaches the detector and because less gets scattered backwards, it lacks some of the features more specific to the fiber gyro. Among these, a high index difference between core and cladding is needed to reduce the extra loss caused by tight winding. Its surface should also be particularly free from defects which make a permamently wound fiber prone to static fatigue and breakage. An aspect which presently receives a lot of attention is that of the stable propagation of a given state of polarization [36], a must in interferometry to prevent fading of the signal. While active schemes [37] have been demonstrated to compensate for the random changes associated with the polarization degeneracy of standard fibers, they are costly to implement and can be dispensed of if polarization conserving fiber is used.

The coil itself may prove to be a problem. Present fiber diameters are such that the volume occupied by a few thousand turns cannot be neglected. The pointing stability required from strap-down gyroscopes will certainly require special winding and potting techniques. The coil need also be protected from magnetic fields and from thermal transients, two recognized sources of non-reciprocities.

The "heart" of the F.O.R.S., the assembly of two beam-splitters, a spatial filter, a polarizer and a modulator, is of course the main topic of technological concern. Once the reciprocal configuration is adopted, the requirements for the gyro heart are very straightforward. It should allow as much as possible of the useful optical signal to ultimately reach the detector, and it should lend itself to non-labour intensive mass production. The only initial performance is then the rejection of the mode filter, both spatially and in polarization, which should ideally reach the hundred dB range to cease to be a limiting factor.

Today, three technologies are competing : miniaturized discrete or bulk optics, all-fiber-optics components and integrated optics. Each one has its own set of advantages but none is today able to solve satisfactorily all of the problems.

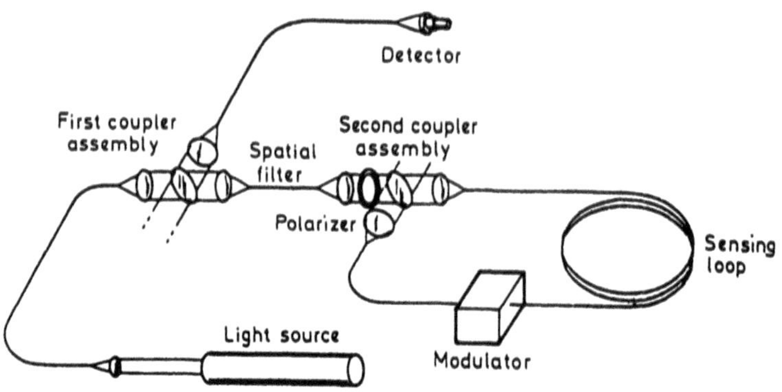

Fig. 16 Bulk optics implementation

The bulk optics approach (Fig. 16), the use of small lenses, mirrors and cube-beam-splitters, requires little progress in technology [20][15]. The components involved have been fabricated industrialy for years, to very high standards and at quite low cost. A definite edge of the bulk optic approach at this time is the availability of acousto-optic modulators with good efficiency and a very high level of fundamental and unwanted side-band rejection, which make quantized closed loop operation possible. Spatial filtering is conveniently performed by a short length of single mode fiber which, combined with a birefringent prism polarizer, provides on adequate level of rejection. The assembly of all these components however, because of the numerous degrees of freedom involved, and of the corresponding number of fine, yet stable, adjustments, is quite labor intensive. The stability and reliability of the completed device is also questionnable. Its size obviously cannot be made indefinitely small, although inch-cube assemblies have been fabricated.

Non-ruggedized breadboard gyroscopes exhibiting drifts on the order of one degree per hour have been demonstrated using this technology.

Because feeding light from free space into a single mode fiber optic requires touchy adjustments, corresponds to a relatively large optical power loss and induces an undesirable amount of back-reflection, it has been proposed [38][39] to construct a fiber optic gyroscope in which the optical wave would be continuously guided in a fiber from the source to

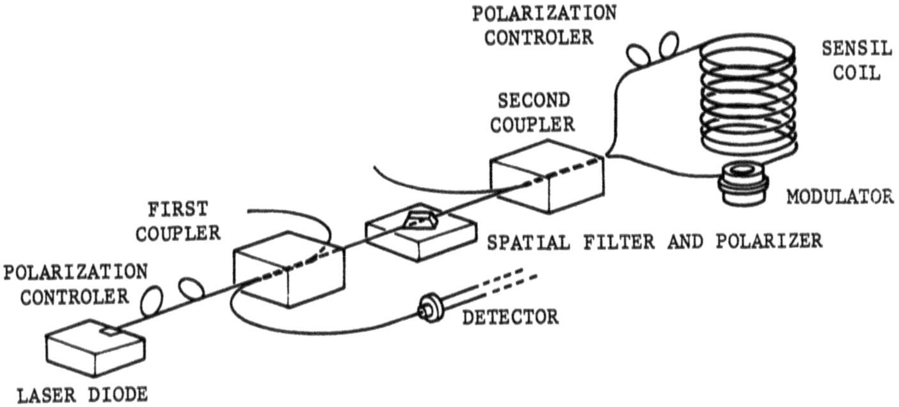

Fig. 17 All-fiber implementation (courtesy of Stanford University)

the detector. It turns out that all the functions required in the minimum configuration can be implemented in an all-fiber form (Fig. 17). As an example, directional coupling between two fibers can be achieved by polishing off a portion of their claddings and bringing their cores together close enough for the evanescent tail of their guided-waves to overlap. These fiber components exhibit extremely low insertion losses. They can be efficiently spliced together to form the interferometer configuration or, better still, be fabricated along an uninterrupted length of fiber also used for the sensing loop. The outstanding performance of the critical element, the polarizer, together with the large optical power ultimately reaching the detector and the low amount of back-reflected light, makes this technology capable of the best drift and random walk figures, on the order of 0.1 degree per hour or better. The main drawback of this approach is the multiplicity and the precision of the operations involved in the construction of the fiber devices which will make their mass production costly. The lack of a solution for a frequency shifter, and the low bandwidth of the piezo-electric all-fiber phase modulator renders the closed-loop operation of the all-fiber gyroscope difficult.

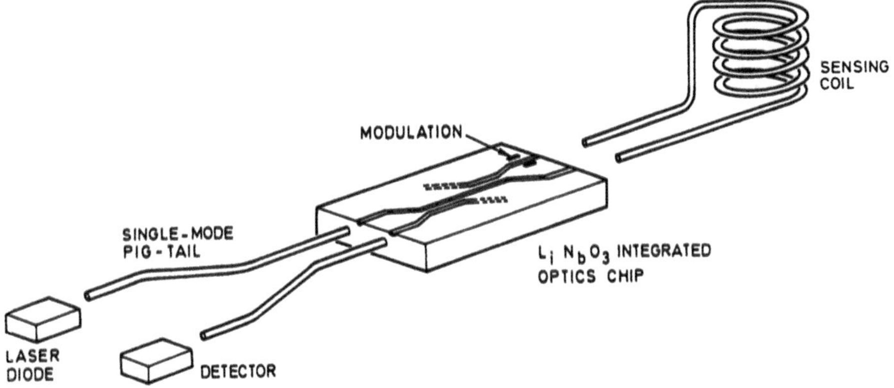

Fig. 18 Integrated optics implementation

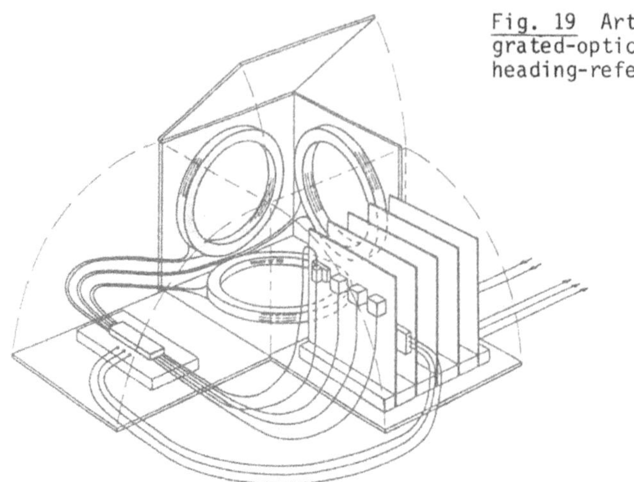

Fig. 19 Artist view of a fiber/integrated-optics strap-down attitude-and-heading-reference-system ($10\times10\times10$ cm^3)

The integrated optics approach [34][40] consists of fabricating a single planar device implementing all the functions required from the heart of the gyroscope in the form of single mode strip wave-guides (Fig. 18). This technique combines the diversity of active functions possible in bulk optics with the advantages of an all guided implementation. The fabrication process, micro-photolithography, similar to that of integrated electronics, is ideally suited to high volume production. The technology, however, is quite difficult to master, and although several teams are nearing that stage, an integrated fiber optic gyroscope remains to be demonstrated. One of the problems to be solved is that of efficient collective (i.e. without individual adjustments) coupling of several fibers to the optical chip. The throughput losses of the integrated optic circuit are expected to remain higher than that of the competing technologies for some time, limiting at present this type of F.O.R.S. to the 1 to 10 degrees per hour class range. It is nonetheless widely considered to be the low-cost gyro of the future (Fig. 19).

Problems in Fiber optic Rotation Sensors

So far we have been concerned primarily with the basic principles of fiber rotation sensors with emphasis on the measurement of small non-reciprocal phase shift in a multiturn fiber interferometer. In this section we discuss briefly a number of error sources that can influence the performance of the fiber gyro.

Fig. 20 lists several sources of error that must be dealt with in order to achieve the predicted performance. Perhaps the major source of noise is backscattering within the fiber [21] and at interfaces, particularly in a set-up that employs discrete components. To overcome this problem researchers have used broadband lasers, [17][22] frequency jittered lasers [18], phase modulators [23], even light emitting diodes (LED) [24]. By destroying the temporal coherence of the light source the detection system becomes sensitive only to the interference between waves that followed identical paths. Thus, interference due to backscattering will, in principle, average to zero.

- RAYLEIGH SCATTERING IN FIBER
- SCATTERING FROM INTERFACES
- POLARIZATION EFFECTS
- PRESENCE OF HIGHER ORDER MODES
- TEMPERATURE GRADIENTS
- NONIDEAL MODULATORS
- NONIDEAL POLARIZERS
- INTENSITY DEPENDENT NONRECIPROCITY
- LIGHT SOURCE PROBLEMS
- MEASUREMENT SYSTEM PROBLEMS
- STRESS INDUCED EFFECTS
- MAGNETIC FIELD EFFECT
- ETC.
- ETC.

Fig. 20 A few problems in fiber gyros

All the fiber gyros under study so far have used single mode fibers. Care has to be taken to check that higher order transverse modes are highly attenuated.

A number of error sources are related to temperature gradients [28], nonideal polarizers [29], nonideal modulators, stress induced effects, external magnetic field effects [30] and electronics problems in the measurement system.

A very basic sources of nonreciprocal phase shift has been uncovered, namely that due to unequal intensities propagating along opposite directions in the fiber [31]. This is a nonlinear optical effect based on four-wave mixing that takes place in a fiber having a third order nonlinear susceptibility. Such an intensity-induced nonreciprocity may be reduced by maintaining equal intensities in the counterpropagating beams.

References

1. D.H. Eckhardt, Proc. Soc. Photo-Opt. Instrum. Eng. 157, 172 (1978).
2. M.O. Scully, in Proceedings of the Fifth International Conference on Laser Spectroscopy, H. Walther and K. Rothe, eds. (Springer-Verlag, Berlin, 1979); M.P. Haugan, M.O. Scully, and K. Just, Phys. Lett. 77A, 88 (1980).
3. J.H. Simpson, Astron. & Aeron. 2, #10, p. 42 (1964).
4. G. Sagnac, Compt. Rend. 157, 708 (1913).
5. E.J. Post, Rev. Mod. Phys, 39, 475, (1967); H.J. Arditty and H.C. Lefevre, Optics Lett. 6, 401 (1981).
6. A.H. Rosenthal, J. Opt. Soc. Am. 52, 1143 (1962).
7. V. Vali and R.W. Shorthill, Appl. Opt. 15, 1099 (1976).
8. S. Ezekiel and S.R. Balsamo, Appl. Phys. Lett. 30, 478, (1977).
9. G.A. Sanders, M.G. Prentiss, and S. Ezekiel, Opt. Lett. 6, 569, (1981).
10. T.A. Dorschner, H.A. Haus, M. Holz, I.W. Smith and H. Stata, IEEE. J. Quant. Electron, QE-16, 1376, (1980).
11. J.L. Davis and S. Ezekiel, Proc. Soc. Photo-Opt. Instrum. Eng. 157, 172 (1978).
12. A. Yariv, "Introduction to Optical Electronics," Holt, Rinehart and Winston, (1976).
13. G. Pircher, M. Lacombat and H. Lefevre, Proc. Soc. Photo-Opt. Instrum. Eng. 157, 212, (1978).
14. W.C. Davis, W.L. Pondrom and D.E. Thompson, Proc. of International Conf. on Fiberoptic Rotation Sensors, M.I.T. (Nov. 1981), Springer-Verlag (this volume).

15. H. Arditty, M. Papuchon, C. Puech and K. Thyagarapan in Digest or Topical Meeting on Integrated and guided-wave Optics, O.S.A. Meeting, Incline Village, NV, (1980).
16. R. Ulrich, Opt. Lett. 5, 173, (1980).
17. R.A. Bergh, H.C. Lefevre and H.J. Shaw, Opt. Lett. 6, 502, (1981).
18. J.L. Davis and S. Ezekiel, Opt. Lett. 6, 505, (1981).
19. D.E. Thompson, D.B. Anderson, S.K. Yao and B.R. Youmans, Appl. Phys. Lett. 33, 940, (1978).
20. R.F. Cahill and E. Udd, Opt. Lett. 4, 93, (1979).
21. C.C. Cutler, S.A. Newton and H.J. Shaw, Opt. Lett. 5, 488, (1980).
22. R. Ulrich, Opt. Lett. 5, 173, (1980).
23. K. Bohm, P. Russer, E. Weidel and R. Ulrich, Opt. Lett. 6, 64, (1981).
24. K. Bohm, P. Marten, K. Petermann, E. Weidell and R. Ulrich, Electron. Lett. 17, 352, (1981).
25. R. Ulrich, Proc. of International Conf. on Fiberoptic Rotation Sensors, M.I.T. (Nov. 1981), Springer-Verlag (this volume).
26. G. Schiffner, W.R. Leeb, H. Krammer and J. Wittman, Appl. Opt. 18, 2096, (1979).
27. I.P. Kaminow, J. Quant. Electron. Q.E. 17, 15, (1981).
28. D.M. Shupe, Appl. Opt. 19, 654, (1980).
29. E.C. Kintner, Opt. Lett. 6, 154, (1981).
30. K. Bohm, K. Petermann and E. Weidel, Proc. of International Conf. on Fiberoptic Rotation Sensors, M.I.T. (Nov. 1981), Springer-Verlag (this volume).
31. S. Ezekiel, J.L. Davis and R.W. Hellwarth, Proc. of International Conf. on Fiberoptic Rotation Sensors, M.I.T. (Nov. 1981) Springer-Verlag (this volume
32. R. Goldstein and W.C. Goss, Proc.Soc.Photo-Opt.Instrum,Eng.157,122(1978).
33. M.N.McLandrich & H.F.Rast, Proc.Soc.Photo-Opt.Instrum.Eng.157,127 (1978).
34. M.Papuchon and C.Puech, Proc.Soc. Photo-Opt.Instrum.Eng. 157,218 (1978).
35. H. Arditty, H.J. Shaw, M. Chodorow and R. Kompfner, Proc. Soc. Photo-Opt. Instrum. Eng. 157, 138, (1978).
36. A. Simon and R. Ulrich, Appl. Phys. Lett. 31, 517, (1977).
37. R. Ulrich, Appl. Phys. Lett. 35, 840, (1979).
38. S.K. Sheem and T.G. Giallorenzi, Opt. Lett. 4, 29, (1979).
39. R.A. Bergh, H.C. Lefevre and H.J. Shaw, Opt. Lett. 6, 198 (1981).
40. W.C. Goss and R. Goldstein, Opt. Eng. 18, 9, (1979).

Fiber-Optic Rotation Sensors. Bibliography

H.J. Arditty

Thomson CSF, Laboratoire Central de Recherches, BP 10, Domaine de Corbeville
F-91401, Orsay, France

The present list of citations is included in this proceeding for the convenience of the reader. It is only a chronological selection of titles extracted from our collection of papers on Optical Rotation Sensing.

No attempt on ordering by subject has been made, on the assumption that the titles given were self-explanatory. Papers dealing only with the laser gyroscope per se have been systematically removed as beeing beyond the scope of this conference. Patents, reports and industry reviews, normally not cited in bibliographies, have been included here to provide a more complete picture of the F.O.R.S. history and present status.

[1] F. HARESS
"Die Geschw. d. Lichtes in bewegten Körpern".
Thesis dissertation, University of Jena, (1912).

[2] G. SAGNAC
"L'éther lumineux démontré par l'effet du vent relatif d'éther dans un interféromètre en rotation uniforme".
C.R. Acad. Sci., 95, 708-710, (1913).

[3] G. SAGNAC
"Sur la preuve de la réalité de l'éther lumineux par l'expérience de l'interférographe tournant".
C.R. Acad. Sci., 95, 1410-1413, (1913).

[4] A. WALLACE
"Electromagnetic wave gyroscopes or angular velocity measuring systems".
U.S. Patent 3102953, September 3, 1963, filed July 9, 1958.

[5] A.H. ROSENTHAL
"Regenerative circulatory multiple beam interferometry for the study of light propagation effects".
J.O.S.A., 52, 1143-1148, (1962).

[6] W.M. MACEK and D.T.M. DAVIS
"Rotation rate sensing with travelling-wave ring lasers".
Appl. Phys. Lett. 2, 67-68, (1963).

[7] D.J. Mc LAUGHLIN
"Optical gyroscope having means for resolving ambiguities of velocity and direction".
U.S. Patent 3512890, May 19, 1970, filed July 27, 1965.

[8] B. EICHENAUER and K.J. BRAUSER
"Procédé et dispositif de mesure précise des vitesses angulaires utilisant l'effet Sagnac".
F.R.G. Patent 16733838, November 20, 1967.
French Patent 1592633, November 13, 1968.

[9] G. PIRCHER and G. HEPNER
"Perfectionnements aux dispositifs du type gyromètre à laser".
French Patent 1563720, December 8, 1967.

[10] E.J. POST
"Sagnac effect".
Rev. Mod. Phys., 39, 475-494, (1967).

[11] R.B. BROWN
"N.R.L. Memorandum Report 1871".
N.R.L. (1968).

[12] E.J. POST
"Interferometric path-length changes due to motion".
J.O.S.A., 62, 234-239, (1972).

[13] E. GITELSON and H. SAKAI
"A polarization Sagnac interferometer".
Annual Meeting of the Optical Society of America.
San Francisco, Ca., U.S.A., 44, October 17-20, 1972.

[14] P. SCHLEGEL
"Flying clocks and the Sagnac effect".
Nature, 248, 180, (1973).

[15] K.P. KOVACS
"Sagnac's experiment".
Fiz. Sz. 321-351, (1974).

[16] R.G. NEWBURGH, Ph. BLACKSMITH, A.J. BUDREAU and J.C. SETHARES
"Acoustic and Magnetic surface wave ring interferometers for rotation rate sensing".
Proceedings of IEEE, 62, 1621-1628, (1974).

[17] A. ASHTEKAR and A. MAGNON
"The Sagnac effect in general relativity".
J. Math. Phys., 16, 341-344, (1975).

[18] V. VALI, R.W. SHORTHILL, R. GOLDSTEIN and R.S. KROGSTAD
"Laser gyroscope".
U.S. Patent 4013365, March 22, 1977, filed June 11, 1975.

[19] Editor
"Fiber beam-paths expected to enhance laser gyro's accuracy and slash cost".
Laser Focus, 11, 6, June, 1975.

[20] P. HARIHAPAN
"Sagnac or Michelson-Sagnac interferometer".
Appl. Opt., 14, 2319-2321, (1975).

[21] V. VALI, R.W. SHORTHILL, R. GOLDSTEIN and R.S. KROGSTAD
"Double optical fiber waveguide ring laser gyroscope".
U.S. Patent 4120587, October 17, 1978, filed November 24, 1975.

[22] G. LIANIE and D. WHICKER
"Electromagnetic phenomena in rotating media".
Arch. Ration. Mech. and Anal., 57, 325-362, (1975).

[23] V. VALI and R.W. SHORTHILL
"Fiber laser gyroscopes".
Proceedings of SPIE, 77, 110-115, (1976).

[24] S. EZEKIEL
"Laser gyroscope".
U.S. Patent 4135822, January 23, 1979, filed May 6, 1976.
French Patent 2350583, May 5, 1977.

[25] V. VALI and R.W. SHORTHILL
"Fiber ring interferometer".
Appl. Opt., 15, 1099-1100, (1976).

[26] V. VALI and R.W. SHORTHILL
"Simulated Brillouin scattering ring laser gyroscope".
U.S. Patent 4159178, June 26, 1979, filed November 24, 1976.

[27] V. VALI and R.W. SHORTHILL
"Ring interferometer 950 m long".
Appl. Opt. 16, 290, (1977).

[28] S. EZEKIEL and S.R. BALSAMO
"Passive ring resonator laser gyroscope".
Appl. Phys. Lett. 30, 478-480, (1977).

[29] S.R. BALSAMO and S. EZEKIEL
"New approach for laser gyroscopes".
Proceedings of IEEE, 1062-1064, May 17-19, 1977.

[30] S.R. BALSAMO and S. EZEKIEL
"New laser gyroscope using a passive ring resonator".
Proceedings of IEEE J. Quantum Electronics, QE-13, 13-14, (1977).

[31] C.M. REDMAN
"Endless fiber interferometer rotary motion sensor".
U.S. Patent 4133612, January 9, 1979, filed July 6, 1977.

[32] A. SIMON and R. ULRICH
"Evolution of polarization along a single-mode fiber".
Appl. Phys. Lett. 31, 517-520, (1977).

[33] V. VALI, R.W.SHORTHILL and M.F. BERG
"Fresnel-Fizeau effect in a rotating optical fiber ring interferometer".
Appl. Opt. 16, 2605-2607, (1977).

[34] P. LECLERC, C. PUECH, M. PAPUCHON and P. LALLEMAND
"Interferometric laser gyroscope".
French Patent 7735039, November 22, 1977.
U.S. Patent 4265541, May 5, 1981, filed November 17, 1978.

[35] G. SCHIFFNER
"Interferometer with a single-mode waveguide coil".
F.R.G. Patent 2804103, January 31, 1978.
U.S. Patent 4259016, March 31, 1981, filed January 11, 1979.

[36] J.M. MARTIN and J.T. WINKLER
"Fiber optic laser gyro signal detection and processing technique".
Proceedings of SPIE 139, 98-102, (1978).

[37] V. VALI, R.W. SHORTHILL and M.F. BERG
"Fresnel drag in single-mode optical fiber waveguides".
Proceedings of SPIE 139, 88-92, (1978).

[38] V. VALI, M.F. BERG and R.W. SHORTHILL
"Fresnel drag effects in fiber optical gyroscope".
Proceedings of IEEE 18, 2, March, 1978, Naecon.

[39] G. SCHIFFNER and D. ROSENBERGER
"Interferometer with a coil composed of a single-mode waveguide".
F.R.G. Patent 2814476, April 14, 1978.
U.S. Patent 4302107, November 24, 1981, filed March 19, 1979.
French Patent 2422142, March 28, 1979.

[40] J.P. HUIGNARD, P. LECLERC, J.C. CARBELES and A. NAPPO
"Structure optique compacte à source intégrée".
French Patent 4818, May 26, 1978.

[41] G. PIRCHER
"Optical fiber interferometric gyrometer with polarization switching".
French Patent 7820540, July 10, 1978.
U.S. Patent 4286878, September 1, 1981, filed July 9, 1979.

[42] G. PIRCHER
"Gyromètre à laser".
French Patent 4850, July 7, 1978.

[43] ROCKWELL int.
"Gyroscope interférométrique utilisable avec une source lumineuse à faible cohérence".
U.S. Patent 936678, filed August 23, 1978.
French Patent 2434371, August 13, 1979.

[44] ROCKWELL int.
"Gyroscope interférométrique".
U.S. Patent 936680, filed August 23, 1978.
French Patent 2434370, August 13, 1979.

[45] D.E. THOMPSON, D.B. ANDERSON, R.R. AUGUST and S.K. YAO
"Interferometer gyro using heterodyne phase detection without severe light source coherence requirements".

U.S. Patent 4208128, June 17, 1980, filed August 23, 1978.
French Patent 2434370, August 13, 1979.

[46] S. EZEKIEL and G.E. KNAUSENBERGER, edt.
"Laser inertial rotation sensors".
Proc. Soc. Photo-opt. instrum. eng. 157, (1978).

[47] R.J. FREDRICKS and W.A CARRINGTON
"Sources of error in a discrete component Sagnac optical rate sensor".
Proceedings of SPIE 157, 50-67, (1978).

[48] S. EZEKIEL, J.A. COLE, J. HARRISON and G. SANDERS
"Passive cavity optical rotation sensor".
Proceedings of SPIE 157, 68-72, (1978).

[49] T. REED, M. MAKLAD, G. BICKEL and A. ASAM
"Low-loss single-mode fibers (laser gyros)".
Proceedings of SPIE 157, 89-94, (1978).

[50] E. GARMIRE
"Optical waveguides for laser gyro applications".
Proceedings of SPIE 157, 95-99, (1978).

[51] V. VALI and M.F. BERG
"Non reciprocity noise in fiber gyroscopes and measurement of fiber dispersion".
Proceedings of SPIE 157, 120-121, (1978).

[52] R. GOLDSTEIN and W.C. GOSS
"Fiber Optic Rotation Sensor (FORS), laboratory performance evaluation".
Proceedings of SPIE 157, 122-126, (1978).

[53] M.N. Mc LANDRICH and H.E. RAST
"Fiber interferometer gyroscope".
Proceedings of SPIE 157, 127-130, (1978).

[54] J.L. DAVIS and S. EZEKIEL
"Techniques for shot-noise-limited inertial rotation measurement using a multiturn fiber Sagnac interferometer".
Proceedings of SPIE 157, 131-136, (1978).

[55] D.E. THOMPSON and D.B. ANDERSON
"Sagnac effect fiber ring gyro with electronic phase sensing".
Proceedings of SPIE 157, 137, (1978).

[56] H. ARDITTY, H.J. SHAW, M. CHODOROW and R. KOMPFNER
"Re-entrant fiber optic approach to rotation sensing".
Proceedings of SPIE 157, 138-148, (1978).

[57] L. SHIH-CHUN
"Sensitivity limit of an optical fiber gyroscope".
Proceedings of SPIE 157, 149-163, (1978).

[58] W.K. STOWELL, R.W. Mc ADORY and R. ZIERNICKI
"Air force applications for optical rotation rate sensors".
Proceedings of SPIE 157, 166-171, (1978).

[59] G. PIRCHER, M. LACOMBAT and H.C. LEFEVRE
"Preliminary results obtained with a Fiber Optic Rotating Sensor (FORS)".
Proceedings of SPIE 157, 212-217, (1978).

[60] M. PAPUCHON and C. PUECH
"Integrated optics. A possible solution for the fiber gyroscope".
Proceedings of SPIE 157, 218-219, (1978).

[61] H. ARDITTY
"L'effet Sagnac, le gyromètre à fibre optique".
SEE - Journée d'étude : La navigation à inertie, 15 novembre 1978.

[62] P. SCHNERB
"Dispositif optique de mesure de la vitesse angulaire d'un mobile".
French Patent 7832317, 15 novembre 1978.

[63] D.E. THOMSON, D.B. ANDERSON, S.K. YAO and B.R. YOUMANS
"Sagnac fiber-ring interferometer gyro with electronic phase sensing using a Ga Al As laser".
Appl. Phys. Lett. 33, 940-941, (1978).

[64] Mc DONNELL DOUGLAS Corp.
"Dispositif optique de détermination de rotation, applicable notamment à un gyroscope optique".
U.S. Patent 967267 filed december 7, 1978.
French Patent 2443668, december 2, 1979.

[65] W.C. GOSS and R. GOLDSTEIN
"Optical gyroscope system".
U.S. Patent 4280766, July 28, 1981, filed december 22, 1978.

[66] D.J. OLKOWSKI and Ch. R. HOLLAND
"Evaluation of errors in a passive ring resonator laser gyroscope".
Res. and Dev. Rep. U.S. AD-A064.046/6GA, december 1979.

[67] A. CLARENCE and J.R. ROBINSON
"Technology program spurs missile intercept andvances".
Aviation week 108, 108-111, (1978).

[68] R. ULRICH and M. JOHNSON
"Procédé et dispositif pour la mesure de taux de rotation par l'effet Sagnac".
F.R.G. Patent P29013885, filed January 15, 1979.
French Patent 2446481, January 15, 1980.

[69] W.C. GOSS and R. GOLDSTEIN
"Fiber Optic Rotation Sensor (FORS) signal detection and processing".
Optical Eng. 18, 9-13, (1979).

[70] M. LACOMBAT and H.C. LEFEVRE
"Gyromètre interférométrique à fibre optique particulièrement adapté à la mesure de faibles vitesses de rotation".
French Patent 50237, February 23, 1979.

[71] SIEMENS
"Fibre optique clivable et méthode de fabrication de celle-ci".
F.R.G. Patent 2909356, March 9, 1979.
Euro. Patent 0015533, March 3, 1980.

[72] S.K. SHEEM and T.G. GIALLORENZI
"Single-mode fiber optical power divider : encapsulated etching technique".
Opt. Lett. 4, 29, (1979).

[73] R.F. CAHILL and E. UDD
"Phase nulling fiber optic laser gyro".
Opt. Lett. 4, 93-95, (1979).

[74] L. SHIH-CHUN and T.G. GIALLORENZI
"Sensitivity analysis of the Sagnac effect optical fiber ring interferometer".
Appl. Opt. 18, 915-931, (1979).

[75] W.R. LEEB, G.S. SCHIFFNER and E. SCHEITERER
"Optical fiber gyroscopes : Sagnac or Fizeau effect ?".
Appl. Opt. 18, 1293-1295, (1979).

[76] M. LACOMBAT, H.C. LEFEVRE and G. PIRCHER
"Dispositif compact de couplage optique et gyromètre interférométrique à fibre optique comportant un tel dispositif".
French Patent 50334, May 2, 1979.

[77] R.F. CAHILL and E. UDD
"Phase nulling optical Gyro".
Proceedings of IEEE, 8-13, (1979), Naecon.

[78] R. ULRICH and M. JOHNSON
"Fiber ring interferometer : polarization analysis".
Opt. Lett. 4,152-154, (1979).

[79] Proceedings of IEEE/OSA. Conference on laser engineering and applications. Digest of technical papers. Washington, DC, USA.
May 30-June 1, 1979.

[80] S. EZEKIEL
"New perspectives in optical gyroscopes".
Proceedings of IEEE, Laser eng., (1979).

[81] W.C. GOSS and R. GOLDSTEIN
"Fiber optic rotation sensor technology".
Proceedings of IEEE, Laser eng. (1979).

[82] R.H. POOL and G.W. SELLERS
"Interferometer gyroscope having related detector linearity requirements".
U.S. Patent 4273444, June 16, 1981, filed June 13, 1979.

[83] G. SCHIFFNER, W.R. LEEB, H. KRAMMER and J. WITTMAN
"Reciprocity of birefringent single-mode fibers for optical gyroscopes".
Appl. Opt., 18, 2096-2097 (1979).

[84] H.C. LEFEVRE
"Gyromètre interférométrique à fibre optique".
Thesis dissertation, University of Paris-Orsay, (1979).

[85] R.W. SHORTHILL, G.J. MORRIS and L.D. WEAVER
"Optical fiber gyroscope".
Automatic Control in Space. Proceedings of the 8th IFAC Symposium
Pergamon Press, Oxford, England, (1980).

[86] D.E. THOMPSON, R.R. AUGUST and D.B. ANDERSON
"Interferometer gyroscope formed on a single plane optical waveguide".
U.S. Patent 4273445, June 16, 1981, filed July 30, 1979.

[87] R. GOLDSTEIN and W.C. GOSS
"Fiber Optic Rotation Sensor (FORS) laboratory performance evaluation".
Opt. Eng., 18, 381-383, (1979).

[88] A. KURIYAGAWA and S. MORI
"Ring laser and ring interferometer in accelerated systems".
Phys. Rev. D, 20, 1290-1293, (1979).

[89] NORTHROP
"Thin film laser gyro".
U.S. Patent 77381, filed September 20, 1979.
Euro Patent 0026066, September 10, 1980.

[90] T.L. WILLIAMSON and D.A. WILLIE
"Passive fiber optic gyro study".
Res. and Dev. Rep. U.S., AD-A079.471/9, October 1979.

[91] R. ULRICH
"Polarization stabilization on single-mode fiber".
Appl. Phys. Lett., 35, 840-842, December 1, 1979.

[92] H.J. ARDITTY, M. PAPUCHON and C. PUECH
"Dispositif interférométrique en anneau et son application à la détection d'effets non-réciproques".
French Patent 2471595, December 14, 1979.

[93] E. UDD and R. CAHILL
"Gyro development progress".
Proceedings, Laser 1979, December 21, 1979, Orlando.

[94] W.T. BRUCKETT, J. MARTIN, J.M. HOLMES and E.L. JOHNSON
"Digital fiber optic rate sensor experimental model".
Proceedings, Laser 1979, December 21, 1979, Orlando.

[95] L.D. WEAVER
"Geophysical fiber interferometer gyroscope".
Res. and Dev. Rep. U.S., AD-A092.913/3, December 31, 1979.

[96] U.U.R.I.
"Fiber Ring Optical Gyroscope (FROG)".
Scient. and Tech. Rep. U.S., N80-28690, December 31, 1979.

[97] R.A. MOTES
"Compact passive gyroscope feasibility investigation".
Res. and Dev. Rep. U.S., AD-A080.155/5, December 1979.

[98] S. EZEKIEL
"Passive ring resonator laser gyroscope".
Res. and Dev. Rep. U.S., AD-A061.649/6, January 3, 1980.

[99] T.G. GIALLORENZI, W.K. BURNS and J. DONOVAN
"Fiber optic sensors".
Topical meeting of integrated and guided-wave optics, incline, NV, USA, January 28-30, 1980.

[100] H.J. ARDITTY, M. PAPUCHON, C. PUECH and K. THYAGARAPAN
"Recent developments in guided-waves optical rotation sensors".
Topical meeting of integrated and guided-wave optics, incline, NV, USA, January 28-30, 1980.

[101] T. FINDAKLY and E. GARMIRE
"Low loss optical waveguides fabricated by solid state electro-thermal diffusion of metals into glass".
Topical meeting of integrated and guided-wave optics, incline, NV, USA, January 28-30, 1980.

[102] N.V. KRAVTSOV and E.G. LARIONTSEV
"Optical non-reciprocity in media having a non-linear refractive index".
Sov. Phys. Tech., Phys., 25, 108-110, (1980).

[103] J.L. LESNES
"Dispositif de détection d'ondes électromagnétiques, typiquement lumineuses, en quadrature de phase avec balance hétérodyne. Application aux vélocimètres Doppler et gyromètres de Sagnac".
French Patent 2475220, february 1, 1980.

[104] D.M. SHUPE
"Thermally induced non-reciprocity in the fiber optic interferometer".
Appl. Opt., 19, 654-655, (1980).

[105] Editor
"Fiber gyro demonstrate high sensitivity".
Electro-optical and laser technology, March 7, 1980.

[106] W.C. GOSS, R. GOLDSTEIN, M.D. NELSON, H.T. FEARNEHAUGH and O.G. RAMER
"Fiber optic rotation sensor technology".
Appl. opt., 19, 852-858, (1980).

[107] T. ORMOND
"Fiber optics flashes : control link, gyro, computer link".
E.D.N., 25, 97, (1980).

[108] B.M. ELSON
"Sensor, spectrum analysis developped".
Aviation week, 81-82, March 31, 1980.

[109] W.C. GOSS
"Fiber optic rotation sensors".
Proceedings of SPIE, 224, 38-45, (1980).

[110] J.L. LESNE and B. DESSUS
"Interféromètre en anneau à fibre optique pour la mesure du courant électrique".
Conférence Européenne d'optique 1980. "Horizon de l'optique", Pont-à-Mousson - France (Poster session), 22-25 avril 1980.

[111] C.G. YOUNG
"Gyroscopes and gyro-stabilized systems. Citation from the engineering index data base (N81-10351) from the NTIS data base (N81-10350) from the international aerospace abstracts data base (N81-10354)".
Scien. and Tech. Reports U.S., (1980).

[112] C.G. YOUNG
"Laser gyroscopes, citations from the engineering index data base (N81-10356) from the NTIS data base (N81-10353), from the international aerospace abstracts data base (N81-10355)".
Scien. and Tech. Reports U.S., (1980).

[113] D.A. JACKSON, A. DANDRIDGE and S.K. SHEEM
"Measurement of small phase shifts using a single-mode optical fiber interferometer".
Opt. Lett., 5, 139-141, (1980).

[114] R. ULRICH
"Fiber optic rotation sensing with low drift".
Opt. Lett., 5, 173-175, (1980).

[115] A. DANDRIDGE, A.B. TVETEN, R.O. MILES and T.G. GIALLORENZI
"Laser noise in fiber optic interferometer systems".
Appl. Phys. Lett., 37, 256, (1980).

[116] P.J. THOMAS, H.M. VANDRIEL and G.I.A. STEGEMAN
"Possibility of using an optical fiber Brillouin ring laser for inertial sensing".
Appl. Opt., 19, 1906-1908, (1980).

[117] R.A. MOTES and S.R. BALSAMO
"A compact passive ring laser gyroscope design".
CPEM Digest, 263-265, June 23-27, 1980.

[118] Editors
"Le gyromètre à fibre optique développé par Quantel et Sfena".
Air et Cosmos, 34, 12 juillet 1980.

[119] Editors
"Quantel et Sfena travaillent à la mise au point d'un gyromètre à fibres optiques".
Electronique Actualité, (1980).

[120] R. ULRICH and S.C. RASHLEIGH
"Beam to fiber coupling with low standing wave ratio".
Appl. Opt., 19, 2453-2456, 15 juillet 1980.

[121] Editors
"Promising advances in a Sagnac type laser gyro...".
Aviation week, 61, July 21, 1980.

[122] S.K. YAD
"Guided-wave optical interferometer using periodic structures".
Proceedings of SPIE, 239, 104-109, (1980).

[123] A.N. GUR'YANOV, D.D. GUSOVSKII, G.G. DEVYATYKH, E.M. DIANOV
A.Ya. KARASIK, V.A. KOZLOV, V.B. NEUSTREV and A.M. PROKHOROV
"Sagnac effect in a fiber optic interferometer".
J.E.T.P. Lett., 32, 221-224, (1980).

[124] D.A. JACKSON, R. PRIEST, A. DANDRIDGE and A.B. TVENTEN
"Elimination of drift in a single-mode optical fibre interferometer using a piezo-electrically streched coiled fiber".
Appl. Opt., 19, 2926, (1980).

[125] R.F. CAHILL and E. UDD
"Solid-state phase nulling optical gyro".
Appl. Opt., 19, 3054-3056, (1980).

[126] P. RUSSER
"Fiber optic rotation sensor".
Symposium on gyro technology-DGON, September 1980, Stuttgart.

[127] J.S. HUNTER and W.W. STRIPLING
"Present and future trends in army inertial sensor concepts".
Symposium on gyro technology, DGON, September 1980.

[128] G. SCHIFFNER
"An optical rotation sensor based on the Sagnac effect using a fibre optics ring interferometer".
NTG-Fachber, 75, 191-193, October 2-4, 1980.

[129] M. PAPUCHON, H.J. ARDITTY and C. PUECH
"Jonction optique hybride et application à un dispositif interférométrique en anneau".
French Patent 8021677, October 10, 1980.

[130] S.K. SHEEM
"Fiber optique gyroscope with [3 × 3] directional coupler".
Appl. Phys. Lett., 37, 869-871, (1980).

[131] R.A. BERGH, H.C. LEFEVRE and H.J. SHAW
"Single-mode fiber optic polarizer".
Opt. Lett., 5, 479-481, (1980).

[132] S.C. RASHLEIGH and W.K. BURNS
"Dual input fiber optic gyroscope".
Opt. Lett., 5, 482-484, (1980).

[133] C.C. CUTLER, S.A. NEWTON and H.J. SHAW
"Limitation of rotation sensing by scattering".
Opt. Lett., 5, 488-490, (1980).

[134] R. BERGH, G. KOTLER and H.J. SHAW
"Single-mode fiber optic directional coupler".
Elect. Lett., 16, 260-261, (1980).

[135] H.C. LEFEVRE
"Single-mode fractional wave devices and polarization controllers".
Elect. Lett., 16, 778-780, (1980).

[136] K. HOTATE and Y. YOSHIDA
"Rotation detection by optical fibre laser gyro with easily introduced phase-difference bias".
Elect. Lett., 16, 941-942, (1980).

[137] M.O. SCULLY
"Laser gyro theory extension".
Res. and Dev. Report U.S., AD-A096.145/8, December 1980.

[138] M. CHODOROW, G.A. PAVLATH and H.J. SHAW
"Research on new approaches to optical systems for inertial rotation sensing".
AFOSR-76-3070, Ginzton LAB Report 3094, (1980), Stanford University.

[139] R.A. MOTES and S.R. BALSAMO
"A compact passive ring laser gyroscope design".
Conference on precision E.M. measurements, 263-266, (1980).

[140] J.A. BUCARO
"Optical fiber sensors".
Laser and Electro-optical system, IEEE, 64, (1980).

[141] P.R. BALL, B. CULSHAW, D.E.N. DAVIES, T.J. HALL and S.A. KINGSLEY
"Fiber optics and some coherent applications".
Laser and Electro-optical system, IEEE, 64, (1980).

[142] G. SCHIFFNER
"Optical fibre gyroscope based on the Sagnac effect".
Siemens Forsch, 9, 16-25, (1980).

[143] D.M. SHUPE
"Fiber resonator gyroscope : sensitivity and thermal non-reciprocity.
Appl. Opt., 20, 286-289, (1981).

[144] Editor
"Optical fibres : also for transducers".
Mes. Regul. Autom., 46, 49-51-53-55-57, January 1981.

[145] M.S. ZUBAIRY, M.O. SCULLY and K. JUST
"Beam width correction in high precision Sagnac interferometry".
Opt. Commun., 36, 175-177, (1981).

[146] K. BOHM, P. RUSSER, K. WEIDEL and R. ULRICH
"Low noise fiber optic rotation sensing".
Opt. Lett., 6, 64-66, (1981).

[147] H.J. ARDITTY, Y. BOURBIN, M. PAPUCHON and C. PUECH
"Etude de faisabilité d'un senseur de courant utilisant une fibre optique comme capteur".
THCSF Report 1752, March 1981.

[148] E.C. KINTNER
"Polarization control in optical fiber gyroscopes".
Opt. Lett., 6, 154-156, (1981).

[149] S.K. SHEEM and W.K. BURNS
"Optical fiber gyroscope with [3 × 2] guided-wave beam splitters".
Proceedings of IOOC 81, 117-118, (1981).

[150] R. BERGH, H.C. LEFEVRE and H.J. SHAW
"All single-mode fiber gyroscope".
Proceedings of IOOC 81, 116-117, (1981).

[151] S.O. RASHLEIGH and W.K. BURNS
"Alternating current operation of the dual-input fiber optic gyroscope (use of Sagnac interferometer)".
Proceedings of IOOC 81, 116-119, (1981).

[152] H.J. ARDITTY, Y. BOURBIN, M. PAPUCHON and C. PUECH
"Current sensor using state-of-the-art fiber optic interferometric techniques".
Proceedings of IOOC 81, 128-130, (1981).

[153] C.H. BULMERS and R.P. MOELLER
"Fiber coupled phase shifter for use in optical gyroscopes".
Proceedings of IOOC 81, 131-132, (1981).

[154] K. BOHM, P. MARTEN, K. PETERMANN, E. WEIDEL and R. ULRICH
"Low drift fiber gyro by using a depolarizer and a superluminescent diode". Proceedings of 100C 81, Paper TUL7-1, (1981), Post Deadline.

[155] J. HAAVISTO and G.A. PAJER
"Characteristics of thin film resonators".
Proceedings of IOOC 81, (1981).

[156] M. PAPUCHON
"Present applications of integrated optics".
Proceedings of IOOC 81, Paper TUI-1, (1981).

[157] T.G. GIALLORENZI
"Fiber optic sensors".
Opt. and Laser Tech., April 13, 1981.

[158] R.A. BERGH, H.C. LEFEVRE and H.J. SHAW
"All single-mode fiber optic gyroscope".
Opt. Lett., 6, 198-280, (1981).

[159] C.A. VILLARUEL and R.P. MOELLER
"Fused single-mode fiber access couplers".
Elect. Lett., 17, 243, (1981).

[160] K. BOHM and P. MARTEN
"Low drift fibre gyro using a superluminescent diode".
Elect. Lett., 17, 353-354, (1981).

[161] J. GOSCH
"Fiber optics point the way to solid-state gyro".
Electronics X, 54, 73-74, (1981).

[162] H.J. ARDITTY, M. PAPUCHON and C. PUECH
"Fiber optic rotation sensor, toward an integrated device, a review".
CLEC 81, 100, (1981).

[163] S.A. NEWTON, C.C. CUTLER and H.J. SHAW
"Reduction of backscattering error in an optical fiber gyro".
CLEO 81, 126, (1981).

[164] G. PAVLATH and H.J. SHAW
"Unpolarized reciprocal operation of fiber-optic rotation sensors".
CLEO 81, 126-128, (1981).

[165] S.K. SHEEM
"Optical fiber interferometer with [3 × 3] directional couplers : an analysis".
J.A.P., 52, 3865-3872, (1981).

[166] J. ANANDAN
"Sagnac effect in relativistic and non-relativistic physics".
Phys. Rev. D, 24, 338-346, (1981).

[167] Ch. J. BORDE
"Phase conjugate optics and application to interferometry and to laser gyroscope".
Quantum optics, experimental gravitation and measurement theory.
NATO AST series (edited by P. MEYSTRE and M.O. SCULLY),
Plenum Press, (1982).

[168] H.J. ARDITTY and H.C. LEFEVRE
"Sagnac effect in fiber gyroscopes".
Opt. Lett., 6, 401-403, (1981).

[169] X. WEEGER
"Les fibres optiques sont aussi des capteurs".
Le Monde, 23 septembre 1981.

[170] H.J. ARDITTY, Y. BOURBIN, M. PAPUCHON and C. PUECH
"Un capteur ampèremétrique à fibre optique".
Rev. Tech. Thomson-CSF, 13, 521-539, (1981).

[171] Y. JIANG, C. FAN and X. YANG
"The fundamental precision limit of optical gyroscope".
Laser J, 8, 34-41, (1981).

[172] G.L.COTTAM
"The fibre optic rate sensor".
Symposium on gyroscope technology-DGON, September 1981, Stuttgart.

[173] B. CULSHAW and I.P. GILES
"Coherent optical fibre gyroscope".
Symposium on gyroscope technology-DGON, Stuttgart, September 1981.

[174] K.V. BARON and K. WICKERT
"Laser gyroscope heading reference for vehicle location by dead reckoning".
Symposium on gyroscope technology-DGON, Stuttgart, September 1981.

[175] B. NOTTBECK and G. SCHIFFNER
"The fibre optic rotation sensor : an analysis of sensitivity limiting effects".
Symposium on gyroscope technology-DGON, Stuttgart, September 1981.

[176] E. SCHLEMPER
"Advantages of frequency modulators in fibre gyroscope".
Symposium on gyroscope technology-DGON, Stuttgart, September 1981.

[177] Editors
"Sensor family developed at naval research lab".
Aviation week, 115, 55-56, 1981.

[178] J.G. HANSE, R.B. SMITH and G.L. MITCHELL
"Dual polarization fiber optic gyro".
Res. and Dev. Rep. U.S., AD-A106-731/3, October 30, 1981.

[179] M.O. SCULLY, M.S. ZUBAIRY and M.P. HAUGAN
"Proposed optical test of metric gravitation theories".
Phys. Rev. A, 24, 2009-2016, (1981).

[180] R.A. BERGH, H.C. LEFEVRE and H.J. SHAW
"All single-mode fiber optic gyroscope with long-tem stability".
Opt. Lett., 6, 502-504, (1981).

[181] J.L. DAVIS and S. EZEKIEL
"Closed loop, low-noise fiber optic rotation sensor".
Opt. Lett., 6, 505-507, (1981).

[182] G.A. SANDERS, M.G. PRENTISS and S. EZEKIEL
"Passive ring resonator method for sensitive inertial rotation measurement in geophysics and relativity".
Opt. Lett., 6, 569-571, (1981).

[183] J.J. ROLAND and G.P. AGRAWAL
"Optical gyroscopes".
Optics and laser tech., 13, 239-244, (1981).

[184] I.P. GILES, B. CULSHAW and D.E.N. DAVIES
"Optical fibre gyroscope".
University College London annual review, October 1981.

[185] T. WEARDEN
"Research moves ahead in optical sensing systems".
IFOC fiber optics, 2, 9-11, (1981).

[186] C.H. BULMER and R.P. MOELLER
"Fiber gyroscope with non-reciprocally operated, fiber coupled. LINBO3 phase shifter".
Opt. Lett., 6, 572-574, (1981).

[187] Ph. J. KLASS
"Work progresses on fiber optic gyros".
Aviation week, 115, 64-68, (1981).

[188] K. PETERMANN
"A new measurement method for rotary movements. Fibre gyro uses the Sagnac effect".
Funkschau, 25-26, 82-84, (1981).

[189] A.E. KAPLAN and P. MEYSTRE
"Enhancement of the Sagnac effect due to non-linearly induced non-reciprocity".
Opt. Lett., 6, 590-592, (1981).

[190] K. HOTATE, Y. YOSHIDA, M. HIGASHIGUSHI and N. NIWA
"Fiber optic laser gyro with easily introduced phase difference bias".
Appl. Opt., 20, 4313-4318, (1981).

[191] H.J. ARDITTY, M. PAPUCHON and C. PUECH
"Dispositif de mesure de la vitesse d'un fluide et son application à la mesure de débit".
French Patent 8105133, March 1981.

[192] R.A. BERGH, C.C. CUTLER, H.C. LEFEVRE, S.H. NEWTON, G.A. PAVLATH and H.J. SHAW
"The all-fiber gyroscope : a practical alternative for rotation sensing".
Topical meeting on integrated and guided-wave optics, WC-2, Asilomar CA, January 1982.

[193] H.J. ARDITTY, M. PAPUCHON, S. VATOUX, Y. BOURBIN and C. PUECH
"Fiber optic gyroscope with a multifunction integrated optical circuit".
Topical meeting on integrated and guided-wave optics, PDP-10, Asilomar CA, January 1982.

[194] J.P. JOLUET
"Fibres optiques : les réalisations gagnent l'avionique".
Inter Elect., 341, 24-26, (1982).

[195] D.E.N. DAVIES
"Making measurements with light".
IEE Proceeding A, 129, 16-23, (1982).

Part 2

Theoretical Considerations

2.1 Sagnac Effect in a Medium

Theoretical Basis of Sagnac Effect in Fiber Gyroscopes

H.J. Arditty

Thomson CSF, Laboratoire Central de Recherches, BP 10, Domaine de Corbeville
F-91401 Orsay, France

H.C. Lefèvre

Edward L. Ginzton Laboratory, Stanford University, Stanford, CA 94305, USA

Simple kinematics considerations give the first order value of the Sagnac phase shift in a rotating vacuum interferometer [1]. It is more complicated to treat the problem when the waves travel within a dielectric medium, as in the case of a fiber gyroscope. Then both Fizeau and Doppler effects occur and compensate the influence of the index of refraction. The results appear to be independent of the properties of the medium [2]. They depend only on the light frequency, the length of the closed loop and its radius. But for the waveguide configuration of fiber rotation sensors where modal dispersion occurs, an electrodynamic approach is more rigorous.

With the kinematics method, we can consider a rotating interferometer in a motionless frame. In electrodynamics, we have to solve the propagation equation in the rest frame of the system where the boundary conditions are time independent. Also we have to find the new coordinates of the rotating frame $\mathcal{F}_R$ from a relativistic point of view. Electrodynamics laws have simple expressions in an inertial frame but can be extended with some modifications to any system of curvilinear coordinates as, in particular, those of a rotating frame. The main problem is to find the expression for the constitutive tensor which connects the contravariant components of the derived field tensor $(\vec{H},\vec{D})$ to the covariant components of the macroscopic field $(\vec{B},\vec{E})$ [3]. This relation involves the properties of the medium and the non-inertial properties of the rotating frame. With this constitutive relation we can find the exact expression for the propagation equation in the rotating frame. This can be solved with a method of perturbation from the solutions for a motionless waveguide in an inertial frame $\mathcal{F}_I$. This leads to the complete expressions for the amplitude and the phase of the counterpropagating modes. Looking at the interferences in a waveguide ring interferometer we find the Sagnac phase shift given by this electrodynamics approach.

Now, what are the coordinates we can use in a rotating frame? Special relativity says that locally the vacuum light velocity is constant and that the coordinates must be measurable with experiments relative to the frame. For a translation, it is well known that a Lorentz transformation has to be used. But what transformation has to be used for a rotation? Locally the rotation of a portion of a circle is completely equivalent to the translation of a line segment. Then we can use a Lorentz transformation with cylindrical coordinates (r, θ, z, t) :

$$r_R = r_I , \quad z_R = z_I$$

$$\theta_R = \gamma(\theta_I - t_I) , \quad t_R = \gamma\left(t_I + \frac{r^2\Omega}{c^2}\theta_I\right)$$

where $\gamma = \left(1 - \frac{r^2\Omega^2}{c^2}\right)^{-1/2}$.

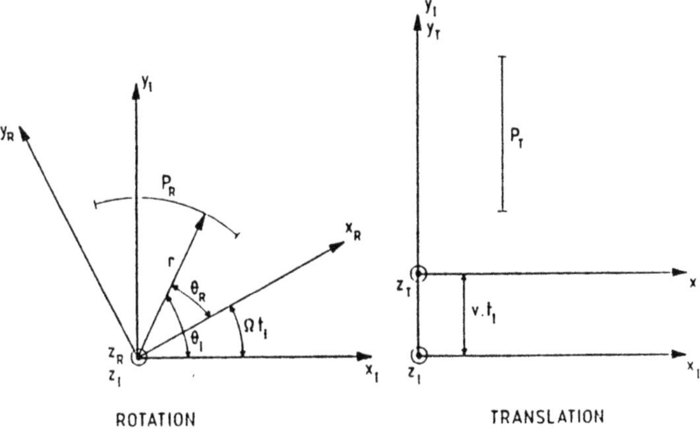

Fig.1 Local equivalence between rotation and translation

But t_R depends on θ_I which is periodic. If we consider a full circle this periodicity will appear in t_R . That means that if we synchronize the time with the light propagation around the circle, the result will depend on the number of turns of propagation. The rotating frame is not inertial any more if we consider a full circle. The solution is to remove the θ_I dependance of t_R . Instead of considering that the synchronization depends on the light path used, we have to say that the time has only one value at one point, and if we send light around a closed loop there is a non-inertial effect which is the Sagnac effect. The new coordinate $t'_R = \gamma t_R$ does not depend on θ_I and can be measured with experiments relative to the rotating frame. t'_R and θ_R still depend on $\gamma(r)$ which is a second order term in $r\Omega/c$.

In any case, a relativistic length contraction occurs and the ratio between the circumference $\mathscr{C}$ of a full circle and its radius r is not 2Π anymore [4], but:

$$\frac{\mathscr{C}}{r} = 2\Pi\left(1 - \frac{r^2\Omega^2}{c^2}\right)^{-1/2} .$$

The coordinate θ_R is the ratio of the length of a portion of a circle over the length of the radius. But it is also possible to give another definition to the angular coordinate. It can be considered as the ratio of the length of a portion of a circle over the length of the full circle multiplied by 2π. Then:

$$\theta'_R = \theta_I - \Omega t_I .$$

Similar considerations can be applied to the temporal coordinates. The expression t'_R is actually the local time at a certain point of the rotating frame. But on the axis of rotation this local time is the time in the inertial frame. This time can be called central time [5]. Sending cylindrical waves from the axis allows one to synchronize all the points of the rotating frame to the central time because locally the direction of propagation of the synchronizing signal is orthogonal to the direction of the movement, and therefore is not affected by it.

In conclusion, a Lorentz transformation cannot be used to define the new coordinates of a rotating frame because the time must have only one value at a certain point. But two different sets of coordinates are measurable with experiments relative to the rotating frame and could be suitable to use. They differ only by the coefficient γ which is a second order term in $r\Omega/c$. The first order result is the same. For simplicity we are going to use the usual Galilean transformation of the coordinates:

$$r_R = r_I , \qquad z_R = z_I ,$$

$$\theta_R = \theta_I - \Omega t_I , \qquad t_R = t_I .$$

This result could seem obvious, but we think that it is interesting to see why the same local special relativity principle leads to different results for translation and rotation because of the particular geometry of a rotating frame of reference.

The expression of electrodynamics laws can be given with four-dimensional tensorial notation in a pseudo-euclidian frame [6]. The Maxwell-Gauss and Maxwell-Ampere laws connect the derived field $G^{\mu\nu}$ to the current four-vector J^ν with:

$$\partial_\mu G^{\mu\nu} = \mu_0 J^\nu .$$

The relationship between the macroscopic field $F_{\sigma\rho}$ and the four-potential ϕ_ρ is:

$$F_{\sigma\rho} = \partial_\sigma \phi_\rho - \partial_\rho \phi_\sigma .$$

The Lorentz gauge can be expressed by

$$\partial_\rho \phi^\rho = 0$$

and the linear constitutive relation is:

$$G^{\mu\nu} = \chi^{\mu\nu\rho\sigma} F_{\rho\sigma}$$

where $\chi^{\mu\nu\rho\sigma}$ is the constitutive tensor. In the case of a non-euclidian frame these expressions can be extended using the determinant g of the metric tensor $g_{\mu\nu}$ [6]. They become

$$\begin{cases} \partial_\mu \sqrt{-g}\, G^{\mu\nu} = \sqrt{-g} \cdot \mu_o \cdot J^\nu , \\ F_{\sigma\rho} = \partial_\sigma\phi_\rho - \partial_\rho\phi_\sigma , \\ \partial_\rho \sqrt{-g}\, \phi^\rho = 0 , \\ G^{\mu\nu} = \chi^{\mu\nu\rho\sigma} F_{\rho\sigma} . \end{cases}$$

For a rotating frame $g_R = -r^2$. Computing these different equations in a dielectric medium where $J^\nu = 0$, the propagation equation of the potential is found to be

$$\begin{cases} \partial_\mu \left[\sqrt{-g_R}\, \chi^{\mu\nu\rho\sigma} (\partial_\sigma\phi_{R\rho} - \partial_\rho\phi_{R\sigma})\right] = 0 , \\ \partial_\rho \left[\sqrt{-g_R}\, g_R^{\rho\sigma} \phi_{R\sigma}\right] = 0 . \end{cases}$$

The problem is to determine the value of the constitutive tensor $\chi_R^{\mu\nu\sigma\rho}$ in a rotating frame with a comoving linear dielectric medium. We have to consider a point M of the medium and its proper frame $\mathscr{F}_P$. It is an inertial frame in translation with a speed $\vec{v} = \vec{\Omega} \times \vec{r}(M)$ [Figure 2].

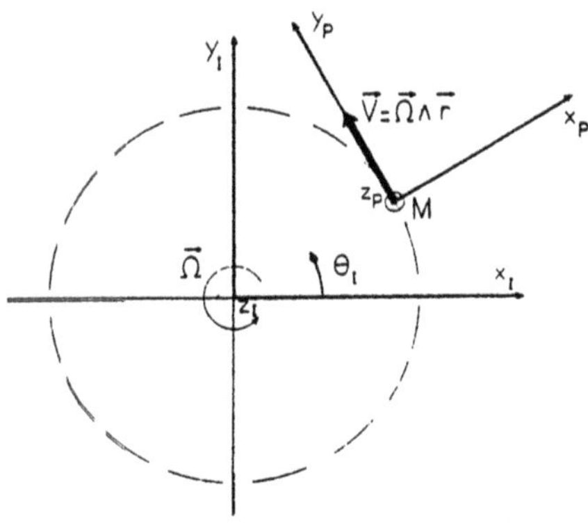

Fig.2 Proper rest frame $\mathscr{F}_p$ of the medium

The axis z_p which is parallel to the axis of rotation z_I, is the instantaneous axis of rotation of the medium in this frame $\mathscr{F}_p$. Therefore, at point M, the medium is at rest in an inertial frame $\mathscr{F}_p$, and thus, the constitutive tensor can be expressed only with the classical dielectric permittivity ε and magnetic permeability μ :

$$\begin{cases} G_P^{23} = \mu^{-1}\mu_o^{-1} F_{P23} , \\ G_P^{13} = \mu^{-1}\mu_o^{-1} F_{P13} , \\ G_P^{12} = \mu^{-1}\mu_o^{-1} F_{P12} , \\ G_P^{14} = -\varepsilon\, \mu_o^{-1} F_{P14} , \\ G_P^{24} = -\varepsilon\, \mu_o^{-1} F_{P24} , \\ G_P^{34} = -\varepsilon\, \mu_o^{-1} F_{P34} . \end{cases}$$

The components of $G_R^{\alpha\beta}$ and $F_{R\gamma\delta}$ in $\mathcal{F}_R$ are connected to the components of $G_P^{\mu\nu}$ and $F_{P\rho\sigma}$ in $\mathcal{F}_P$ through the relations of transformation of coordinates applied twice because these fields are second rank tensors:

$$G_R^{\alpha\beta} = a_\mu^{\ \alpha} \, a_\nu^{\ \beta} \, G_P^{\mu\nu} ,$$
$$F_{P\rho\sigma} = \bar{a}_\rho^{\ \gamma} \, \bar{a}_\sigma^{\ \delta} \, F_{R\gamma\delta} .$$

That leads to:

$$\chi_R^{\alpha\beta\gamma\delta} = a_\mu^{\ \alpha} \, a_\nu^{\ \beta} \, \chi_P^{\mu\nu\rho\sigma} \, \bar{a}_\rho^{\ \gamma} \, \bar{a}_\sigma^{\ \delta}$$

and the constitutive relation in $\mathcal{F}_R$ becomes:

$$\left\{ \begin{aligned}
G_R^{23} &= \mu_o^{-1}\mu^{-1}r^{-2}\gamma^{-2}F_{R23} - \mu_o^{-1}\mu^{-1}\,\frac{\Omega}{c}\,F_{R34} , \\
G_R^{13} &= \mu_o^{-1}\mu^{-1}\,F_{R13} , \\
G_R^{12} &= \mu_o^{-1}\mu^{-1}r^{-2}\gamma^{-2}F_{R12} + \mu_o^{-1}\mu^{-1}\,\frac{\Omega}{c}\,F_{R14} , \\
G_R^{14} &= -\mu_o^{-1}\left[\varepsilon + (\varepsilon+\mu^{-1})\frac{r^2\Omega^2\gamma^2}{c^2}\right]F_{R14} + \mu_o^{-1}\,\mu^{-1}\,\frac{\Omega}{c}\,F_{R12} , \\
G_R^{24} &= -\mu_o^{-1}\,\varepsilon\, r^{-2}F_{R24} , \\
G_R^{34} &= -\mu_o^{-1}\left[\varepsilon + (\varepsilon+\mu^{-1})\,\frac{r^2\Omega^2\gamma^2}{c^2}\right]F_{R34} - \mu_o^{-1}\,\mu^{-1}\,\frac{\Omega}{c}\,F_{R23} .
\end{aligned} \right.$$

In addition to the usual relationship which depends on the coefficients μ and ε of the medium, and to first order terms in r, because of the use of cylindrical coordinates, there are cross-terms which depend on Ω/c and which explain the Sagnac effect. From this result we can find the expression for the propagation equation.

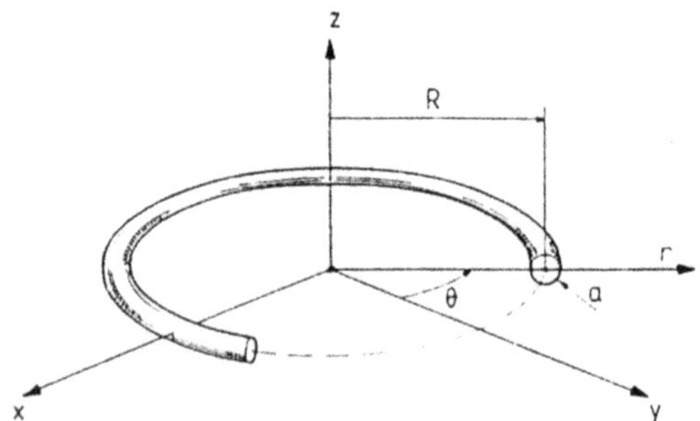

Fig.3 Toroidal waveguide

The calculation is simplified if we consider a toroidal dielectric waveguide where the radius of curvature R is much larger than its lateral dimensions [Figure 3]. For a straight single mode fiber in an inertial frame, the equation of propagation of the amplitude A of the mode has the usual form:

$$\frac{\partial^2 A}{\partial x^2} + \frac{\partial^2 A}{\partial y^2} + \frac{\partial^2 A}{\partial z^2} - \frac{n^2}{c^2}\frac{\partial^2 A}{\partial t^2} = 0 \tag{1}$$

where the index of refraction is $n = \sqrt{\varepsilon\mu}$. An approximate solution is a pseudo-Gaussian mode [7]:

$$A_1(x,y,z,t) = A_0 \exp - \frac{x^2+z^2}{2w_0^2} \exp(i\,\omega t - iky)$$

where k is the propagation constant, ω the light frequency and w_0 the mean mode radius which is a function of the radius of the core and the normalized frequency V of the mode [7]. Now, in a first approximation, the expression of the Laplacian operator in cylindrical coordinates gives:

$$\frac{\partial^2 A}{\partial r^2} + \frac{1}{r^2}\frac{\partial^2 A}{\partial \theta^2} + \frac{\partial^2 A}{\partial z^2} - \frac{n^2}{c^2}\frac{\partial^2 A}{\partial t^2} = 0 . \tag{2}$$

This propagation equation corresponds to the symmetry of a toroidal wave-guide and differs only slightly from the propagation equation (1) to which is applied the change of variable $\theta = y/R$ and $r = R + x$:

$$\frac{\partial^2 A}{\partial r^2} + \frac{1}{R^2}\frac{\partial^2 A}{\partial \theta^2} + \frac{\partial^2 A}{\partial z^2} - \frac{n^2}{c^2}\frac{\partial^2 A}{\partial t^2} = 0 . \tag{3}$$

Thus equation (3) has a solution A_3 directly derived from the solution A_1 of equation (1):

$$A_3(r, \theta, z, t) = A_1(r-R, R\cdot\theta, z, t) .$$

Therefore:

$$A_3(r, \theta, z, t) = A_0 \exp\left(- \frac{(r-R)^2 + z^2}{2w_0^2}\right)\exp(i\,\omega t - ikR\theta) .$$

The solution A_2 of equation 2 can be deduced from A_3 with a method of perturbation which yields, for a mode propagating through a toridal single mode fiber [7]:

$$A_2(r, \theta, z, t) = A_0 \exp - \frac{(r-R-\Delta R_c)^2 + z^2}{2w_0^2} \exp(i\omega t - ikR\theta) .$$

The term ΔR_c represents the centrifugal mode shift due to the fiber curvature and is equal to $[n^2 w_0^4 \omega^2/Rc^2]$ [7] [Figure 4].

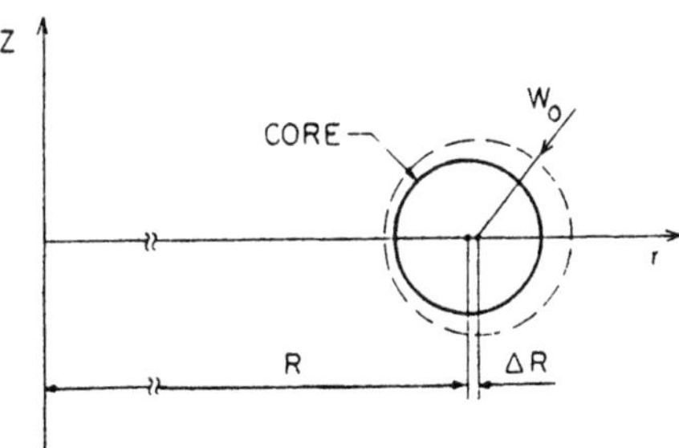

Fig.4 Radial shift of the pseudo Gaussian mode

When the waveguide is rotating about its axis of curvature, the propagation equation in its rest frame deduced from our relativistic electrodynamics approach is, to first order in $r\Omega/c$:

$$\frac{\partial^2 A}{\partial r^2} + \frac{1}{r^2}\frac{\partial^2 A}{\partial \theta^2} + \frac{\partial^2 A}{\partial z^2} - \frac{n^2}{c^2}\frac{\partial^2 A}{\partial t^2} + \frac{\Omega}{c^2}\frac{\partial^2 A}{\partial \theta \partial t} = 0 . \tag{4}$$

The additional derivative $\partial^2 A/\partial\theta\partial t$ depends on the direction of propagation through the first order derivative $\partial/\partial\theta$ and a non-reciprocal behavior occurs between counterpropagating modes. The solution of equation (4) can be found with a method of perturbation from the solutions of equation (2). Looking for solutions of a kind:

$$A_4(r, \theta, z, t) = A_0 \exp - \frac{(r - R - \Delta R_c - \Delta R_R)^2 + z^2}{2w_0^2} \exp(i\omega t - ikR\theta + i\Delta K\theta)$$

we found that the value of the additional radial shift ΔR_R due to rotation is

$$\Delta R_R = \frac{4\omega\, w_0^4\, k\,\Omega}{c^2} .$$

The perturbation ΔK on the propagation constant is:

$$\Delta K = \frac{\omega (R + \Delta R_c)^2}{c^2}\Omega .$$

The radial shift of the mode ΔR_R depends on the respective directions of the light propagation and the waveguide rotation through the sign of the product $k\,\Omega$. Numerically it appears to be negligible when compared to ΔR_c . For $R = 10$ cm, $w_0 = 3$ µm, $\omega = 2 \times 10^{15}$Hz, $n = 1.45$ and $\Omega = 1$ rad s^{-1} , we obtain

$$\Delta R_c = + 0.2 \ \mu m \quad \text{and} \quad \Delta R_R = \pm 6 \times 10^{-11} \ \mu m .$$

The problem is different for the perturbation of the phase. The relative change in the phase $\Delta\phi/\phi$ due to rotation is also very small but, since the phase is cyclical its absolute value $\Delta\phi$ is measured with respect to 2Π in the interference and the value $\Delta\phi/2\Pi$ accumulates with the length of propagation path. After N turns of counterpropagation with the respective constant $[+|k| - \Delta K/R]$ and $[-|k| - \Delta K/R]$ the phase difference becomes:

$$\Delta\phi_R = 4\,N\,\Pi\,\Delta K$$

$$\Delta\phi_R = \frac{4\,N\,\Pi\,(R + \Delta R_c)^2 \omega\,\Omega}{c^2} .$$

This result depends only on the light frequency ω , the vacuum light velocity c , the rotation rate Ω , the number of turns N and the area $\mathcal{A} = \Pi(R + \Delta R_c)^2$ defined by the radius of curvature $R + \Delta R_c$ corresponding to the maximum of the mode. The indices of refraction of the core and the cladding, the phase or group velocities, the dispersion of the medium or of the waveguide have no influence.

We have treated only the case of a toroidal fiber bent in a plane orthogonal to the axis of rotation, but this approach gives, through a more complicated calculation, a generalized result using the equivalent area vector $\vec{\mathcal{A}}$ enclosed by a closed path of arbitrary shape and a rotation pseudo-vector $\vec{\Omega}$

$$\Delta\phi_R = \frac{4\,N\,\omega}{c^2}\,\vec{\mathcal{A}}\cdot\vec{\Omega}\ .$$

$\Delta\phi_R$ appears to be the flux of $\vec{\Omega}$ through the closed path. Now, assuming that the value of the Sagnac phase shift $\Delta\phi_R$ can be measured with good precision, the stability of the scale factor will depend only on the stability of $\vec{\mathcal{A}}$ and ω .

References

1 E.J. Post, Rev. Mod. Phys., 39, 475 (1967)

2 H.J. Arditty and H.C. Lefevre, Optics Letters, 6, 401, 1981

3 J.L. Anderson and J.W. Ryon, Physical Review, 181, 1765, (1969)

4 L.D. Landau and E.M. Lifshitz, The Classical Theory of Fields, Pergamon Press, New York (1975)

5 H. Arzelies, Cinematique Relativiste, Ed. Gauthier-Villars, Paris (1955)

6 See for example M.A. Tonnelat, Principles of Electromagnetic Theory and Relativity, Gordon and Breach, London (1966)

7 W.A. Gambling, H. Matsumura and C.M. Ragdale, Electronics Letters, 14, 130, (1978)

2.2 Reciprocal Operation of the Sagnac Interferometer

2.2.1 Polarization Problems

Polarization and Depolarization in the Fiber-Optic Gyroscope

R. Ulrich

Technische Universität Hamburg-Harburg
D-2100 Hamburg 90, Fed. Rep. of Germany

Abstract. From the scattering matrix of a general N-mode fiber-optic ring interferometer conditions are derived for gyro operation with stable zero point and stable scale factor. The discussion includes reciprocity, polarization filtering, stray light suppression, polarization adjustment and control, birefringence effects in optical fibers, nonreciprocity by magnetic field, polarization-holding fibers, and polarization scrambling by a depolarizer.

1. Introduction

The fiber-optic gyroscope in the form of a SAGNAC interferometer is a remarkable optical device by the fact that the SAGNAC phase shift $2\phi_S$ is independent of the refractive index of the optical fiber [1]. Therefore, this interferometer represents an *absolute* physical instrument. It compares directly the angular rotation Ω of the fiber coil with an effective angular rotation rate Ω_o of the light in that fiber coil. Rewriting the well-known SAGNAC formula, we see that

$$\phi_S = 2\pi\Omega/\Omega_o \tag{1}$$

with the effective rotation rate

$$\Omega_o = c/\bar{R} . \tag{2}$$

Here c is the light velocity in free space, and $\bar{R} \equiv (P_c/\lambda)R_c$ appears as an effective radius of the fiber coil, equal to the actual coil radius R_c times the number (P_c/λ) of free-space wavelengths λ that fit in the total perimeter P_c (fiber length) of the fiber coil. This number is large, of the order $P_c/\lambda \sim 10^9$, thus explaining the high possible sensitivity of the SAGNAC interferometer.

In the gyroscope, where the phase shift $2\phi_S$ must be measured with high accuracy, some fundamental problems arise, e.g. photon noise, phase bias, backscatter, and polarization. The latter problem is inherently related to the questions of stability and sensitivity of the fiber-optic SAGNAC interferometer. They will be discussed in this contribution. We start the discussion from the viewpoint of electromagnetic network theory [2]. The fiber and the directional coupler are treated as general *multimode* optical devices in Sec.2, and the optical properties of a general multimode fiber ring are stated. Section 3 deals with those conditions under which such a ring can operate as a gyro with stable zero under varying conditions.This treatment of a gyro with N modes comprises the desired polarization properties of the gyro with a "single"-mode fiber as the special case N = 2. Moreover, this multimode treatment may be used to account for stray light and multimode-propagation (sometimes confused with nonreciprocal propagation), and it shows most clearly the need for a single-mode spatial filter, having equally high quality as the polarization filter. Section 4 contains a summary of birefringence effects in single-mode optical fibers. Their understanding is necessary to reduce fluctuations in the interference contrast of the SAGNAC interferometer, to explain the magnetic field sensitivity of the fiber-optic gyro, and also for the adjustment and control of the polarization conditions in a fiber gyroscope. In Sec.5, finally, the methods are discussed by which it is possible to achieve operation with stable polarization conditions. In that section, the use of a polarization scrambler (depolarizer) is shown to yield a particularly simple and stable configuration of a gyroscope.

Most of these topics have been discussed already at several places in the literature [3 - 11] or are addressed by other authors in this volume. Therefore, the goal of the present contribution is to give a closed-form, coherent presentation of the various problems involved.

2. General, N-mode Fiber Loop

An optical fiber is a passive, linear electromagnetic network [2]. With suitable (mutually orthogonal, normalized) wave functions defined in either end plane of the fiber (i = 1,2, see Fig. 1), the fields at the fiber for a given optical frequency ω are described completely by the set of complex modal amplitudes $a_{ij}(\omega)$ of all waves entering the fiber and $b_{ij}(\omega)$ of all waves leaving the fiber, with j = 1...N. Here N is the total number of modes involved. It is well known [11,12] that we must use N = 2 to describe the light guided in the core of a so-called "single"-mode fiber, e.g. with j = 1 referring to the x-polarized HE_{11} mode, and j = 2 to the y polarized mode.

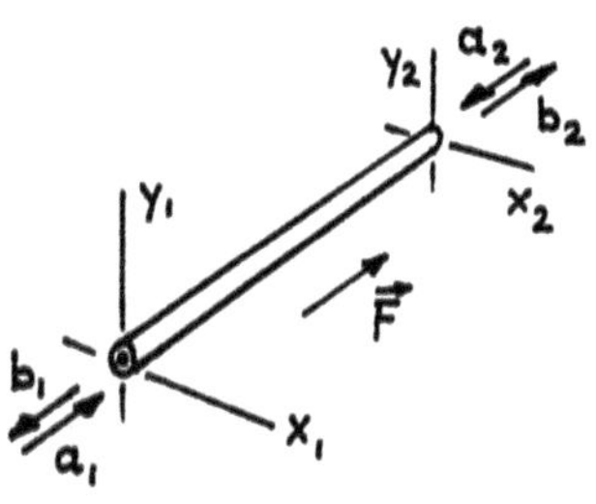

Fig.1 General fiber with input waves a_i and output waves b_i at both ends

Regarding the high accuracies (e.g. 10^{-6} rad) desired in measuring the SAGNAC phase shift it is reasonable, however, to include into our treatment also light which may be guided in the cladding of the fiber. Such light is generally considered as stray light, and it may be suppressed more or less completely by a cladding mode stripper. We may include this light here very simply, essentially without complicating our mathematical formalism in any way, by identifying N with the total number of core *plus* cladding modes that exist in the fiber. With that generalization we are able, beyond treating polarization, to show quantitatively to what level ($\simeq 10^{-12}$) the cladding modes must be suppressed. Thus, N may be a number of the order of 10^3.

At sufficiently low light intensities, linear relationships must exist between all input and output amplitudes. In order to express them in a simple form, we combine the N amplitudes a_{ij} and b_{ij} at either fiber end to form the (column) vectors

$$a_i = \{a_{i1}, a_{i2}, \dots a_{iN}\}, \quad (3)$$

$$b_i = \{b_{i1}, b_{i2}, \dots b_{iN}\}. \quad (4)$$

They are represented in Fig. 1. Ignoring here the problem of backscattering, we have

$$b_2 = \vec{F} a_1, \qquad 1 \rightarrow 2 \quad \text{transmission}, \quad (5)$$

$$b_1 = \overleftarrow{F} a_2, \qquad 1 \leftarrow 2 \quad \text{transmission}. \quad (6)$$

The matrices $\vec{F}(\omega)$ and $\overleftarrow{F}(\omega)$ are general N×N transmission matrices describing the properties of the fiber, including all possible perturbations. We consider them as given quantities throughout this section. They are in general analytical functions of the optical frequency ω.

If we focus our attention only on the polarization properties of the fiber (N = 2), the vectors a_i and b_i are simple Jones vectors, and likewise $\vec{F}$ and $\overleftarrow{F}$ are the Jones matrices of the fiber.

2.1 Reciprocity

General network theory [2] shows that the 1 → 2 and 2 → 1 transmission matrices are interrelated if the fiber is time-invariant and not exposed to a magnetic field. Under those conditions, the fiber is reciprocal and

$$\vec{F} = \overleftarrow{F}^T \quad (7)$$

where the symbol T denotes transposition. The reason for (7) is that $\vec{F}$ and $\overleftarrow{F}$ are parts of the general scattering matrix of the fiber. We shall discuss the significance of this condition of reciprocity below in Sec.3. In the most general case, i.e. even when the reciprocity condition (7) is violated, we can always split the pair of matrices $\vec{F}$ and $\overleftarrow{F}$ into their nonreciprocal and reciprocal parts

$$\vec{F} = \vec{F}_R + \vec{F}_{NR} , \tag{8}$$

$$\overleftarrow{F} = \overleftarrow{F}_R + \overleftarrow{F}_{NR} , \tag{9}$$

where

$$\vec{F}_R = \overleftarrow{F}_R^{\,T} = \frac{1}{2}(\vec{F} + \overleftarrow{F}^T) , \tag{10}$$

$$\vec{F}_{NR} = - \overleftarrow{F}_{NR}^{\,T} = \frac{1}{2}(\vec{F} - \overleftarrow{F}^T) . \tag{11}$$

The condition (7) of reciprocity is equivalent then to the requirement that $\vec{F}_{NR} = 0$.

2.2 Directional Coupler

The directional coupler (or beamsplitter) of the fiber-optic SAGNAC interferometer is described in a similar way. It is sufficient here to consider the fields at only three of its four arms (Fig. 2), assuming that the fourth arm is matched. Using the same set of wave functions as with the fiber, the most general coupler may be characterized by four matrices $\vec{C}_{\parallel}$, $\overleftarrow{C}_{\parallel}$, $\vec{C}_X$, and $\overleftarrow{C}_X$. Here the subscripts $\parallel$ and X refer to straight-through propagation and to cross-coupling, respectively. The arrows → and ← refer to forward and backward propagation according to Fig. 2.

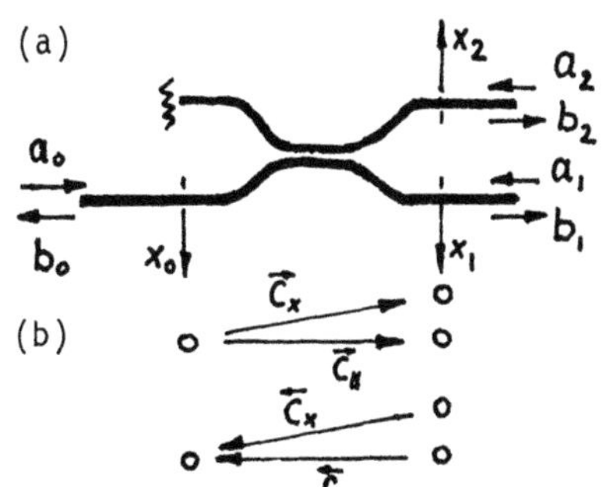

Fig.2 (a) General directional coupler, (b) representation of the coupling matrices $C_{\parallel}$ and C_X

$$b_1 = \vec{C}_{\parallel} a_o , \tag{12}$$

$$b_2 = \vec{C}_X a_o , \tag{13}$$

$$b_o = \overleftarrow{C}_{\parallel} a_1 + \overleftarrow{C}_X a_2 . \tag{14}$$

These coupler matrices may be identical to the propagation matrix of a short section of fiber (e.g. in the fiber-optic directional coupler [9]). However, the X matrices must also account for the inversion of the directions of the x_1 and x_2 axes (Fig.2) that results from bending the fiber of Fig. 1 into a loop. To each pair of forward/backward coupler-matrices, the same comments about reciprocity apply and relations equivalent to (8)-(11) hold as for the pair of fiber matrices, $\vec{F}$ and $\overleftarrow{F}$. In particular, if the directional coupler is time-invariant and not exposed to a magnetic field, the nonreciprocal parts of C and C_X vanish, so that

$$\overleftrightarrow{C} = \overleftrightarrow{C}^T \qquad \text{for X and } \| \, . \tag{15}$$

2.3 General Fiber Ring as a SAGNAC Interferometer

Fiber and directional coupler are connected now in the fashion of Fig. 3 to form a general, N-mode fiber ring. This ring can be operated as a SAGNAC interferometer in various configurations, differing by the choice of input and output ports. It can be shown under very general conditions that in each of those configurations the N-mode fiber ring will act as a gyroscope in the following sense. If, for a moment, we include again the fourth arm of the directional coupler into the discussion, there exist 2N modes as possible input ports, and the same number of possible output ports. If light is coupled into any one of those input ports or into a linear combination thereof, and if the output power p_o from an arbitrary combination of output ports is detected, that power will vary sinusoidally with the rotation rate Ω of the loop,

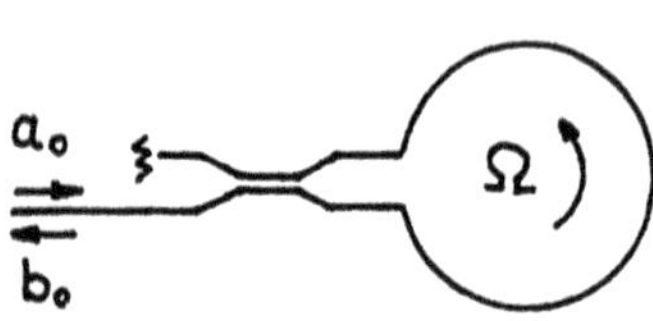

Fig.3 General fiber ring, operating in reflection as a SAGNAC interferometer

$$p_o(\Omega) = \bar{p} + \tilde{p} \cos 2(\phi_S - \phi_o) \, . \tag{16}$$

The SAGNAC phase shift $\phi_S(\Omega)$ has been defined earlier by (1). Moreover, $\bar{p}$ is the mean power level, and $\tilde{p}$ the maximum power variation. In other words, no matter how simply or how sophisticated the interferometer is built, it will always produce a rotation-dependent signal (16) with the correct periodicity.

This signal $p_o(\Omega)$ is generally not very useful, however. Its zero point, defined here (in the absence of any phase biassing means) by the maximum of p_o, is offset by some angle ϕ_o from the condition of zero rotation. This offset, as well as the signal amplitude (scale factor) $\tilde{p}$, depends critically upon the properties of fiber and of directional coupler in almost all of the possible configurations. Therefore, these quantities will generally vary with changing ambient conditions (temperature, pressure) of the fiber, possibly resulting in an unstable and unreliable gyroscope. For this reason, no general proof of the validity of (16) is given here. Yet, for the special configuration of Fig. 3, the relation (16) is derived below in Sec. 3.1.

3. Conditions for a Stable Zero Point of the Interferometer

Comparing all of the mentioned possible input/output configurations, one unique set of operating conditions can be identified which guarantees a stable zero point of the interferometer in any kind of ambient situation. These are the conditions:

(I) The fiber loop must be operated in reflection.
(II) A single-transverse-mode, single-polarization filter must be inserted in the common input/output arm.
(III) The fiber, the directional coupler, and the mentioned filter must be reciprocal in the sense of (7) and (15) when they are at rest ($\Omega = 0$).

In the following, it will be shown that the simultaneous fulfilment of these conditions is sufficient for an absolutely stable zero point ϕ_o and that, moreover, $\phi_o = 0$ holds in this simple arrangement without a phase bias.

It must be emphasized, however, that these conditions are not necessary ones. It is possible of course (at least in principle) to construct a stable interferometer, employing any one of the mentioned configurations, if absolutely stable components are used. Yet, the significance of conditions (I)-(III) is that they yield a stable zero point even with (very slowly) varying components, which means a gyro with vanishing drift.

3.1 Fiber Loop Operating in Reflection

To discuss now the meaning and consequences of the conditions (I)-(III) we note first that condition (I) requires operation of the fiber ring in the very configuration shown in Fig. 3, i.e. without using the fourth arm of the directional coupler. We can calculate the general reflection matrix $R(\Omega)$ for this fiber ring. It relates the output amplitudes b_{oj} of the N-mode fiber ring to the input amplitudes a_{oj}. We combine (5), (6), and (12)-(14), and we introduce explicitly the SAGNAC phase shift $2\phi_S$ between the two counter-rotating light waves. Splitting this phase evenly between those two waves, we obtain

$$b_o = Ra_o \quad , \tag{17}$$

where R is the scattering matrix of the fiber ring,

$$R = \vec{G}\exp(i\phi_S) + \overleftarrow{G}\exp(-i\phi_S) \, , \tag{18}$$

$$\vec{G} = \overleftarrow{C}_X \vec{F} \vec{C}_{\parallel} \, , \tag{19}$$

$$\overleftarrow{G} = \overleftarrow{C}_{\parallel} \overleftarrow{F} \vec{C}_X \quad . \tag{20}$$

Thus, the pair of matrices $\overleftarrow{G}$ and $\vec{G}$ characterize the clockwise and counterclockwise travelling waves in the ring at rest, as seen from the $x_o y_o$ reference plane. Using (18) we can evaluate now directly the total power p_o reflected by the rotating SAGNAC ring of Fig. 3 when it is excited with an arbitrary input wave a_o. Denoting by $*$ the complex conjugate and by $\dagger$ the hermitean conjugate, we find

$$p_o = |b_o|^2 = b_o^{\dagger} b_o = a_o^{\dagger} R^{\dagger} R a_o \tag{21}$$

$$= \bar{p} + p_1 \exp(-2i\phi_S) + p_1^{*}\exp(2i\phi_S) \tag{22}$$

$$= \bar{p} + 2|p_1|\cos 2(\phi_S - \phi_o) \, . \tag{23}$$

Here, we have abbreviated

$$a_o^\dagger (\vec{G}^\dagger \vec{G} + \overleftarrow{G}^\dagger \overleftarrow{G}) a_o = \bar{p} , \tag{24}$$

$$a_o^\dagger \vec{G}^\dagger \overleftarrow{G} a_o = p_1 = |p_1| \exp(2i\phi_o) . \tag{25}$$

We recognize that the reflected power (23) has indeed the form (16) mentioned earlier. Moreover, (25) indicates that the phase ϕ_o at zero rotation rate depends in general both upon the kind of excitation of the ring (a_o) and upon the properties (e.g. birefringence) of the fiber through $\vec{G}^\dagger \overleftarrow{G}$. As a consequence, the zero point of the interferometer may still drift if only condition (I) is satisfied.

3.2 Fiber Ring Operating With a Filter

This problem is removed by the condition (II). It requires that both the input wave a_o and the output wave b_o be filtered by the same single-mode, single-polarization filter P, as indicated schematically in Fig. 4. We describe the properties of such a filter in forward and backward propagation by a pair of matrices, $\vec{P}$ and $\overleftarrow{P}$, respectively. They are particularly simple for an ideal filter that transmits only the j = 1 mode, i.e. in a single-mode fiber only core-guided light of x-polarization. In that case we have $\vec{P} = \overleftarrow{P} = P_{11}$, representing an N × N matrix that contains a 1 in the top left corner and zeros everywhere else. In the general case, corresponding e.g. to an elliptical-polarization filter of oblique azimuthal orientation, we may represent the ideal filter P by a sequence of three elements: a general (unitary) mode transformer (e.g. birebringent plate) U, the above-mentioned filter P_{11}, and the inverse transformer $U^{-1} = U^\dagger$. We anticipate here already part of condition (III) by postulating the filter (and therefore U) to be reciprocal. Such a filter has the following properties

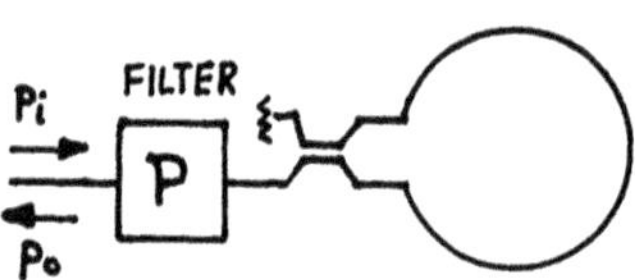

Fig.4 General fiber ring interferometer, operating through the single-mode, single-polarization filter

$$\vec{P} = U P_{11} U^\dagger = \vec{P}\vec{P} = \vec{P}^\dagger = \overleftarrow{P}^T = \overleftarrow{P}^* . \tag{26}$$

The single polarized mode passed by this filter is characterized by the first column of U. This filtered mode may coincide with any one of the fiber modes used in the present representation, or it may be a linear combination of them.

Using these properties, we calculate the power p_o reflected by a rotating SAGNAC fiber ring which includes the filter (Fig. 4), assuming excitation with an arbitrary wave a_o.

We find

$$\begin{aligned} p_o &= |\overleftarrow{P} R \vec{P} a_o|^2 = a_o^\dagger \vec{P} R^\dagger \overleftarrow{P} R \vec{P} a_o \\ &= p_i \, T_{Pa_o} \, |r^{(P)}|^2 . \end{aligned} \tag{27}$$

Here $p_i = |a_o|^2$ is the optical input power, and $T_{Pa_o} = a_o^{\dagger} \vec{P} a_o / p_i$ is the power transmission factor of the filter P for the particular mode distribution and polarization of the input wave a_o. This factor could include, for example, coupling or splice losses at the input of the loop of Fig. 4. The third factor appearing in (27) is

$$r^{(P)} = [U^T R U]_{11} \tag{28}$$

$$= 2[U^T \vec{G}_R U]_{11} \cos \phi_S + 2i[U^T \vec{G}_{NR} U]_{11} \sin \phi_S .$$

Here the pair of matrices, $\vec{G}$ and $\overleftarrow{G}$, have been split into their reciprocal and nonreciprocal parts in analogy to (8) - (11), and the symbol $[\ ]_{11}$ identifies the top left matrix element (i,j=1,1). We may interpret $r^{(P)}$ as the $[\ ]_{11}$ element of a projection of the ring matrix R on the filter matrix P. In the same sense, we abbreviate $[U^T \vec{G}_R U] = g_R^{(P)}$ and

$$[U^T \vec{G}_{NR} U] = g_{NR}^{(P)} ,$$ and we obtain

$$p_o = 4p_i T_{Pa_o} \Big[|g_R^{(P)}|^2 \cos^2 \phi_S \tag{29}$$

$$+ |g_{NR}^{(P)}|^2 \sin^2 \phi_S + 2\mathrm{Im}\{g_R^{(P)} g_{NR}^{(P)*}\} \sin \phi_S \cos \phi_S \Big] .$$

Here, Im { } indicates the imaginary part. The result (29) is still of the general form (16) with $\phi_o \neq 0$ because of the presence of the terms containing $\sin \phi_S$. We recognize, however, that those terms are associated only with the nonreciprocal part $g_{NR}^{(P)}$ of $\vec{G}$. They vanish if condition (III) is satisfied, i.e. if all components of the gyro loop (fiber, directional coupler, and filter) are reciprocal. In that case, a condition corresponding to (7) holds for each component and, therefore, also for $\vec{G}$, and we have $\vec{G}_{NR} = 0$.

3.3 The Ideal Gyro

For the ideal gyro, satisfying all three conditions (I)-(III), the reflected power becomes simply

$$p_o = 4p_i T_{Pa_o} |g^{(P)}|^2 \cos^2 \phi_S . \tag{30}$$

The subscript R has been omitted here from $g^{(P)}$ because $\vec{G} = \vec{G}_R$ anyway. We compare this result (30) with (16). The zero point of the gyro, given gy the maximum of the power (30), coincides exactly with zero rotation rate ($\phi_S = 0$), independent e.g. of the birefringent properties of the fiber (contained in $g^{(P)}$) or of the polarization of the input light. Therefore, the zero offset vanishes, $\phi_o = 0$, and this gyro has no drift of its zero.

This result may appear somewhat trivial because we had postulated by condition (III) time-invariant, i.e. stable components. However, the result (30) still represents an important advantage in comparison to the cases described by (16), (23), or (29). In all those cases there exists generally some permanent offset ($\phi_o \neq 0$), whereas a deviation from (30) can exist only momentarily while a component is actually changing.

The factor $|g^{(P)}|^2$ in Eq. (30) characterizes the combination of directional coupler and fiber loop as seen through the filter P when the gyro is reciprocal (loop at rest, no magnetic field). In particular, this factor contains the birefringence properties of coupler and loop and, therefore, is central to this paper.

The definition $g^{(P)} = [U^T\vec{G}U]_{11}$ indicates that this factor describes the following process. Light passes from the source through the filter $\vec{P}$ and may then be characterized by a vector Uu_1. Here u_1 is a unit column vector, $u_1 = \{1,0,0,....\}$ denoting the linear horizontal polarization in the fundamental mode which emerges from the (fictive) polarizing part P_{11} of the general polarizer (26). This light flux is normalized to unit power at this point. The unitary matrix U transforms u_1 into the true output state of $\vec{P}$. This light Uu_1 passes through the directional coupler $\vec{C}_\parallel$, propagates around the loop $\vec{F}$ in the c.c.w. direction, and passes through the coupler $\overleftarrow{C}_X$ again. Along its way, the optical power may have been scattered into a number of modes, differing in spatial distributions and polarizations. This mixture of modes is characterized by $\overleftarrow{C}_X \vec{F} \vec{C}_\parallel Uu_1 = \vec{G}Uu_1$. Returning again to the filter $\overleftarrow{P}$, this light is transformed to $U^T \vec{G}Uu_1$ and falls on P_{11}. By multiplication with $u_1^\dagger$ we find the amplitude passing through P_{11} in the form $u_1^\dagger U^T\vec{G}Uu_1 = g^{(P)}$. The amplitude of the other beam, propagating c.w. around the loop, is attenuated by the same factor.

The magnitude of $|g^{(P)}|^2$ depends through $\vec{G}$ upon the intermodal scattering of the directional coupler and the fiber loop, on their birefringence, and on their losses. The factor is generally largest when the light returning to P_{11} after the roundtrip is still propagating in the fundamental mode and arrives with the polarization state u_1 passing P_{11}. If the light is scattered into other modes or has modified polarization, the power transmission of P_{11} is reduced by a factor $0 \leq T_{PG} \leq 1$. In the limit of weak scattering and nondichroic losses we may express the power transmissions of fiber and directional coupler by the factors $T_\parallel = [\vec{C}_\parallel^\dagger \vec{C}_\parallel]_{11}$, $T_X = [\overleftarrow{C}_X^\dagger \overleftarrow{C}_X]_{11}$, and $T_F = [\vec{F}^\dagger \vec{F}]_{11}$. For an ideal lossless 3dB directional coupler, $T_\parallel = T_X = 1/2$ and $4\, T_\parallel T_X \to 1$. With these factors we may rewrite (30)

$$p_o = p_i \, T_{pa_o} \, (4\, T_\parallel T_X) \, T_F \, T_{PG} \cos^2\phi_s. \tag{31}$$

In order to discuss the influence of fiber birefringence on T_{PG} (i.e. on the scale factor of the gyro) we consider the ideal gyro at rest, $\Omega = 0$, and ignore scattering. The fiber loop is characterized then by the symmetric Jones matrix

$$R_o \equiv R(\phi_s{=}0) = \vec{G} + \overleftarrow{G} = \vec{G} + \vec{G}^T. \tag{32}$$

Postulating further that fiber and couplers are non-dichroic, two real eigenvectors of R_o exist, and we recognize that the polarization eigenstates of R_o are *linear* in this ideal case.

As a consequence, it is then possible to use a linear polarizer in the filter P. By adjusting its azimuth one can find two orthogonal directions at which P agrees with one of the polarization eigenstates of R_o, and T_{PG} assumes a relative maximum. Generally, however, these maxima are <u>not</u> absolute maxima ($T_{PG} \to 1$), even in the case of perfect spatial overlap between the returning light and the filter. The reason is that the magnitude of the eigenvalues of R_o (the amplitude reflectivity of the loop) depends on the birefringence of the loop, i.e. on $\vec{G}$. In the worst possible case ($\vec{G}$ skew symmetric) these eigenvalues may even vanish, $R_o \to 0$. The absolute maximum, $T_{PG} \to 1$, requires $\vec{G}$ to be symmetric, i.e. identical linear polarization eigenstates for both the c.w. and c.c.w. propagating beams. This problem will be discussed further in Sec. 5.

When the gyro is rotating, the eigenstates of R are changed. As the filter P remains fixed, the fraction of power is reduced which is transmitted by P after the roundtrip of the light through the rotating ring, thus explaining the factor $\cos^2\phi_S$ in (30). For a rotation rate $\Omega = \Omega_o/4$, for example, we have $\phi_S = \pi/2$, and the reflection matrix (18) becomes skew-symmetric, $R^T = -R$. Such a matrix has vanishing eigenvalues. Therefore, the reflected power is zero in this case, in accordance with (30).

It is also interesting to compare, at an arbitrary position z along the fiber, the polarization states of the two counterpropagating light waves in an ideal ring interferometer. We split the fiber at z so that two fiber sections $\vec{F}_1$ and $\vec{F}_2$ are formed, with $\vec{F}_2\,\vec{F}_1 = \vec{F}$. The amplitudes of the two waves arriving at z are

$$b_+ = \vec{F}_1\,\vec{C}_{\parallel}\vec{P}a_o. \tag{33}$$

$$b_- = \overleftarrow{F}_2\,\vec{C}_X\,\vec{P}a_o\;. \tag{34}$$

Here, the subscripts + and - refer to the c.c.w. and c.w. propagating waves, respectively. Assuming that the filter P and the fiber ring (R) are perfectly matched ($T_{PG} \to 1$), the polarization $\vec{P}a_o$ at the input to the ring is linear, ($\vec{P}a_o$ is real), and it is the same again after the roundtrip of the light along $\vec{G}$ or $\overleftarrow{G}$. Therefore, ignoring here all losses, $\vec{G} = \overleftarrow{C}_X\,\overleftarrow{F}_2\,\vec{F}_1\,\vec{C}_{\parallel}$ is an identity transformation in the ideal gyro, and

$$\vec{F}_1\,\vec{C}_{\parallel} = (\overleftarrow{C}_X\,\vec{F}_2)^{-1} = \overleftarrow{F}_2^{\,*}\,\vec{C}_X^{\,*}\;. \tag{35}$$

We have used here the fact that $\vec{F}_2$ and $\overleftarrow{C}_X$ are unitary under the assumptions of losslessness and reciprocity. Combining (33)-(35) and remembering that Pa_o is real, we find, for any z,

$$b_+ = b_-^{\,*}\;. \tag{36}$$

This means that the two counterpropagating waves in an ideal, optimally adjusted ring interferometer have *identical polarizations* everywhere, i.e. that the azimuths and ellipticities of their polarization ellipses coincide, and that they have the same sense of rotation (e.g. both left elliptical) relative to their respective directions of propagation. When referred to the same direction (e.g. 1 → 2 on the fiber), these rotations have opposite senses, of course. For this statement about the identity of polarizations the assumption (III) of reciprocity is the key condition.

From these considerations, in particular from (30) and (31), we recognize that conditions (I) - (III) do guarantee a stable zero point, but that they cannot guarantee a stable signal p_o (scale factor) for a given rotation rate. In fact, under certain conditions of birefringence or mode coupling in the fiber, the light returning from the ring to $\tilde{P}$ may have a polarization orthogonal to that of $\tilde{P}$. In that case, $T_{PG} \to 0$ and the signal fades. The technical possibilities of avoiding such a fading will be discussed below in Sec. 5. Here, we proceed by deriving some estimates on how well the conditions (I) - (III) must be satisfied.

3.4 Gyro With Reciprocity Error

Nonreciprocal effects may arise in the optical system of a fiber gyro due to temporal variations, e.g. during a change of the temperature distribution [13], and also by the action of a magnetic field like the earth magnetic field [14]. Without going into details here, we can estimate the effect of such nonreciprocities on the indication of the fiber gyro by means of (29). We note first that (29) can be rewritten in the form (16), exhibiting a zero offset

$$\phi_o \approx \mathrm{Im}\left\{ g_R^{(P)} \; g_{NR}^{(P)*} \right\} \Big/ \left| g_R^{(P)} \right|^2 , \qquad (37)$$

$$|\phi_o| \leq \left| g_{NR}^{(P)} \Big/ g_R^{(P)} \right| . \qquad (38)$$

In deriving Eq. (37) it has been assumed that the nonreciprocal part of G is small compared to the reciprocal one. Under the same assumption, line (38) gives an upper limit for the possible nonreciprocity offset ϕ_o. The actual magnitude and the sign of that offset depend on the phase of $g_{NR}^{(P)}$ relative to $g_R^{(P)}$.

The best fiber-optic gyros reported to date [15,16] measure the SAGNAC phase shift ϕ_S to an accuracy of the order of 10^{-6} rad. At least the same accuracy should be specified concerning a possible zero offset from nonreciprocal effects. Hence it should be required for such a gyro that the relative nonreciprocity (38) of the fiber ring must be smaller than 10^{-6}. With respect to magnetic-field-induced nonreciprocity, this can be achieved by shielding [14], by spatial or temporal averaging (see Sec.5 below), or by a combination of these methods.

3.5 Polarization Filtering

The discussions in Sec. 3.2 and 3.3 have shown the need for a filter P which passes only one spatial mode in one polarization. In practice, this spatial filtering can be realized conveniently by a short section of single-mode fiber [5]. The polarization filter may be a bulk polarizer arranged in series with that fiber, but preferably it is integrated directly into that fiber [15]. The specifications of this filter P are extremely demanding, as they affect directly the stability of the zero point of the interferometer and, therefore, the drift and the minimum detectable rotation rate of the interferometer.

The consequences of imperfect polarization discrimination in the filter P have been discussed by KINTNER [6]. Following his argument, we represent an imperfect polarization filter by a matrix $P = \vec{P} = \overleftarrow{P} = P_{11} + \varepsilon P_{22}$, where P_{22} contains a 1 only at the position $i,j = 2,2$ and zeros everywhere else, so that the small parameter ε is the *amplitude* extinction ratio of that polarizer. In the arrangement of Fig. 4, the reflected amplitude becomes, instead of (27),

$$\begin{aligned} b_o &= P\ R\ Pa_o \\ &\approx P_{11}\ R\ P_{11}a_o + \varepsilon(P_{22}\ R\ P_{11} + P_{11}R\ P_{22})a_o \end{aligned} \tag{39}$$

and in components

$$b_{o1} = r_{11}\ a_{o1} + \varepsilon\ r_{12}\ a_{o2}\ , \tag{40}$$

$$b_{o2} = \varepsilon\ r_{21}\ a_{o1}\ . \tag{41}$$

Here r_{ij} are the components of the matrix R. We have neglected terms of order ε^2 in (39) - (41). We do so again when we evaluate the reflected power p_o. We assume $\vec{G} = \overleftarrow{G}^T$ as reciprocal now. Denoting its components by g_{ij} we obtain

$$\begin{aligned} p_o &= |b_{o1}|^2 + |b_{o2}|^2 \approx |b_{o1}|^2 \\ &= \big|(g_{11}\ a_{o1} + \varepsilon\ g_{12}\ a_{o2})\ \exp(i\phi_S) \\ &\quad + (g_{11}\ a_{o1} + \varepsilon\ g_{21}\ a_{o2})\ \exp(-i\phi_S)\big|^2\ . \end{aligned} \tag{42}$$

Representing a fiber ring operating with more than a single mode, this expression corresponds again to the form (16). By inspection of (42) it can be seen that the zero point is generally offset from its ideal value ($\phi_S = 0$) by an angle ϕ_o, with an upper limit $\phi_{o,max}$

$$|\phi_o| \le \phi_{o,max} = \varepsilon \left|\frac{g_{12} + g_{21}}{g_{11}}\right| \cdot \left|\frac{a_{o2}}{a_{o1}}\right|\ . \tag{43}$$

The actual offset ϕ_o may vary in magnitude and sign within the limit (43), because the g_{ij} are components of the Jones matrix of the fiber ring and, therefore, may slowly vary with changing ambient conditions.

In order to keep the resulting rate error of the gyro small, the value of $\phi_{o,max}$ must be kept below the noise-equivalent SAGNAC phase determined by other perturbations, e.g. photon noise. As mentioned, this limit lies presently near 10^{-6} rad. Hence, to reach that limit, the right-hand side of (43) must be kept below 10^{-6} or - 120dB. Various steps are possible to achieve this, according to the three factors there.

A first step is to use a polarized source and align it so that most of its power passes through the filter P. This is equivalent to minimizing the crosspolarized amplitude (a_{o2}) falling on the imperfect filter. With some care, a ratio $|a_{o2}/a_{o1}| < 10^{-1}$ is possible, corresponding to -20dB. As a second step, the off-diagonal elements of the Jones matrix G must be minimized. This adjustment is equivalent to maximizing g_{11} and, consequently, the power falling on the detector. It is achieved by hand-operated polarization controllers [15] or by an active polarization control [5]. These techniques will be discussed below in Sec.5. They permit an amplitude ratio $|(g_{12}+g_{21})/g_{11}|$ reaching -20 to -40dB under optimum conditions.
Thirdly, and most importantly, a good polarization filter is necessary, characterized by a small extinction ratio ε. BERGH et al. [15] estimate a -95dB extinction for their in-line fiber-optic polarizer. Thus, the required total reduction of $\phi_{o,max}$ to better than -120dB is possible.

Beyond these three steps mentioned, the perturbations resulting from an imperfect polarization filter can be further reduced by temporal or wavelength averaging, e.g. by a suitable polarization scrambler. The principle here is to make the offset ϕ_o vary rapidly *and uniformly* through all possible values. As these values are generally distributed symmetrically about $\phi_o = 0$, the averaged output power p_o may appear essentially free of any offset of its zero point. Details of this method are discussed below in Sec. 5.

3.6 Spatial filtering

The preceeding discussion about the suppression of light in the cross-polarized mode applies, in perfect analogy, also to the suppression of light in any other spatial mode of the fiber, i.e. to stray light propagating in the cladding of the fiber, and to stray light reaching the detector by any other way except through the mode selected by P. The origin of either kind of stray light may be scattering from discrete discontinuities (e.g. at fiber end faces, at surfaces of a lens or polarizer etc., in the directional coupler, at splices), or it may be Rayleigh scattering distributed along the fiber. If such stray light reaches the detector and interferes with the signal light, it causes an offset ϕ_o of the zero point as discussed in detail in connection with Eq. (16). This kind of offset should not be confused with the nonreciprocity offset (37), however. The "multimode offset" (16) discussed here can exist even in an interferometer which is perfectly reciprocal in the sense of (7) if the filter P is insufficient.

To function as a spatial mode filter, the filter fiber must be operated well within its single-mode frequency range (i.e. at sufficiently low V-number), and most importantly it must be equipped with a highly efficient cladding mode stripper. The required discrimination is identical to the value (e.g. -120dB) derived above for the polarization filter. The reason is that an argumentation exactly analogous to (39) - (43) may be applied to any one of the cladding modes instead of the cross-polarized core-mode labelled above by $j = 2$. The result (43) must then be interpreted as follows. The parameter ε is the single-pass amplitude extinction ratio of the particular cladding mode in the mode stripper. The coefficients g_{12} and g_{21} correspond to the amplitude scattering ratio between the main mode (j=1) at the input to the fiber ring (without P), and the particular cladding mode at the output of the ring. The ratio $|a_{o2}/a_{o1}|$ describes here how well the power available from the source is coupled into the main mode. In this context, the significance of a high coupling efficiency does not lie primarily in the increased detector signal. Rather, a high efficiency η is advantageous here because the complement $(1-\eta)$ of the source power is going mainly into the cladding modes, thus causing a poor ratio $|a_{o2}/a_{o1}|$.

In practice, the situation with cladding modes is somewhat more complicated than outlined above because of the presence of many ($N \simeq 10^3$) cladding modes. Their large numbers aggravate the offset problem but, on the other hand, they tend to interfere with random phases and therefore cause some averaging. In any case, the main contribution to spatial filtering must come from the attenuation of the cladding modes by the mode stripper. According to the earlier considerations, it must suppress all cladding modes by approximately 120 dB or better in a high quality gyroscope. This high extinction can be reached by choosing a sufficiently long fiber and a well matched absorptive coating around the cladding of the filter fiber.

4. Birefringence Effects in Optical Fibers

Ideally, the ring interferometer (Fig. 4) would be equipped with a fiber in which the polarization injected through the filter P is maintained along the entire fiber length and through the directional coupler, so that the output light from the ring can pass with maximum efficiency through P again. In practice, however, the fiber is subject to various anisotropic influences causing birefringence. They may modify the state of polarization (SOP) of the light along the fiber and at the output, thus generally reducing T_{PG} and possibly even causing a complete fading of the signal.

These birefringence effects will be discussed now. Their understanding is necessary to avoid, as far as possible, all disturbing birefringences in the construction of fiber-optic gyroscopes. On the other hand, some of these birefringence effects are produced deliberately, in a controlled manner (see Sec. 5.1) to cancel the disturbing birefringences existing unintentionally in practical ring interferometers.

We may distinguish between internal birefringence effects which are already present in a straight isolated piece of fiber, and birefringence effects due to external forces or fields acting on the fiber. There are two internal effects [11]. Deviation of the fiber core from perfect circular symmetry is a geometrical anisotropy, causing birefringence particularly in fibers with a high core/cladding index difference and correspondingly small core diameter [17]. The "slow" axis of this birefringence coincides with the direction of maximum polarizability of the cross-section of the core i.e. with the direction of its maximum transverse dimension. The other internal effect is anisotropic mechanical stress, frozen into the fiber during its fabrication, and causing birefringence by the elasto-optic effect in the fiber material [30]. In this case the slow axis coincides with the direction of minimum transverse compressive stress.

Both internal effects usually exist unintentionally already in a fiber with a nominally round and stress-free core. Typically, their magnitude in such a fiber is characterized by a phase difference (specific retardation) of $\beta \approx 2\pi$ rad/m between the fast and slow eigenstates of polarization, corresponding to a polarization beat length $L_p \equiv 2\pi/\beta$ of the order of 1m. However, both effects have also been utilized deliberately to produce fibers with the highest possible birefringence. In those fibers beat lengths of less than 1mm have been reached. Experimentally, it is possible to distinguish both effects in a given piece of fiber by their different wavelength dependences [18].

A number of the external birefringence effects are sketched in Fig. 5. The simplest one is caused by a force Q acting with parallel jaws transversely on the fiber [19]. The stress-induced birefringence β is proportional to the force per unit fiber length, Q/ℓ, as indicated in Fig. 5. When the fiber is pressed into a V groove of angle 2γ, the reactive forces from the walls induce additional birefringence which partially cancels that of the primary force Q. Ignoring friction one would expect [20] perfect cancellation. Actually, however, birefringence is only reduced to 0.2...0.3 of its value at $2\gamma = 180^{\circ}$. In a freely bent fiber, the outer part is under tensile stress and is pressed, therefore, laterally on the inner, compressed part. The birefringence induced by that lateral pressure varies quadratically with the curvature of the fiber [21]. If, in addition, axial tension is applied to the fiber, e.g. by winding it under tension onto a drum, an extra birefringence effect arises from the reactive force of the drum [22]. This extra effect is proportional to the curvature and to the relative fiber elongation $\varepsilon = \Delta\ell/\ell$ produced by the axial tension. In all these cases (a) - (d), the induced birefringence is a linear one, with the fast axis along the direction of maximum compressive stress, i.e. along the vertical direction of Fig. 5.

A circular birefringence effect (optical activity) is induced in a fiber when it is elastically twisted [23]. This effect is directly proportional to the twist rate τ and is reciprocal in the sense of (7). The only nonreciprocal effect of Fig. 5 is the circular birefringence which results when an axial magnetic field is acting on the fiber [24], as shown in Fig. 5f. This is the well-known Faraday effect, linear in the magnetic field strength H. The last effect of Fig. 5 is the electro-optic Kerr effect, inducing a (reciprocal) linear birefringence in the fiber by application of a transverse electric field [25]. In silica, this effect is a quadratic one.

In a real fiber-optic interferometer, most of these effects are simultaneously present to a certain extent, together with the internal birefringence of the fiber. Their combined action can be understood in detail by local superposition of the individual effects and subsequent integration, either by matrix methods like those used in the present discussion, or (depending upon personal taste and mental mobility) by the Poincaré forma-

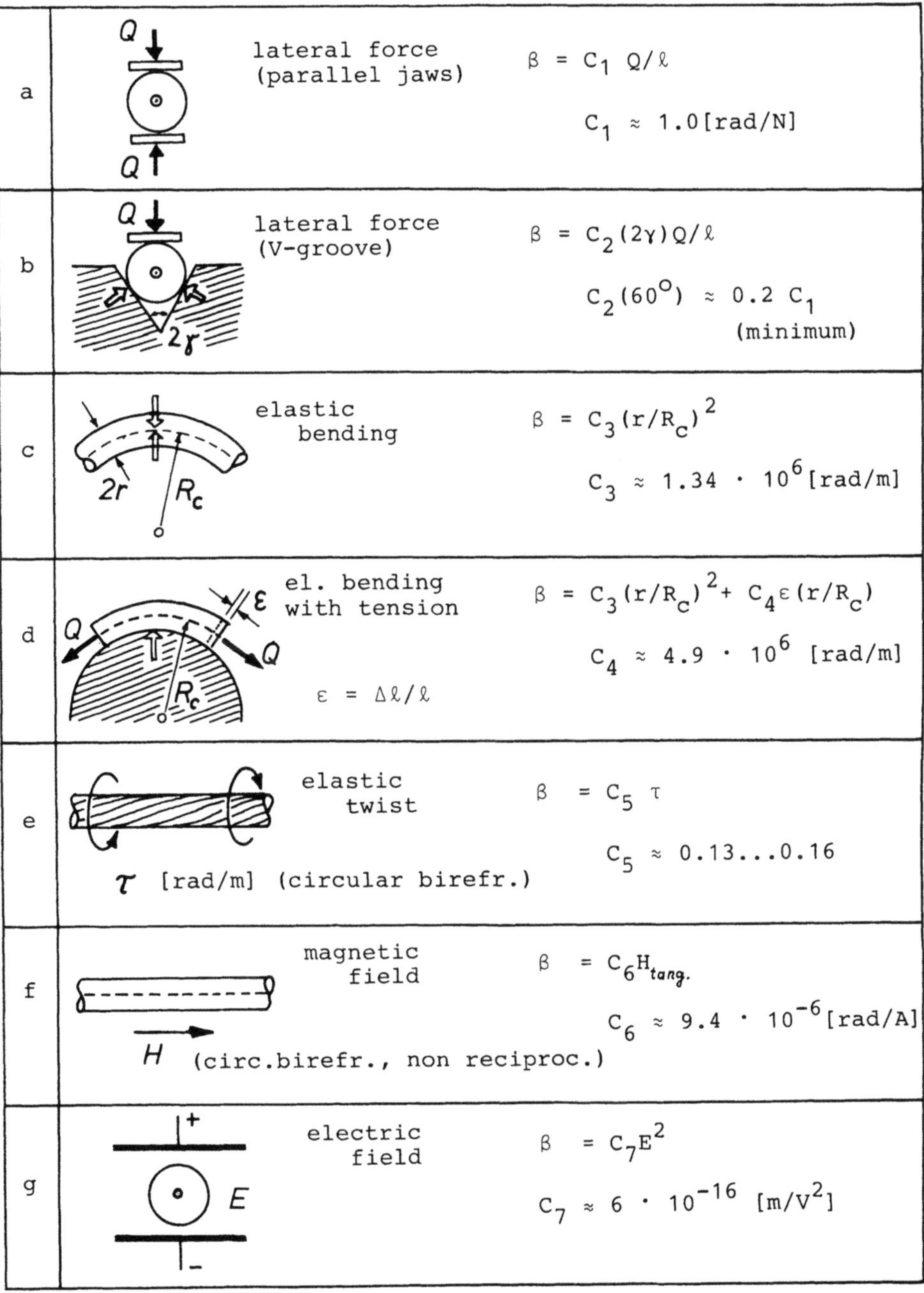

Fig.5 Schematic representation of linear and circular birefringence effects in single-mode optical fibers. The numerical values are given for a silica fiber at a wavelength of 0.633 [μm]. $\beta = |k_{slow} - k_{fast}|$

lism [18, 23, 12]. Generally, the individual birefringence effects tend to mix and to influence each other, resulting in a complicated evolution of polarization along the fiber and in a complicated dependence of the overall polarization properties ($\vec{G}$, $\vec{G}$) on optical frequency, ambient temperature and pressure, magnetic field, age, vibration, etc.

As an illustration of this dependence Fig. 6 shows a record of the output polarization state at the end of a 500 m long single-mode fiber which was heated slowly from 25°C to 68°C. This fiber had a nominally round stress-free core, a soft silicone coating, and was wound loosely on an aluminum drum of 30 cm diameter. In a fiber-optic SAGNAC interferometer equipped with that fiber and not having any polarization control, the output power p_o at constant rotation rate might vary in proportion to the distance of the actual polarization state in Fig. 6 from the center of this Figure, because the factors $|g^{(P)}|^2$ in (30) or T_{PG} in (31) are typically proportional to that distance. Therefore, Fig. 6 indicates clearly the need for some kind of polarization control in the fiber-optic gyro to keep T_{PG} constant or at least nonvanishing. This problem will be discussed in Sec. 5.

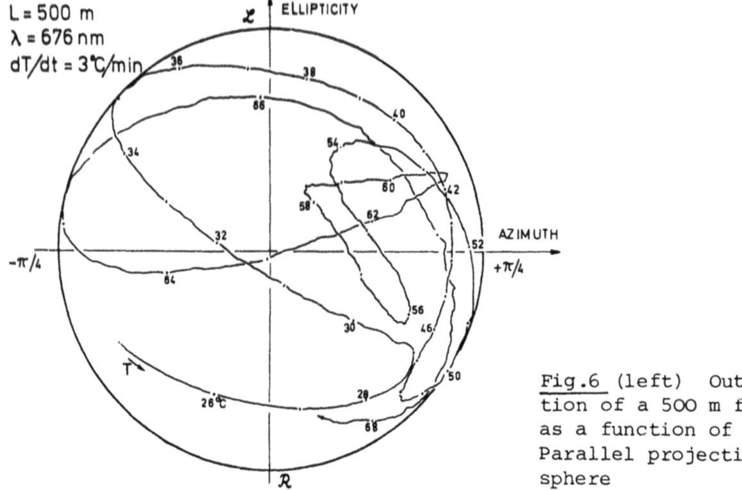

Fig.6 (left) Output polarization of a 500 m fiber, recorded as a function of temperature. Parallel projection of Poincaré sphere

5. Polarization Control

In the basic configuration of the SAGNAC interferometer discussed here, polarization control should generally be provided in at least two places, corresponding to the two factors T_{Pa_o} and $|g^{(P)}|^2$ in (30). At these places, indicated by A and B in Fig. 7, adjustments of polarization are necessary for optimization of those factors, e.g. in the form of (fiber-) elements with adjustable birefringence which modify the polarization of the light passing through.

The first control (A) is arranged somewhere between the light source and the filter P . This control serves to adjust the polarization arriving at the filter for maximum transmission, thus optimizing the factor T_{Pa_o} . For the light returning from the filter to the detector, this control remains without consequences, because the detector is polarization-insensitive. Instead of arranging an explicit polarization control (A) it is alternatively possible to rotate the laser or the polarizer [5] until $T_{Pa_o} \to 1$.

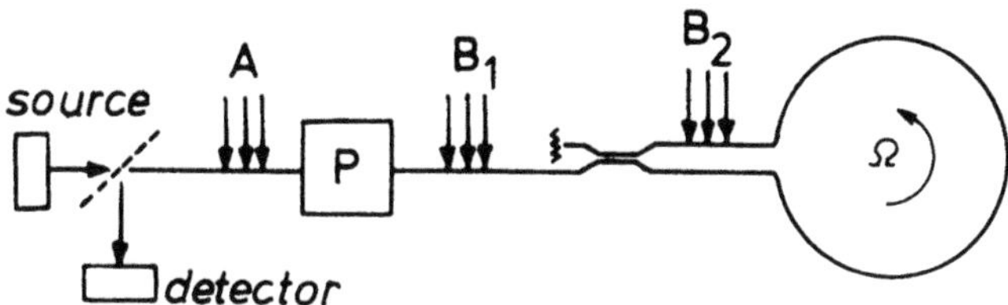

Fig.7 (top) Arrangement of polarization controls for optimization of the filter transmission

A second control is required to optimize the factor $|g^{(P)}|^2$. As discussed earlier, this optimization is required to adjust the azimuth of the linear polarization eigenstate of R_o to that of the filter P and, more importantly, to maximize the amplitude reflectivity of the loop, i.e. the eigenvalues of R_o. Two basically different positions for this second control appear possible, marked by B_1 and B_2 in Fig. 7. However, as will be shown now, these two positions are not at all equivalent for the polarization control.

At B_1, in the common input/output path of the loop, insertion of a general birefringent element modifies the polarizations of both the c.w. and c.c.w. beams at all points in the loop, permitting adjustment of any one of them to any desired state. However, the variations of both beams are correlated in such a way as to always maintain linearly polarized eigenstates of the loop (including B_1). Only the azimuths of these eigenstates can be adjusted by a control at B_1. The optimization of the eigenvalues of R_o is not possible at B_1. To show these facts, we denote by $\vec{B}_1$ and $\overleftarrow{B}_1$ the transfer matrices of a birefringent element inserted at position B_1. Assuming this element as lossless and reciprocal, $\vec{B}_1$ is unitary and $\overleftarrow{B}_1 = \vec{B}_1^T$. For the gyro at rest, the new reflection matrix of the loop is

$$R_1 = \overleftarrow{B}_1 R_o \vec{B}_1 = \vec{B}_1^T R_o \vec{B}_1 .$$

As R_1 is symmetric and nondichroic like R_o, its eigenstates are linear too, possibly with rotated azimuth, regardless of the choice of $\vec{B}_1$. Moreover, the eigenvalues of R_1 are equal to those of R_o except for an unimportant common phase factor. Consequently, it is not possible by adjustments at the position B_1 to reach the absolute maximum ($T_{PG} \to 1$) of the amplitude reflectivity of the loop.

The situation is different at position B_2, inside the loop. By inserting there a general adjustable birefringent element (lossless, reciprocal) we modify $\vec{G}$ directly. We can transform $\vec{G}$ into all other possible transfer matrices describing the same spatial spreading of the light, but differing in polarization. In particular, we are able to adjust $|[U^T \vec{G} U]_{11}|^2 = |g^{(P)}|^2$ to its absolute maximum and thus obtain $T_{PG} \to 1$. Such an adjustment includes both the azimuthal alignment of the eigenstate of R_o with the polarizer P and the optimization of the amplitude reflectivity (eigenvalues of R_o) of the loop. In contrast to a control at B_1, however, the control at B_2 does not modify both the polarizations of the c.w. and of the c.c.w. beams. Rather, at any given point in the loop, the polarization of only one of these beams is affected by B_2.

We conclude that the control at B_2, inside the loop, is necessary and sufficient to achieve optimum polarization conditions ($T_{PG} \to 1$), whereas control at B_1 is not sufficient. Yet, a control at B_1 may be necessary (in addition to a control at B_2) for other reasons, e.g. if a particular input polarization into the fiber is required.

The degree to which an optimization of T_{Pa_o} and $|g^{(P)}|^2$ is required, depends upon the way in which the gyro is biased and on how the rotation rate Ω is evaluated. If the gyro is of the open-loop type and Ω is evaluated directly [5,7-9] from a measurement of p_o, a very high degree of polarization control is required. In that case, any deviation in T_{Pa_o} or $|g^{(P)}|^2$ would represent an equivalent error in the scale factor of the gyro. For this reason, closed-loop gyro systems [16,26] are advantageous if a high stability of the scale factor is desired. In those systems, where the SAGNAC phase is compensated by e.g. a servo-controlled frequency shift between the counterpropagating waves, the interferometer operates essentially continuously at the zero point defined earlier as the maximum of p_o. For a reliable identification of that maximum it is not necessary to maintain the factors T_{Pa_o} and $|g^{(P)}|^2$ very near their optimum. Rather, a signal drop by e.g. 3dB is tolerable there, possibly causing a slight deterioration of the noise equivalent rotation rate. Hence, in the closed-loop systems, a relatively coarse polarization control appears sufficient.

5.1 Adjustable Birefringent Elements

For the controlled adjustment of the polarization, any one of the reciprocal birefringence effects of Fig. 5 is suitable, at least in principle. In practice only two effects have found application, i.e. the lateral pressure and the bending effect. Birefringence control by lateral pressure is extremely simple. The fiber is squeezed by a lateral force (Fig. 8a) which is adjustable either by hand, by an electromagnet [27,28,9], or by piezoelectric control [29]. A force of up to ~ 3N is sufficient to induce up to 180° phase delay. It is an advantage of this method that it can operate relatively fast ($\leq$ 0.1 ms), permitting highly accurate control. Moreover it can be employed even with fibers of the shortest beat lengths. A disadvantage is perhaps that by excessive pressure the bare fiber may be damaged. This problem is avoided by the arrangement of three loops (Fig. 8b) in azi-

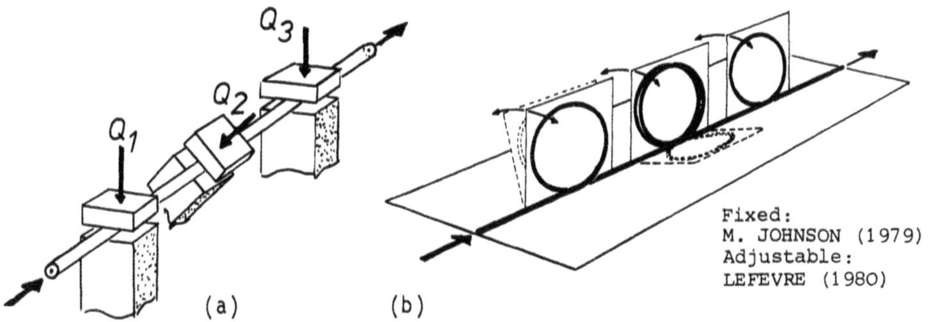

Fig.8 (left) Polarization control by (a) adjustable lateral pressure and (b) birefringent loops of adjustable azimuth [10]

muthally adjustable planes [9,15]. Here, the coating of the fiber need not be removed. The bending radius of the loops is arranged so that they act e.g. as ($\lambda/4$) retarders.

With either method, a set of three adjustable elements is required to permit the most general polarization transform, i.e. to convert a given pair of polarization states into another pair [27]. The simpler problem of transforming one given state into another one can generally be solved by just two controls.

5.2 Servo Control of Polarization

The large possible variations of polarization with temperature shown in Fig. 6 indicate that the polarization control must be readjusted after even minor temperature changes (< 1°C). Therefore, an automatic, servo-controlled polarization adjustment has been suggested [28,5]. Its principle is sketched in Fig. 9. An auxiliary directional coupler at one end of the ring fiber branches off a small fraction of the light which has propagated ccw through the ring. The polarimeter measures the polarization of that light and generates two electrical signals, characterizing e.g. azimuth and ellipticity of the polarization. In the servo unit, these signals are compared with those characteristic for optimum transmission by the filter $\hat{P}$. Suitable error signals are sent to the two polarization controllers, e.g. in the form of two electromagnetic fiber squeezers, which restore the polarization to its optimum.

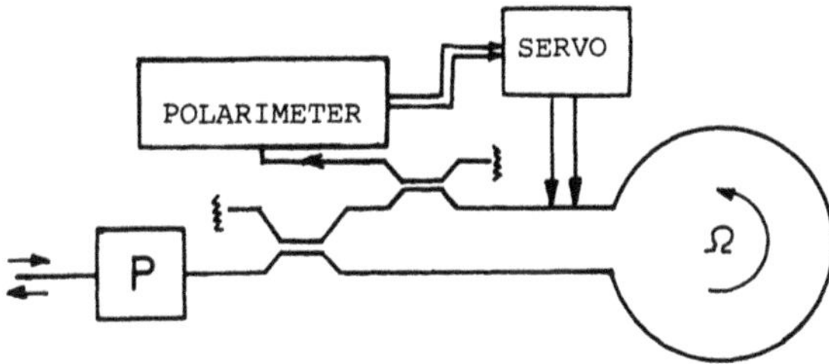

Fig.9 (top) Schematic of an automatic, servo-controlled adjustment of polarization

Such an automatic control has the advantage of fast and accurate optimization of T_{PG} and, therefore, is suitable for a direct-measuring (open-loop) gyro.

5.3 Polarization-Holding Fibers

Fibers which can accurately maintain the polarization of the guided light under varying ambient conditions would be ideal for the construction of a fiber-optic SAGNAC interferometer. After inital adjustments of polarization at the positions A and B of Fig. 7, no further polarization control would be necessary. For a closed-loop system, the requirements on the polarization-holding properties would not even be very strict, as discussed earlier.

Such polarization-holding fibers have been under development for several years [11]. They are all based on the principle of providing a high, uniform internal birefringence β_i in the fiber. The magnitude of this β_i must far exceed all irregular birefringence effects which result from external perturbations (Fig. 5) of the fiber, e.g. from coating, coiling, clamping, vibration, magnetic or electric fields. Moreover it is important that the polarization beat length $L_p = 2\pi/\beta_i$ of these fibers is much shorter than the characteristic lengths of those perturbations [11]. Under these conditions, the fiber will maintain either of its two polarization eigenstates.

Various schemes are investigated to provide the required high birefringence. As mentioned, a high linear birefringence exists in fibers with an elliptical core of small dimensions and high index difference [17], and in fibers with highly anisotropic internal stress [30]. Alternatively it has been suggested [24] polarization-holding may be obtained by the circular birefringence induced in an elastically twisted fiber, and such fibers are now under development [31].

To date, no SAGNAC interferometer with a polarization-holding fiber has been described in the literature. Therefore, it is presently only possible to speculate about the performance of such a gyroscope. In doing so, the first question is how well such a fiber can actually hold polarization. Although the high internal birefringence is generally advantageous for polarization-holding, the birefringence value β_i is not, by itself, a measure for the polarization-holding quality of the fiber. Rather, the magnitude and distribution of perturbations must also be taken into account. This is done by specifying the h-parameter of the fiber [11,32]. It describes the average rate at which power launched into one polarization eigenstate leaks out into the other, cross-polarized state. A systematic study of the h-parameters of a variety of highly birefringent fibers [33] has shown that the best available fibers have $h \lesssim 10^{-4}\ m^{-1}$. It means that after 1000 m fiber length only ~ 10 % of the power is in the cross-polarization state, 90 % remaining in the original one. These numbers are statistical averages, and in any individual fiber the values may deviate considerably. Nevertheless we recognize that such a fiber would be sufficiently close to the ideal fiber mentioned earlier, obviating the need for continuous or repeated polarization control.

Another important aspect for the application of highly birefringent fibers in a gyro is the expectation that their polarization-holding property makes them less sensitive also to magnetic fields. An ideal SAGNAC interferometer, having uniform polarization along the entire length of the fiber, is completely insensitive to arbitrary magnetic fields. The reason is that the net Faraday effect, (integrated along the fiber) vanishes because

$$\oint H_o ds = 0 \tag{44}$$

where H_o is the magnetic field vector and ds a line element of the fiber. However, in a real gyro using a conventional fiber, the polarization is not uniform due to the various birefringent perturbations. Therefore, the contributions to the Faraday effect from various line elements do not cancel completely. It has been demonstrated recently [14] that the resulting nonreciprocal zero-offset (37) from the earth magnetic field strongly contributes to the instability and drift of a fiber gyro with a conventional fiber which has a beat length of the order of 1 m. In a detailed analysis [34] of this instability the expectation value of the nonreciprocal offset (37) has been calculated for the limiting cases of a fiber with low or with high birefringence β_i

$$\langle \phi_o^2 \rangle^{1/2} \approx C_6 H_o (P_c/\ell_\alpha)^{1/2} \kappa^{-1} \quad \text{if } \beta_i \ll \kappa , \tag{45}$$

$$\langle \phi_o^2 \rangle^{1/2} \approx C_6 H_o (P_c/\ell_\alpha)^{1/2} \beta_i^{-1} \quad \text{if } \beta_i \gg \kappa . \tag{46}$$

The constant C_6 is the Verdet constant (Fig. 5f), P_c is the fiber length, and ℓ_α is a correlation length along which the azimuth of the birefringence of the fiber is correlated. Moreover, $\kappa = 1/R_c$ is the curvature of the fiber on the SAGNAC coil. Comparing these equations we recognize a strong superiority of the highly birefringent fiber. Firstly, the offset (46) with a highly birefringent fiber is smaller than (45) by the factor β_i/κ. For a fiber with e.g. a beat length of 3 mm on a coil of 100 mm diameter this factor is 200. Secondly, the correlation length ℓ_α will approach the fiber length P_c in the highly birefringent fiber, as discussed above. (In the ideal gyro, $\ell_\alpha \to \infty$.) In the case of a conventional fiber, however, ℓ_α is typically only of the order of 10 m. Assuming a fiber length P_c = 250 m, we thus obtain another factor $(P_c/\ell_\alpha)^{1/2} \approx 5$ in favor of the polarization-holding fiber. Hence, the total sensitivity to magnetic fields may be expected to be reduced by 3 orders of magnitude, possibly permitting operation of the gyro without magnetic shielding.

In absolute terms, Eq. (44) yields an average nonreciprocal offset of $\sim 1.4 \cdot 10^{-4}$ rad for ℓ_α = 10 m and the experimental parameters of [34] (P_c = 800 m, H_o = 16 A/m, $\kappa = 10\ m^{-1}$). This offset agrees within a factor of 2 with the experimentally observed value.

5.4 Polarization Averaging (Depolarization)

The preceeding schemes of polarization control are based on the idea of achieving stable deterministic polarization conditions, guaranteeing optimum overlap of the polarizer state with an eigenstate of $\overleftarrow{G}$, and $T_{PG} \to 1$. An alternative possibility is to permit rapid statistical fluctuations of the polarization, but in such a way that a stable mean value $\langle T_{PG} \rangle$ results. The average must be taken over the integration time of the gyro, typically $t_o = 0.1$ s. According to this scheme, all possible polarization conditions are scrambled uniformly, so that the overlap varies over its full range, $0 \leq T_{PG} \leq 1$, yielding in the average $T_{PG} = 0.5$. The advantage of this scheme is that (slow) variations of the birefringence of the fiber do not affect the average $\langle T_{PG} \rangle$ if the statistical distribution of the polarization was truly uniform. The price paid for this independence of T_{PG} from the ambient conditions of the fiber is a 3dB reduction in the output power p_o.

As a first possibility of realizing this scheme it appears conceivable to operate the polarization control B_1 or B_2 in Fig. 7 sufficiently fast in a random fashion, e.g. by driving it from a high-pass filtered noise generator. In practice, however, such an arrangement of temporal polarization averaging may cause problems because it violates the conditions (III), given earlier, which required a time-invariant fiber ring. The only apparent exception where such an arrangement might work is to use a polarization controller B_2 placed exactly at the midpoint of the fiber, in close analogy to the scheme of Ref. [35] for the reduction of temporal coherence of the light.

A time-invariant averaging of polarization is achieved by inserting a LYOT depolarizer somewhere (e.g. at the position B_2 in Fig. 7) into the fiber ring and operating the interferometer with a broad-band source [8]. The depolarizer causes a rapid dependence of the matrices $\overleftarrow{G}(\omega)$ and $\overrightarrow{G}(\omega)$ on optical frequency. This variation can be characterized by a bandwidth $\delta\omega$. If the bandwidth $\Delta\omega$ of the source is much larger, $(\Delta\omega >> \delta\omega)$, efficient frequency averaging of T_{PG} can be obtained. The advantage of this scheme is that the fiber ring can be perfectly time-invariant and reciprocal.

The LYOT depolarizer is shown schematically in Fig. 10a. It consists of two birefringent plates (e.g. calcite) whose thicknesses are chosen in the ratio $s_2 : s_1 = 2 : 1$ and whose optical axes include a 45° angle. When the plates are sufficiently thick (e.g. $s_2 = 4$ mm), this combination possesses a large birefringence, varying rapidly as a function of the optical frequency ω. As an illustration, Fig. 10b gives a POINCARÉ representation of this frequency dependence. For one particular input state of polarization, the variation of the output state vs. optical is shown. On the POINCARÉ sphere, the output state runs along a banana-shaped trajectory which contains the input state. This trajectory is a three-dimensional LISSAJOUS figure. For the mentioned integer thickness ratio $(s_2 : s_1)$ the figure is a *closed* curve. It is centered exactly at the center point of the POINCARÉ sphere which represents unpolarized light. The birefringence $\overleftarrow{G}(\omega)$ or $\overrightarrow{G}(\omega)$ of a general fiber ring including this depolarizer is described by a similar type of trajectory but with general orientation. When the birefringence of the fiber changes due to varying ambient conditions, that orientation changes, too. The centroid remains invariant, however.

The optical bandwidth $\delta\omega$ characterizing one complete revolution around the trajectory is $\delta\omega = 2\pi c/(n_1-n_2)s_2$ where n_1, n_2 denote the slow and the fast refractive index of the crystal plates. This bandwidth corresponds typically (calcite, s_2 = 4 mm, λ = 800 nm) to an optical wavelength interval $\delta\lambda$ = 1 nm. The superluminescent diode used as the source in the interferometer of Ref. [8] has a smooth spectrum of width $\Delta\lambda \approx 8$ nm. Consequently, the state of polarization arriving from the ring at P performs ~ 8 full revolutions on the POINCARÉ sphere within the optical bandwidth. On average, then, that arriving state appears unpolarized, and $T_{PG} = 0.5$. With a precision depolarizer this average is stable and permits operation of the interferometer as an open-loop gyro, evaluating the rotation rate Ω directly from a measurement of p_o [8]. No adjustment (B_1 or B_2) of polarization is required, and no other polarization control. The use of the broad-band superluminescent diode in this interferometer has the added advantage of reduced fluctuations of p_o from interferences with backscattered light. Using this scheme a gyro was demonstrated whose open-loop scale factor was stable to better than ± 2 % over a 2 hr period [8]. Considerable improvements appear possible by increasing the ratio $\Delta\lambda/\delta\lambda$ and by stabilizing the laser power.

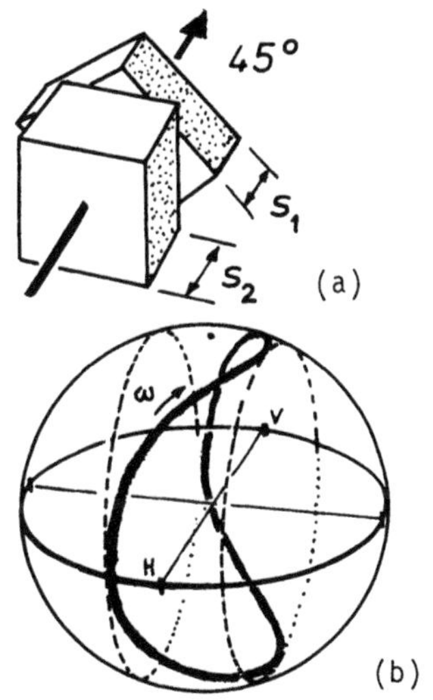

Fig.10 LYOT depolarizer

(a) Arrangement of crystal plates
(b) Output trajectory on POINCARE sphere

Other arrangements for depolarizing the light are conceivable, e.g. spreading the output polarization not along a line trajectory, but rather over the entire surface of the POINCARÉ sphere [36]. This could be achieved by a LYOT depolarizer with a non-commensurable ratio $s_2 : s_1$. In that case, however, a *two-dimensional* statistical averaging must be performed, so that a larger relative source bandwidth $\Delta\omega/\delta\omega$ is required to achieve the same independence of T_{PG} from variations of the fiber birefringence (with $\delta\omega$ determined again by the thickest plate).

6. Conclusions

The discussion has shown that the polarization conditions in a fiber-optic ring interferometer affect both the stability of the zero point of the gyro and the magnitude of its output signal p_o. With respect to the stability of zero, the polarization problem represents but a special case (N = 2) of the general N-mode ring interferometer. It is possible to obtain an intrinsically stable zero-point by operating the interferometer in reflection through a common single-mode, single-polarization filter and by guaranteeing a sufficiently high degree of reciprocity, as expressed by (37). The specifications of the filter are very demanding for a good gyro, concerning its polarization filtering as well as its suppression of higher order modes (stray light). Although the discussion here dealt only with the basic fiber loop having no phase bias, these results apply generally to all fiber-optic ring interferometers, regardless of their biasing scheme. A variety of internal and external effects cause a real optical fiber to have a birefringence which may vary slowly in time. Therefore, suitable arrangements are required to keep the rotation-dependent output signal $\tilde{p}$ stable or at least nonvanishing. Besides passive or active deterministic stabilization of polarization, the described frequency averaging scheme of polarization by a depolarizer permits a particularly simple construction of a fiber gyro, giving stable zero and stable signal $\tilde{p}$ with an ordinary fiber and without any further polarization control.

Acknowledgement

The author wishes to thank K. Petermann of AEG Telefunken, Ulm, for numerous interesting and fruitful discussions.

References

1. H.J. Arditty and H.C. Lefèvre, Opt. Lett. 6, 401 (1981)
2. H.J. Carlin, Proc. IEEE 55, 482 (1967)
3. R. Ulrich and M. Johnson, Opt. Lett. 4, 152 (1979)
4. G. Schiffner et al., Appl. Opt. 18, 2096 (1979)
5. R. Ulrich, Opt. Lett. 5, 173 (1980)
6. E. Kintner, Opt. Lett. 6, 154 (1981)
7. K. Böhm, P. Russer, E. Weidel, R. Ulrich, Opt. Lett. 6, 64 (1981)
8. K. Böhm, P. Marten, K. Petermann, R. Ulrich, Electron. Lett. 17, 352 (1981)
9. R.A. Bergh, H.C. Lefèvre, and H.J. Shaw, Opt. Lett. 6, 198 (1981)
10. H.C. Lefèvre, Electron. Lett. 16, 778 (1980)
11. I.P. Kaminow, IEEE Journ. QE-17, 15 (1981)
12. R. Ulrich, Opt. Lett. 1, 109 (1977)
13. D.M. Shupe, Appl. Opt. 19, 654 (1980)
14. K. Böhm, K. Petermann, and E. Weidel, "Fiber Gyro Performance in the Presence of External Magnetic Fields", Intl. Conf. on Fiberoptic Rotation Sensors, Cambridge 1981 (this volume).
15. R.A. Bergh, H.C. Lefèvre, and H.J. Shaw, Opt. Lett. 6, 502 (1981)
16. J.L. Davis and S. Ezekiel, Opt. Lett. 6, 505 (1981)
17. R.B. Dyott, J.R. Cozens, and D.G. Morris, Electron. Lett. 15, 380 (1979)
18. W. Eickhoff, Y. Yen, and R. Ulrich, Appl. Opt. 20, 3428(1981)
19. Y. Namihira, M. Kudo, Y. Mushiako, Trans. Inst. Chem. Eng. Japan, 60 C, 391 (1977)
20. A. Kumar and R. Ulrich, Opt. Lett., Dec. 1981(to be publ.)
21. R. Ulrich, S.C. Rashleigh, and W. Eickhoff, Opt. Lett. 5, 273 (1980)

22. S.C. Rashleigh and R. Ulrich, Opt. Lett. 5, 354 (1980)
23. R. Ulrich and A. Simon, Appl. Opt. 18, 2241 (1979)
24. S.C. Rashleigh and R. Ulrich, Appl. Phys. Lett. 34, 768(1979)
25. A. Simon and R. Ulrich, Appl. Phys. Lett. 31, 517 (1977)
26. R.F. Cahill and E. Udd, Appl. Opt. 19, 3054 (1980)
27. M. Johnson, Appl. Opt. 18, 1288 (1979)
28. R. Ulrich, Appl. Phys. Lett. 35, 840 (1979)
29. F. Mohr, "Polarization Control for an Optical Fiber Gyroscope", Intl. Conf. on Fiberoptic Rotation Sensors, Cambridge 1981 (this volume)
30. R.H. Stolen, V. Ramaswamy, P. Kaiser, and W. Pleibel, Appl. Phys. Lett. 33, 699 (1978)
V. Ramaswamy, R.H. Stolen, M.D. Divino, and W. Pleibel, Appl. Opt. 18, 4080 (1979)
31. F. Gauthier, J. Dubos, S. Blaison, and H.J. Arditty, "Attempts to Fabricate Circular Polarization Conserving Fibers", Intl. Conf. on Fiberoptic Rotation Sensors, Cambridge 1981 (this volume)
32. S.C. Rashleigh, W.K. Burns, R.P. Moeller, and R. Ulrich, Opt. Lett., (to be published Jan. 1982)
33. S.C. Rashleigh, W.K. Burns, N.J. Marone, and R. Ulrich, "Characterization of Polarization Holding in Birefringent Single-Mode Fibers", Intl. Conf. on Fiberoptic Rotation Sensors, Cambridge 1981 (this volume)
34. R. Ulrich, (to be published)
35. R.A. Bergh, H.C. Lefèvre, and H.J. Shaw, "All Single-Mode Fiber Gyroscope", Third Intl. Conf. on Integrated Optics and Optical Fiber Communication, San Francisco, 1981
36. R.B. Smith, C.H. Anderson, and G.L. Mitchell, "Numerical Modeling of Dual Polarization Interferometric Gyros and Sensors", Intl. Conf. on Fiberoptic Rotation Sensors, Cambridge 1981 (this volume)

Polarization Problems in Optical Fiber Gyroscopes

E.C. Kintner

The Charles Stark Draper Laboratory, Inc.
Cambridge, MA 02139, USA

The importance of polarization control is becoming increasingly recognized in the development of optical fiber gyroscopes [1,2]. Optical gyroscopes depend essentially on the establishment of a single optical path (optical mode), whose length, measured optically in opposite directions, can be taken to indicate rotation rate. However, even a so-called single-mode optical fiber typically provides two optical modes (polarization states) which may couple to each other in a linear but otherwise unpredictable manner. If one of the two polarization states is not suppressed, then it becomes impossible to define (to the accuracy required) which optical path the two counterpropagating beams actually are taking, and therefore the drift stability of the gyro is compromised.

The degree of polarization required to achieve a specified level of gyro drift stability may be inferred through the use of the Jones calculus applied to a nominal model for an optical fiber gyro. Fig.1 shows a model which typifies many current gyro prototypes. In Fig.1a, the polarized output of the light source is divided into two beams, each of which travels in opposite directions through an optical fiber coil. Then the two beams are recombined, and through the use of a beamsplitter the resulting interference pattern is directed to an optical detector. The sensing length of the instrument is the round-trip distance of the optical beams from the beamsplitter through the fiber and back to the beamsplitter. The effective difference between the lengths measured optically in opposite directions indicates the rotation rate. Fig.1b indicates that this Sagnac interferometer, in which the two optical paths share the same physical path but in opposite directions, may be characterized by a conventional interferometer with seperate but equal optical paths.

Let the polarization state of the optical source be given by the Jones vector

$$\vec{a} = \begin{pmatrix} a_A \\ a_B \end{pmatrix} = \begin{pmatrix} |a_A| e^{i\alpha_A} \\ |a_B| e^{i\alpha_B} \end{pmatrix},$$

and that at the detector by

$$\vec{d} = \begin{pmatrix} d_A \\ d_B \end{pmatrix} .$$

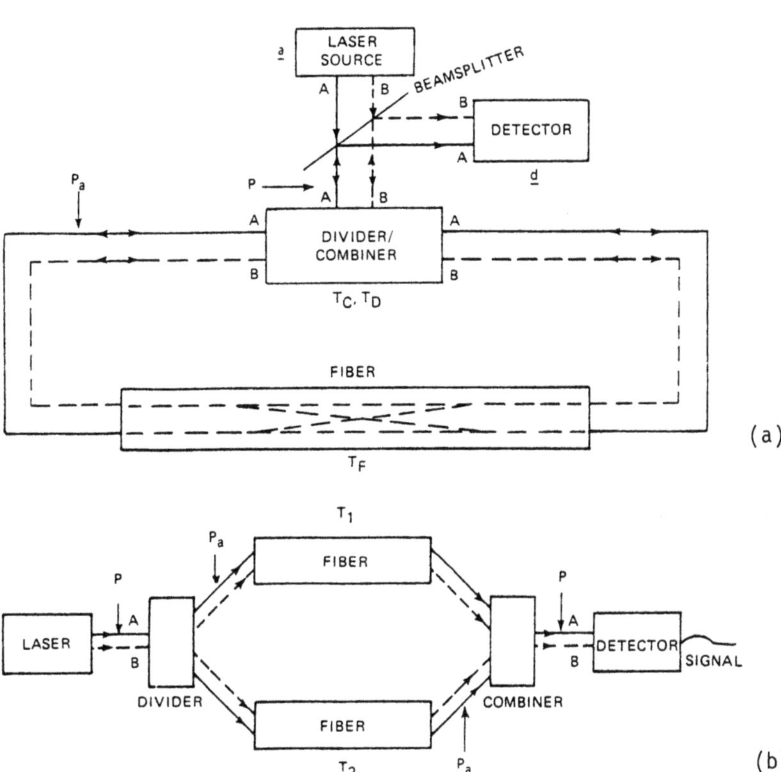

Fig.1 Schematic diagram of fiber interferometer gyro; (a) physical arrangement (b) arrangement for analysis

The propagation of each beam through the gyro can be described by the transformations

$$\vec{d}_1 = \frac{1}{\sqrt{2}} T_{C1}\, T_{F1}\, T_{D1}\, \vec{a} = \frac{1}{\sqrt{2}} T_1\, \vec{a} \quad \text{(clockwise)} ,$$

$$\vec{d}_2 = \frac{1}{\sqrt{2}} T_{C2}\, T_{F2}\, T_{D2}\, \vec{a} = \frac{1}{\sqrt{2}} T_2\, \vec{a} \quad \text{(counter-clockwise)}$$

where the subscripts D, F, and C indicate divider, fiber, and combiner, respectively. Each transformation takes the matrix form

$$T = \begin{pmatrix} |t_{AA}|e^{i\tau_{AA}} & |t_{AB}|e^{i\tau_{AB}} \\ |t_{BA}|e^{i\tau_{BA}} & |t_{BB}|e^{i\tau_{BB}} \end{pmatrix}.$$

If appropriate precautions are taken, the two opposite optical paths are reciprocal; that is

$$T_2 = T_1^t$$

where the superscript t indicates the matrix transpose. In this case, the optical system can be characterized by one matrix, $T_1 = T$, where T is the matrix defined above.

To suppress the second polarization channel, an imperfect polarizer

$$T_p = \begin{pmatrix} 1 & 0 \\ 0 & \varepsilon \end{pmatrix},$$

with extinction ratio $E = \varepsilon^2$, can be inserted at point P in Fig.1.

The two oppositely-directed beams can now be characterized by the modified transformations

$$\vec{d}_1 = \frac{1}{\sqrt{2}} T_p \; T \; T_p \, \vec{a} \;,$$

$$\vec{d}_2 = \frac{1}{\sqrt{2}} T_p \; T^t \; T_p \, \vec{a}.$$

The optical amplitude at the detector is the sum of the amplitudes d_1 and d_2, together with a phase shift ϕ due to rotation:

$$\vec{d} = \vec{d}_1 + \vec{d}_2 e^{i\phi}.$$

The electrical output of the detector is proportional to the optical intensity, which is the sum of the squared moduli of the two polarizations:

$$\begin{aligned} i \propto I &= |d_A|^2 + |d_B|^2 \\ &\simeq \tfrac{1}{2}\{|t_{AA}|^2 |1 + e^{i\phi}|^2 |a_A|^2 \\ &\quad + 2\varepsilon \mathrm{Re}[t_{AA}^* \, (t_{AB} + t_{BA} e^{i\phi})(1 + e^{i\phi})^* \, a_A^* \, a_B]\} \end{aligned}$$

The first term in braces represents the ideal gyro output, while the term linear in ε is an error term whose effect is determined by both the magnitudes and phases of the complex quantities contained within this term. Because it is difficult to control the phases characterizing the fiber (τ_{AB} and τ_{BA}), controlling magnitudes is obviously the first step in reducing this error term. A little algebra will show that in the worst case, where the phases are most unfavorable, the error in the estimate of the phase shift due to polarization is given by

$$|\phi_\varepsilon| \simeq \frac{\varepsilon|a_B|(|t_{AB}| + |t_{BA}|)}{|a_A|\ |t_{AA}|} .$$

Significantly, the error ϕ_ε is proportional to the amplitude-extinction ratio ε, not the intensity-extinction ratio $E = \varepsilon^2$, even though the beams traverse the polarizer twice.

The above equation should not be taken as a strict constraint on the performance of the polarizer alone. Rather, it specifies a budget for the polarization performance of the optical system as a whole, including the source polarization (expressed by the ratio a_B/a_A) and the fiber polarization stability (expressed by the ratio $(|t_{AB}| + |t_{BA}|)/|t_{AA}|$). Much depends on careful optical alignment to minimize these ratios.

Nevertheless, unless these ratios can be considerably reduced in practical applications, the performance demands placed on the polarizer are serious. To realize this, assume for the moment that the other ratios in the polarization budget are of order unity, and consider the following example. A prototype fiber interferometer gyro recently built has a radius R of 14cm and a fiber length L of 1 km, and operates at a wavelength λ of 846 nm. The scale factor for this gyro is then

$$K = \frac{4\pi RL}{c\lambda} = 7 \text{ deg/(deg/sec)}.$$

The very modest goal for drift stability for this prototype gyro is Ω_ε = 0.01 deg/sec. Through the polarization constraint, this drift specification implies

$$\varepsilon \simeq \phi_\varepsilon = K\Omega_\varepsilon = 1.2 \times 10^{-3} \text{ rad}.$$

Since for this special case the polarizer provides the only polarization control in the gyro, the extinction ratio of this polarizer ($E = \varepsilon^2$) must approach 10^{-6} even for this relatively undemanding application. This is comparable to the specifications for the best polarizers commercially available.

This example should be taken as an illustration of the importance of polarization control for optical fiber gyroscopes and of the utility of the Jones calculus in analyzing the polarization problem. The nominal gyro model posed here can be modified to correspond to different prototype configurations. Further, probabilistic models can be exercised to indicate performance in more realistic circumstances. While the polarization problem is clearly challenging, numerous possibilities can be created and tested to overcome it.

References

1. Kintner, E.C., "Polarization Control in Optical-Fiber Gyroscopes", Opt. Lett. 6, 154-156, (1981)
2. Ulrich, R., "Fiber-optic rotation sensing with low drift", Opt. Lett. 5, 173-175, (1980)

Scattering Matrix Analysis on the Use of a Wide-Band Laser Source in a Passive Fiber Rate Sensor

R.J. Fredricks

Advanced Development Operations, Lear Siegler, Inc., Instrument Division
Grand Rapids, MI 49508, USA

Summary

Recent results obtained by various laboratories experimenting with passive fiber optic rotation sensors have suggested that a wide-band laser source, provided it is not so broad that dispersion effects in sensor elements external to the fiber ring, such as the input-output coupler, become significant, can appreciably reduce low frequency random noise at the sensor output. This low-frequency noise is due to the time variation of the elements comprising the fiber ring's scattering matrix model at a given source wavelength.

A question has arisen, additionally, as to whether the use of a fluctuating, or random, state of polarization (SOP) at each such source mode may further reduce the low-frequency noise. In this theoretical analysis we show that the use of a random SOP for each mode is not only unnecessary but, in general, gives rise to a nonreciprocal phase (NRP) shift between the countertraveling beams in the fiber ring at each mode. We find instead that the use of a nonrandom SOP at each source wavelength via suitably oriented polarizers and analyzers eliminates the NRP.

Finally, we show that the resultant intensity signal at the sensor photodetector exhibits a $1/\sqrt{N}$ suppression of the noise amplitude when a wide-band source with N modes is employed. This will occur provided there is sufficient spacing between the various modes so that the scattering matrix elements become uncorrelated with each other from one wavelength to the next.

1. Introduction

In this presentation we shall

a. Formulate the scattering matrix analytical model for a single wavelength laser source and a passive Sagnac Ring of single mode optical fiber. This formulation is an extension of one used by SCHIFFNER et al. [1].

b. Show the origin of a nonreciprocal phase when the transverse components of the input light are completely time uncorrelated. Compare this to the case of completely correlated, i.e., polarized, input light. The correlation time interval involved here is the detector integration time.

c. Extend the analytical model to a wide-band laser source such as a multimode laser diode with N axial modes. A continuous broadband light source such as a superluminescent laser diode will give similar results; however, the analysis becomes more involved with integrals replacing summations and so such sources will not be treated here.

d. Statistically model the time fluctuations in the scattering matrix elements for each discrete frequency and show a reduction in the variance of the intensity bias noise when the sensor output signals due to N modes are incoherently summed, such reduction going as 1/N.

We shall make the following assumptions for simplicity:

a. The fiber ring uses single mode optical fiber, but not necessarily polarization preserving fiber.

b. A one frequency homodyne sensor configuration employing a laser diode is used.

c. The radiation at the various axial modes of a wide-band source is mutually uncorrelated over the detector's integration time. Typically, such mode spacing is on the order of 1Å and the detector's bandwidth is only 100-200 MHz.

d. Additional improvements in sensor noise reduction are possible via nonreciprocal phase modulation and sophisticated signal processing schemes but these techniques will not be analyzed here.

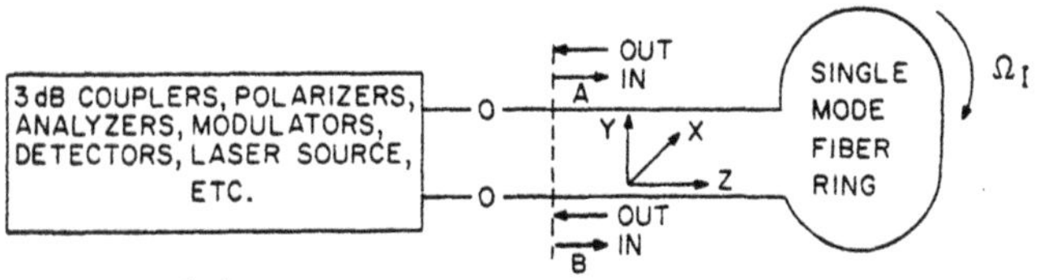

$\lambda = \lambda^{(n)}$ for nth axial mode

Ω_I = inertial rate

Fig.1 Four Port Scattering Matrix Model for the Basic Fiberoptic Rate Sensor

$$\begin{bmatrix} E_{xA}^{(n)} \\ E_{yA}^{(n)} \\ E_{xb}^{(n)} \\ E_{yB}^{(n)} \end{bmatrix}_{OUT} = \begin{bmatrix} S_{11}^{(n)} & 0 & S_{13}^{(n)^+} & S_{14}^{(n)^+} \\ 0 & S_{22}^{(n)} & S_{23}^{(n)^+} & S_{24}^{(n)^+} \\ S_{12}^{(n)^-} & S_{23}^{(n)^-} & S_{33}^{(n)} & 0 \\ S_{14}^{(n)^-} & S_{24}^{(n)^-} & 0 & S_{44}^{(n)} \end{bmatrix} \cdot \begin{bmatrix} E_{xA}^{(n)} \\ E_{yA}^{(n)} \\ E_{xB}^{(n)} \\ E_{yB}^{(n)} \end{bmatrix} \quad (1)$$

where

$$S_{13}^{(n)^+} = S_{13}^{(n)} \exp\{+j\, \Theta_s\} \ , \tag{2}$$

$$S_{13}^{(n)^-} = S_{13}^{(n)} \exp\{-j\, \Theta_s\} \tag{3}$$

and where

$$\Theta_s = \frac{8\pi A}{C}\, \Omega_I \tag{4}$$

with similar meanings for the + and - superscripts appearing on the other terms in $S^{(N)}$ when a rate is present. In the model the S_{14} and S_{23} terms represent cross coupling of x and y polarized components. Also the S_{11}, S_{22}, S_{33}, and S_{44} diagonal terms represent like polarized backscatter and, for simplicity, we neglect any cross-backscatter effects.

We shall model the statistics of the $S_{ij}^{(n)}(t)$ as follows:

a. Each $S_{ij}^{(n)}(t)$ is really a random time process.

b. Randomness is due to

□ External temperature and pressure changes on fiber ring,

□ Laser frequency fluctuations. Each mode is not a single frequency fixed in time, but a homogeneously broadened line whose center frequency is not constant.

c. We will sample the photodiode at time t and, hence, the $S_{ij}^{(n)}$ matrix elements should be viewed as complex random variables. For N laser modes (with N>>1) the set $\{S_{ij}^{(n)}\}$ with ij fixed constitutes a statistical ensemble.

d. We will, furthermore, assume that the frequency separation of the modes is sufficient so that ensemble member $S_{ij}^{(n)}$ is statistically uncorrelated with member $S_{ij}^{(m)}$ for all $m \neq n$.

e. Uncorrelation → $\overline{S_{ij}^{(n)}\ S_{ij}^{(m)}} = \overline{S_{ij}^{(n)}} \cdot \overline{S_{ij}^{(m)}}$

where the bar denotes a statistical or ensemble expectation. (This is considerably weaker than statistical independence which states that the joint probability function of the two random variables is the product of the two marginal probability distributions.)

f. We will also assume that the $S_{ij}^{(n)}$, $S_{ij}^{(m)}$, are identically distributed for all n,m..... This is only for simplicity here, as the essential assumption is the uncorrelation condition above.

g. While the exact distributions need not be known, it is reasonable to assume that for i=j the absolute value functions $|S_{11}^{(n)}|$, $|S_{22}^{(n)}|$, $|S_{33}^{(n)}|$, $|S_{44}^{(n)}|$ representing backscattered signal amplitude are Rayleigh distributed; i.e., $F_{|S_{ij}^{(n)}|}\{x\} = \frac{x}{(\alpha^2)} \exp\{-x^2/(2\alpha^2)\}$ for $x \geqq 0$ (5)

$= 0$

with mean value proportional to α and variance proportional to α^2 where F is the probability density.

The square functions $|S_{11}^{(n)}|^2$, etc., would thus be expected to be chi squares of order two distributed, i.e., an exponential distribution proportional to $\exp\{-x^2\}$.

2. Necessary Condition for Phase Reciprocity

We shall "derive" this condition by two examples employing our scattering model. First we consider the extreme case of completely unpolarized input light (i.e., an input SOP uniformly distributed over the Poincarè sphere) at each mode n. The question of whether such an (ideal) source could actually be used to advantage in reducing both the number of sensor elements and the fiber scattering noise has already surfaced a few times at this conference and this example should serve to address that question. Next we obtain the no iRP condition for a basic sensor wherein the input light at each mode d represents a single point on the Poincarè sphere. This is the more conventional approach.

2.1 Case I - X,Y Polarization Components of Input Radiation Perfectly Uncorrelated over Photodiode Response Time

For the following discussion refer to fig.2.

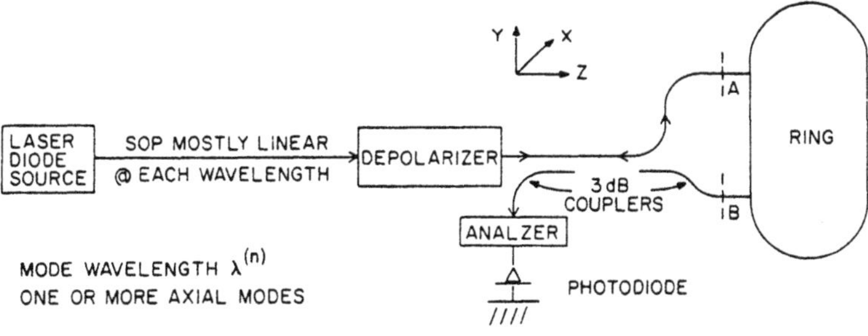

Fig.2 "Ideal" Unpolarized Source Used in Fiberoptic State Sensor

a. Assuming 3 dB couplers $D_{xA}^{(n)} = E_{xB}^{(n)} \triangleq E_x^{(n)}$; $E_{yA}^{(n)} = E_{yB}^{(n)} \triangleq E_y^{(n)}$ at ring input.

b. At mode (n), $\langle E_x^{(n)} \cdot E_y^{(n)} \rangle = 0$; $\langle |E_x^{(n)}|^2 \rangle = \langle |E_y^{(n)}|^2 \rangle = 1$ due to action of depolarizer where $\langle\ \rangle$ denotes time average.

c. All modes will be assumed equal in intensity. Component intensities are all normalized to unity.

d. Let us place the analyzer at 45° with respect to x,y. (This direction is arbitrary as the input light is unpolarized.)

e. Consider the output amplitude, $E_{45°}$, at the photodiode for $\lambda = \lambda^{(n)}$:

$$E_{45°}{}^{(n)} = [1 \;\; 1 \;\; 1 \;\; 1]\,[S^{(n)}] \begin{bmatrix} E_x{}^{(n)} \\ E_y{}^{(n)} \\ E_x{}^{(n)} \\ E_y{}^{(n)} \end{bmatrix}_{IN} . \tag{6}$$

f. Therefore,

$$E_{45°}{}^{(n)} = A^{(n)} + B^{(n)} \exp\{+j\theta_s/2\} + C^{(n)} \exp\{-j\theta_s/2\} \tag{7}$$

where

$$A^{(n)} = [S_{11}{}^{(n)} + S_{33}{}^{(n)}]\; E_x{}^{(n)} + [S_{22}{}^{(n)} + S_{44}{}^{(n)}]\; E_y{}^{(n)}, \tag{8}$$

$$B^{(n)} = [S_{13}{}^{(n)} + S_{23}{}^{(n)}]\; E_x{}^{(n)} + [S_{24}{}^{(n)} + S_{14}{}^{(n)}]\; E_y{}^{(n)}, \tag{9}$$

$$C^{(n)} = [S_{13}{}^{(n)} + S_{14}{}^{(n)}]\; E_x{}^{(n)} + [S_{24}{}^{(n)} + S_{23}{}^{(n)}]\; E_y{}^{(n)} . \tag{10}$$

g. Unless $S_{23}{}^{(n)} = S_{14}{}^{(n)}$, we will not have $B^{(n)} = C^{(n)}$ and an effective NRP will exist that is not rate related.

h. To see this consider $I^{(n)} \sim \langle|E_{45°}{}^{(n)}|^2\rangle$ where

$$\begin{aligned} \langle|E_{45°}{}^{(n)}|^2\rangle = {} & \langle|A^{(n)}|^2\rangle + \langle|B^{(n)}|^2\rangle + \langle|C^{(n)}|^2\rangle \\ & + \langle A^{(n)}C^{(n)*} + B^{(n)}A^{(n)*}\rangle \exp\{+j\theta_s/2\} \\ & + \langle C^{(n)}A^{(n)*} + A^{(n)}B^{(n)*}\rangle \exp\{-j\theta_s/2\} \\ & + \langle B^{(n)}C^{(n)*}\rangle \exp\{+j\theta_s\} \\ & + \langle C^{(n)}B^{(n)*}\rangle \exp\{-j\theta_s\} . \end{aligned} \tag{11}$$

i. The terms independent of θ_s constitute an intensity bias, those involving $\theta_s/2$ are due to coherent backscatter effects and those involving θ_s are the desired rate-dependent output signal.

j. If $B^{(n)} = C^{(n)}$, the desired signal term becomes $\sim 2|B^{(n)}|^2 \cos\{\theta_s\}$ and has a peak at $\theta_s = 0$.

k. However if, say, $B^{(n)} = C^{(n)} \exp\{j\varepsilon\}$, then the signal term goes as $\cos\{\theta_s + \varepsilon\}$ and hence peaks at $\theta_s \neq 0$. Furthermore, if ε changes with time, the location of the intensity peak plotted vs θ_s also changes. This NRP effect is independent of the backscatter and can be shown to persist at all orientations of the output analyzer.

l. While use of multiple axial modes will tend to average out the NRP-caused peak shifts, why not eliminate the NRP entirely by forcing $B^{(n)} = C^{(n)}$ at each $\lambda^{(n)}$? This will happen if the ring input radiation and recombined output radiation are characterized by the same location on the Poincarè sphere at the splitting and reuniting element in the sensor as we shall next show.

2.2 Case II: X,Y Polarized Components of Input Radiation Perfectly Correlated over Detector Response Time

For the following discussion please refer to fig.3.

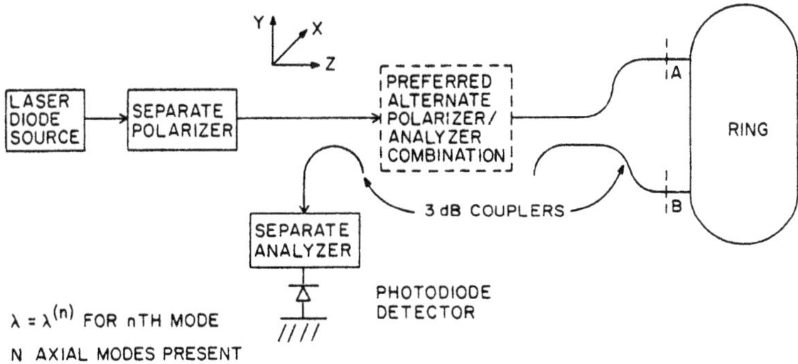

Fig.3 Basic Sensor Configuration Using Well-polarized Light at each Source Line

a. We will first show the NRP at each mode vanishes if the polarizer and analyzer have the same orientation with respect to the X,Y axes. (We neglect any circular birefringence or twist in the pigtails to the 3-dB couplers. The preferred arrangement precludes these problems from arising and, of course, guaranteesthe same orientation of polarizer and analyzer.)

b. We will then show a reduction in scattering noise arises as the n modes are summed to give the resultant intensity at the photodiode.

2.2.1 Subcase A: Polarizer and Analyzer not Parallel ("Separate" Elements Assumed)

a. Example:

45° orientation for input polarizer ($E_{xA}^{(n)} = E_{yA}^{(n)} = E_{xB}^{(n)} = E_{yB}^{(n)} = 1$),

0° orientation for output analyzer (only $E_{xA}^{(n)}$, $E_{xB}^{(n)}$ passed).

b. Therefore,

$$E_{0^\circ}^{(n)}|_{output} = [1 \quad 0 \quad 1 \quad 0] \quad [S^{(n)}] \begin{bmatrix} 1 \\ 1 \\ 1 \\ 1 \end{bmatrix}_{\text{Normalized input}} , \tag{12}$$

$$\triangleq A^{(n)} + B^{(n)} \exp\{+j\theta_s/2\} + C^{(n)} \exp\{-j\theta_s/2\} \tag{13}$$

where

$$A^{(n)} = S_{11}^{(n)} + S_{33}^{(n)}, \tag{14}$$

$$B^{(n)} = S_{13}^{(n)} + S_{14}^{(n)}, \tag{15}$$

$$C^{(n)} = S_{13}^{(n)} + S_{23}^{(n)} . \tag{16}$$

c. Again, unless $S_{14}^{(n)} = S_{23}^{(n)}$, an NRP will exist on the signal term at the photodiode. Rayleigh's Reciprocity Theorem does not require this to be true in general.

2.2.2 Subcase B: Polarizer and Analyzer Parallel

a. This condition will force $B^{(n)}$ to equal $C^{(n)}$ and again we use an example to show this.

b. Example: Assume a 45° orientation for both input polarizer and output analyzer. Then neglecting the $\sqrt{2}$ factors we have

$$E_{45^\circ}^{(n)}|_{out} = [1 \quad 1 \quad 1 \quad 1] \quad [S^{(n)}] \begin{bmatrix} 1 \\ 1 \\ 1 \\ 1 \end{bmatrix}_{\text{Normalized input}} \tag{17}$$

$$= A^{(n)} + B^{(n)} \exp\{+j\theta_s/2\} + C^{(n)} \exp\{-j\theta_s/2\} . \tag{18}$$

c. Now, however,

$$A^{(n)} = S_{11}^{(n)} + S_{22}^{(n)} + S_{33}^{(n)} + S_{44}^{(n)} , \tag{19}$$

$$B^{(n)} = S_{13}^{(n)} + S_{23}^{(n)} + S_{24}^{(n)} + S_{14}^{(n)} = C^{(n)} . \tag{20}$$

d. Since $B^{(n)} \equiv C^{(n)}$, no NRP exists regardless of the actual values the scattering matrix terms assume.

e. In general, we can show the NRP will vanish if the polarizer and analyzer are like oriented. Since the use of a single polarizer/analyzer guarantees this it should be used instead of separate devices.

f. However, low frequency variations in the magnitudes and phases of the scattering matrix elements will produce fluctuations in the signal scale factor and, more importantly, in the coherent backscatter and intensity bias terms. Depending on the signal processing scheme used the variations in coherent backscatter and bias may be mistaken for a false rate.

g. Hence, we need to reduce the variance of these fluctuations with respect to the desired signal term. Intensity averaging over N modes can effect this reduction as we next show.

3. Reduction in Output Noise via Multimode Source

a. For simplicity we assume that all N modes are input to the ring as linearly polarized light with azimuths at 45° with respect to X and Y axes.

b. Expanding $I^{(n)} \sim |E_{45°}{}^{(n)}|^2$ for the example in section 2.2.2 we have

$$\begin{aligned} I^{(n)} \sim\ & |S_{11}^{(n)} + S_{22}^{(n)} + S_{33}^{(n)} + S_{44}^{(n)}|^2 \\ & + 2\,|S_{13}^{(n)} + S_{23}^{(n)} + S_{24}^{(n)} + S_{14}^{(n)}|^2 \\ & + 4\,\mathrm{Rel}\,\{(S_{11}^{(n)} + S_{22}^{(n)} + S_{33}^{(n)} + S_{44}^{(n)}) \\ & + (S_{13}^{(n)} + S_{23}^{(n)} + S_{24}^{(n)} + S_{14}^{(n)})^*\} \cos\{\theta_s/2\} \\ & + 2\,|S_{13}^{(n)} + S_{23}^{(n)} + S_{24}^{(n)} + S_{14}^{(n)}|^2 \cos\{\theta_s\} \end{aligned} \tag{21}$$

c. In the above we omit the short time average braces. The first line of terms represents the bias intensity (including backscatter power), the second line represents the coherent backscatter and the third line the desired signal.

d. For N such modes the photodiode senses I_{total} where

$$I_{total} = \frac{1}{N} \sum_{n=1}^{N} I^{(n)} \tag{22}$$

and the $\frac{1}{N}$ is for renormalization.

e. We shall assume that at each (n) the complex backscatter amplitude coefficients, $S_{11}^{(n)}$ are not only uncorrelated but are zero mean. Furthermore, we assume equal power at each mode. These are not essential, but do simplify the following expansions. Therefore,

$$I_{BIAS} = \frac{1}{N} \sum_{n=1}^{N} [|S_{11}^{(n)}|^2 + |S_{22}^{(n)}|^2 + |S_{33}^{(n)}|^2 + |S_{44}^{(n)}|^2]$$

$$+ \frac{2}{N} \sum_{n=1}^{N} |S_{13}^{(n)} + S_{23}^{(n)} + S_{24}^{(n)} + S_{14}^{(n)}|^2$$

$$= I_{bias1} + I_{bias2} \; . \tag{23}$$

f. Most of the bias noise comes from the incoherent backscatter amplitude coefficients in I_{bias1} which we have assumed to be identically Rayleigh distributed.

g. Each $|S_{ii}^{(n)}|^2$ is therefore chi squared of degree 2 and has a positive mean value, M, and a variance, σ^2.

h. Then, since the mean value of a sum of random variables equals the sum of the means and since the variance of the sum equals the sum of the individual variances, provided the variables are uncorrelated, we have

$$\overline{I_{bias1}} \sim \frac{1}{N} NM = M; \tag{24}$$

$$\mathrm{Var}\{I_{bias1}\} \triangleq \overline{(I_{bias1} - \overline{I_{bias1}})^2} \sim \frac{1}{N^2} N\sigma^2 = \frac{\sigma^2}{N} \; . \tag{25}$$

i. Note that the variance of I_{bias1} goes down as $\frac{1}{N}$.

j. Similarly, the variance of I_{bias2} is reduced by $\frac{1}{N}$ even though $S_{13}^{(n)}$, $S_{23}^{(n)}$, $S_{24}^{(n)}$, $S_{14}^{(n)}$ are not mutually uncorrelated nor zero mean.

k. To see this, let

$$S_{1234}^{(n)} \triangleq S_{13}^{(n)} + S_{23}^{(n)} + S_{24}^{(n)} + S_{14}^{(n)} \; . \tag{26}$$

Then, $S_{1234}^{(n)}$ is a random variable and hence possesses a mean and a variance. If we assume $S_{1234}^{(m)}$ is uncorrelated with $S_{1234}^{(n)}$ for $m \neq n$ and that both are identically distributed, then the above reduction in variance must occur.

l. Similarly, for the coherent backscatter term we find that its mean value approaches zero. This is because

$$\overline{S_{11}^{(n)} + S_{22}^{(n)} + S_{33}^{(n)} + S_{44}^{(n)}}$$

equals zero and the $S_{ii}^{(n)}$ are uncorrelated with $S_{ij}^{(n)}$ $i \neq j$. More importantly, we also find a $\frac{1}{N}$ suppression in the variance of this term.

m. Finally, the signal term is given by

$$I_{sig} = \frac{2}{N} \cos \{\Theta_s\} \sum_{n=1}^{N} |S_{13}^{(n)} + S_{23}^{(n)} + S_{24}^{(n)} + S_{14}^{(n)}|^2 \quad . \tag{27}$$

As N becomes large, then

$$I_{sig} \rightarrow \overline{I_{sig}} \tag{28}$$

where

$$\overline{I_{sig}} = 2 \cos\{\Theta_s\} \overline{|S_{13}^{(n)} + S_{23}^{(n)} + S_{24}^{(n)} + S_{14}^{(n)}|^2} \quad . \tag{29}$$

n. Thus, the desired signal term grows as $\sqrt{N}$ with respect to the standard deviation of the fiber noise terms and the sensor's signal-to-noise ratio is improved by the same factor as N becomes larger.

o. We conclude that the use of a like oriented polarizer/analyzer combination at each wavelength in a Sagnac Effect fiberoptic sensor will minimize nonreciprocal-phase and that the use of a wideband source with many axial modes will reduce the noise arising from time variations in the fiber scattering matrix elements.

4. Discussion

In order to secure the full 1/N reductions in the intensity bias and scale factor noise variances we must have the $S_{ij}^{(n)}$ uncorrelated over n for each ij. A common $\exp\{j \frac{2\pi}{\lambda_n} \eta L\}$ phase factor in the transmission coefficients, where η is core index and L ring length does not produce the desired bias and scale factor smoothing. This is because $S_{1234}^{(n)}$ appears in I_{BIAS2} and in I_{SIG} as a square magnitude and hence a common phase term among the scattering matrix transmission coefficients cancels at the sensor output. Furthermore, we need the backscatter amplitudes $s_{ii}^{(n)}$ (i=1 to 4) to be uncorrelated with the cross and bar state transmission coefficients for the x and y polarized light components in order to minimize coherent backscatter.

From Mie diffraction theory we know that both the electric and magnetic partial waves scattered via inhomogeneities in the fiber are rapidly oscillating functions of the partial wave order number and of the optical wavelength. Hence, for the 1Å mode spacings assumed it is quite reasonable to state that the backscatter coefficients indeed are uncorrelated with each other and with the transmission coefficients at a given source wavelength line and furthermore are mutually uncorrelated from one such line to the next. Thus, the coherent backscatter false signal and the backscatter bias noise should indeed see a 1/N reduction in variance.

The random variable $S_{1234}^{(n)}$ requires somewhat more careful consideration. AEG Ulm[2] succeeded in effectively ensuring that the $S_{1234}^{(n)}$ are indeed uncorrelated over n by using a different SOP in the ring at each source wavelength. This caused a different complex mixing of the transmission coefficients at each λ_n instead of the simple sum of terms characteristic of the 45° linear SOP that we assumed. In turn then, the sum over N modes of the $|S_{1234}^{(n)}|^2$ power terms was precluded from vanishing at any time and the variance of the sum was reduced.

The AEG researchers utilized a Lyot depolarizer in one arm of the fiber ring to uncorrelate the $S_{1234}^{(n)}$. However, if a single mode fiber possesses sufficient noncircular core birefringence, β_c, stress birefringence, β_s, and/or optical activity resulting from elastic twist, and provided the fiber length is sufficiently great, then the fiber ring itself should serve to effect the same result as AEG's Lyot depolarizer. A recent paper by ULRICH[3] as well as his presentation at this conference indicates that a typical fiber coil in lengths of 200 meters or so can spread the output SOPs over a good portion of the Poincarè sphere given a wide band source of bandwidth 10 nm or greater. If the wavelength dependent spreading from a given input polarization (say 45° linear) forms at least a great circle over the Poincarè sphere, it would appear that the needed decorrelation of the $S_{1234}^{(n)}$ should be achieved without recourse to any additional optical elements. This should be a topic for further investigation among fiber gyro research workers and by the single mode fiber supply houses themselves.

5. Acknowledgments

I wish to thank Dr. S. Rashleigh of NRL and Dr. R. Ulrich of the Technische Universität, Hamburg, for their technical advice and comments on the subject matter treated here. Also, I wish to express my gratitude to Dr. H. Arditty of Thomson-CSF, Paris, for his encouragement in the preparation of this paper.

References

1 G. Schiffner, et al, "Reciprocity of Birefringent Single-Mode Fibers for Optical Gyros", Applied Optics, Vol. 18, No. 13, 1 July 1979

2 K. Böhm, P. Marten, K. Petermann, E.Weidel, R. Ulrich, "Low Drift Fiber Gyro by Using a Depolarizer and a Superluminescent Diode", Paper TuL7-1, Third International Conference on Integrated Optics and Optical Fiber Communication, San Francisco, CA, April 1981

3 W. Eickhoff, Y. Yen, R. Ulrich, "Wavelength Dependence of Birefringence in Single-Mode Fiber", Applied Optics, Vol. 20, No. 19, 1 October 1981

Numerical Modeling of Dual Polarization Interferometric Gyros and Sensors

R.B. Smith, C.D. Anderson, and G.L. Mitchell

Honeywell, Inc., Minneapolis, MN 55413, USA

Introduction

In the research and development of fiber gyros there is a need for a capability to analyze and simulate the optical system. In order to predict performance, sensitivity, and the effect of changing parameters and imperfect components, some form of quantitative calculations would be very useful. Analytical methods quickly become intractable as the complexity of the system increases. For even the simplest fiber gyro configuration, for several optical paths, and for a few optical components, it is practically impossible to reduce a complete analytical description to a useful form. An accurate numerical simulation would be invaluable for low cost exploration of different designs, sensitivity calculations, and performance predictions.

We have been developing a program for such a simulation, and believe that it will be a useful tool in such work. The novel features of the program are:

1. It is applicable to arbitrarily large systems, with multiple optical paths and a large number of optical components.
2. The optical system is described by a sequence of input lines, not by the program logic. The program does not need to be changed for each optical structure to be studied.
3. Different types of "experiments" can be run, each one again determined by single input lines.
4. The program is easily expandable to include additional optical elements or experiments.

Method: Jones Vector-Matrix Representation

The optical system is modeled using the well-known Jones vector-matrix representation for the wave amplitude propagation along multiple optical paths. The effect of optical components on the light is represented by matrix operations. Wave amplitudes are represented by complex two-vectors for the horizontal and vertical (x and y) components, here denoted as Ax and Ay. Optical components are represented by 2 x 2 complex matrices whose diagonal elements may be viewed as propagation factors for each of the x and y wave components; the off-diagonal elements are coupling or conversion coefficients between the Ax and Ay. Standard matrices are well-known for most conventional optical components such as polarizers, isotropic or birefringent path lengths, wave plates, retarders, rotators, etc.

For modeling interferometric systems it is essential to retain, along each path, information on the total wave amplitude intensity and the absolute phase along the path. This is done by including a common complex propagation factor associated with each two-vector. Conventionally, the Jones representation assumes normalized vectors and matrices such that the total power in a path is irrelevant, and any phase factor common to both Ax and Ay is assumed to be factored out (zero-length components). With such an assumption it is not possible to compare (add, combine, or couple) the waves in two paths. Thus, the present method may be described as the Jones vector-matrix method, enhanced by preservation of the absolute phase and total amplitude along each path.

Beam splitters, combiners, and distributed directional couplers are also necessary optical components. These are slightly less familiar to the usual descriptions of the Jones method, but the form of the required 2 x 2 and 4 x 4 matrices is well-known.

The vector-matrix representation is ideally suited to numerical work and to computer programs. The mathematical quantities are easily calculated for any component and values of the parameters which describe it, e.g., path lengths, orientation, reflection and transmission coefficients, coupling, etc. The complexity of a system lies in the large number of its components and the multiplicity of interconnecting paths, rather than in the basic components themselves. This problem can be handled by a good computer program structure.

Program Structure

Logically there are three main parts to the program. The first part reads the input lines, one per optical component, and builds up to a description of the optical system. Pointer vectors and lists are built up which describe the operations to be carried out, in which order, and on which paths. The operations are simply the multiplication of the complex wave amplitude of each path by the 2 x 2 matrix. This phase of the program execution may be described as a "compilation" of the optical system, to be executed later any number of times.

The compilation of the system description from the input lines is done only once. But execution for numerical output may be carried out any number of times to give the output data points vs. some changing parameter such as wavelength, a rotation rate, or path length, etc.

The heart of the program is a subroutine containing a large number of small blocks of code for matrices and multiplications. A block of lines, from a few to perhaps 20, exists for each possible component. During execution, for any fixed set of parameter values, and progressing from input to output, the logical flow is to skip to each block in order, executing only those operations called for. This is done very efficiently following the pointer lists made up during the build mode. It is done repeatedly with changing parameter values during any one "experiment."

The third part of the program is another subroutine which conducts the "experiments" or simulation runs. It reads additional input lines, one or more per experiment, which describe the parameters to be varied and the values to be used, and then calculates the "data points." The detector outputs are printed or stored for each data point. Any number of different experiment runs can be carried out by successive input lines, all for the same optical system, without repeating the system description.

Program Input: Available Components and Experiments

The structure of the optical system and the types of numerical experiments to be simulated are controlled by the list of input lines. One type of input line defines an optical component. It specifies: a mnemonic for the component; the path or paths of which it is a part; the sense of propagation direction; and up to four real quantitative parameters for the component. A different type of input line chooses the type of numerical experiment to be performed: it specifies the particular optical component(s) and parameter(s) to be varied, and defines the list or range of values to be used, as well as the disposition of the generated output.

Standard optical components include: path lengths, both isotropic and birefringent with arbitrary orientation; polarizers for linear, circular, and arbitrary elliptical polarization; Faraday cell; beam splitters and combiners, both polarizing and nonpolarizing; polarization rotators; waveplates and retarders of quarter-, half- and arbitary length and any orientation; distributed couplers in terms of coupling coefficient, coupling length and phase mismatch; and beam splitters/combiners.

The list of available optical components also includes non-standard components such as detectors of total intensity, or intensity in specific polarizations, and a Stokes parameter detector, allowing the polarization state to be plotted on the Poincare sphere. These detectors, which may be positioned at any point in the optical system, do not affect beam propagation. Other non-standard components are a non-reciprocal (rotation) simulator and a "component" which specifies the excitation state of the input beam; for gyros a Sagnac path length is sensitive to rotation rate and direction of propagation.

Additional pseudo-components are available which can duplicate or symmetrically reflect a previously input component (or set of components) to a position later in the chain. One novel input "component" with mnemonic FLOP can unfold a gyro structure. Any number of components, preceding the midpoint of the fiber-coil or a circular path, are reflected about the midpoint; path numbers are interchanged and the sense of propagation is reversed. This economizes the input required, and guarantees duplication of the components and their parameters in the counter-propagation paths. The latter guarantee is important in experiments which vary a parameter of the same component in both counter-propagating paths of the system.

Various types of experiments may be performed by varying: 1) any one parameter of one or several duplicated components; 2) the relative frequency of the input light beam; and 3) any two parameters in one (or duplicated) components. Parameters may be varied by specifying a list of values, or by incrementing over a range.

Examples

Examples are presented in this section to illustrate the modeling capabilities:

1. Dual Polarization Gyro.
2. Narrow-band Birefringent Depolarizer.
3. Four Sections of Birefringent Fiber.
4. Basic All-Fiber Gyro, Effect of Coupler Drift.

These results are important because they represent an examination of fundamental gyro error sources. The first example is central to the dual polarization gyro; the others are important to recently-developed high-performance gyro configurations.

Example 1. Dual Polarization Gyro

Figure 1 shows the gyro structures as it can be modeled, both in its folded and unfolded form. All significant optical components are included. Input excitation is at +45 deg polarization, and separate detectors for x and y polarizations are used at the output. The wavelength used was $0.633 \cdot 10^{-3}$mm.

Polarization-preserving fiber with a $\Delta n = 0.15 \times 10^{-3}$ was used for the coil of 100 meters. Because a model is not yet included for a path which is both birefringent and rotation sensitive, the length was partitioned into alternating segments of birefringent fiber and nonreciprocal fiber, for a total

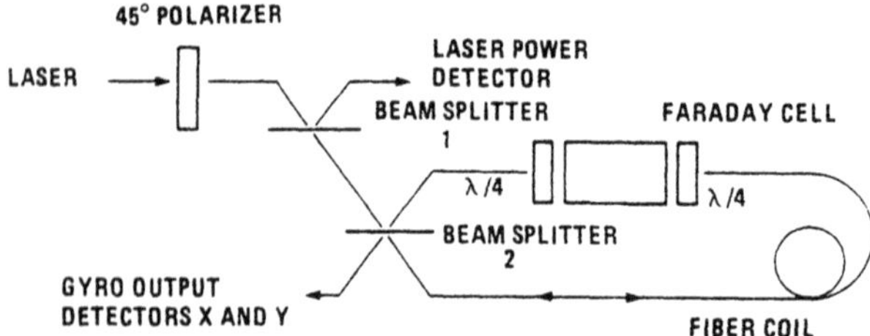

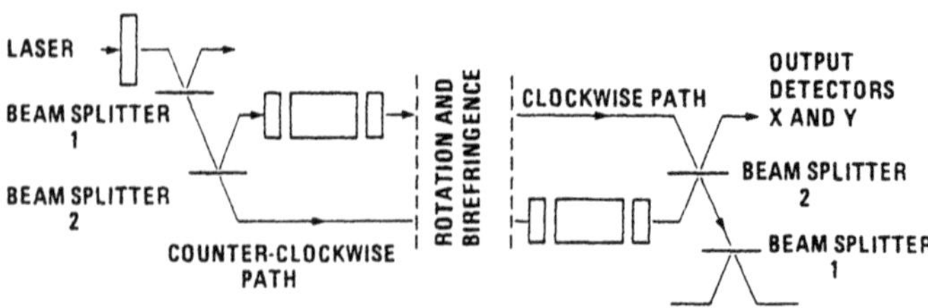

Figure 1. Dual Polarization Gyro as Modeled. Top, Physical Form. Bottom, Unfolded Form

of 50 meters of each type (with twice the retardation coefficients appropriate for a 100m length). To simulate fiber twist or misalignments, the orientation of the 3 birefringent sections could be set at ± a few degrees.

Figure 2 shows the output from the x and y polarization detectors vs. rotation rate in deg/sec. Scale factor and drive for the Faraday cell were adjusted for the quadrature operating point, and the fast axis of all the birefringent fiber sections was aligned to 0 deg. The output is as expected, with the rate signal from the two detectors having opposite sign with respect to the direction of orientation. (The two polarizations are biased to complementary quadrature points.) The rate signal is the difference between the x and y detectors. An arbitrary scale factor was used to make this model sensitivity agree with the experimentally-observed 80 deg/sec for 2π phase shift in the detector outputs.

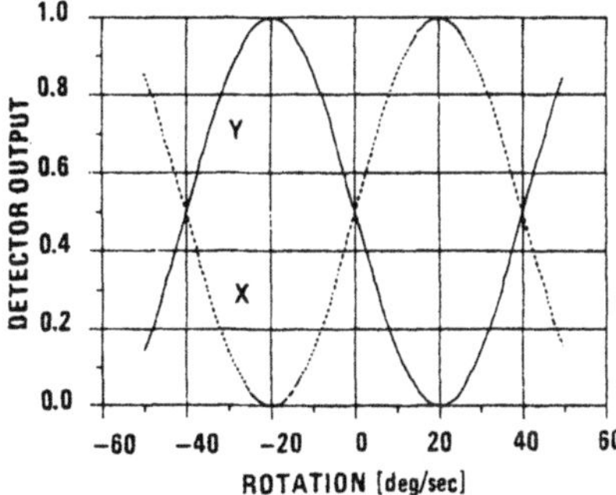

Figure 2. Dual Polarization Gyro Operation. X and Y Polarized Detectors vs. Rotation Rate in deg/sec.

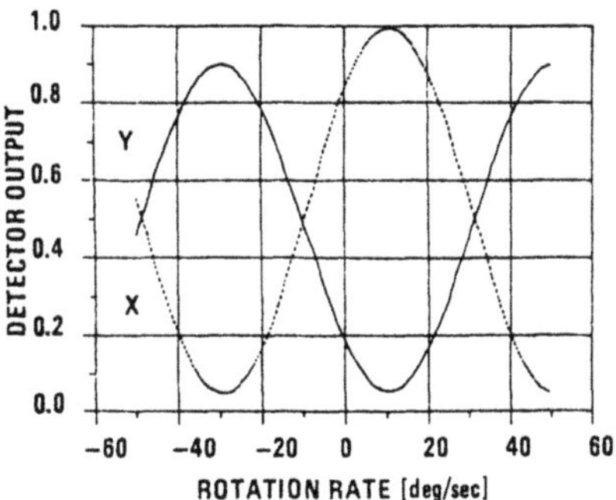

Figure 3. Dual polarization Gyro Operation with Fiber Splice Misalignment

To simulate possible angular misalignment of the fiber, the three sections were then oriented at +5, +7, and -8 deg respectively, with the results shown in Fig. 3. Most notable is the shift from equal detector output, appearing as a drift to an equivalent +10 deg/sec rotation rate. Also, the curves are no longer symmetric--most notably, the curve for the x polarized output.

A different experiment is shown in Figure 4, where the scale factor of the Faraday cell (Verdet constant) was varied over a relative range of $\pm 5 \cdot 10^{-4}$. This was intended to simulate the effects of a temperature variation of, say, ±5 deg, assuming a temperature coefficient of 10^{-4}/deg. This demonstrates the effects of small changes, as evidenced by the expanded scales needed. The rotation rate was 0.002 deg/sec, and phase bias was at quadrature at the 0.15 point on the horizontal axis, resulting in 0.5 output for both detectors. The change in detector output with this "drift" in the Faraday cell was identical; the two curves have a constant separation. With respect to the sense of direction of rotation, the sense of detector drift is opposite for the two outputs. The rotation signal, the difference between the two detectors, does not drift. This directly demonstrates the advantage of the dual polarized gyro.

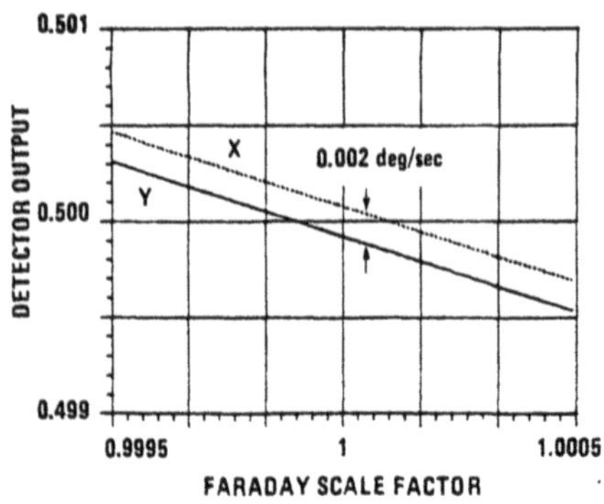

Figure 4. Dual polarization Gyro Outputs vs. Phase Bias Drift Showing Bias Stability

When the two detector outputs are not equal, the compensation for drift in the phase bias may not be as good as in the above case. For example, at the zero rotation rate in Fig.3 the signal inequality is large, equivalent to a large bias due to the fiber misalignment. An experiment equivalent to that of Fig.4 was run, and showed that the slope vs. Faraday scale factor was different for the two detectors; compensation was not perfect.

Example 2. Narrow-Band Birefringent Depolarizer

Rather than preserving a polarization state, it is sometimes desirable to depolarize a light source. This has meaning only for non-monochromatic light, and the bandwidth over which depolarization can be obtained is an important question. As an example, in the use of superluminescent diodes as sources for fiber gyros, a depolarizer over a 1% bandwidth is desirable.

Figure 5 shows a possible depolarizer using two lenths of quartz. For narrow-band performance, the pieces need to be rather long, depending on the degree of birefringence. One example given in the literature states that the longer piece should be twice as long as the short piece. It can be shown that this is a poor choice; rather, the two lengths should have a ratio not near the ratio of small integers. The total length, in number of retardation wavelengths at center band, should be many times the inverse relative bandwidth.

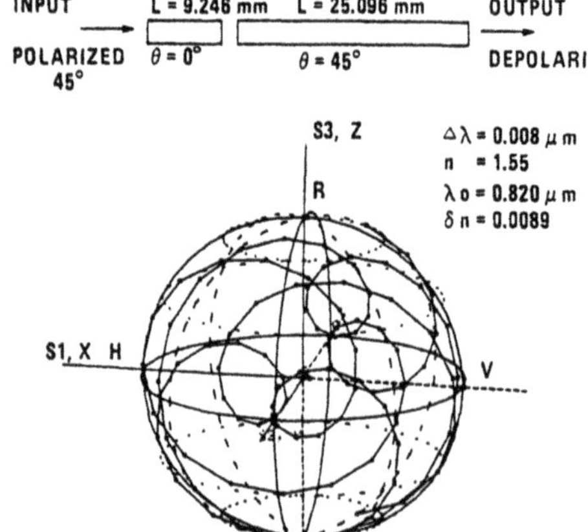

Figure 5. Narrow-Band Two-State Birefringent Depolarizer. Locus of Polarization States on Polarization States on Poincaré Sphere over 1% Spectral Range

The Poincare sphere is very useful to exhibit the spreading of the polarization states by a depolarizer. In a good depolarizer, as the frequency of the light is varied within the bandwidth, the state of polarization should be redistributed widely and uniformly over the surface of the sphere. The lower part of Fig.5 shows the points on the sphere which were obtained using the computer program. (This model requires a simple list of about 10 input lines.) The relative frequency was varied over the range of 0.995 to 1.005 using 100 points. That is, the total spectral range displayed is one percent, and the individual points shown are spaced at 10^{-4}. Good distribution over the sphere is evident, although there is a somewhat greater number of points (loops) near the ±45 deg linear polarized points P and Q. Different choices of parameters for best distribution on the sphere are very easily explored using the computer program. Design iterations are efficient because of the flexible program structure.

Example 3. Four Sections of Birefringent Fiber

The results from a simple example of birefringent fiber sections misoriented in angle at their junctions are shown in Figs. 6 and 7. Four sections of fiber with a linear birefringence of $\delta n = 2 \cdot 10^{-7}$ were used (beat length of 2.5 meter, at $\lambda = 0.5 \cdot 10^{-3}$mm). Lengths were 9.5, 9.09, 9.72, and 9.45 meters; the orientations (fast axis) were all nominally 45 deg, but actually misoriented to 41, 47, 50, and 45 degs respectively. Input light polarized at 45 deg is nominally aligned with the fast axis of the fiber, and should emerge as 45 deg polarized light. But the fiber misorientation will result in deviations from pure 45 deg polarization, as is clearly shown in the two figures where the relative frequency was varied over the range 0.95 (0.001).05.

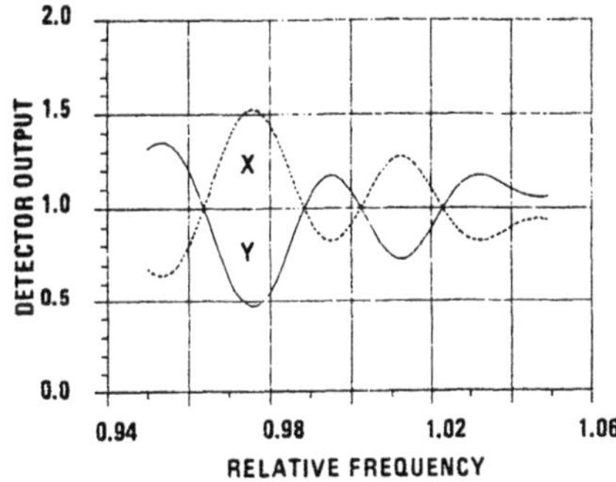

Figure 6. Output from X and Y Polarization Detectors vs. Relative Frequency

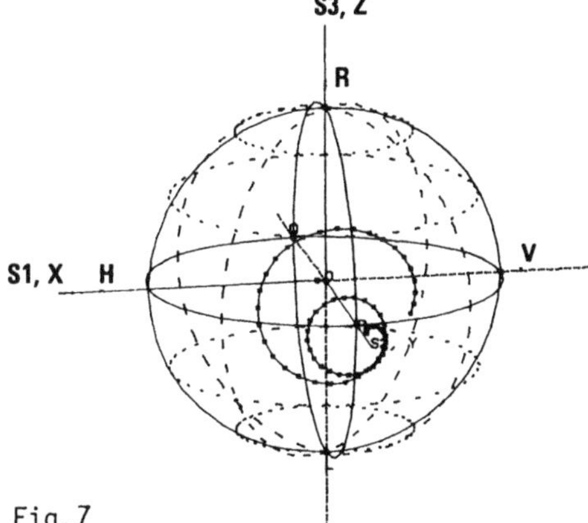

Fig.7

Figure 7. Output from Four Sections of Misaligned Birefringent Fiber vs. Relative Frequency. Locus of Polarization State on Poincaré Sphere are indistinguishable

Figure 6 shows the output from x and y polarized detectors. The sum of the two is constant (energy conservation); where the two are equal, the azimuth of the polarization is 45 deg, though there will generally be some ellipticity. More clearly displaying the departure from the +45 deg polarization state is the locus shown on the Poincaré sphere in Fig. 7. The locus spirals around the point P, or Stokes parameter axis S2. Over a wider frequency band the spiral path repeats periodically. Projecting this locus onto the S1 axis from H to V gives the resolution into x and y polarization shown in the previous figure.

Example 4. Temperature-Induced Bias Errors in Gyros Using Distributed Couplers

The basic all-fiber gyro consists of a distributed 3dB fiber coupler, used for a beam splitter/combiner, plus the rotation sensing fiber coil. This may be modeled quite simply, with a single non-reciprocal phase retarder in the path to establish the quadrature operating point. Two unfolded paths are used for the clockwise and counterclockwise beams respectively, with a single detector for the output of each path, giving complementary rotation signals. Only a single polarization is used.

An experiment of interest is how the output signal might drift with, say, temperature changes in the distributed coupler. It is not obvious whether the drift would be equivalent to, or be interpreted as, a rotation signal, but this is easily modeled. Figure 8 shows the results of varying the coupling coefficient of the distributed coupler (this is also equivalent to varying the effective coupler length or the frequency of the light). The consequence is that the light is not equally divided between the two paths. It is the coupling coefficient which is varied here, not the power division ratio. Because of the nature of the distributed coupler, a phase shift also accompanies the power division change; this is a phase change that may also appear as a gyro bias error.

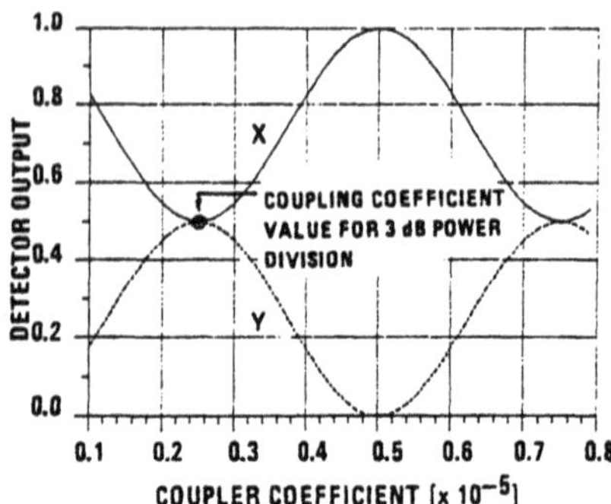

Figure 8. Output from Clockwise and Counterclockwise Beams in Basic Fiber Optic Gyro Operating at Quadrature. Coupling Coefficient of Distributed Beam Splitter/Combiner is Varied

With the choices of wavelength and coupler length used, equal power division required a coefficient of 0.25 x 10^{-5} per unit length. This point is shown in Fig.8. At quadrature and zero rotation rate, the detectors for the two outputs are equal. Along the horizontal axis, the coupling coefficient is varied over a wide range. The detector outputs vary, and are indistinguishable from a rotation rate. To first order, very near the 3dB point, the effect vanishes, but the second order effect, for changes of even one percent in the coefficient, still can be large compared to the rotation rates to be measured.

6. Discussion

The program is just reaching fruition, and these examples only hint at the principal power of the program that is, the ability to do simulations of arbitrarily complicated optical systems with multiple paths and many optical components.

The limitations of the program are those imposed by the Jones method. Only monochromatic steady state can be treated at present. Consequently it is not useable for time modulated systems which are becoming important for fiber gyros. For quasi-monochromatic excitation (e.g. semiconductor laser LED's etc.) it is, in fact, possible to extend the Jones formulation and the program. It is only necessary to replace the complex two-vector of the wave amplitudes by their coherency matrix. This is a trvial change in the program; the same Jones matrices are used for the optical components. It is conceivable that the program can be extended to treat the time modulated cases but this would probably require considerably more complexity to keep track of transit time effects. The method is also not applicable to resonator structures; those become boundary value problems.

In actual practice an important limitation will be lack of adequate quantitative data to use with the program for realistic modeling. Many optical components are less than ideal, and too little is known about the parameters. Examples are the polarization effects in beam splitters and distributed couplers, and twist or misalignment in birefringent fibers. The modeling can be as good as the optical parameters are well known.

Acknowledgements

This work was supported by the Naval Research Laboratory and the Honeywell Systems and Research Center.

2.2.2 Other Reciprocity Considerations

Reciprocity Properties of a Branching Waveguide

H.J. Arditty, M. Papuchon, and C. Puech

Thomson-CSF, Laboratoire Central de Recherches, BP 10, Domaine de Corbeville
F-91401 Orsay, France

After a review and synthesis of the properties of the single-mode dielectric branching waveguide, we propose a particularly simple and attractive integrated-optics device, the "double Y", which fulfils all the requirements for reciprocal operation of a fiber-optic rotation sensor.

It is now recognized that, in order to obtain photon-noise-limited drift performance, a Sagnac interferometer rotation sensor must be operated in a strictly reciprocal fashion [1, 2]. Although many configurations can and have been made to operate in a fairly stable fashion, the achievement of reciprocity requires an implementation rather more complex than that of the classical Sagnac interferometer. This arrangement, which is independent of the technology used, is called the "minimum configuration" [3].

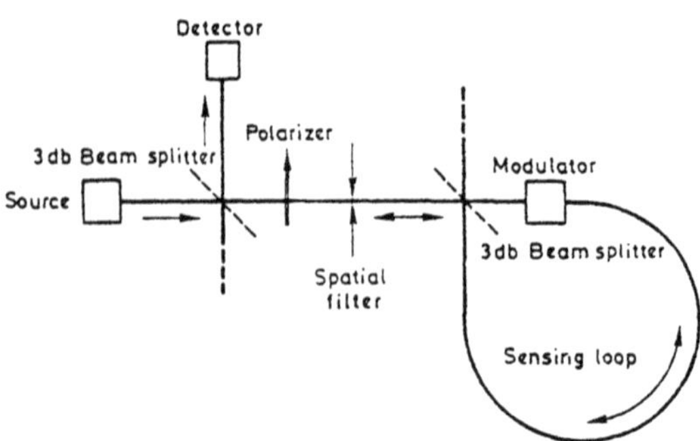

Fig.1 Reciprocal operation of a fiber-optic rotation sensor: the "minimum configuration"

Apart from the light source, the fiber loop and the detector, the minimum configuration involves two 3 dB couplers, a spatial filter, a polarizer and a light modulator (Fig. 1). As limited as this list may seem, the construction of such a configuration using discrete optical elements is quite involved, mainly because of the number of degrees of freedom and of the corresponding adjustments (Fig. 2). Furthermore, the light travels serially twice through most of these components, so that any loss of the individual components quickly leads to an unacceptable overall transmission loss for the system.

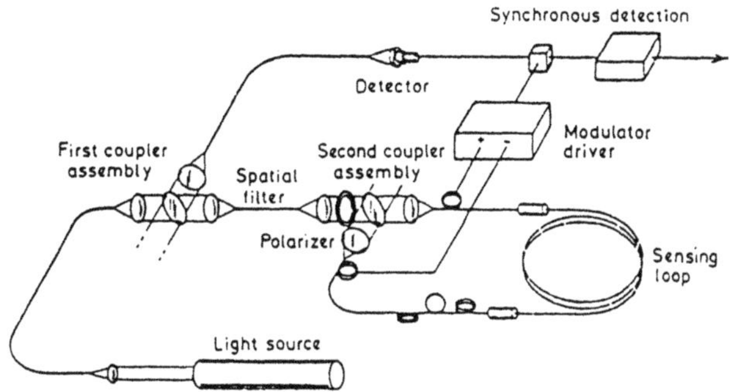

Fig. 2 The minimum configuration implemented using conventional discrete optics

The possibilities offered by integrated-optics for the construction of the central functions of the fiber-optic rotation sensor (FORS) have been pointed out very early. It clearly presents potential mass reproducibility at low cost and benefits from an unusual optical stability due to the sharp reduction of the number of degrees of freedom associated with the use of single-mode waveguides.

The utilization of integrated-optics to implement the minimum configuration functions looks even better [4]. One of the sources for drift in the fiber gyroscope arises from the non-reciprocal coherent interference between light travelling the direct path and light undergoing spurious reflections. In the integrated form, these are virtually eliminated by the absence of sharp interfaces between individual components.

Furthermore, even though integration of several different optical functions on one substrate is one of the goals of integrated-optics, most demonstrated applications to date have consisted of a single component. The minimum configuration FORS may well be the first application where complex integration is really meaningful, yet simple enough to be feasible with today's state-of-the-art technology.

Fig. 3.a depicts a possible implementation of the FORS core functions in integrated form. The beam-splitting devices used are "cobra" type directional couplers adjusted for 50 % splitting ratio. The phase modulation is achieved by taking advantage of the electro-optic properties of the substrate material used, $LiNbO_3$ in this case. Spatial filtering is automatically performed by the use of single mode waveguides. The critical function of the polarizer can be fulfilled in a number of ways, the one used here being differential loss induced by plasmon coupling to a metal layer.

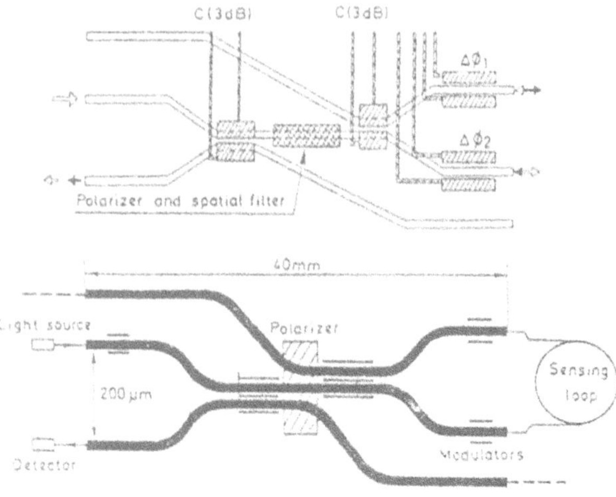

Fig. 3a Straightforward integrated-optics implementation using "cobra" 3 dB directional couplers

Fig. 3b Scale representation of the actual integrated optical circuit (50/1 vertical/horizontal expansion)

It turns out, however, that the complexity and performance requirements of even this simple configuration are pushing the limits of the present integrated-optics technology. Because of the high optical losses associated with sharp waveguide bends, the only practical topology is the in-line arrangement. The length of the individual components, even restricted to the minimum, together with the necessity to separate the output waveguides far enough for parallel fiber pigtail butt coupling, add up to a very long device, in excess of 40 mm, both potentially lossy and in the upper range of practical micron lithography.

Furthermore, the precision in the prediction of the coupling length of the directional couplers does not permit the deterministic achievement of the 3 dB splitting ratio, so that an electro-optic tuning has to be implemented, which may even have to be actively stabilized to compensate for several drift mechanisms

Such a device was however fabricated (Fig. 3.b) with relative success but was never used in a FORS becaused of the problems evocated above [5].

We propose here a novel, much simplified, integrated-optics implementation based on a combination of properties of a simple device, the branching single-mode dielectric waveguide or "Y" junction.

Those properties have been known for years and have been studied and described at length [6][7][11][12][13][14], but have been periodically forgotten or ignored since, and their synthesis into a useful proposition, namely the close analogy between the "Y" junction and a 3 dB quadrupole, has never been pointed out in connection with the reciprocal operation of an interferometer.

The "Y" junction is defined as the smooth or adiabatic branching of a truly single-mode dielectric waveguide into two geometrically symmetrical single-mode waveguides (Fig. 4). It can be fabricated in a planar technology as shown or in a number of other approaches, including fiber-optics, for which the following analysis is equally valid.

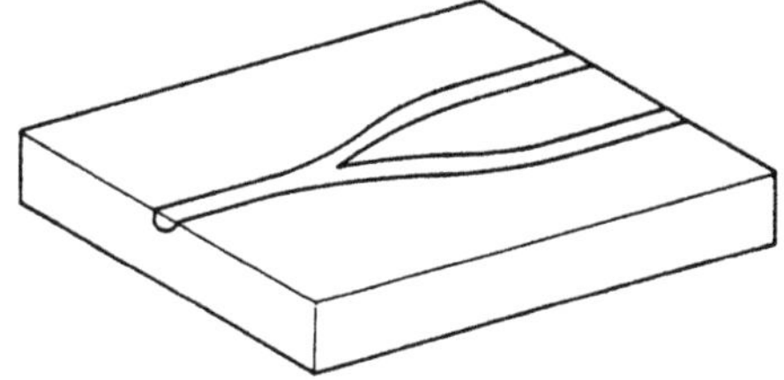

Fig. 4 The branching single mode waveguide or "Y" junction

Propagation of light in this device in the direct, or branching direction, is quite straight forward : light entering through the common waveguide is split evenly into the two output waveguides. The splitting ratio is strictly 50 % because it derives directly from the absolute geometrical symetry of the branch and does not rest on some fine-tuned equilibrium between two competing processes (Fig. 5). A fraction of the incident light may also be radiated at or near the junction, and recent work show that this loss may not be reduced indefinitely, but the experimentation shows that, for all practical purposes, it can be held arbitrarily low.

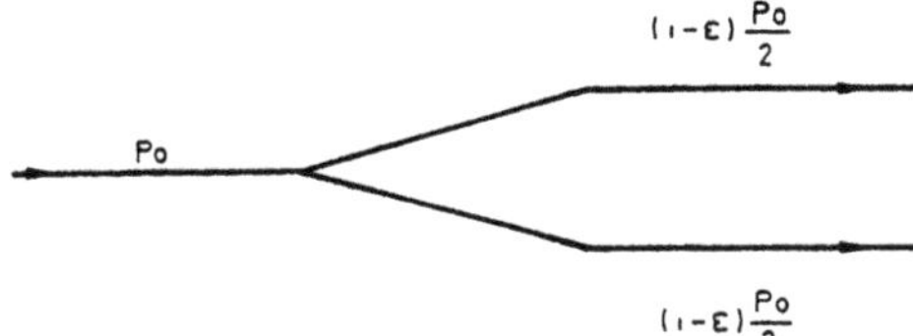

Fig. 5 The power-splitting property

More intriguing is the behaviour of the "Y" junction in the reverse direction. Much insight can be gained by the analysis of its operation in a now classical integrated-optics device : the Mach-Zehnder electro-optic amplitude modulator (Fig. 6.a).

In the amplitude modulator, the incident light is split evenly by a "Y" junction used in the direct direction. After a short path in two independent waveguides, the two waves are recombined in a "Y" junction used in the reverse direction. An electro-optic diffential phase shift is introduced between the two arms of the device.

When the two waves are recombined in phase, they interfere constructively to yield a maximum output power. When the phases differ by π, they interfere destructively and the output power is zero. In the general case, the output is the raised cosine function of the phase difference.

The above analysis must of course satisfy the conservation of energy. What happens to the energy not transmitted to the output part is quite obvious in the traditional optics implementation of the Mach-Zehnder interferometer (Fig. 6.b) : it is swerved to exit through another port of the device.

Because no absorbtion loss mechanism is available, the integrated optics version must also radiate the unused energy, even though the spare exit part is not readily apparent at the recombining "Y" junction. In fact the ways this happen may differ, depending on the characteristics of the junction. In the case of a waveguide with perfectly conductive walls, the only alternative for the energy is to reverse direction and be retro-reflected into the two input waveguides [8][9][10] (Fig. 6.c).

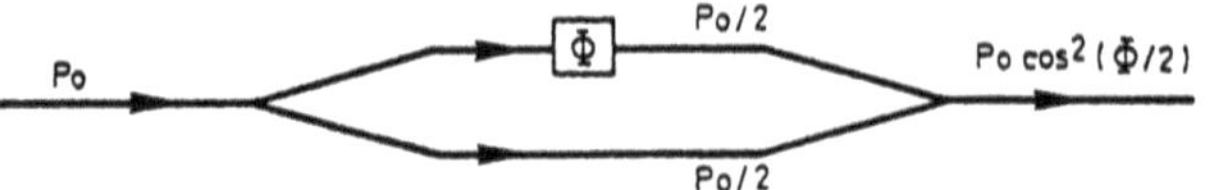

Fig.6a The Mach-Zehnder integrated interferometer: an electro-optic amplitude modulator

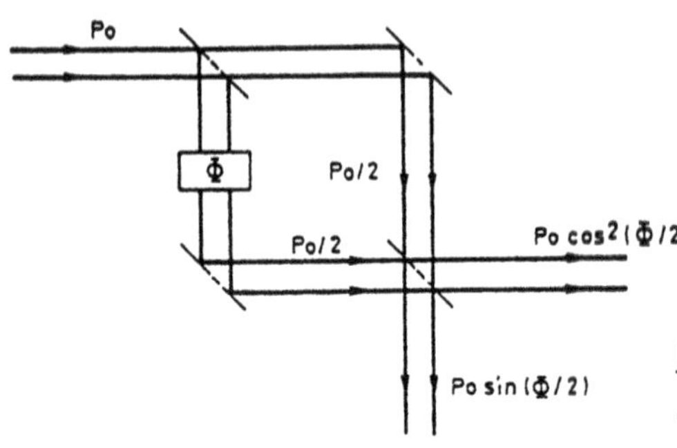

Fig.6b The conservation of energy in the discrete optics implementation of the Mach-Zehnder interferometer

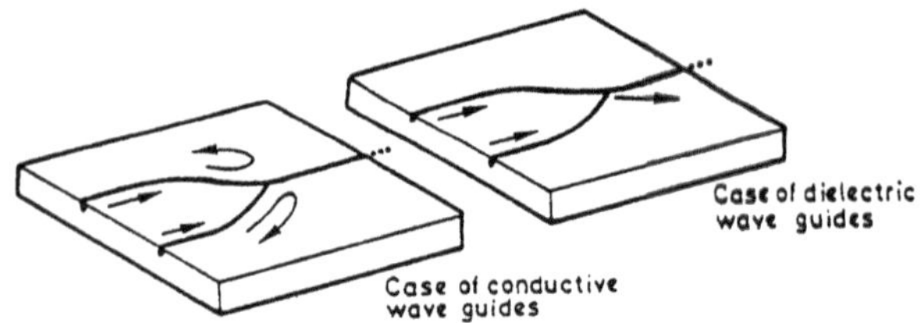

Fig.6c Mechanisms for the exit of energy in conditions of destructive interference

For the case of interest here, that of an adiabatic branch of strictly dielectric waveguide, the energy is radiated at a shallow angle into the substrate. The mechanism of this behaviour is depicted in Figs. 7a and 7b: when the two input waveguides are fed in phase, the branch region carries a symetric mode, adiabatically transformed from a double-hump to a single-hump distribution, which efficiently couples to the output waveguide.

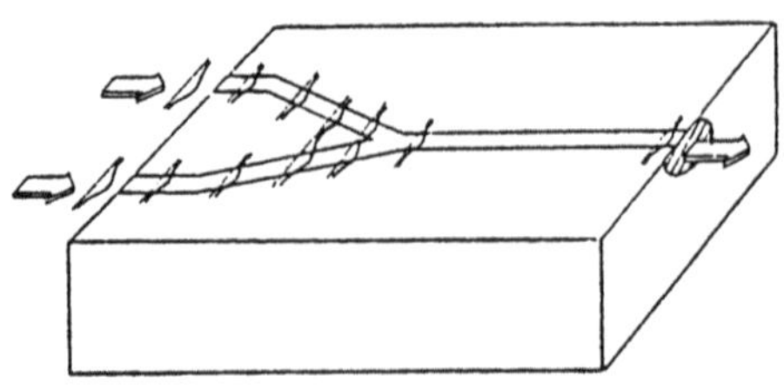

Fig.7a Excitation of the symmetric guided mode by inputs in phase

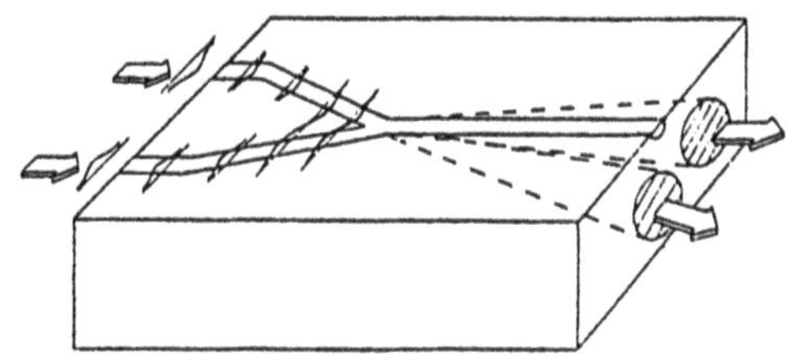

Fig.7b Excitation of the anti-symmetric radiation mode by inputs in opposition

When the input waveguides are fed by waves out-of-phase by π, it is an anti-symmetric mode of the structure that is excited. The overlap integral of that mode with that of the output waveguide remains identically zero throughout the adiabatic transition so that no coupling into this waveguide occurs. The energy is radiated in the form of a two-lobe (anti-symmetric) pattern. The direction of radiation is that which satisfies K-matching, a shallow angle into substrate, and the divergence arises from diffraction due to the finite width of the radiating structure.

Hence the single-mode dielectric branching waveguide is actually a four-port device, and, eventhough the fourth port is not guided and thus usually unusable, the consequences are many-fold.

One obvious, but too often overlooked, consequence is that the symetrical single-mode "Y" junction, used in reverse and excited through one of the two waveguides only, exhibits on intrinsic 3 dB minimum "loss". This property is of course readily accepted of a 50 % traditional beam splitter (Fig. 8.a). The modal explanation is given, for the guided-wave device, in Fig. 8.b. The single input excites, in the branch region, a linear superposition of the symmetric mode, which couples into the output waveguide, and of the anti-symmetric mode which radiates.

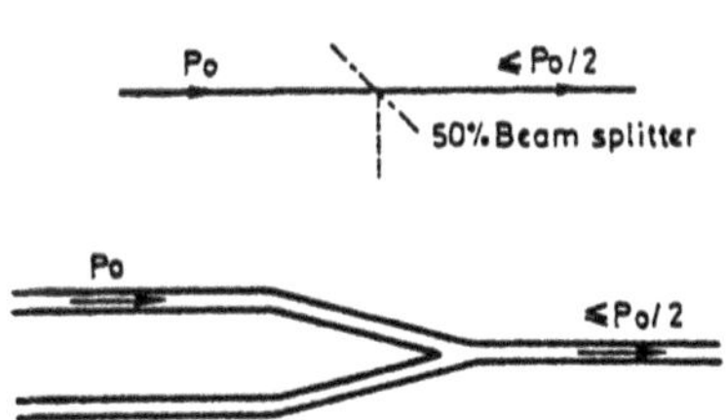

Fig.8a An often forgotten intrinsic limit of the symmetrical single-mode

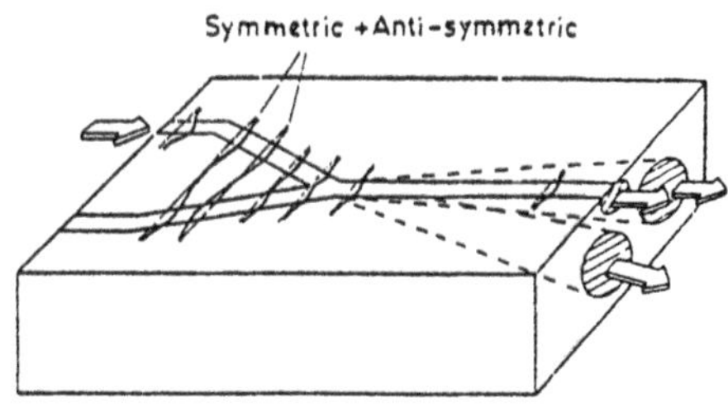

Fig.8b The mode-coupling mechanism of the 3dB intrinsic radiation loss

The even split in power between the two phenomena arises from time reversal consideration (reciprocity) applied to the direct power splitting ratio discussed earlier. It certainly contradicts the gut-feeling that waveguide should behave like plumbing, but, no matter how careful the design of the recombination region may be, the 50 % "loss" will occur. Its necessity can be traced to the fundamental limit set by the second law of thermodynamics.

The similarity of the "Y" junctions with a conventional 3 dB coupler having been emphasized, a subtle but important difference should be pointed out : if both devices are four-part devices, their fundamental symmetries, however, differ. A conventional beam splitter exhibits a true symmetry axis normal to its plane. One "input" port and one "output" port are symmetrical twins while the other "input" port forms another pair with the other "output" port. Although this is not generally true, another symmetry can exist through the plane of the structure if care is taken to construct the device with identical coatings on both sides. In this case the two input ports form a pair and the two output ports form the other (Fig. 9). In either case, no interesting consequence arises from these symmetries, and the same holds for most other four port devices, including directional couplers.

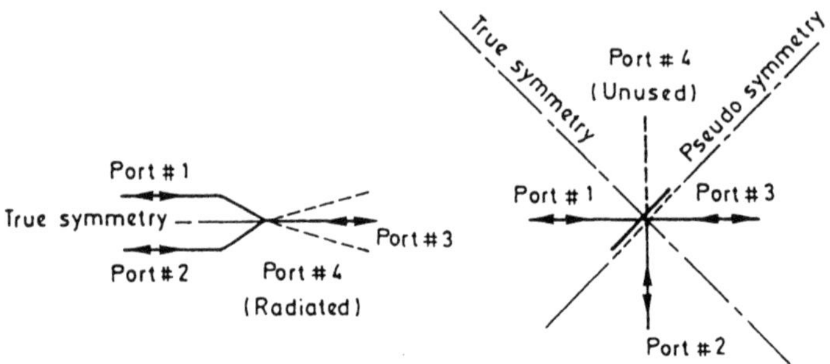

Fig. 9 Equivalence and differences between a "Y" junction and a 3 dB beam-splitter :

- the geometrical symmetry of the "Y" junction guarantees an even split.
- No useful property derives from the different symmetry of the traditional beam-splitter

On the contrary, the symmetry plane of the "Y" junction makes one "input" port the twin of the other but dissects each of the "output" ports, each one being its own twin. This is the origin of the intrinsic stability of the 3 dB splitting ratio, a very desirable feature for a FORS interferometer.

Coming to the point of this analysis, the observation that, in the minimum configuration, two four-port 3 dB devices are needed but that, fortunately, in each case one of the ports is not, and should not be, used brings us to propose the FORS integrated circuit depicted in Fig. 10, which we call the "double Y" configuration. In spite of the striking simplification of its topology and of the absence of any form of splitting ratio control, this configuration strictly adheres to the "minimum configuration" dogma and guarantees perfectly reciprocal, hence drift free, operation.

This design leads to a very simple fiber-optic gyroscope (Fig. 11.a). Such an integrated circuit has been fabricated successfully and is presently being tested in the above FORS configuration. Results of this experiment will be reported in the near future.

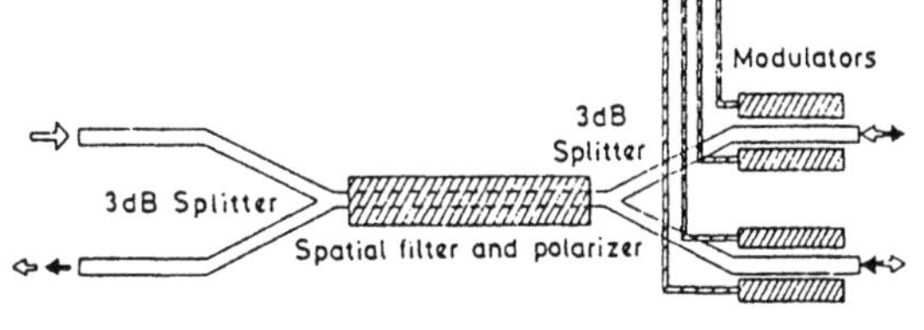

Fig. 10 The "minimum configuration" implemented using two "Y" junctions

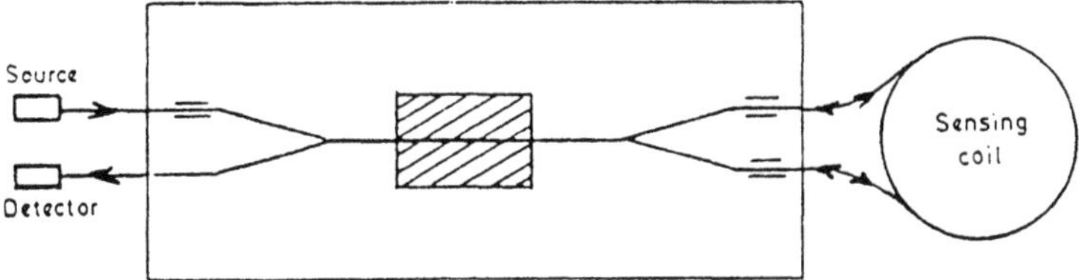

Fig. 11a The proposed integrated fiber-optic rotation sensor using the "double Y" configuration

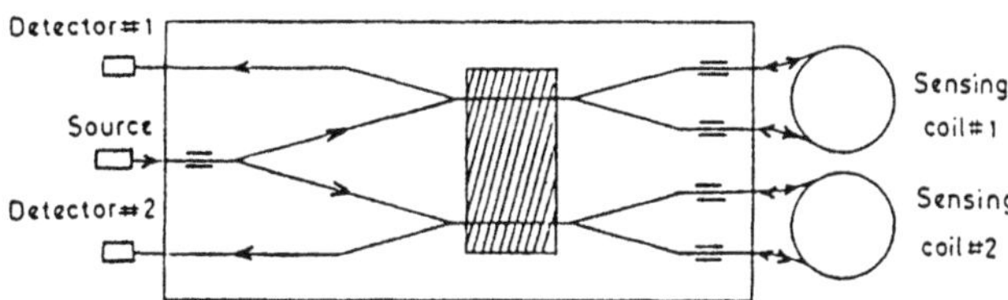

Fig. 11b A two-axis version using a single light source

Because of the high cost of present single transverse mode light sources and of the pigtailing and packaging operations, it is economically meaningful to build two-axis rotation sensors. The configuration proposed in Fig. 11.b takes advantage of the simplicity of the topology of the "double Y" configuration to achieve that goal at minimum cost in technology complexity and performance loss.

In conclusion, the synthesis of simple properties of the single-mode dielectric branching waveguide, all well established but some periodically forgotten, has led to the design of a particularly simple and attractive integrated-optics device, the "double Y", which fulfills all requirements for the reciprocal operation of a fiber-optic rotation sensor. This development increases the advantages offered by the integrated-optics approach and strongly enhances its credibility as the privileged technology for the mass fabrication of high performance fiber-optic gyroscopes.

References

[1] R. Ulrich, "Fiber-Optic rotation sensing with lowdrift", Opt. Lett. 5, 173-175 (1980)

[2] H.J. Arditty, Y. Bourbin, M. Papuchon and C. Puech, "Current sensor using state-of-art fiber optic interferometric techniques", IOOC'81, April 27-29, 1981 (WL-3)

[3] S. Ezekiel and H.J. Arditty, "Fundamentals of fiber-optics rotation sensors", this volume

[4] H.J. Arditty, M. Papuchon and C. Puech, "Fiber-optic rotation sensor : towards an integrated device", CLEO'81, June 10-12, 1981 (THE-1)

[5] M. Papuchon, "Integrated Optics", this volume

[6] H. Yajima, "Dielectric thin-film optical branching waveguide", Appl. Phys. Lett. 12, 647-649 (1973)

[7] W.K. Burns and A.F. Milton, "Mode conversion in planar dielectric separating waveguides", IEEE J. Quantum Electron. QE-11, 32-39 (1975).

[8] W. Streifer, D.R. Scifres and R.D. Burnham, "Integrated interferometric reflector", Appl. Phys. Lett. 30, 521-523 (1977)

[9] D.R. Scifres, W. Streifer and R.D. Burnham, "Semiconductor lasers with integrated interferometric reflectors", Appl. Phys. Lett. 30, 585-587 (1977)

[10] W. Streifer, D.R. Scifres and R.D. Burnham, "Erratum : Integrated interferometric reflector", Appl. Phys. Lett. 31, 309 (1977)

[11] H. Yajima, "Coupled mode analysis of dielectric planar branching waveguides", IEEE J. Quantum Electron. QE-14, 749-755 (1978)

[12] H. Sasaki and I. Anderson, "Theoretical and experimental studies on active Y junctions in optical waveguides", IEEE J. Quantum Electron. QE-14, 883-892 (1978)

[13] W.K. Burns and A.F. Milton, "An active analytic solution for mode coupling in optical waveguide branches", IEEE J. Quantum Electron. QE-16, 446-454 (1980)

[14] H. Sasaki and N. Mikoshiba, "Normalized power transmission in single mode optical branching waveguides", Electron. Letter. 17, 136-138 (1981)

Multimode Fiber Gyroscopes

G.A. Pavlath and H.J. Shaw

Edward L. Ginzton Laboratory, Stanford University, Stanford, CA 94305, USA

We have recently shown [1] that the basic reciprocity requirements [2,3] for sensitive fiber gyros can be met in single mode systems in which the usual polarized light source is replaced by an unpolarized source, and that certain other operational features also result which are of interest. In the present paper we extend the analysis of both polarized and unpolarized gyros to the case of gyros using multimode fibers, and show that there are interesting features in these cases also.

The electric field in the multimode fiber is expanded in the modes of the fiber as

$$E(x,y,z) = \sum_{k=1}^{N} a_k(z) M_k(x,y) \tag{1}$$

where $M_k(x,y)$ is the kth mode and $a_k(z)$ is a complex number representing the amplitude and phase of the kth mode. For quasi-monochromatic light, the correlation between the modes must be specified along with the power in the mode. This is accomplished mathematically by a generalized coherence matrix [4] which is $N \times N$. The $k\ell^{th}$ term is

$$J_{k\ell} = \langle a_k(z) a_\ell^*(z) \rangle \tag{2}$$

where the asterisk (*) and the brackets (< >) denote the complex conjugate and the infinite time average respectively. The generalized coherence matrix has the same properties as the 2 × 2 coherence matrix decribed in Ref. [4].

The optical fiber is modeled as a 2N port linear system. The ports are identified with the fiber modes at either end of the fiber. The fiber is described by two transfer matrices T^{21} and T^{12} which characterize propagation from end 1 to 2 and vice versa. The $k\ell^{th}$ element of these matrices is

$$T^{21}_{k\ell} = r^{21}_{k\ell}\, e^{j(\phi^{21}_{k\ell} + \phi_s/2)} , \tag{3a}$$

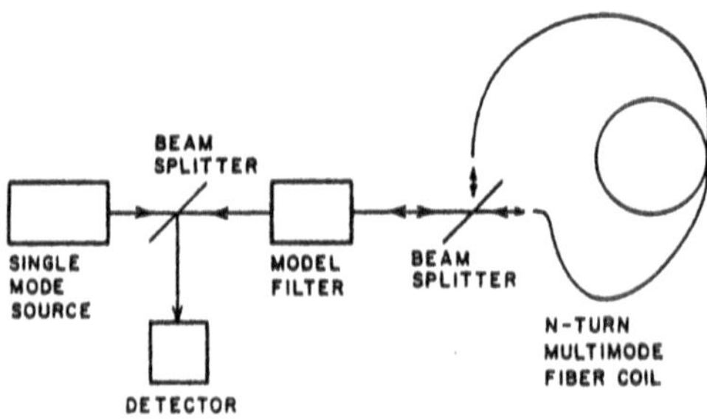

Fig. 1a Multimode polarized, fiber optic gyroscope configuration

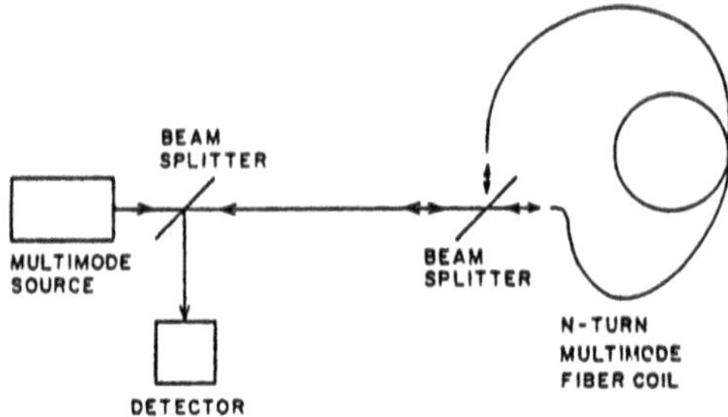

Fig. 1b Multimode, unpolarized, fiber optic gyroscope configuration

$$T_{k\ell}^{12} = r_{k\ell}^{12}\, e^{j(\phi_k^{12} - \phi_s/2)} \tag{3b}$$

where ϕ_s is the Sagnac phase shift, $r_{k\ell}$ is the fraction of the light incident at one end in mode ℓ which exits the other fiber end in mode k, and $\phi_{k\ell}$ is the phase shift along the path connecting modes ℓ and k. Reciprocity is assumed to hold at rest thus $T_{k\ell}^{12} = T_{\ell k}^{21}$. The transfer matrices are unitary if the fiber is lossless. A transfer matrix for the gyroscope can be written in terms of the transfer matrices of the components. The $T_{k\ell}$'s are environmentally sensitive.

We consider two configurations of multimode fiber gyros, one using polarized light (Fig. 1a) and the other using unpolarized light (Fig. 1b), where polarized and unpolarized are defined in terms of the general coherence matrix. For polarized light all of the power is in one mode (i.e., $J_{k\ell} = \delta_{mk}\delta_{m\ell}$), while for unpolarized light the power is evenly divided among the modes and the modes are uncorrelated (i.e., $J_{k\ell} = 1/N\, \delta_{k\ell}$).

The normalized power assuming ideal components is calculated to be

$$I_{\text{polarized}} = 1/2(r_{mm})^2(1 + \cos\phi_s), \tag{4a}$$

$$I_{\text{unpolarized}} = 1/2\left[1 + \left\{\frac{1}{N}\sum_{k=1}^{N}(r_{kk}^{12})^2 - \frac{1}{N}\sum_{k=1}^{N}\sum_{\substack{\ell=1\\ \ell\neq k}}^{N}(r_{k\ell}^{12})^2\cos(\phi_{k\ell}^{12} - \phi_{\ell k}^{12})\right\}\cos\phi_s\right]. \tag{4b}$$

The above formalism reduces to the case for single mode fibers by setting $N = 2$.

Response curves (I versus ϕ_s) for multimode fibers are shown in Fig. 2 for various values of $T_{k\ell}$. Note that the response curves are sinusoidal with the positions of the maxima and minima of each curve independent of $T_{k\ell}$. This indicates reciprocal behavior. For the polarized gyro, the value of the minima are fixed while the value of the maxima depends on $T_{k\ell}$. In the unpolarized gyro, the average value of the curves is constant, while the maxima and minima expand or contract uniformly about the average value depending on $T_{k\ell}$. Thus the behavior of polarized and unpolarized multimode gyros has the same basic characteristics as for polarized and unpolarized single mode gyros.

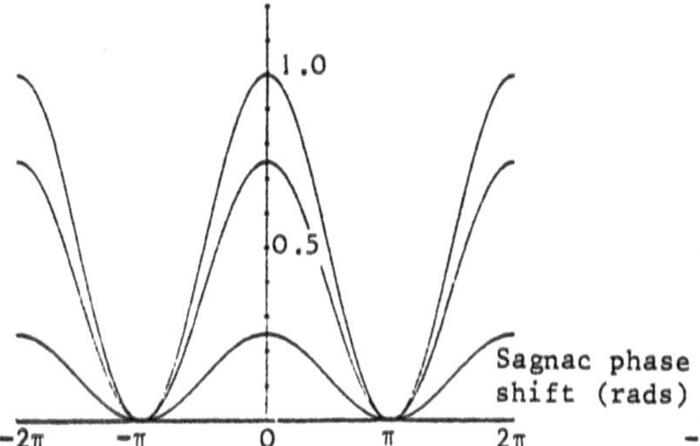

Fig. 2a Response curves of polarized, multimode fiber optic gyroscope for various values of the scattering coefficients ($T_{u\ell}$)

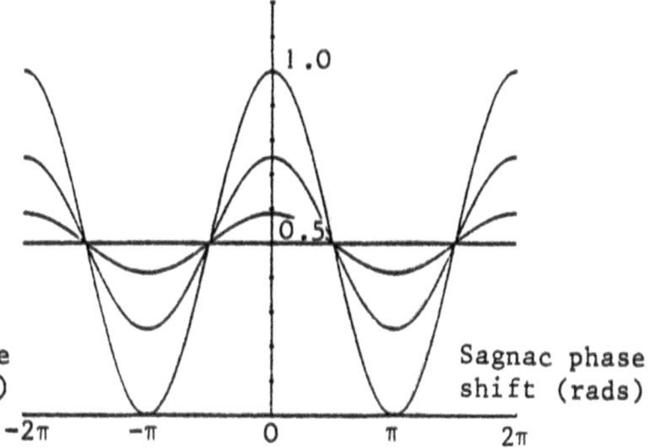

Fig. 2b Response curves of unpolarized, multimode fiber optic gyroscope for various values of the scattering coefficients ($T_{u\ell}$)

For the polarized gyro, Fig. 1 a , the insertion loss can be large if the optical fiber scatters the incident power uniformly over the N modes. However, individual modes are sometimes well preserved, which would result in reduced insertion loss. The insertion loss of the unpolarized gyro, Fig. 1 b , is expected to be low due to the absence of mode selective components. All N modes are detected. The large numerical apertures of multimode fibers allow more optical power to be coupled into them than into single mode fibers. This can be important in reducing the effects of detector noise. Low cost LED's can be used with multimode fibers. Relaxed alignment tolerances and cheaper fiber are further advantages. Also coherent backscattering and environmental sensitivity are expected to be lower because of averaging over the modes.

In conclusion, we see that reciprocal operation of gyros with multimode fiber is theoretically possible in both polarized and unpolarized forms and that there are some prospective advantageous tradeoffs over single mode gyros.

References

1 G. Pavlath and H.J. Shaw, "Unpolarized Reciprocal Operation of Fiber-optic Rotation Sensors," Technical Digest, Conference on Lasers and Electro-Optics, (Washington, D.C., 1981) 126-127

2 H. Arditty, H.J. Shaw, M. Chodorow, and R. Kompfner, "Re-Entrant Fiber-optic Approach to Rotation Sensing," Proceedings of the Society of Photo-Optical Instrumentation Engineers, Vol. 157 (August 1978) 138

3 R. Ulrich and M. Johnson, Optics Letters 4, (1979) 152

4 M. Born and E. Wolf, *Principles of Optics*, Pergamon Press, 5th Edition, (1975), 544-553

This work was supported by the Air Force Office of Scientific Research under Contract F49620-80-C-0040.

Part 3

Building Blocks

3.1 Integrated-Optic Components

Integrated Optics

M. Papuchon

Thomson-CSF, Laboratoire Central de Recherches, BP 10, Domaine de Corbeville
F-91401 Orsay, France

Introduction

Work in integrated optics has been continuing since 1965 when the microwave concepts were combined with photolithographic techniques to build thin film waveguides for applications in the infrared [1]. In fact, the name "Integrated Optics" was given to this new technology four years later [2] and since this date, the field and the number of involved researchers have been continuously growing. The basic element of the integrated optic technology is the optical waveguide which in many cases is of the single mode type. Many different materials are used to create the desired structure starting from passive glass or plastic to active materials such as ferroelectric, magnetooptic or semiconductors. In fact, two types of integrated optics are still competing : the hybrid and monolithic technology. If any material can be used in the first one, the candidates for the last one are obviously semiconductors like Gallium Arsenide or Indium Phosphide where different types of functions can be realized : lasers, amplifiers, modulators, detectors...

In fact, depending on the application, it can be preferable to choose the most suitable material and to connect the different waveguides together via single mode optical fibers.

In this paper, some of the applications of hybrid integrated optics will be described. This will include some devices for optical communications systems, optical processing, non linear optics interaction and obviously for this new but already growing field, optical fiber sensors.

Practically all the examples that will be given here use waveguides realized with the so called Ti:diffusion technique in $LiNbO_3$ [3] which has permitted the realization of the best electrooptical devices available to day.

Optical communications applications

The field of integrated optics began practically at the same time as the optical fiber communication programs. So it was believed that integrated optical waveguides would play an important role in transforming point to point links into "active" optical links. Unfortunately most of the active integrated optical devices are of the single mode type and altough it would have been possible to use them at the beginning ofthe link before entering the multimode fiber it was necessary to wait for the advent of high data rate single mode fiber systemsto think really about using integrated optical circuits.

Among the number of devices which have been proposed and experimentally tested we have chosen to describe two of them which are very significant of what can be achieved with integrated optics.

In the case of high data rate optical fiber links, although it seems possible to modulate directly the injected current in semiconductor lasers , this becomes difficult as the bandwidth is increased over one or several gigahertz. In this case, external modulation could be preferred for amplitude modulation but also especially for phase modulation if heterodyne systems are going to emerge.

Using integrated optic technology and the electrooptical effect, very efficient modulators can be realized the bandwidths of which are only limited by the care with which driving electrodes are realized. Ultimately the bandwidth of the device will be limited by the velocity mismatch existing between the optical guided wave and the modulating electrical signal.

A good example of such a device is shown in figure 1 where the configuration of an integrated Mach-Zehnder interferometer has been realized with Ti:$LiNbO_3$ waveguides to obtain an intensity modulator. In this case, the electrode configuration has been carefully designed in order that an electrical travelling wave type interaction may occur [4]. In this experiment, the length of the arms of the interferometer were 6 mm and the driving voltage 8,8 V. The achieved bandwidth was of the order of 12 GHz.

In fact, if amplitude modulators are important for optical communications, other devices may be of great interest. This is the case for electrooptical switches which permit light to be switched from one waveguide to another using an electrical signal. These have been demonstrated with directional couplers in particular using $LiNbO3$ (see for example [5,6]) showing command voltages of few volts and crosstalks below - 30 dB. The directional coupler principle can be used to realize wavelength multiplexer/demultiplexer.

A device of this type is shown in figure 2 [7]. It presents the additional advantage of being insensitive to the input polarization which could be important as standard single mode fibers do not preserve a particular type of polarization. The working principle is based on a combination of a directional coupler and a TE $\longleftrightarrow$ TM mode converter. The TE (TM) wave in one waveguide is coucoupled to the TM (TE) wave of the other waveguide via an off-diagonal electrooptic coefficient and evanescent wave coupling. Lithium Niobate is strongly birefringent and, as a consequence, the TE $\longleftrightarrow$ TM interaction is not phase matched. This problem is solved by using periodic electrodes whose periods (Λ) satisfy:

$$\frac{2\pi}{\lambda}\left(\frac{\beta}{K_0 TE} - \frac{\beta}{K_0 TM}\right) = \frac{2\pi}{\Lambda} .$$

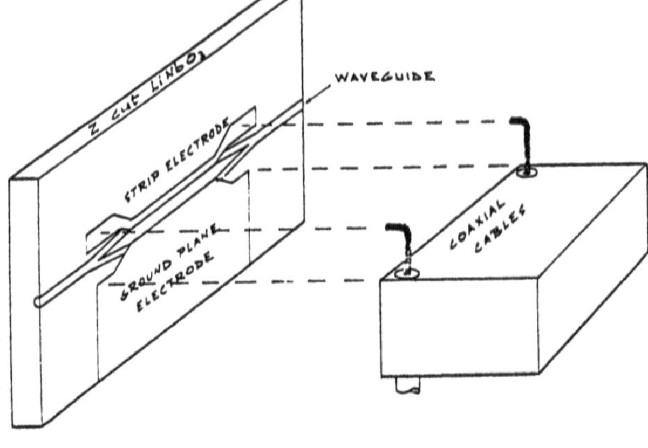

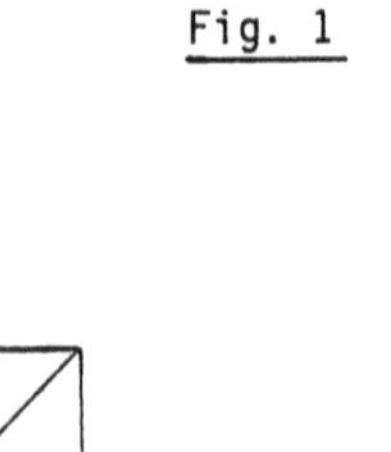

Fig. 1

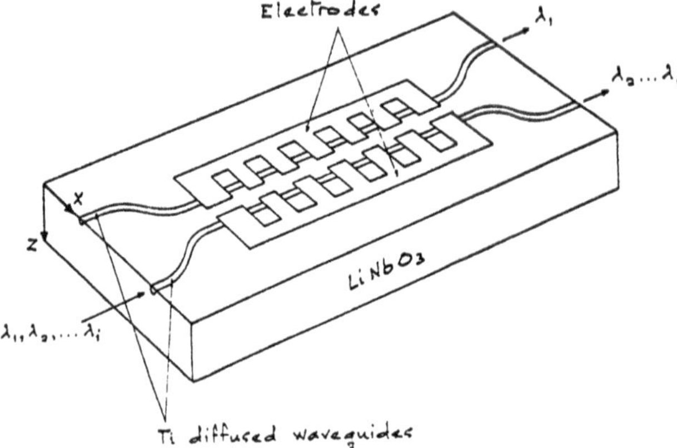

Fig. 2

This relation is wavelength dependent and so a filter can be realized. To overcome direct TE (TM) $\leftrightarrow$ TE (TM) evanescent coupling between the waveguides, these have been fabricated with different refractive indices.

A peak filter efficiency of 75% for a bandwith of 5 Å has been achieved.

Optical data processing

Another field of application for integrated optics is the so called optical data processing.

The most well known device is certainly the RF spectrum analyzer, the operation of which has been recently demonstrated in an integrated optic form [8]. It uses optical planar waveguides in which the acoustooptic interaction permits diffraction of the guided light in different directions depending on the frequency of the acoustic surface wave. Several types of lenses have been proposed and tested such as Luneburg, geodesic or Fresnel lenses. In fact the experimental demonstration has been performed using geodesic lenses with $LiNbO_3$ as the substrate and a GaAs laser as the source. A resolution of 8 MHz has been obtained over a bandwith of 300MHz and with a dynamic range of the order of 25 dB.

Other examples of optical circuits for signal processing have been implemented. The two circuits that we are going to describe are intended to be used aboard spacecraft. The first one is shown on figure 3. It is a part of a much more complex system, realized to preprocess the data coming from a multichannel sensor before sending the information to the earth.

In this system different types of functions are needed such as DC level control, amplitude control, A/D conversion etc.

The circuit shown here is capable of performing some of them (like DC level control, amplitude control) while driven directly by the on board computer via a parallel digital word [9]. The working principle lies in the fact that the electrooptically induced phase shift is proportional to the length of the interaction region. If the command electrode is divided in to several parts having lengths equal to L, 2L, 4L, 8L ..., for the same applied voltage, the induced phase shifts will be Ø, 2Ø, 4Ø, 8Ø ... If the phase modulator is a part of an integrated Mach Zehnder interferometer, a digitally driven amplitude modulator is realized. In this configuration the required command voltage for a logic "1" is 3 V and the modulation efficiency was measured to be better than 95%.

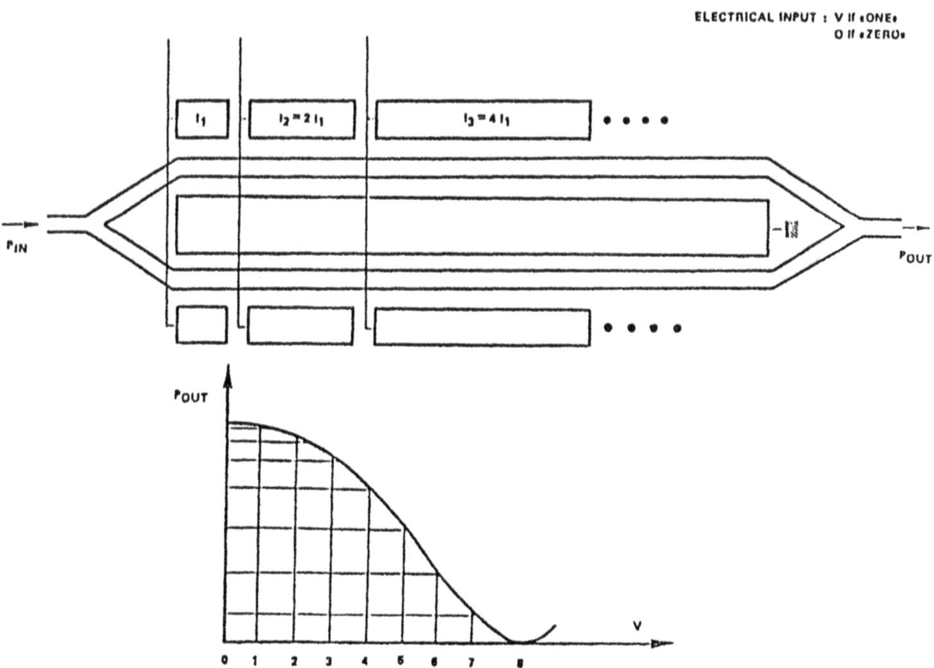

Fig. 3

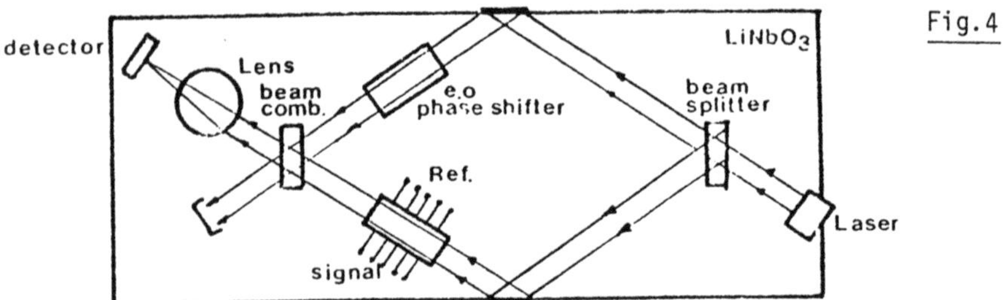

Fig.4

Another type of circuit for on board processing is shown in figure 4.

It is based on the principle of an integrated interferometer formed on a planar $LiNbO_3$ waveguide. In one arm a set of electrodes is used to adjust the nominal phase shift between the two arms. In the other arm two sets of electrodes are used, one for introducing an externally stored voltage set. The electrodes are arranged so that the nominally incident plane wave upon the electrode region is distorted by the signal voltages when the signal and reference voltages are equal. The match between a signal set and a given reference set is signaled by the null resulting from destructive interference (adjusted with the electrodes in the other arm between the two beams).

Non linear optic interaction

Since the beginning of integrated optics, it has been recognized that the very high energy density achievable in single mode optical waveguides would permit the observation of non linear optical effects with moderate incident powers.

In fact work began very early by investigating second harmonic generation using non linear waveguides or substrates [11, 12, 13]. Recently very high conversion efficiencies (of the order of 25%) have been obtained. One of the advantages of non linear optics is the ability to generate new optical frequencies in a very broad range. This has been demonstrated recently in an integrated form by operating a parametric oscillator in Ti : $LiNbO_3$ waveguides pumped with a dye laser [14]. The next goal should be to demonstrate parametric oscillation with a semiconductor laser as the pump thus leading to a miniature tunable source in the infrared.

Optical fiber sensors - the fiber gyroscope

Optical sensors and especially those using fibers formed a rapidly growing field. Some classical examples are the hydrophone, current sensor and the gyroscope. In general, interferometric techniques are used to detect the phase shifts induced by the external parameter to be measured. That means that beam combiners, dividers, lenses and some time active devices such as phase modulators, switches or frequency shifters are needed. All these types of optical devices can be realized in a compact way using integrated optic technology thus leading to a very small and stable heart of the interferometer. This is particularly true for the optical gyroscope using passive Sagnac interferometers where the phase shifts induced by rotation onto the two counterpropagating waves are very small.

In fact, several types of passive Sagnac interferometers can be used and the corresponding integrated circuits reflect at the same time the type of inter-

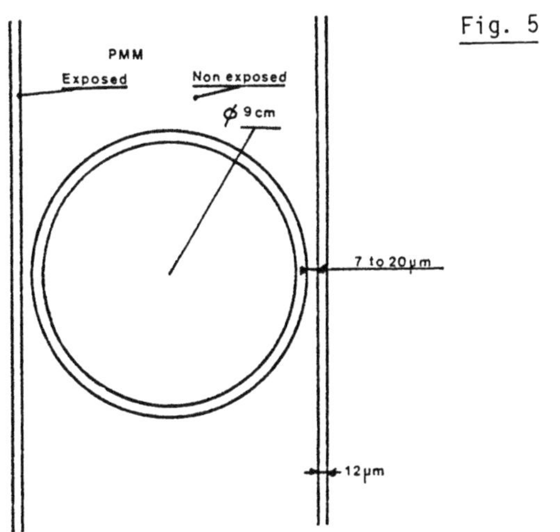

Fig. 5

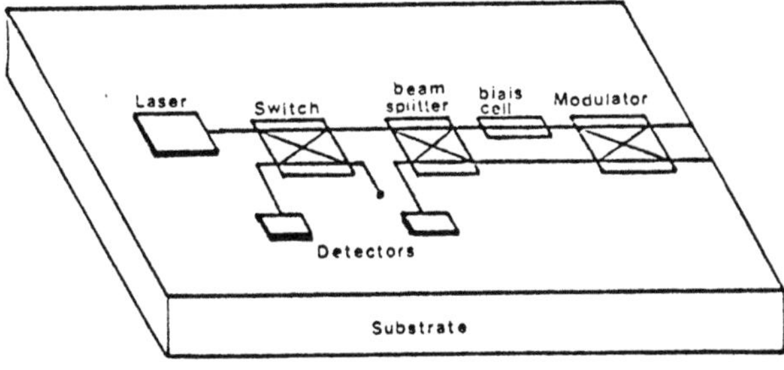

Fig. 6

ferometer used and the techniques employed to extract the weak Sagnac signal from the background. Figure 5 shows a part of a circuit [16] based on the principle of the resonant Sagnac interferometer [17]. The ring forms a travelling wave resonator for the two counterpropagating waves (the ring is fed with the directional couplers). When the interferometer is at rest the two resonant frequencies are identical. When the interferometer is rotating the two frequencies are no longer the same and the difference is directly related to the rotation rate.

Another type of integrated circuit is shown in figure 6 and is intended to be used with a non resonant fiber gyro [18]. In this case the so-called fiber coil flipping technique is used to reverse the sign of the Sagnac signal leaving the same background. This permits the use of synchronous detection to extract the interesting signal from the background.

The last integrated circuit shown in figure 7 is also intended to be used with a non resonant optical fiber gyro.It comprises the two beam splitters (3 dB directional couplers) needed to ensure reciprocal operation of the system and polarizer and phase shifters used to modulate the Sagnac signal in a reciprocal way to permit the use of synchronous detection techniques [19]. In fact simpler circuits can be used [20] to achieve the same functions without the need of command electrodes for the beam splitters. In addition

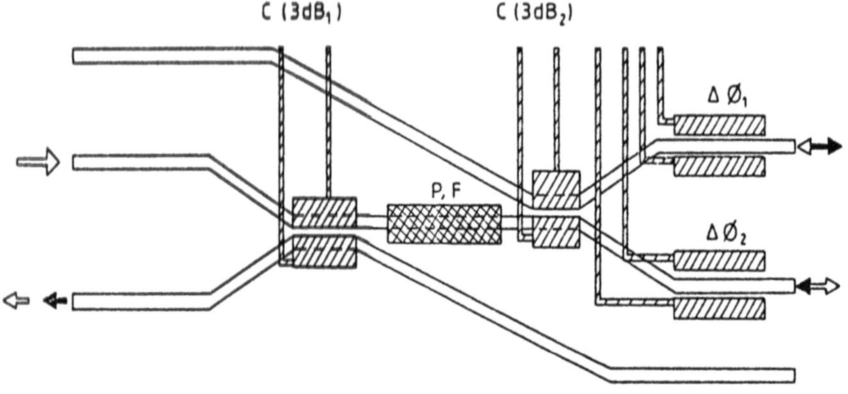

FIG7

these new circuits are shorter than the previous one and so present lower propagation losses and are easier to fabricate.

There is no doubt that these circuits are under test in different laboratories and that the operation of an integrated optic gyroscope will be reported soon.

Conclusion

In this paper we have reviewed some interesting possibilities of integrated optics technology in fields as different as optical communications, data processing, non linear optics and optical sensors especially the fiber optic gyroscope. It is clear that the integrated optical circuits are now ready for application and that in the next few years they will appear in various optical systems where compactness, insensitivity to vibrations etc. will be important.

References

1) D.B. Anderson and R.R. August, Proc. IEEE 54, 657 (1966)
2) S.E. Miller, Bell System Tech. J. 48, 2059 (1969)
3) R.V. Schmidt and J.P. Kaminow, Appl. Phys. Lett. 25, 458 (1974)
4) T. Sueta and M. Izutsu, paper TUM2 IOOC'81, San Francisco (1981)
5) M. Papuchon et al., Appl. Phys. Lett. 27, 289 (1975)
6) R.V. Schmidt and H. Kogelnik, Appl. Phys. Lett. 2, 503 (1976)
7) R.C. Alferness, IEEE JQE - 17, 946 (1981)
8) T.R. Ranganath, T.R. Joseph and J.Y. Lee, paper WH3, IOOC'81, San Francisco (1981)
9) M. Papuchon, C. Puech and A. Schnapper, Elect.Lett. 16, 142 (1980)
10) C.M. Werber et al., Paper 13.6.1. Optical communication conference Amsterdam (1979) -
11) P.K. Tien, R. Ulrich and R.J. Martin, Appl. Phys. Lett. 7, 447 (1970)
12) Y. Suematsu et al.,IEEE JQE-17, 222 (1974)
13) C. Deverdun and P. Lallemand, Rev.Tech. TH.CSF 6, 1030, (1974)

14) W. Sohler and H. Suche, Appl. Phys. Lett. 33, 518 (1978)
15) W. Sohler and H. Suche, Paper WB1, IOOC'81 San Francisco
16) J.H. Haavisto and G.A. Pajer, Opt. Lett., 5, 510 (1980)
17) J.L. Davis and S. Ezekiel, Proc. SPIE, 157, 131 (1978)
18) W.C. Goss, R. Goldstein, M.D. Nelson, H.T. Fearnehaugh and O.G. Mauer, Appl. Opt. 19, p.852 (1980)
19) H. Arditty, M. Papuchon, C. Puech and K. Thyagarajan, Topical meeting on integrated and guided wave optics,Lake Tahoe (1979)
20) M. Papuchon, H. Arditty, S. Vatoux and C. Puech,"Reciprocity Properties of a branching waveguide",this conference

Progress on Integrated Optic Waveguide Devices for Fiber Gyro Applications

O.G. Ramer

Hughes Research Laboratories, 3011 Malibu Canyon Road, Malibu, CA 90265, USA

1. Introduction

It is presently widely believed that the production future of fiber optic gyros depends on the development of integrated optic circuits [1] to replace the various bulk optic components and labor intensive components presently utilized in all waveguide systems. Before this can become practical, the fabrication of low-loss integrated optical circuits and the low-loss coupling of single-mode fiber to $LiNbO_3$:Ti in-diffused channel waveguides are also required technologies [2]. This paper reports progress in the areas of fiber-to-chip coupling (Sections II and IV), circuit losses and device fabrication (Section II). The demonstrated fiber-to-channel waveguide coupling efficiencies and measured circuit losses due to waveguide bends and offsets are in agreement with theoretical estimates. A four- port optical switch has been pigtailed with an approach that can be extended to the coupling of fibers to numerous devices on a single chip, as will be required for gyro triads.

2. Experimental Fiber-to-Channel Waveguide Coupling Losses

To optimize the single-mode fiber to integrated-optic channel waveguide coupling, Z-cut $LiNbO_3$:Ti in-diffused channel waveguide samples with different waveguide widths, Ti depths, and diffusion times were fabricated. All samples were diffused at 1000°C. In the fiber-to-channel waveguide coupling, an ITT single-mode fiber with a 4.5 μm core coupled to a laser diode with a wavelength of 0.83 μm was used. The waveguide samples contained straight waveguides and the following fabrication parameters were used.

1) Ti thickness of 430 Å and 470 Å.
2) Diffusion times of 4, 6, and 8 h.
3) Waveguide width from 1 μm to 10 μm in 0.5 μm steps.

A schematic of the experiment used in the evaluation is shown in Fig.1. The TE polarization was maintained by elimination of any evidence of the planar Li_2O out-diffusion mode known to have a TM orientation in Z-cut $LiNbO_3$. The ITT fiber was epoxied into the Si V grooves, and the fiber and Si ends were polished. The positioning of the polished Si face against the polished $LiNbO_3$ chip edge ensured consistent measurement from waveguide sample to sample. The Si fiber holder was attached to a micromanipulator. Glycerol was used as an index matching medium between the fiber and $LiNbO_3$. The detected light has been corrected by the following values to obtain the coupling coefficient: lens loss of 1.1 dB (measured value), propagation loss of 1 dB (assumed from an independent measurement on planar guides with losses of 0.8-0.9 dB/cm), and reflection loss of 0.5 dB (theoretical value). The values obtained for the TE polarization versus waveguide width are plotted in Fig.2; values have been plotted only for single-mode guides. The 430 Å samples had excellent waveguides for all three diffusion times. The 470 Å samples appeared incompletely diffused for the 4 and 6 h diffusions. The 8 h diffusion samples produced excellent waveguides.

At the 5 µm waveguide width, the 430 Å sample gave a coupling loss below 0.6 dB. The 470 Å, 8 h diffusion sample gave the lowest coupling loss of 0.4 dB for a wide range of waveguide widths. These values must be viewed as lower bounds since they are corrected measurements based on an assumed propagation loss; however, we can be assured of a coupling loss of less than 1 dB if the 0.8-0.9 dB/cm planar guide losses are taken as a lower bound. Measurements for the TM polarization gave similar losses.

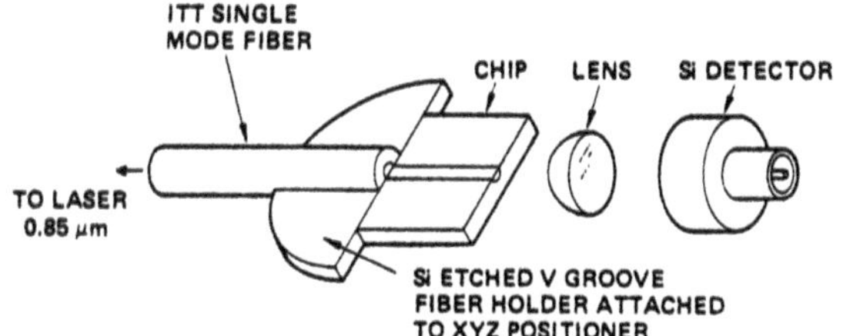

Figure 1. Schematic of experimental arrangement used in fiber-to-fiber coupling measurement

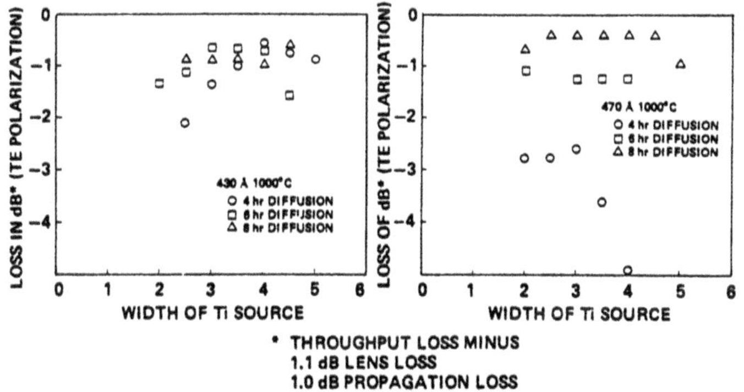

Figure 2. Data from the coupling loss measurements for Ti thicknesses of 430 Å and 470 Å

3. Waveguide Losses

In addition to simple propagation loss (∿1dB/cm), there are other sources of loss associated with an optical circuit that do not exist in straight waveguides. The two primary sources are the following:

1) Waveguide bends and displacement,
2) metal electrode placement on waveguides.

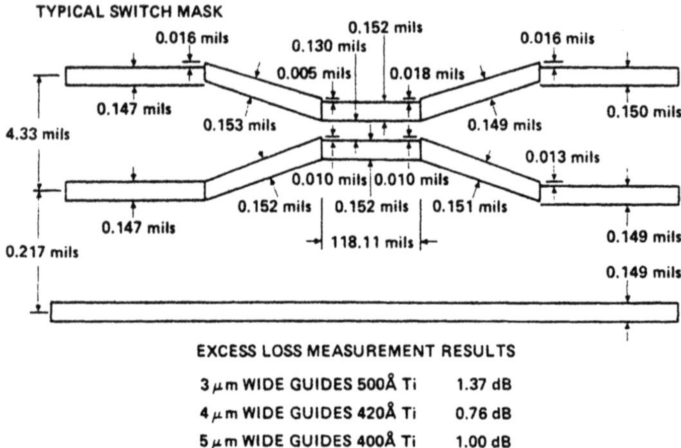

EXCESS LOSS MEASUREMENT RESULTS

3 μm WIDE GUIDES 500Å Ti	1.37 dB
4 μm WIDE GUIDES 420Å Ti	0.76 dB
5 μm WIDE GUIDES 400Å Ti	1.00 dB

Figure 3. Sources of additional loss in integrated optic circuits are related to mask making accuracy and the requirement for waveguide bends

In the typical switch waveguide mask (see Fig.3), there are two additional sources of loss that do not exist with straight waveguides. These are the bends and waveguide displacement at the connection between angled and non-angled sections. We have analyzed what the expected loss should be for a bend and a displacement between straight guides. The initial calculations were done using the mode profiles obtained from the Marcatili [4] dielectric waveguide and analysis. The second analysis was done using Gaussian profiles: this analysis gives the following simple formulas that are dependent on the Gaussian waist. For the bend, the transmission is given by [2,3]

$$\varepsilon = e^{-k^2 w^2 \sin^2\theta/4} \qquad (1)$$

where k is the propagation constant and θ is the bend angle in the waveguide. For the displacement, the transmission is given by

$$\varepsilon = e^{-\delta^2 w^2} \qquad (2)$$

where δ is the displacement between waveguides. These equations allow predictions in general trends, but a more detailed analysis is required to include the waveguide parameters such as Δn, aspect ratio, and refractive index distribution. The more detailed analysis comes by using the Marcatili waveguide modes.

On our switch masks, which have four 0.5° bends and 0.4 μm displacements at the waveguide interconnections, the expected additional loss from Equations (1) and (2) s ∿1 dB. The displacements are due to limitations imposed by current mask-making technology; high resolution E-beam or ion-beam technology can improve on this in the future. This additional loss has been measured at 0.83 μm using Z-cut $LiNbO_3$:Ti in-diffused samples with straight guides and switch patterns showing excellent transmission. These three samples were diffused for 6 hr and have the following:

1) 3 μm wide Ti sources 500 Å thick
2) 4 μm wide Ti sources 420 Å thick
3) 5 μm wide Ti sources 400 Å thick.

The throughput for the switch patterns is consistently lower than for the straight guides. The excess losses associated with the TE polarization caused by the bends in the samples were 1.37, 0.76, and 1 dB for the samples given above, respectively. This is believed to be in excellent agreement with the theory.

The second source of additional loss associated with an optical circuit is the placement of electrodes on the chip such that the metal interferes with the light. The electrodes placed directly on the $LiNbO_3$ waveguides as required for the Z-cut $LiNbO_3$ switch are a source of propagation loss (∿1 dB/mm for the TE mode and >10 dB/mm for the TM mode). In general, with Z-cut $LiNbO_3$, the TM mode is not observed when the electrodes are in place. Therefore, it is desirable (and, in fact, necessary for low-loss chips) to isolate the electrodes from the $LiNbO_3$ surface with a low-loss dielectric if the Z-cut configuration is to be used. A buffer layer of 2500Å of sputter-deposited SiO_2 reduced the propagation losses for both polarizations. Switch operation using the TM polarization with the larger r_{33} electrooptic coefficient was possible.

4. Fiber-to-Chip Coupling

This section discusses the fiber pigtailing of a Z-cut $LiNbO_3$ chip containing a four-section Δβ-reversal switch. The switch was fabricated using 4 μm wide strips of 470 Å thick Ti diffused 8 h at 1000°C. In the interaction region, the guides were separated by 3 μm for 8.2 mm. The total chip length was 2.5 cm. A 2500 Å SiO_2 buffer layer was sputter-deposited on the $LiNbO_3$ before deposition of the electrodes.

The technique of permanently attaching single-mode fibers to the integrated-optic chip is illustrated in Fig.4 with a photograph of a completed device. To hold the single-mode fibers, V grooves are etched in (100)-oriented Si wafers with KOH. The photolithographic masks made for both this process and the waveguide fabrication have matching center-to-center spacings of 110 μm. After the fibers are positioned in the grooves, the Si and fibers are polished such that the fiber and Si edge are aligned. The fixture holding the $LiNbO_3$ chip is designed such that the Si chip can be adjusted into place with external micropositioners, epoxied into place, and have the positioner removed.

However, this process does not leave everything in an optimum position; in some cases, total decoupling has been observed. Therefore, we have designed into the fixture the ability to adjust the platform to which each Si chip has been attached independently for the $LiNbO_3$ sample. The primary cause of the misalignment from removing the micropositioner is the stress placed on the Si chip by the curing of the epoxy.

The device shown in Fig.4 has a total throughput loss of 7.8 dB, divided as follows:

1) 2.5 dB propagation loss (1-dB/cm loss)
2) 1 dB bend and offset loss (circuit losses, see Section 3)
3) 0.35 dB reflection loss due to lack of antireflection coatings at fiber $LiNbO_3$ interfaces
4) 3.9 dB fiber-to-chip coupling loss.

This latter loss is the remainder after the three preceding losses have been subtracted from the total loss. This larger than expected coupling loss is believed to be associated with the precision with which the fibers can be positioned with the present fixture or possibly with the concentricity of the fiber core. This statement is based on the following observations. First, the total throughput was maximized with both input fibers excited.

Figure 4. Photograph of fiber-pigtailed $LiNbO_3$ chip with four section $\Delta\beta$-reversal switch

After blocking either laser source, the remaining throughput could then be increased by ~10 percent. Unblocking the second laser source then decreased the overall throughput. Second, the total loss at one point is believed to have been closer to 6 dB before the fibers were permanently attached to the chip.

5. Conclusion

We have investigated the losses associated with an integrated optic circuit and the fiber pigtailing of the integrated optic chips. The experimental results presented for the fiber-to-channel waveguide coupling coefficient show potential coupling losses of less than 1.5 dB/chip. The circuit losses are dependent on the mask-making capability and on the circuit complexity. Theoretical and experimental data are in good agreement for the single device circuit discussed in this paper (i.e., ∿1 dB). Finally, we have presented a promising technique for permanently coupling fibers to chips. Coupling losses for the present device are very good (∿3.9 dB total), but it is expected that future improvements will reduce this loss by 2.4 dB.

References

1. W. C. Goss, R. Goldstein, M. D. Nelson, H. T. Fearnehaugh, and O. G. Ramer,"Fiber-optic rotation sensor technology", Applied Optics 19, p 852 (1980).
2. L. D. Hutcheson, I. A. White, and J. J. Burke."Comparison of bending losses in integrated optical circuits",Opt. Lett., Vol. 5, p. 262 (1980)
3. O. G. Ramer, C. Nelson, and C. Mohr, "Experimental Integrated Optic Circuit Losses and Fiber Pigtailing of Chips",IEEE Journal of Quantum Electronics, Q.E. 17, p 970 (1981)
4. E. A. J. Marcatili, "Dielectric rectangular waveguide and directional coupler for integrated optics", Bell System Tech J. Vol. 48, p 2071 (1969)

Guided-Wave Electrooptic Modulators

F.J. Leonberger

Lincoln Laboratory, Massachusetts Institute of Technology
Lexington, MA 02173, USA

In the development of fiber-optic rotation sensors, a variety of optical modulators are needed. It would be desirable to monolithically integrate such modulators on a chip to provide reliable, compact and low-drive power operation. In this paper, progress in guided-wave electrooptic modulators is reviewed. Devices in both $LiNbO_3$ and GaAs are described for use in phase, intensity, and frequency modulation. In addition, progress in waveguide bends, essential for monolithically interconnecting the modulators, and in fiber-to-chip coupling is summarized. In each instance, representative examples of the state of the art are given rather than a complete literature review.

Figure 1 is a schematic drawing of the two most common single-mode three-dimensional waveguide structures. In both cases, the indices of refraction are labelled and n_2 is assumed to have the largest index. Also in both cases, the single-mode guides support both a TE- and TM-like mode. The first guide, an embedded stripe, is commonly formed in $LiNbO_3$ by indiffusion of Ti to raise

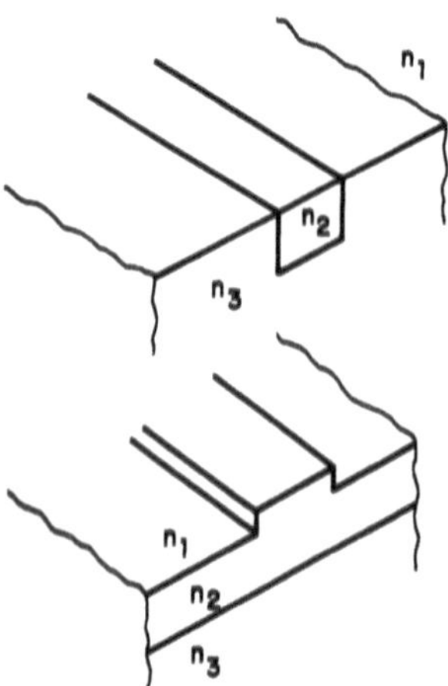

Fig. 1 Schematic drawing of embedded stripe and rib waveguides. In each case, the relative size of the refractive indices is $n_2 > n_3 > n_1$

the index [1,2]. Attenuation in these guides is typically $\leq$ 1dB/cm. Optical power handling densities at visible wavelengths are $\leq$ 50 W/cm^2 due to optical damage effects. Recent work by several workers suggests that the guides can sustain at least an order-of-magnitude higher power density in the 0.85- to 1.5-μm range.

The second guide shown is a rib guide, and is commonly fabricated in semiconductors. In GaAs-based structures, the loss is typically 4 to 6 dB/cm at wavelengths in the 1-um range, but recently a loss of only 2 dB/cm was reported in an oxide-confined structure in which a special epitaxial growth technique was used so that SiO_2 rather than a semiconductor layer formed the region of index n_3 in the figure [3]. An obvious advantage of the semiconductor guides, in spite of their higher propagation loss and their less efficient electrooptic effect (x5 more voltage than $LiNbO_3$ for comparable phase shift in identical guide structures), is the potential to fabricate both lasers and detectors on chip with other guided-wave components.

Figure 2(a) shows a typical phase modulator fabricated in $LiNbO_3$. For electric fields along the c axis, a modulation rate of 0.3 V/rad has been achieved for TE polarization in a structure of 12-pF capacitance [4]. For phase modulation of an arbitrary input polarization, a polarization-independent modulator [5] such as the one shown in Fig. 2(b) can be built. In this case, separate electric fields, parallel and perpendicular to the surface, are provided to control both TE and TM components. This polarization-independent concept has been demonstrated to date only in $LiNbO_3$.

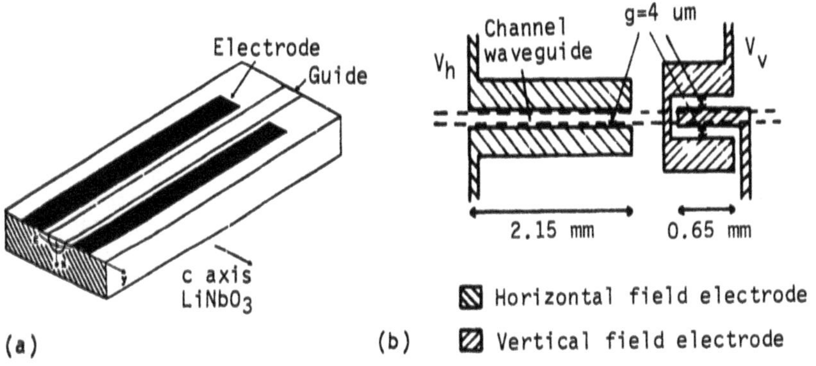

Fig. 2 Schematic drawing of (a) conventional $LiNbO_3$ electrooptic phase modulator and (b) polarization-insensitive phase modulator (from [5])

Several different types of intensity modulators have been demonstrated. Each type has exhibited extinction values of ~ 25 dB. Fig. 3 illustrates an electroabsorption modulator [6]. This device has been demonstrated in GaAs and relies on Franz-Keldysh absorption $\Delta\alpha$ near the band edge at moderate electric fields E ($\Delta\alpha \sim 10^3$ cm for $E = 10^5$ V/cm at 0.2 μm below the band edge). The effect is achieved by reverse biasing the Schottky diode shown in the figure. This device requires that the wavelength of the guided light be near the band edge and, because of the large effect, generally results in very high modulation depth per unit drive power.

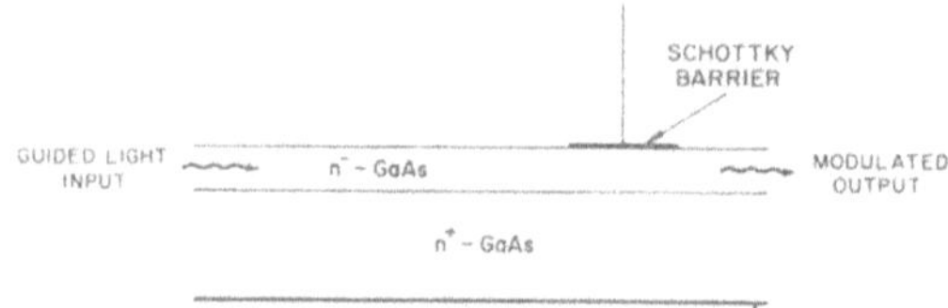

Fig. 3 Schematic drawing of GaAs electroabsorption waveguide modulator. The light is loss-modulated by reverse biasing the Schottky barrier

Figure 4 illustrates a 2-section $\Delta\beta$-reversal directional-coupler switch. This device can be operated as a 2 x 2 switch, a tunable power splitter or as a modulator. Unlike the conventional directional-coupler switch, both states can be electrically tuned by changing the difference in propagation constant $\Delta\beta$ between the two guides. Here $\Delta\beta \sim kV$ where k is an electrooptic coefficient, and V is the applied voltage. Its operation has been described in detail [7], and high-quality devices have been demonstrated in both $LiNbO_3$ and GaAs. This type of modulator has been operated at bandwidths of 3 GHz [8]. The transfer function for a N-section $\Delta\beta$-reversal switch is given by

$$\frac{I}{I_0} = \sin^2 \left[N \sin^{-1} \left(\frac{\sin^2 \left\{ \frac{\pi L}{2 \ell N} \left[1 + \left(\frac{\Delta\beta\ell}{\pi} \right)^2 \right]^{1/2} \right\}}{1 + \left(\frac{\Delta\beta\ell}{\pi} \right)^2} \right)^{1/2} \right] \qquad (1)$$

where I is the output power in the crossover state, I_0 is the input power, L is the electrode length of each section, and ℓ is the coupling length.

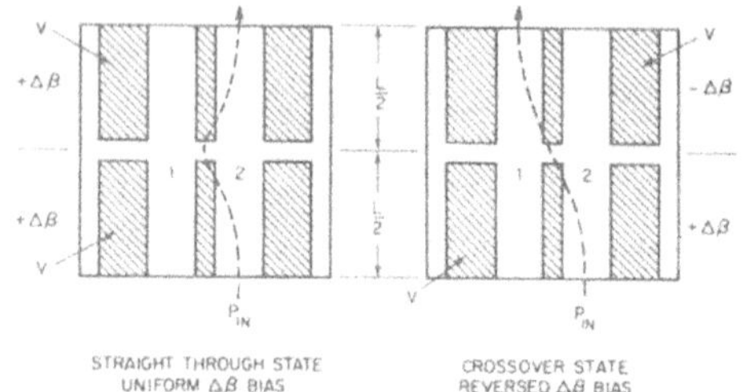

Fig. 4 Schematic drawing of $\Delta\beta$-reversal directional-coupler switch. The light output is switched between guides 1 and 2 by electrooptically altering the sign of $\Delta\beta$ in the second half of the device

Figure 5 illustrates an electrooptic Mach-Zehnder interferometric modulator. In this device, the output phase of the two arms is varied electrooptically and the output Y serves as a mode filter to eliminate the second order mode [9]. This device requires a factor-of-3 less drive power for full modulation than a comparable $\Delta\beta$ switch. Its operation can be described by

$$\frac{I}{I_0} \sim \cos^2 \left(\frac{\Delta\beta L}{2} \right) \qquad (2)$$

where L is the electrode length. This transfer function is periodic without any drop in intensity with values of $\Delta\beta > 2\pi$, as is the case with the directional-coupler devices. The achievement of good performance relies on the ability to make good 3-dB splitters. Devices with Y-junction splitters, as shown in the figure, have exhibited a 25-dB extinction ratio [10], indicating a splitting accuracy of $\lesssim 2.5\%$.

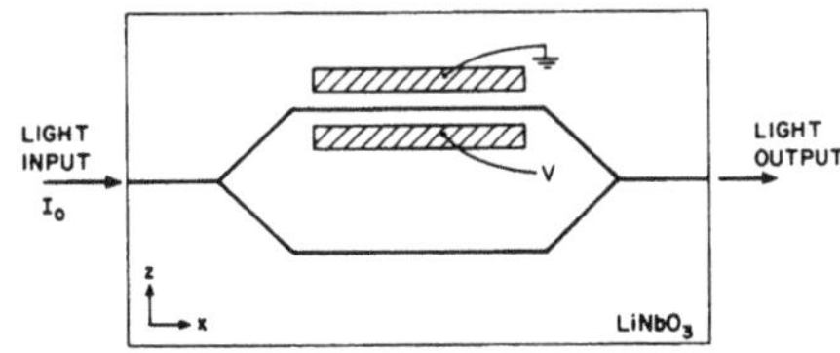

Fig. 5 Schematic drawing of guided-wave Mach-Zehnder interferometer. The relative phase of the light in the two arms at the output Y-junction is electrooptically controlled

An interesting possiblity is the use of a modulator as a frequency shifter. Mach-Zehnder interferometers have been investigated for this purpose because of their relatively simple transfer function and the high degree of output accuracy attainable with both ramp and sinusoidal drive signals. If an rf signal $(\phi_1/2)\cos\omega_1 t$ is applied to a device with ideal Y-splitters the output is FM-modulated with most of the power in the first two sidebands. The relative intensities of the frequency outputs is given by

$$I_0 = I(\Omega_0) = J_0^2(\phi_1/2)[1 + \cos\phi_0] , \tag{3}$$

$$I_1 = I(\Omega_0 \pm \omega_1) = J_1^2(\phi_1/2)[1 - \cos(\phi_0 \pm \theta_1)] , \tag{4}$$

$$I_2 = I(\Omega_0 \pm 2\omega_1) = J_2^2(\phi_1/2)[1 + \cos(\phi_0 \pm 2\theta_1)]. \tag{5}$$

Here Ω_0 is the optical frequency, ϕ_0 is the dc phase tuning of the interferometer and θ_1 is the rf phase difference introduced between the two modulator arms. From these equations and a knowledge of the zeroes of the Bessel functions J_n, one can predict various types of sideband generation. For example by making $\phi_0 = \theta_1 = \pi/2$ and $\phi_1/2 = 2.5$, one can produce a type of single-sideband suppressed-carrier modulation, with 6-dB insertion loss relative to the unmodulated carrier throughout. In this mode of operation the second sidebands are 6 dB down relative to the first sideband.

Frequency-modulation experiments of this type using Mach-Zehnder modulators have been reported by AURACHER and KEIL [11]. Shown in Fig. 6 are some of their data. In Fig. 6(a) there is no phase difference between the arms (θ_1 = 0) and the device is biased at ($\phi_0 = \pi$) to suppress the carrier and generate both of the first sidebands. Fig. 6(b) shows single-sideband suppressed-carrier modulation obtained by setting $\phi_0 = \theta_1 \approx \pi/2$. The other first-order sideband was generated by setting $\theta_1 = -\pi/2$.

Another potentially important modulator is the TE-to-TM converter which allows the polarization of the output wave to be controlled. Due to space limitations, this device will not be reviewed here, but the reader is referred to [12] and [13] for a complete description. Recently the use of a mode-converter in conjunction with two phase modulators has led to the demonstration of an electrooptic generalized polarization transformer [14].

To effectively utilize these modulators and power dividers, it is desirable to monolithically interconnect them with low-loss waveguide bends. Recently, a 1-cm-long 1-cm-radius-of-curvature bend used as part of an 0.63-μm unequal-arm-length interferometric temperature sensor was demonstrated that had a total loss of only ~ 6 dB [15]. The bend was formed by interconnecting straight waveguide sections using coherent coupling [16] between short (~ 200-μm) sections to minimize the loss. This technique is especially amenable to photolithographic pattern generators which typically can expose only rectangles.

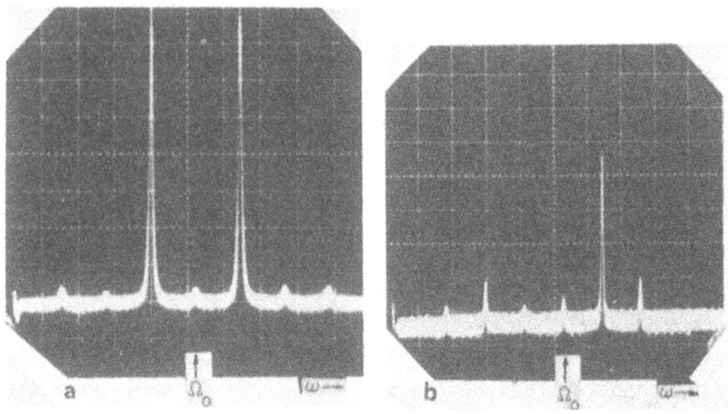

Fig. 6 Output frequency spectra for dc-biased Mach-Zehnder modulator driven with rf sinusoid to suppress the optical carrier frequency (a) both first sidebands generated with $\phi_0 = \pi$ and $\theta_1 = 0$ and (b) one first sideband generated along with minimal second-sideband generation and with $\phi_0 = \theta_1 = \pi/2$ (from [11])

Fig. 7 shows a sketch and photomicrograph of a 1.06-µm single-mode 4-way optical power splitter [10]. Coherent coupling was also used for the bend design of this device. The splitter was 3 mm long, the output waveguides were separated by 100 µm, the splitting uniformity was ± 0.5 dB. This initial device had a 6-dB total insertion loss (i.e. the sum of the outputs of the four guides relative to the output of a straight guide on the same wafer).

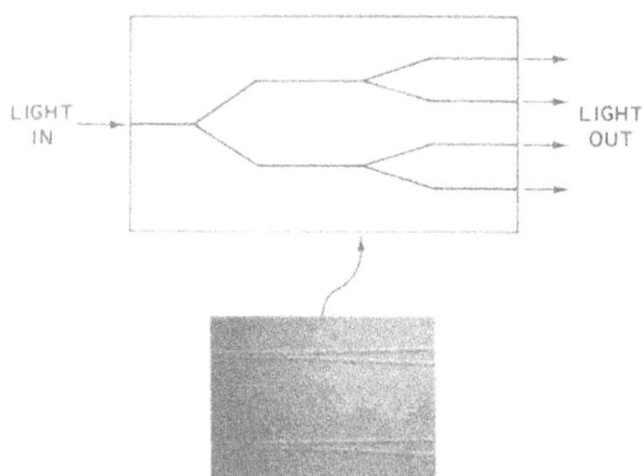

Fig. 7 Single-mode 1.06-µm 4-way optical power splitter. The branching circuit length is 3 mm and the output guides are separated by 100 µm (from [10])

Another fundamental problem is coupling light between single-mode fibers and modulator chips. Fig. 8 illustrates a flip-chip approach developed by BULMER and coworkers [17]. In this approach, V-shaped fiber-alignment grooves are etched in a silicon chip and the $LiNbO_3$ chip aligned to the grooves. A particular advantage of this approach shown in the figure is the use of tapered alignment fibers in cross grooves which allow the fiber height to be positioned to account for any lack of centration of the fiber core. Coupling losses in excess of Fresnel loss as low as ~ 1 dB at each port have been achieved. This number is largely limited by the overlap of the guided modes in the fiber and in the $LiNbO_3$ waveguide. There has been no reported fiber coupling to single-mode semiconductor waveguides.

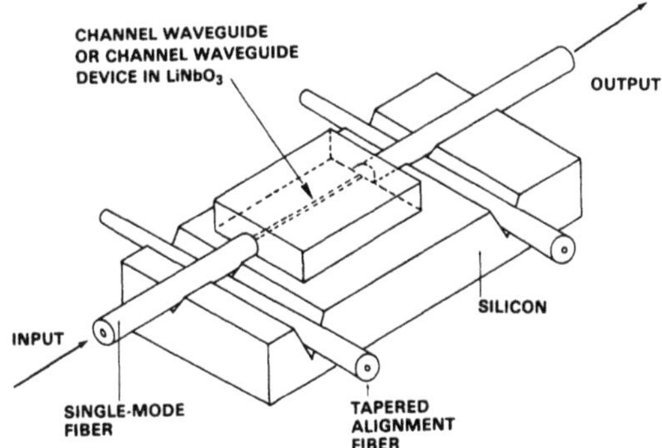

Fig. 8 Schematic drawing of flip-chip grooved-Si method of coupling single-mode fiber to $LiNbO_3$ waveguide chips (from [15])

In summary, representative performance characteristics have been presented for a variety of guided-wave electrooptic modulators for phase, intensity and frequency modulation. The quality of performance of these modulators, in conjunction with the described progress in waveguide bends and fiber-to-chip coupling, makes the devices attractive for use in fiber rotation sensors.

References

1 R. V. Schmidt and I. P. Kaminow, Appl. Phys. Lett. 25, 458 (1974)
2 W. K. Burns, P. H. Hiklein, E. J. West and L. E. Plew, J. Appl. Phys. 50, 6175 (1979)
3 F. J. Leonberger, C. O. Bozler, R. W. McClelland and I. Melngailis, Appl. Phys. Lett. 38, 313 (1981)
4 I. P. Kaminow, L. W. Stulz and E. H. Turner, Appl. Phys. Lett. 27, 555 (1975)
5 C. H. Bulmer and R. P. Moeller, Tech. Digst. Third Int'l. Conf. Integrated Optics and Optical Fiber Comm., San Francisco, CA, April 1981, pa 6 WL5
6 G. E. Stillman, C. M. Wolfe, C. O. Bozler and J. A. Rossi, Appl. Phys. Lett. 28, 544 (1976)
7 H. Koegelnik and R. V. Schmidt, IEEE J. Quantum Electron. QE-12, 396 (1976)
8 R. C. Alferness, N. P. Economou and L. L. Buhl, Appl. Phys. Lett. 38, 214 (1981)
9 See, for example, F. J. Leonberger, C. E. Woodard and D. L. Spears, IEEE Trans. Circts. and Systms. CAS-26, 1125 (1979)
10 F. J. Leonberger, Proceed. SPIE, 269, 64 (1981)
11 F. Auracher and R. Keil, Appl. Phys. Lett. 36, 626 (1980)
12 R. C. Alferness, Appl. Phys. Lett. 36, 513 (1980)
13 O. Mikami, Appl. Phys. Lett. 36, 191 (1980)
14 R. C. Alferness and L. L. Buhl, Appl. Phys. Lett. 38, 655 (1981)
15 L. M. Johnson and F. J. Leonberger, Tech. Digst. Third Int'l. Conf. Integrated Optics and Optical Fiber Comm., San Francisco, CA, April 1981, paper TuM1
16 H. F. Taylor, Appl. Optics 13, 642 (1974)
17 C. H. Bulmer, S. K. Sheem, R. P. Moeller and W. K. Burns, Appl. Phys. Lett. 37, 351 (1980)

3.2 Fiber-Optic Components

Single Mode Fiber Optic Components

R.A. Bergh, M.J.F. Digonnet, H.C. Lefèvre, S.A. Newton, and H.J. Shaw

Edward L. Ginzton Laboratory, Stanford University
Stanford, CA 94305, USA

In the process of studying single-mode fiber optic systems it became rapidly apparent a few years ago that it would be desirable to perform a number of functions with in-line components, which would allow for signal processing without the removal of the optical signal from the fiber. Some advantages of single-mode fiber optic components over conventional bulk optic components would be reduction of insertion loss, high stability, and compactness. They would also be free of difficulties involved in interfacing their integrated optic counterparts to single mode fiber optic systems.

We present in this paper three such components that were recently developed in our laboratory for use in fiber gyroscopes, namely a single mode fiber directional coupler, [1] a single mode fiber polarizer [2] and a polarization controller [3].

1. Fiber Optic Bidirectional Coupler

In the last few years several approaches have been proposed and demonstrated for fabricating fiber optic couplers, the main ones using chemical etching, [4] fusion, [5] and mechanical techniques. [1] The method we preferred to follow is of the last type. The optical fiber is locally stripped of its jacket and bonded in a grooved quartz block or fiber substrate (Fig. 1a). The bottom of the substrate groove is given a certain radius of curvature R which controls the length of the coupling region in the final device. The substrate is first ground, then polished until sufficient cladding material has been removed from the embedded fiber. Two such substrates are then mated together (Fig.1b), and an appropriate index liquid is introduced at the interface to optically match the two fibers. So as to be able to align the fibers with respect to each other the coupler assembly is placed in a fixture (Fig.2) where positioning of the fibers is achieved with two micrometer positioners.

Such couplers were successfully fabricated with different brands of single mode fibers, and found to perform very well and predictably over a broad

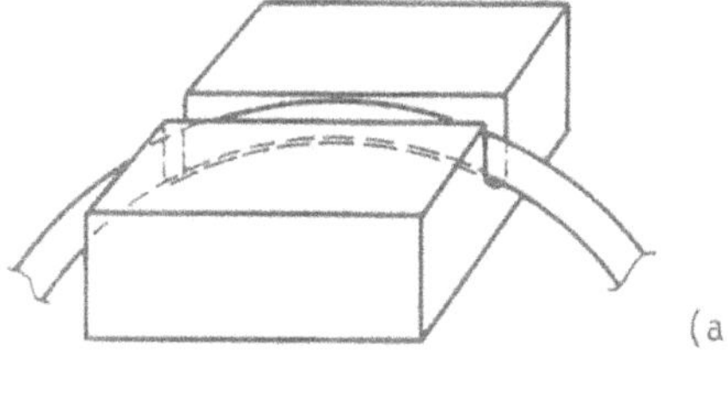

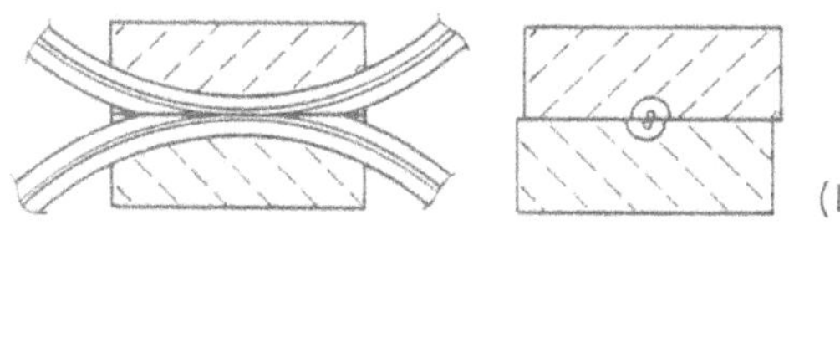

Fig.1 Schematic of single-mode fiber directional coupler; (a) half coupler, (b) assembled coupler

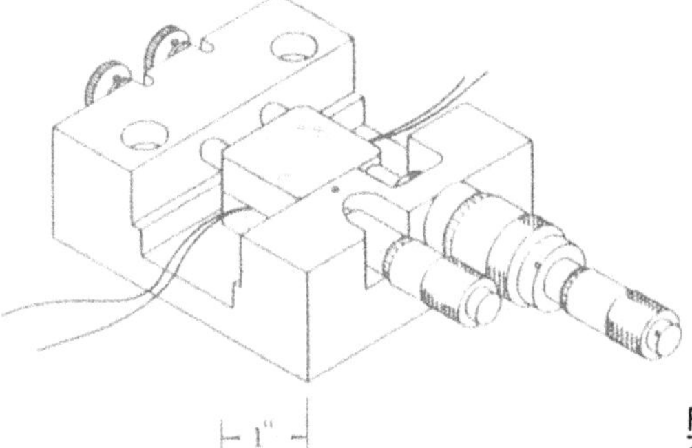

Fig.2 Sketch of directional coupler in adjustable holder

range of wavelengths. They were found to provide coupling ratios anywhere between 0 and 1 depending, among other parameters, on the spacing between the fiber cores. This coupling ratio can be easily tuned by adjusting the fiber spacing, which is conveniently done by offsetting the top fiber with respect to the bottom fiber by means of the larger (differential) micrometer (Fig.2). As was expected on the ground of the near-degeneracy of the two polarization modes in weakly-guiding fibers, these couplers also exhibit a coupling ratio which is very nearly independent of the signal polarization, typically less than a few parts in 10^3. Finally insertion power losses through a typical device was found to be quite low, on the order of 2 to 5% or less.

To better understand and predict the behavior of such optical couplers, we developed a theoretical model [6] on the basis of existing analysis of coupling between two HE_{11} fiber modes [7]. In the simpler case of two identical waveguides parallel over a length L, it is a well-known result that the coupling ratio, defined here as the ratio of the coupled power to the total coupler output, is simply

$$\eta = \sin^2 (cL) \tag{1}$$

where c is a coupling coefficient involving the E-field overlap of the fiber modes, a strong function of the fiber spacing h_0, and L is the coupling length (the length of the device in the case of straight parallel guides). In the present case of a curved geometry, however, the coupling coefficient varies along the coupling region. By working through the new coupled-wave equations it can be shown that the coupling ratio may be written in the same form as in the parallel case, provided the product $c \cdot L$ is replaced by [6]:

$$cL_e = \int_{\substack{\text{coupling}\\ \text{region}}} c(z)\, dz \;. \tag{2}$$

For convenience, we defined the coupling coefficient of a curved-fiber coupler as the coupling coefficient of a parallel fiber coupler with a uniform spacing equal to the minimum fiber spacing h_0 of the curved fiber geometry. The effective interaction length L_e is then given by Eq.(2). It can be shown that with this definition, the interaction length is nearly independent of the minimum fiber spacing as in parallel-fiber couplers, and that it depends mostly on the radius of curvature R of the fibers, increasing proportionally to $\sqrt{R}$ [6]. Typically, it is on the order of 1 mm in a 25 cm radius coupler.

One of the most useful results of this analysis is what we refer to as coupling curves, which display the evolution of the coupling ratio as a function of the fiber center-to-center minimum spacing. Such curves are very useful during coupler fabrication as it is important to know when to stop polishing in order to achieve a given coupling ratio at the upper end of the adjustable range. We show in Fig.3 a theoretical coupling curve, plotted for one of the fiber brands we use in our laboratories (NA ≃ 0.09) at a signal wavelength of 1.15 μm, showing the coupling ratio versus fiber spacing. The coupling ratio increases as the fiber cores are brought closer together, reaches a maximum value of unity, then decreases again in what we call an overcoupling region as some of the coupled power is coupled back into the throughput fiber. The several experimental data points shown were taken at different stages during a coupler fabrication. Since the absolute spacing between the fiber cores, measured by mechanical tools, is known to only ± 1 μm, this error can be cancelled out by translating the experimental points horizontally by a common displacement which gives the best fit to the theoretical curve, and the absolute spacing is then known accurately.

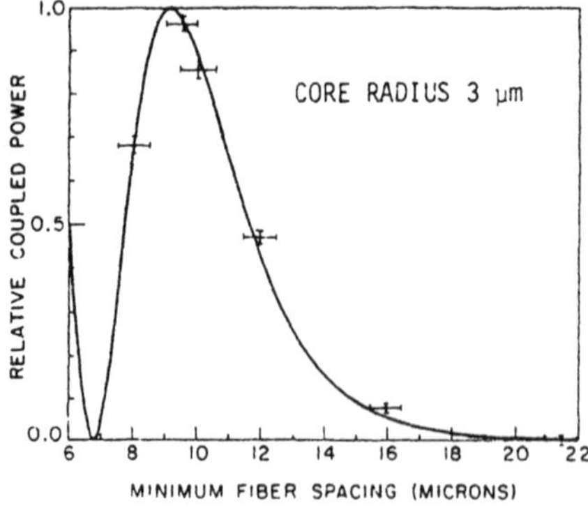

Fig.3 Coupling curves for directional coupler. Solid curve: theoretical; plotted points: experimental

An important feature of these couplers is the tunability of their coupling ratio. The behavior and sensitivity of a typical coupler is shown in Fig.4, which was obtained by motorizing the differential micrometer of a coupler holder (Fig.2) to automatically scan the offset between the fibers while recording both coupler outputs on a dual-trace oscilloscope. The smoothness and symmetry of the resulting curve reflects the good optical quality of the couplers and the ease of adjustment of the coupling ratio.

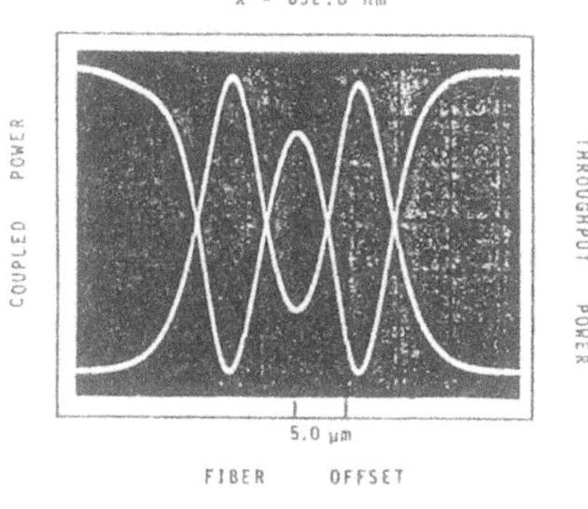

Fig.4 Experimental directional coupler motor driven tuning curves. At the left-hand side of the oscillogram, the lower curve corresponds to coupled power and the upper curve corresponds to throughput power

To compare experimental and theoretical tuning curves a number of couplers were tested by manually scanning the fiber offset and sampling the coupling ratio every one micron or so. For a given signal wavelength and radius of curvature the parameter to be used to fit the theory to the experimental results is the minimum fiber spacing h_0. As one can see from the two examples given in Fig.5 of fiber couplers with respective spacings of 5.4 and 4.8 μm, this single parameter fitting gave a very good agreement between our data and the theoretical prediction. Another interesting property of single-mode fiber couplers is the dependence of the coupling coefficient on the signal wavelength, in total analogy to the phenomena previously observed, for instance, in integrated optic waveguide couplers. Using this effect, one can realize efficient wavelength multiplexing in single-mode fiber de-

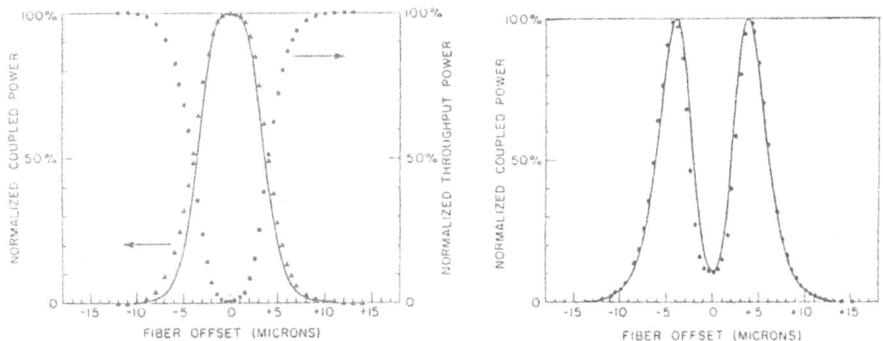

Fig.5 Experimental directional coupler point-by-point tuning curves (plotted points) and theoretical tuning curves (solid curves). Fiber radius of curvature: 25 cm; wavelength: 633 nm

vices [8]. For example, for a coupler which is excited at its two input ports by two separate wavelengths λ_1 and λ_2, a proper choice of parameters can cause λ_1 to be totally coupled while λ_2 is essentially uncoupled, so that both signals are multiplexed into the same fiber at the coupler output.

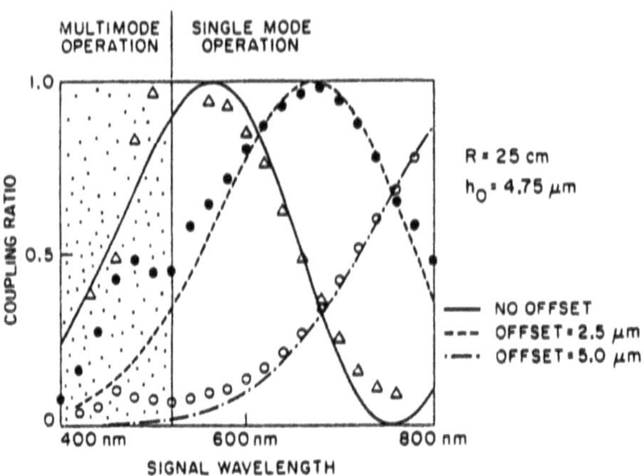

Fig.6 Experimental measurements of wavelength dependence of coupling ratio of a directional coupler

This effect was verified experimentally by measuring the coupling ratio of a 25 cm radius coupler for discrete values of the wavelength over the visible range (Fig.6). The solid line was measured for superposed fibers, while the curves in broken lines show the behavior of the same coupler when fiber offsets of 2.5 μm and 5 μm were applied.

As predicted by the theory an offset reduces the coupling coefficient while the coupling length remains essentially unchanged, which results in a shift of the multiplexing curves towards longer wavelengths.

With this device a resolution of 200 nm was achieved, and it was possible to efficiently multiplex any pair of signals separated by this quantity. Higher resolutions can be achieved by increasing the radius of curvature of the fibers. For example, 20 nm resolution is predicted for a 25 m radius of curvature.

2. Single Mode Fiber Optic Polarizer

If, somewhere along an optical fiber, a section of the cladding is replaced by an appropriate birefringent material, one signal polarization can be caused to couple to radiation modes and be removed from the fiber core while the other polarization remains guided. [2]

In our polarizer we use again a fiber embedded in a substrate (Fig.1a) polished to within one micron or so of the core and cover it with a birefringent crystal, namely potassium pentaborate, having refractive indices which bracket those of the fiber [2] (Fig.7). For the vertical polarization the crystal refractive index is slightly lower than that of the original fiber cladding, and this polarization remains in the fiber while experiencing little loss. The other polarization sees an index slightly larger than that of the fiber and is caused to efficiently couple out of the fiber core. Fig.8 is a schematic view of the fiber polarizer showing the fiber output (vertical polarization) and the radiation mode excited in the crystal that comes out of the device in a semi-conical beam (horizontal polarization).

The fiber polarizer can be easily adjusted by rotating the crystal about the vertical axis until the desired extinction ratio is achieved. As the crystal is rotated the refractive index experienced by the horizontal polarization is changed between two extreme values which are the proper indices of the crystal (n_a = 1.42 and n_b = 1.49 for this particular crystal). As shown in Fig.9, the extinction ratio clearly peaks when the index of the crystal becomes near that of the fiber material ($n \simeq 1.45$). Note that the value of the extinction ratio, as measured by classical means, was in this case as high as 60 dB.

Such a device has been used in a fiber optic gyroscope, in which the sensitivity to rotation can be related to the non-reciprocal phase shift induced by a residual birefringence. The sensitivity that was achievable

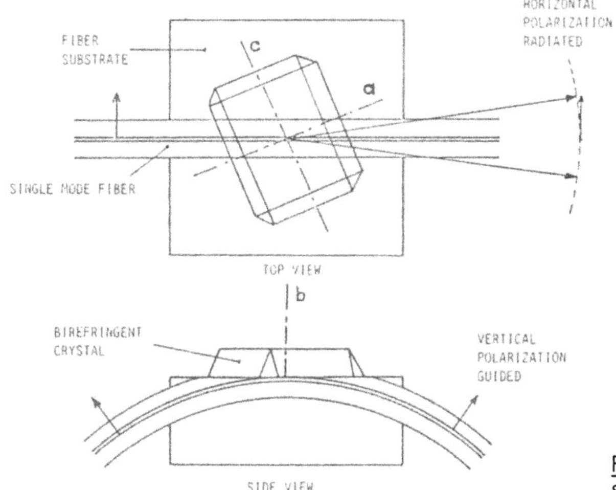

Fig.7 Schematic sketch of single-mode fiber polarizer

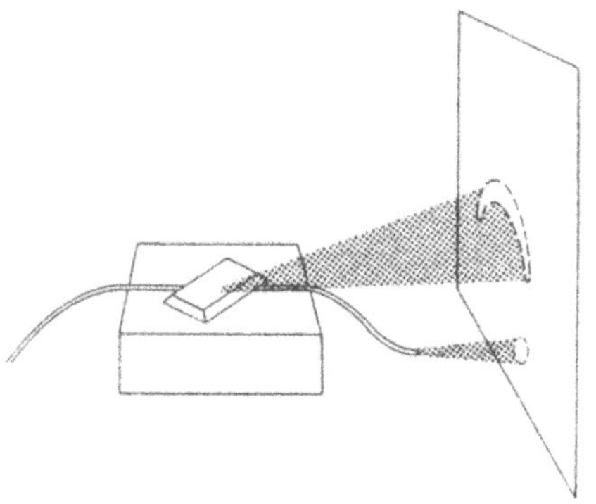

Fig.8 Schematic view of polarizer, showing radiation of guided mode from output fiber (vertical polarization) and radiation from birefringent crystal (horizontal polarization)

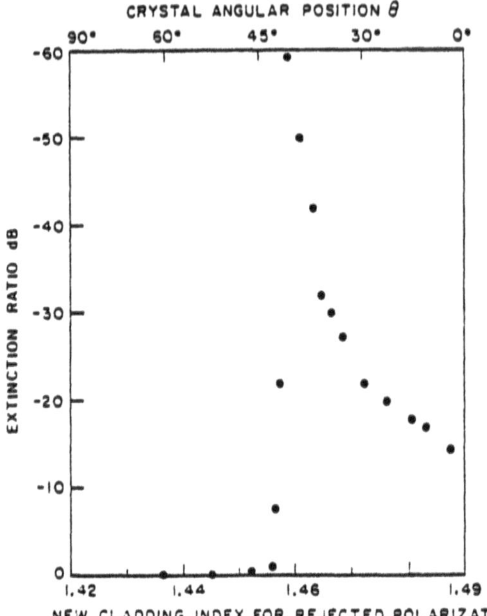

Fig.9 Measured extinction ratio of polarizer as a function of crystal index as seen by the rejected polarization

with this gyro showed that the extinction ratio was on the order of 100 dB which is orders of magnitude better than that specified for bulk optic polarizers. Such high performance was complimented by very little insertion losses, on the order of 1% or less.

3. Fiber Optic Polarization Controllers

In this device, the stress induced by bending the optical fiber is used to generate a predetermined birefringence [3] (Fig.10). Bending the fiber induces a second order stress in the radial direction, which defines a fast extraordinary axis in the plane of the bend, and a slow ordinary axis perpendicular to it. A simple model based on the theory of elasticity is able to predict that in a typical single mode fiber, and for a given wavelength, the photoelastic coefficients of silica are such that devices with $\lambda/4$ and $\lambda/2$ values of differential delay can be realized with 1 or 2 turns of fiber with radius on the order of 1 to 2 cm [3]. By rotating the plane of the loop of such a device one rotates the orientation of the principal axes of

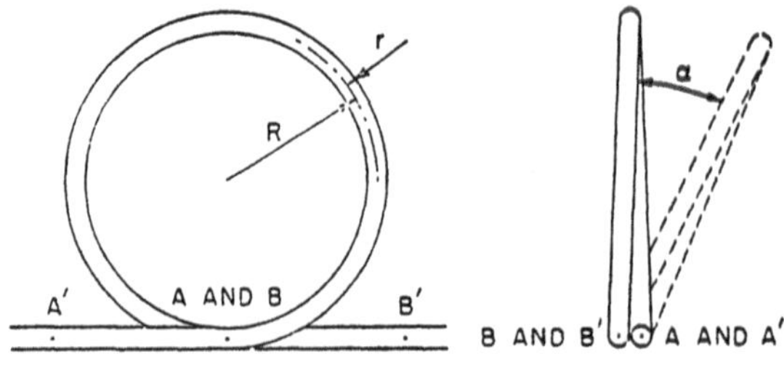

Fig.10 Schematic view of fiber loop used to produce controlled bending birefringence in the construction of single mode fiber polarization controller

the loop with respect to the input light polarization in a manner totally analogous to bulk optic fractional wave plates. The state of polarization (SOP) of the output signal is thus easily adjusted. Fig.11 is a general view of the final device, which consists of two successive $\lambda/4$ plates. Such a device, about 10 cm in length, is able to transform any SOP into any other SOP. Because the radius of the fiber can be arbitrarily chosen, provided an adequate number of turns are wrapped in the coil, bending losses can be kept to a minimum and the overall loss of this fiber polarization controller is practically negligible.

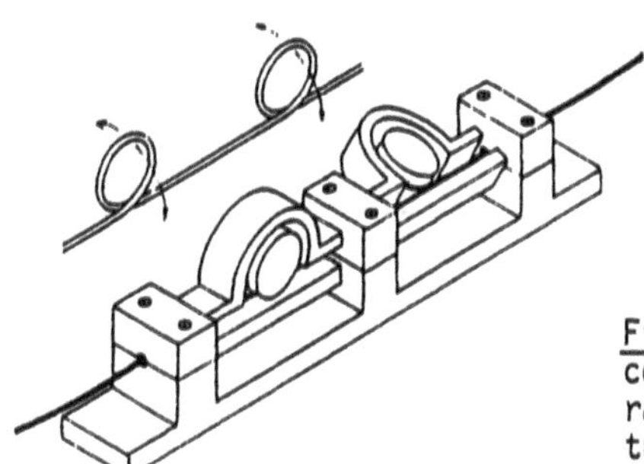

Fig.11 Sketch of complete polarization controller. Fiber diameter = 80 um; radius radius of loops = 0.85 cm; number of turns per loop: 1

4. Conclusion

As a conclusion we should stress again the importance of these components. They exhibit very good performance, introduce negligible back scattering, and all present very little loss. They have been successfully implemented in our laboratory in a cw fiber optic gyroscope with which we have been able to measure state-of-art rotation rates,[9] and in a recirculating pulsed gyroscope, and are currently being used in a number of other applications where they are proving to be extremely useful tools in laboratory prototype devices.

References

[1] R. A. Bergh, G. Kotler, and H. J. Shaw, Electronics Letters 16, (1980), p. 260.

[2] R. A. Bergh, H. C. Lefevre, and H. J. Shaw, Optics Letters 5, (1980), p. 479.

[3] H. C. Lefevre, Electronics Letters 16, (1980), 778.

[4] S. K. Sheem and T. G. Giallorenzi, Optics Letters 4, (1979), p. 29.

[5] T. Oseki and B. S. Kawasaki, Electronics Letters 12, (1976), p. 151.

[6] M. J. F. Digonnet and H. J. Shaw, Journal of Quantum Electronics QE-18, (April 1982).

[7] D. Marcuse, BSTJ 50, (July/Aug. 1971), p. 1791.

[8] O. Parriaux, et al., Applied Optics 20,(1981).

[9] R. A. Bergh, H. C. Lefevre, and H. J. Shaw, "All Single Mode Fiber Gyroscope," these Proceedings

Coupling and Multiplexing Between Single Mode Optical Fibers

O. Parriaux, G. Chartier, and F. Bernoux

Institut National Polytechnique de Grenoble, ENSIEG, BP 46
F-38402 St. Martin d'Hères, France

1. Integrated optics or "all-fiber" approach ?

Perhaps more than in the field of single mode optical communications, single mode fiber applications in sensors will require a whole set of functions that should be performed directly on the optical wave sensing a physical quantity. These are usually expected to be achieved by having recourse to the possibilities offered by integrated optics. However there still are a number of problems associated with the insertion of, say, a lithium niobate chip in a fiber sensor system, such as the difficulty in achieving an easy and straightforward fiber-film connection, the losses of such a junction due to the geometrical mismatch between the two guiding structures and also the additional problem caused by the reflection of coherent light due to the electromagnetic mismatch.

An alternative approach, in performing passive functions at least, is to act on the fiber mode, kept in the core, by playing with the possibilities offered by the mechanism of distributed mode coupling [1], [2], [3]. This, of course, implies a technique giving access to the neighbourhood of the fiber core.

The comparative advantages of the two approaches, as we see them for passive functions, with that of power division as a particular case, are summarized in Table 1.

2. Wavelength selectivity of a fiber coupler

We want to discuss here the possibility of designing "all-fiber" wavelength selective components using the basic coupling configuration as shown in Fig. 1 for the fiber directional coupler discussed in Table 1.

For reaching the fiber core, we developed a polishing procedure allowing a rapid and controllable etching of part of the cladding [4]. Polishing is operated along a plane on a fiber laid on a high curvature radius R convex surface, while the fiber is excited, and it is stopped when the bright core

Table 1 Comparative advantages of integrated optics and the "all-fiber" approach in the case of the power division function

	INTEGRATED OPTICS	ALL-FIBER APPROACH
Insertion loss	3 dB	5 % [2]
Adjustment	3 butt-joints 3 x 3 space variables	1 space variable (relative lateral shift)
Reflections	High Fresnel reflections	negligible
Size	< 1 cm.	~ 1 cm.
Vibration	Immune	Can be made immune
Polarization	Very sensitive	Independent
Multicomponent integration	Possible	Impossible
Cost	Could tend towards zero	Cannot tend towards zero

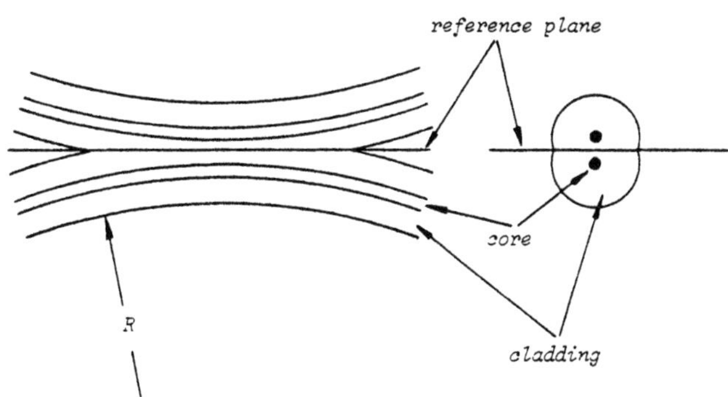

Fig.1 Fiber coupler made of two bent single mode fibers. a_1 , a_2 : core radii and Δn_1, Δn_2: core-cladding index difference

trace appears. A directional coupler is made by setting the polished planes of two such fibers in front of each other, as shown in Fig. 1.

The wavelength selectivity of a coupler made of two identical fibers only originates from the wavelength dependence of the coupling coefficient C, as phase matching between the coupled modes is always achieved. Using the expression for C derived in [5], and taking the two bent fibers as a series of short sections of parallel fibers, the coupled power P_2 into fiber 2 from a normalized unit power launched into fiber 1, is calculated and sketched in Fig. 2 versus the wavelength λ.

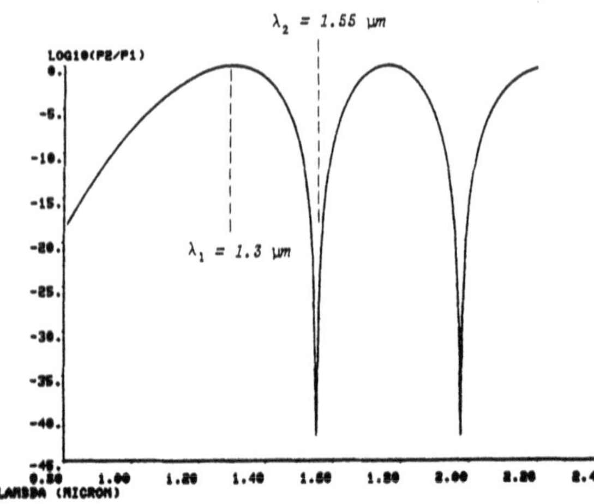

Fig.2 Wavelength dependence of the power P_2 in dB coupled to fiber 2 with $a_1 = a_2 = 2.75\ \mu m$, $\Delta n_1 = \Delta n_2 = 0.01$, when fiber 1 is excited with unit power. $R = 1\ m$

In Fig.2 the core radii $a_1 = a_2$, and the core-cladding index difference $\Delta n_1 = \Delta n_2$ of fibers 1 and 2 were chosen so that the fiber normalized frequency V is around 2 in the wavelength range $1.3 < \lambda < 1.6\ \mu m$. It can be seen from Fig.2 that wavelength multiplexing could be operated in two separate channels. Theoretically there is nothing against defining more channels in the same wavelength range by increasing the coupling coefficient, ie, by bringing the fibers closer. In practice however, it is not so: polishing would be carried on closer to the core, which would considerably increase the losses, particularly of the waves of longer wavelength channels, as their power would travel from one core to the other several times.

3. Improved wavelength selective coupler using fibers of different dispersion characteristics

The wavelength selectivity of a fiber directional coupler could be considerably improved if phase matching can be made to occur at a given wavelength λ_c only. This can be achieved by coupling two fibers having different dispersion characteristics as demonstrated in Fig.3, showing the single mode part of the effective index curves $n_e(\lambda)$ versus λ of two fibers 1 and 2, having $a_1 > a_2$ and $\Delta n_1 < \Delta n_2$. Fig. 3a is for fibers with the same cladding, whereas in b, the cladding of fiber 2 is lowered in order to get a larger intersection angle, ie, a larger phase mismatch between the coupled modes as λ differs from λ_c, for the same core-cladding index differences.

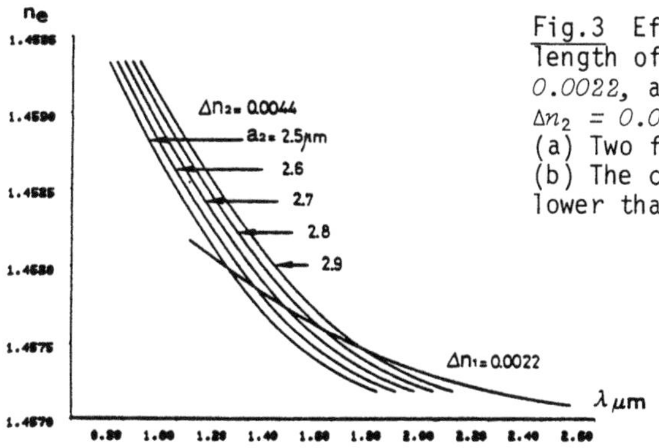

Fig.3 Effective index curves versus wavelength of fiber *1* with $a_1 = 5$ µm, $\Delta n_1 = 0.0022$, and a number of fibers *2* with $\Delta n_2 = 0.01$ and a number of radii a_2
(a) Two fibers have the same cladding
(b) The cladding index of fiber *2* is lower than that of fiber *1*

(a)

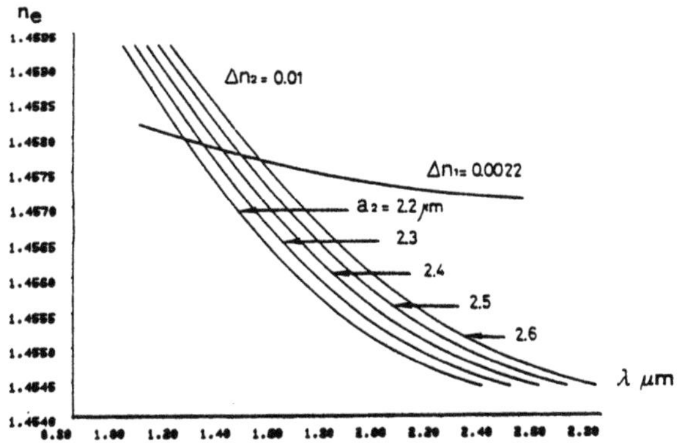

(b)

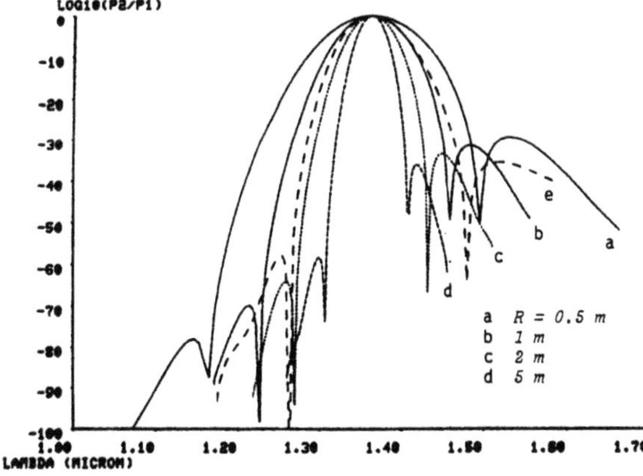

Fig.4 Wavelength dependence of the power P_2 in *dB* coupled into fiber *2* when fiber *1* is excited with unit power. Curves a, b, c and d, with *R* as a parameter, are for fibers with different cladding indices, with $\Delta n_1 = 0.0022$, $\Delta n_2 = 0.01$, $a_1 = 5$ µm, $a_2 = 2.4$ µm. Curve e is for fibers with the same Δn_i as above, but identical cladding index; $a_2 = 1.4$ µm

It naturally proceeds from Fig.3 that, for a given main fiber *1* (with its a_1 and Δn_1), and a given Δn_2, the coupling wavelength λ_c can be chosen by just selecting the proper radius a_2 of fiber *2*. In the coupled wave formalism the amplitude of the coupled wave around $\lambda = \lambda_c$ is governed by the term [6]:

$(1 + (\frac{\Delta\beta}{2C})^2)^{-1/2}$, where $\Delta\beta = \frac{2\pi}{\lambda}(n_{e_1} - n_{e_2})$ is the phase mismatch between the two modes of effective indices n_{ei} . It follows that, for given fibers, ie, for a given $\Delta\beta$, and for a prescribed amount of coupling, the longer the coupling length, the larger the curvature radius R of the coupled fibers, the smaller the coupling coefficient C, the better the wavelength selectivity. Calculations, based on the same decomposition of the bent fibers coupler in a number of short parallel sections [7], lead to Fig.4, showing the effect of R on the coupler bandwidth around $\lambda_c = 1.35\ \mu m$ in the case of two fibers with different claddings. Typically, with $R = 5\ m$, a channel spacing of $35\ nm$ could be expected in this configuration.

As a comparison, curve e shows the bandwidth obtained with two fibers with the same index difference as in curves a, b, c and d, and $R = 5\ m$, but identical cladding.

In Fig.5 are summarized a number of computed results showing the wavelength spacing $\Delta\lambda$ between adjacent channels in a coupler designed with fiber 1 having $a_1 = 5\ \mu m$ and $\Delta n_1 = 0.0022$, versus $(R)^{-1/2}$, with Δn_2 as a parameter, when fiber 2 has a different cladding index. Typically, with $\Delta n_2 = 0.01$ and $R = 5\ m$, one can expect 10 channels between $\lambda = 1.25$ and $1.6\ \mu m$ with $-20\ dB$ of isolation.

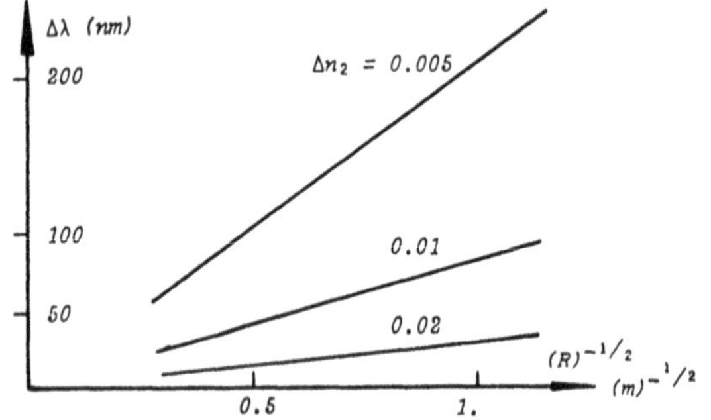

Fig.5 Wavelength spacing between adjacent channels versus $(R)^{-1/2}$, with $\Delta n_1 = 0.0022$, $a_1 = 5\ \mu m$, with Δn_2 as a parameter, in the range $1.3 < \lambda < 1.6\ \mu m$

4. Conclusion

The simplicity of the described coupling design, its very low achievable insertion loss, its very high directivity, make its application very attractive when low wavelength selectivity "all-fiber" components are required.

5. References

1 R.A. Bergh, G. Kottler, H.J. Shaw, Electron. Lett. 16, 260 (1980)

2 R.A. Bergh, H.C. Lefevre, H.J. Shaw, Opt. Lett. 5, 479 (1980)

3 O. Parriaux, S. Gidon, F. Cochet, Proc. 7th ECOC, Copenhagen, P6-1 (1981)

4 O. Parriaux, S. Gidon, A.A. Kuznetsov, Appl. Opt. 20, 2420 (1981)

5 P.D. McIntyre, A.W. Snyder, J. Opt. Soc. Am. 63, 1518 (1973)

6 A. Yariv in *Fiber and Integrated Optics*, Plenum Press, 223 (1979)

7 O. Parriaux, F. Bernoux, G. Chartier, J. Opt. Commun. 2, *to be published* (1981)

Polarization Preserving Single Mode Fiber Optic Coupler

M.D. Nelson and W.C. Goss
Jet Propulsion Laboratory, Pasadena, CA 91109, USA

A technique is described for fabrication of etched single mode fiber optical waveguide couplers which preserve the polarization state to within 10^{-4}. The coupling ratio is tunable over a broad range (0-90%) during fabrication. Back-coupling is less than 10^{-3}, insertion loss is less than 1.5 dB, and coupling ratio thermal coefficient is about 1% per degree C.

INTRODUCTION

Directional couplers promise to be even more useful for optical waveguide systems than for their microwave counterparts. Although principally useful as power dividers in multimode waveguide systems, waveguide couplers may also be used as the splitting and combining elements in single mode interferometers which may have important applications for a variety of sensors [1,2,3,4]. A considerable amount of literature describes the analysis and fabrication of diffused channel waveguide couplers [5,6,7]. Recently attention has been given to processes for fabricating couplers using commercially available single mode glass fiber waveguides [8,9,10]. Discrete non-integrated waveguide couplers (as well as other components) available in a readily connectable form would be very useful for breadboarding single mode optical waveguide systems.

In this paper we describe a technique for fabricating directional couplers. This technique does not require precision lapping or polishing processes and yields a coupler with determinate polarization preserving characteristics. Polarization preservation is important to single mode interferometers. The process involves removal of the majority of the cladding glass by etching, in a manner similar to that described in [8]. However, the fibers are not twisted about each other, but are instead contacted along their length, the fiber axes and the line of contact remaining in a single plane parallel to the mounting base.

This paper presents the results of one phase of research carried out at the Jet Propulsion Laboratory, California Institute of Technology, under contract NAS7-100, sponsored by the National Aeronautics and Space Administration.

THEORY

Directional coupling occurs between single mode waveguides when the waveguides are close enough for the evanescent fields to interact between the waveguides [6,7]. For optical fibers the extent of this coupling depends on how strongly the mode is guided and on the proximity of the cores. A weakly guiding optical fiber with a small numerical aperture has more power distributed in its evanescent fields than a fiber with a larger numerical aperture and thus will couple more for a given core separation. The coupled mode equations for two waveguides in close proximity are [7]

$$\frac{dA(z)}{dz} = j\, \beta_a A(z) + K_{ab} B(z) ,$$

$$\frac{dB(z)}{dz} = j\, \beta_b B(z) + K_{ab} A(z)$$

where A(z) and B(z) are the waveguide mode amplitudes and β_a and β_b are the propagation constants.

The solutions to the coupled mode equations are sinusoids of the form

$$A(z) = \sin\,[\theta(z)] ,$$

$$B(z) = \cos\,[\theta(z)]$$

where $\theta\,(z) = \int_0^L K\,(z)\,dz$.

L is the length over which the waveguides interact. For degenerate waveguides ($\beta_a = \beta_b$ and $K_{ab} = K_{ba}$) K(z) is a constant and there can be total power transfer between the waveguides. The coupling coefficient takes the form

$$K = \frac{k_x{}^2 \zeta}{\beta a \left(1 + k_x{}^2 \zeta\right)}\, e^{-\left(\frac{c}{\zeta}\right)}$$

where k_x is the transverse propagation constant, ζ is the evanescent field penetration, a is the waveguide radius, and c is the core separation. ζ is a polarization dependent quantity.

For single mode fiber interferometers the guided mode must be split into two equal parts (3-dB coupling). For a fixed coupling constant the coupling ratio can be adjusted to 3-dB coupling by choosing the proper interaction length, L, to satisfy the equation

$$L = \frac{n\pi}{4K}$$

where n = 1,2.3...

A plot of the coupling constant versus the core separation for our single mode fiber is shown in Figure 1. Two orthogonal polarizations are shown. "Parallel polarization" (11) indicates the input beam is polarized in the plane defined by the two fiber axes (parallel to the mounting block base).

From Fig. 1 and the equation for interaction length we see that if the cores are separated by a distance of a core diameter, the coupling is strong and many

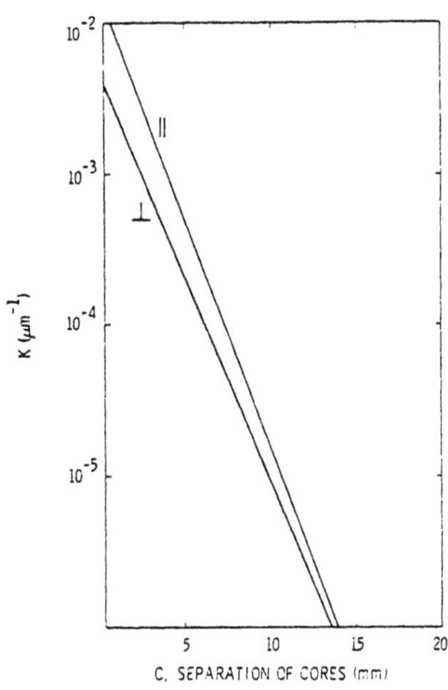

Figure 1. Coupling Constant for 2 Polarizations

cycles of power transfer will occur within a centimeter of interaction length. With these criteria and the need for preserving polarization, we developed the following fabrication technique.

FABRICATION

The optical fiber used is ITT single mode fiber which has a 4-µm diameter core and a numerical aperture of 0.1. A short length of fiber is first folded in the middle. This forms the two legs of the coupler, ensuring that both will have similar propagation constants and facilitating the application of an identical tension in both legs as they are secured to the mounting block. An axial tension of about 0.5 gm is introduced into the fibers through a small pulley placed into the looped end. About 5 cm of Hytrel and RTV jacket is stripped from the centers of these two legs, exposing the glass waveguide. These stripped regions are placed side-by-side on the surface of a block of plexiglass 7 cm long and 2.5 cm wide. A shallow cavity 2.3 cm long by 0.9 cm wide has been milled into the center of the plexiglass coupler block. Once the fibers have been placed on the block they are secured with wax or acrylic cement such that they lie side-by-side, parallel with the plane of the surface and span the cavity. This configuration is shown in Figure 2.

The cavity is filled with 48% HF acid and the fibers are allowed to etch until the diameters are reduced from 80 µm to about 7.5 µm (about 15 min). The etching is stopped by flushing with distilled water.

Once the fibers have dried, a thin coating of an RTV silastic which has an index of refraction slightly below that of the fiber claddings is applied to the etched fibers (Corning Sylgard 184, index = 1.41). The surface tension of the silastic causes the fibers to contact each other and holds them together until polymerization is complete (Fig. 3). The index for the silastic changes very little during the curing process. Figure 4 shows the etched fibers being held together by the RTV.

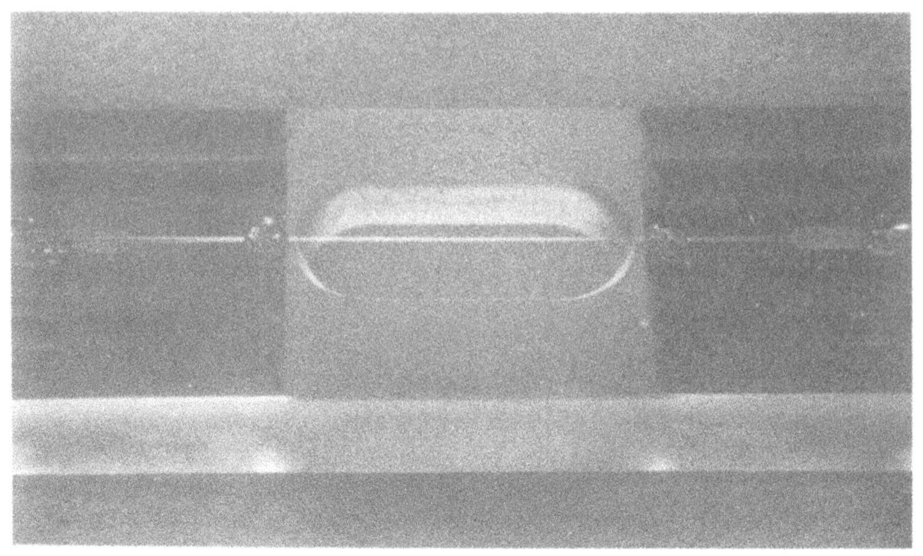

Figure 2. Coupler Block Prior to Etching

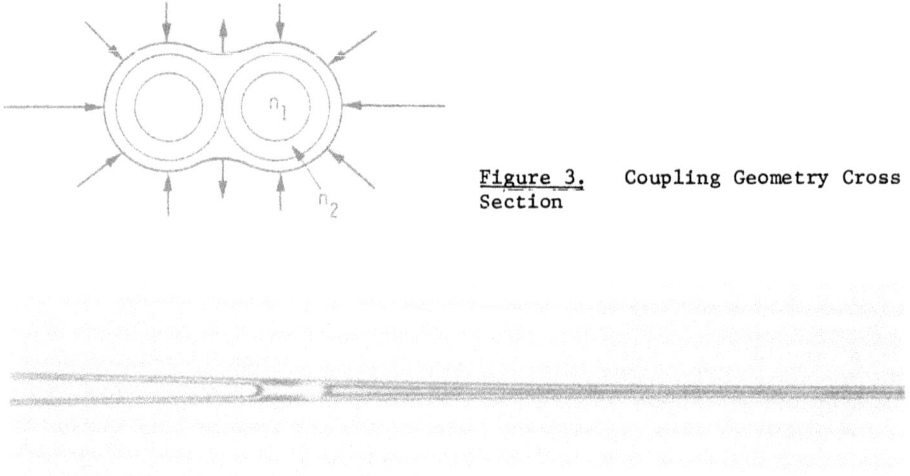

Figure 3. Coupling Geometry Cross Section

Figure 4. Etched Fibers Held Together with RTV

Before the silastic has set up, the power transfer ratio can be adjusted by shortening the interaction length. This is done by inserting a probe between the fibers at one end where the fibers are already separated. Running a probe down the middle of the fibers causes them to start to separate, shortening the inter-

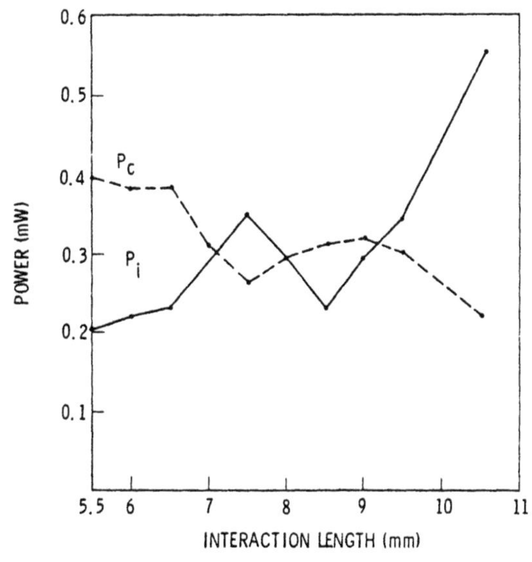

Figure 5. Change in Power Transfer as the Fibers are Separated

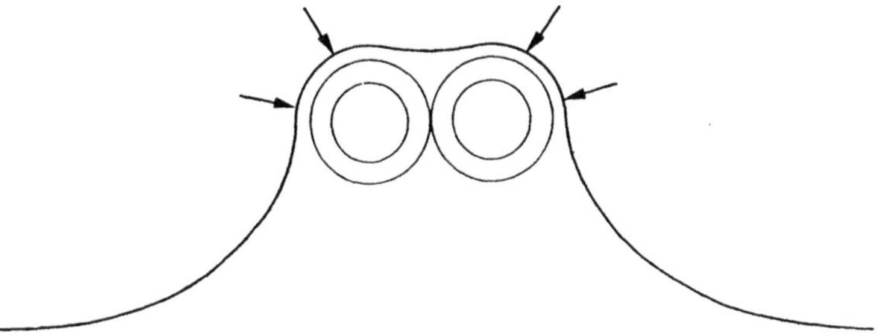

Figure 6. Coupling Geometry Cross Section, Arrows Show Surface Tenson Forces

action length. The probe is stopped when the coupling is as desired. Figure 5 shows the power coupling while separating a test coupler.

An alternative technique involves filling the cavity above the level of the fibers with RTV. By then lowering the level of the RTV, removing the material with an eyedropper, the fibers are again drawn together by surface tension (Fig. 6). As the level of the silastic is lowered, the length of the liquid meniscus which contacts the fibers is shortened, thus reducing the coupling.

Both techniques require total encapsulation after the initial application and curing of the RTV. The thin initial coating does not provide an adequate cladding and so a small proportion of the single mode wave couples into higher order modes. Also when the RTV to air interface is part of the waveguide structure, the guided mode is easily scattered and anti-guided by contaminants. Figure 7 shows a completed coupler that is coupling 60%.

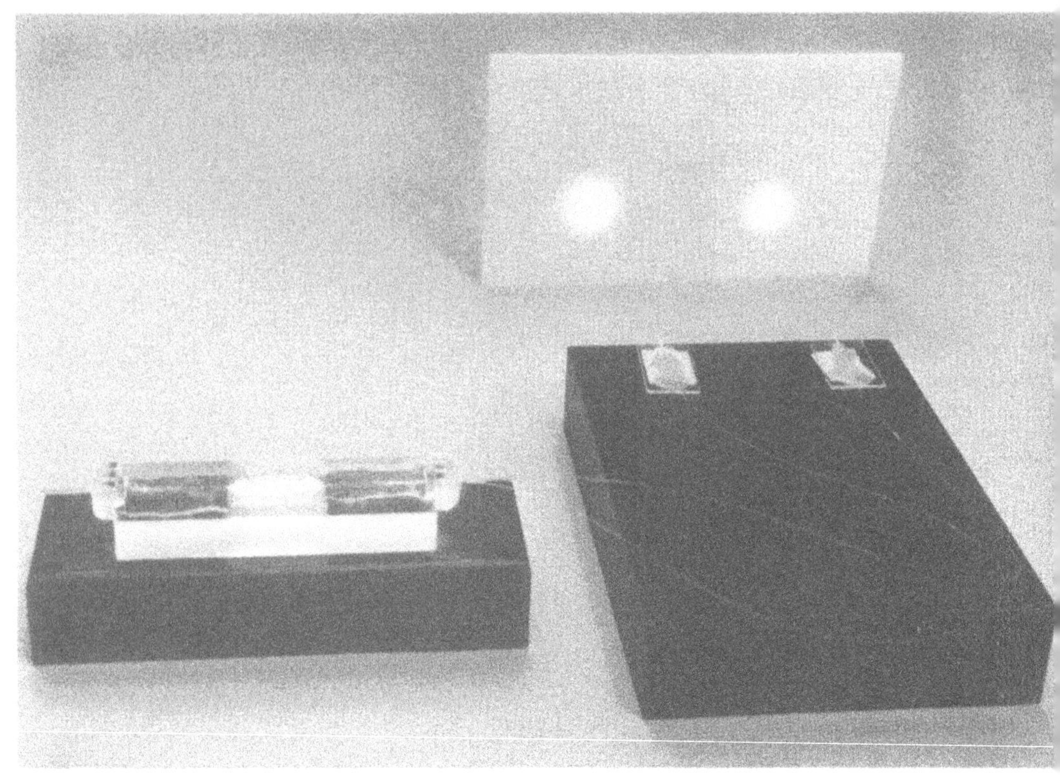

Figure 7. Coupler

RESULTS

Power transfer is easily controlled by the techniques previously described. The insertion loss is typically 50%, although we have seen a loss of only 25%. The etched fibers and RTV coatings form secondary waveguide structures of higher normalized frequency which causes a percentage of the single mode power to be coupled into higher order modes which are subsequently lost by mode stripping. This accounts for only a 5 to 10% loss when the coupler is encapsulated. Contamination on the fibers is the main contributor to loss when present. Backscatter due to the coupler is less than 10^{-3} of the power throughput.

The polarization dependent coupling characteristics of the couplers are shown in Figures 8 and 9. The coupling coefficient is shown to be larger when the input beam is polarized in the plane of the two fibers and less when perpendicular. This is due to the evanescent field penetration into the cladding being greater along the axis of the E vector of the polarized propagating mode. Note in Fig. 9 that for polarization states for polarization approximately parallel and perpendicular to the mounting block base, the couplers preserve linear input to 10^{-4}.

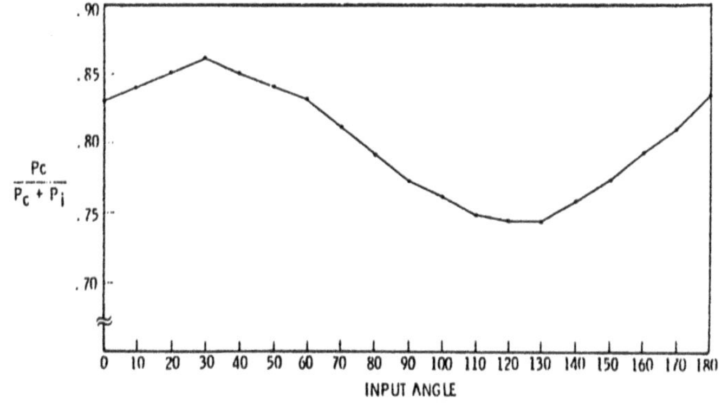

Figure 8. Coupling Dependence on Input Polarization Angle

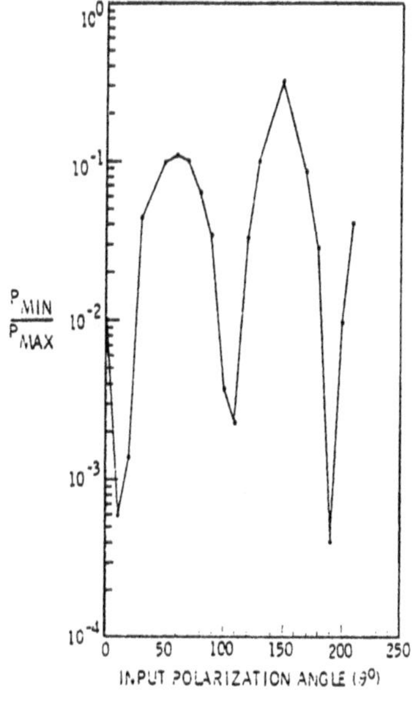

Figure 9. Coupled Fiber Polarization Ratio

Figure 10. Coupling Temperature Dependence

$\sim 0.7\ \%/^{\circ}C$

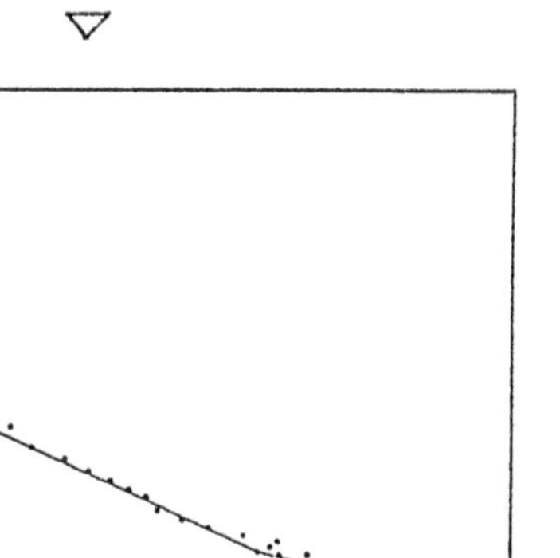

Temperature dependence of the coupling ratio is shown in Figure 10 to be about 0.7% per degree Centigrade. The temperature dependence can be attributed to two factors; thermal expansion of the RTV silastic and mounting substrate, and to a negative thermal coefficient of the index of refraction of the silastic. It should be noted that the temperature dependence provides a method of fine tuning a coupler to a desired power transfer ratio.

CONCLUSION

We have presented a method for fabricating a directional coupler with a single mode fiber. The method is simple and yields an initially tunable directional coupler without high precision lapping and polishing. These devices are expected to be useful in breadboarding single mode fiber systems and interferometric systems where polarization preservation of the split laser beam is of primary concern.

REFERENCES

1. V. Vali and R.W. Shorthill; Appl. Opt 15, 1099 (1976)
2. R.A. Bergh, H.C. LeFerre, H.J. Shaw; Optics Letters 6, 502 (1981)
3. J.A. Bucaro, H.D. Daydy, J. Carome; J. Acoust. Soc. Am 62,1302, (1977)
4. A.B. Tveten, A. Dandridge, C.M. Davis, T.G. Gaillorenzi; Electronics Letters 16, 260 (1980)
5. R.V. Schmidt, H. Kogelnick; Appl. Phys. Lett 28,503 (1976)
6. A.L. Jones; J.O.S.A. 55,261 (1965)
7. E.A.J. Marcatili; B.S.T.J. 48,2071 (1969)
8. S.K. Sheem, T.G. Gaillorenzi; Optics Letters 4 29,(1979)
9. O. Parriaux, S. Gidon, A.A. Kuznetsov; Appl. Opt. 20,2420 (1981)
10. R.A. Bergh, G. Kotler, H.J. Shaw; Electronics Letters, 16,260 (1981)

High Kilohertz Frequency Fiber Optic Phase Modulators

E.F. Carome and G. Adamovsky

Department of Physics, John Carroll University
Cleveland, OH 44118, USA

For several years we have been conducting studies of the behavior of various types of optical fiber sensors, including hydrophones, microphones, accelerometers and pressure transducers [1-5]. Experimentally this has involved mainly the use of Mach-Zehnder fiber interferometer configurations which can be highly sensitive to thermal and other forms of environmental disturbance. It has been convenient, therefore, to employ fiber phase shifters in conjunction with electronic feedback circuitry to maintain the interferometers in quadrature, i.e., with a constant $\pi/2$ phase shift between the optical beams from the two arms, so that they operate at their maximum sensitivity [6,7].

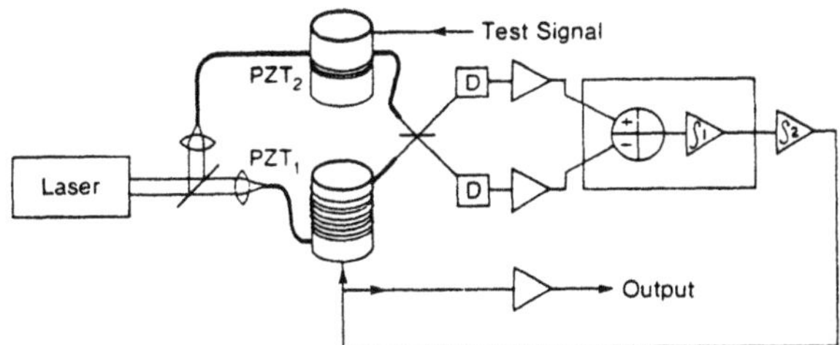

Fig. 1 Schematic diagram of optical fiber Mach-Zehnder test interferometer

A basic optical fiber Mach-Zehnder test interferometer of the type employed for sensor development is shown schematically in Fig. 1. The output from a single mode He-Ne laser is split and injected into the two arms of the interferometer. A short portion of the fiber forming the upper arm is wound around a lead titanate-lead zirconate piezoelectric cylinder PZT_2. By voltage driving this simple test element it is possible to calibrate the interferometer system and simulate fiber index of refraction and length changes that may be introduced by more complex sensor configurations. Two photodetectors are used to monitor the intensity of the centers of the circular interference patterns produced by combining the expanding beams from the output ends of the fibers. The sum and difference of the photodetector outputs are processed to compensate for common mode effects, such as laser intensity fluctuations. These outputs are then integrated to yield a stabilization signal that is applied to the piezoelectric cylinder PZT_1 that is wrapped with a portion of the fiber in the lower arm of the interferometer.

The circuitry in the feedback loop shown in block diagram form in Fig. 1 is such that the voltage applied to PZT_1 is just that which is required to make the optical path of the lower arm track the length changes induced by the test or a sensing element in the upper arm [6]. It should be obvious that, besides holding the interferometer in quadrature, the voltage applied to PZT_1 becomes the sensor output signal in any particular sensor application of this active feedback type stabilization circuit. In most of our sensor development work we have been concerned with detecting signals at frequencies below 5 kHz. Therefore, the frequency response of the stabilization circuit has been made flat from DC to about 6 kHz and then it falls off rapidly as the frequency increases.

The system sketched in Fig. 1 is being used both for sensor development work and also to evaluate the properties of various piezoelectric phase shifters and phase modulator configurations. We are, for example examining the use of thin films of the piezoelectric polymer polyvinylidene fluoride PVF_2 [7]. This material appears to have great promise for application as a thin, piezoelectrically active coating for optical fibers, capable of yielding optical phase shifts of the order of 1 radian per volt per meter of coated fiber. In addition, radially poled PVF_2 coatings should have the capability of generating optical phase shifts without producing stress-induced polarization effects.

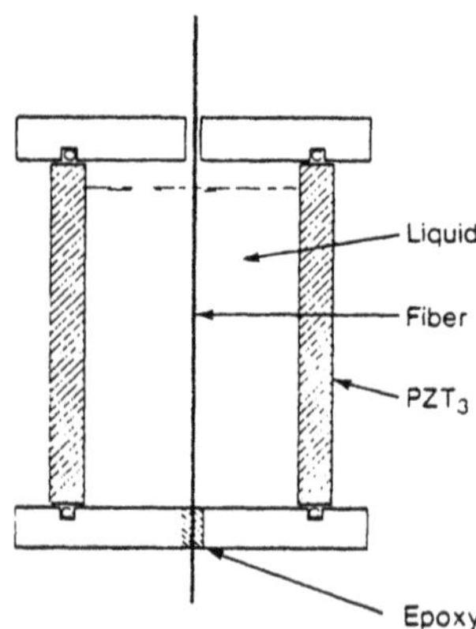

Fig. 2. Diagram indicating configuration of axially symmetric optical phase modulator

While working on coating-type phase shifters and modulators it occurred to us that another relatively simple configuration should have this same capability and thus be of use in various interferometeric sensor applications. One of a number of possible designs is outlined in Fig. 2. A piezoelectric cylindrical shell PZT_3 is capped at bottom and top with plastic plates fitted with O-rings to form liquid tight seals. The optical fiber forming one arm of an interferometer is threaded through small holes in the end caps so that it lies along the axis of the cylinder. The fiber is sealed in the bottom cap with epoxy cement and the shell is filled with liquid.

This is the simplest arrangement for examining the acoustic behavior of such a configuration since the liquid can easily be removed using a hypodermic syringe inserted through a small hole in the top cover plate. In a more practical design the cavity might be filled with a plastic potting material or gel so that the cylinder may be aligned in any orientation and, more importantly, to prevent cavitation effects that can occur in liquids for pressure oscillation in excess of 1 atmosphere.

In considering the acoustic behavior of the device sketched in Fig 2, to a first approximation the liquid column should act as a rigid walled, compressible cylinder with zero shear modulus. It should have a series of axially symmetric radial resonance modes with pressure antinodes on the axis and at the outer walls of the cavity. The pressure antinode at the center should be of high amplitude due to the effective focussing of the acoustic wave on the axis of the cylinder. This focussing effect should enhance the pressure induced optical path length changes in the fiber. Since the modes are axially symmetric the fiber should be subjected to uniform, radial stress changes, thus reducing the possibility of stress-induced birefringence for radially symmetric fibers.

Preliminary data have been obtained on the behavior of a modulator constructed in this fashion using a slightly modified version of the Mach-Zehnder test interferometer sketched in Fig. 1. The optical portion of the test system is shown in more detail in Fig 3; included in the sketch are the usual cladding mode strippers MS. Also shown are the third modulator PZT_3 and the modified photodetection arranagement that is employed to examine this modulator's frequency response.

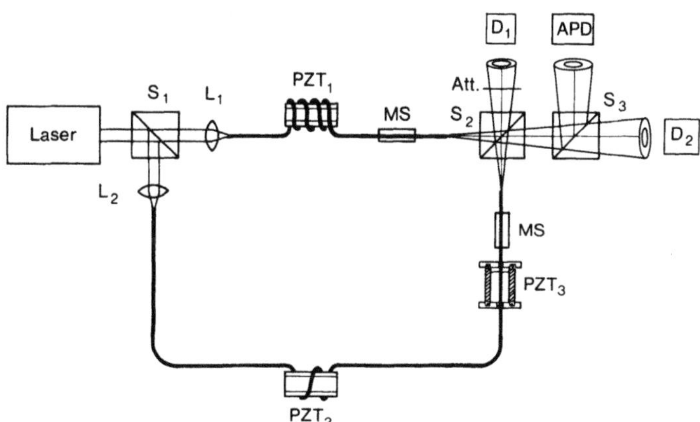

Fig. 3 Schematic diagram of modified optical fiber interferometer

The output signals from the PIN diodes D_1 and D_2 again are fed to the feedback circuit sketched in Fig 1. This locks the interferometer in quadrature, and equalizes all optical path length variations that occur at frequencies below 5kHz. An attenuator, inserted in the path of the beam directed at D_1, balances out the added losses due to the insertion of the beam splitter S_3. This latter element is included so that high frequency phase modulation signals generated by PZT_3 may be detected using the avalanche photodiode APD. Since the interferometer is maintained in quadrature, the fundamental frequency components of these high frequency signals do not randomly fade but remain quite stable at their maxima.

The piezoelectric cylinder PZT_3 used for the measurements reported here is of length 5.08cm, inner diameter 1.90cm, and wall thickness 0.32 cm. The amplified output from a frequency synthesizer was used to excite the cylinder at frequencies in the range 100 to 800 kHz. The output signal from the avalanche photodiode was monitored using a spectrum analyzer. When the frequency was scanned, with the cylinder filled with water, this signal exhibited a number of well defined peaks. Their relative amplitudes, for constant amplitude drive voltage, are shown graphically in Fig. 4. It may be seen that the peaks are spaced, on the average, by 78 kHz, which is exactly the value to be expected for the radial mode spacing of an ideal rigid walled, water filled cylinder; these are the frequencies for which the inside radius of the cylinder is equal to $\lambda/2$, $2\lambda/2$;.., $n\lambda/2$, where λ is the sound wavelength in water and n is an integer. That these peaks do correspond to resonances of the water column was confirmed by removing the water from the cylinder, in which case all of the peaks disappeared. The fact that there are substantial differences in their relative amplitudes is at least partially due to the complexity of the modulator structure, e.g., both the finite length and thickness of the cylinderical shell and acoustic coupling to the end gaps should affect the behavior of the resonator.

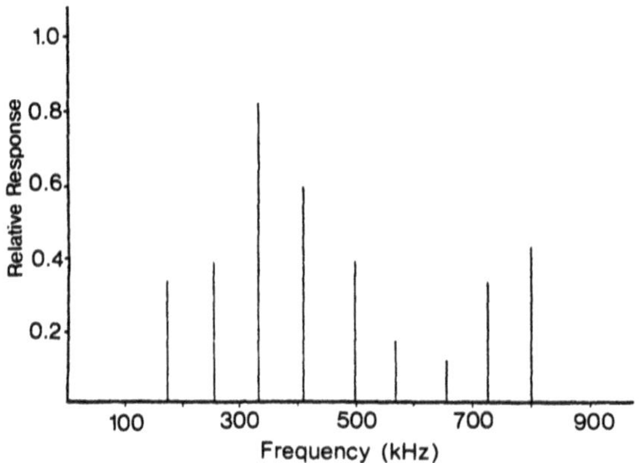

Fig. 4 Graph indicating the relative strength of the various axial modes of the cylindrical phase modulator

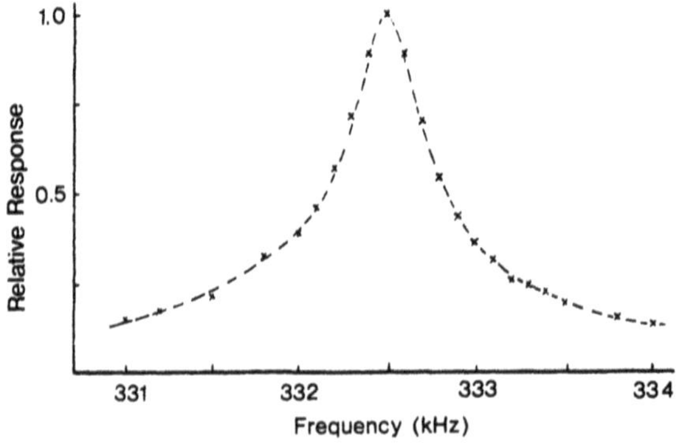

Fig. 5 Graph indicating typical resonance response curve of an axial resonance mode of the cylindrical phase modulator

The individual resonance peaks were examined in detail by taking measurements of the APD relative response as the frequency was varied incrementally. As an example of the results of such measurements, a detailed plot of the peak in the vicinity of 332 kHz is shown in Fig. 5. Note the narrowness of the peaks which corresponds to that of a resonator with a quality factor $Q = \Delta f_{fw}/f_o$ of the order of 600.

Ordinarily to determine the calibration constant of an optical fiber phase modulator we drive it at relatively high amplitudes so that phase shifts of at least 10 radians are produced. This can be done at low frequencies, for example, with the modulators PZT_1, and PZT_2 in Fig. 3, with the stabilization circuit inoperative. In this case one can easily obtain a precise determination of the phase shift per volt by observing the drive voltages that produce nulls in the photodetector output signals at the fundamental and various harmonics of the driving frequency [7]. Since we are able to generate only relatively small phase shifts with PZT_3 this procedure may not be used. Instead we extrapolate from our low frequency calibation results, assuming that the APD frequency response remains flat and that it is linear at low amplitudes.

Using this approach yields a value of 0.01 radian/volt for the phase shift constant for the PZT_3 configuration used in this preliminary study. Assuming that the entire 5.08 cm is equally affected by the acoustic pressure yields a value of 0.002 radians per volt per cm of fiber for this modulator.

In our initial attempt to generate relatively large phase shifts with PZT_3 we have encountered instability problems and overheating of the piezoelectric cylinder. Initially we attributed the instability to the occurrence of cavitation effects along the axis of the cylinder in the vicinity of the fiber. These are to be expected for pressure oscillations in excess of 1 atmosphere. There is some question as to whether we have reached such high pressure levels. Kingsley, for example, has reported detecting phase shifts of the order of 0.05 rad/cm atm for bare silica fibers acoustically irradiated in water at 7.8 MHZ [8]. In addition the fiber employed in our measurements is plastic-coated silica (ITT 1601) for which enhancement of acoustic sensitivity has been observed at low frequencies [9,10]. Thus, at present, there are a number of undetermined factors regarding the results reported here.

We are continuing this studies and hope to resolve these matters and improve the accuracy of our measurements. To reach higher acoustic pressures and thus induce larger phase shifts we plan to go to a pulsed mode of operation so that overheating can be eliminated. We also plan to use various potting techniques and PZT elements of other lengths and diameters.

Acknowledgements

This work has been supported in part by the Office of Naval Research

References

1. Bucaro JA, Dardy HD and Carome EF 1977 Appl. Opt. 16 1761
2. Bucaro JA, Dardy HD and Carome EF 1977 J. Acoust. Soc. Am. 62 1302
3. Bucaro JA and Carome EF, 1978 Appl. Opt. 17 330

4. Carome EF and Satyshur MP 1979 Fiber Optics: Advances in Research and Development (Plenum Publishing) p 657
5. Carome EF and Koo KP 1980 Opt. Lett. 5 359
6. Fritsch K and Adamovsky G 1981 Rev.Sci Instrum. 52 996
7. Carome EF and Koo KP 1980 Ultrasonics Synposium Proceedings (New York; Institute of Electrical and Electronics Engineers) p 710
8. Kingsley SA 1978 (Unpublished; personal communication)
9. Hughes R and Jarzynski J 1980 Appl. Opt. 19 98
10. Hocker GB 1979 Opt. Lett 4 320

Polarization Control for an Optical Fiber Gyroscope

F.A. Mohr

Standard Elektrik Lorenz AG, Hellmuth-Hirth-Str. 42
D-7000 Stuttgart 40, Fed. Rep. of Germany

U. Scholz

Fachhochschule für Technik, D-7300 Esslingen, Fed. Rep. of Germany
(presently at Standard Elektrik Lorenz AG)

It is well known that if birefringent fibers are used in the sensing loop of a fiber gyro, a kind of nonreciprocity arises. To overcome this, the insertion of polarizers in the loop has been proposed (Fig.1) [1] and has become common practice [2] . However, this solution of the problem causes further difficulties: time-varying birefringence of the fiber modulates the transmittance of the sensor-loop. To give an example, assuming a fiber length L = 1000 m, a beat length of the orthogonally polarized modes Λ = 10 m and a temperature coefficient of $\partial\Lambda / \partial\vartheta = 2 \cdot 10^{-3}$/K (data which are common for standard fibers with intrinsic birefringence) one can calculate a temperature change of 10 K to be sufficient to completely block the transmittance. This might be a worst-case example but it illustrates the situation.

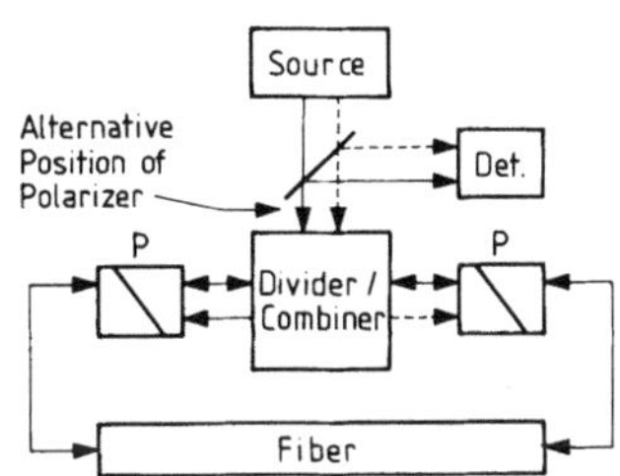

Fig.1. Schematic representation of Sagnac-Interferometer

Therefore, in order to maintain the interferometer signal to noise ratio even under changing environmental conditions the polarization transfer characteristics of the fiber must be stabilized [3] . This can be done by using polarization preserving fibers but these are still rather expensive and have high losses. Alternatively, standard or spun fibers (= fibers with extremely low birefringence) in connection with an active polarization stabilization can be used. We propose a novel method of in-fiber elastooptic polarization stabilization which in comparison to other versions reported so far [4] , [5] has the advantage of vanishing insertion loss.

In order to determine the dynamical requirements which a polarization control has to satisfy we measured the polarization noise spectrum of the output from a 1000 m long fiber excited by a HeNe-laser beam. Fig.2 shows the results for the circularity of the output ellipse (Stokes-parameter $\bar{S}_3$) when the input state of polarization (SOP) was adjusted to give a linearly polarized output beam (mean value of $\bar{S}_3 = 0$). The measurement bandwidth was 10 Hz, the 0 dB-level corresponds to the DC light level. We found that only low frequency noise components well below 500 Hz are present.

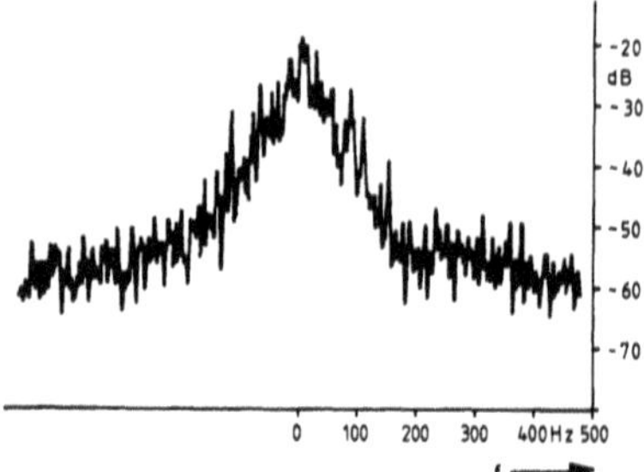

Fig. 2. Polarization noise spectrum at fiber end (S_3). Fiber length L = 1000 m

Our polarization controller is based upon elastooptic polarization modulation in a monomode fiber. A fiber subjected to uniaxial transverse pressure becomes linearly birefringent with the birefringence main axes parallel and orthogonal to the pressure direction and the retardation adjustable via the pressure. In a simple experiment (Fig.3a) we tested the suitability of piezo drives for exerting the required pressure on the fiber. The fiber was clamped between two slightly beveled glass shoes, one mounted on a force transducer, the other on a commercially available piezo drive as used for mechanical translators with 20 μm travel per 1000 V. We measured the circularity of the output polarization ellipse ($\bar{S}_3$) and the force F acting on the fiber when the driving voltage was varied.

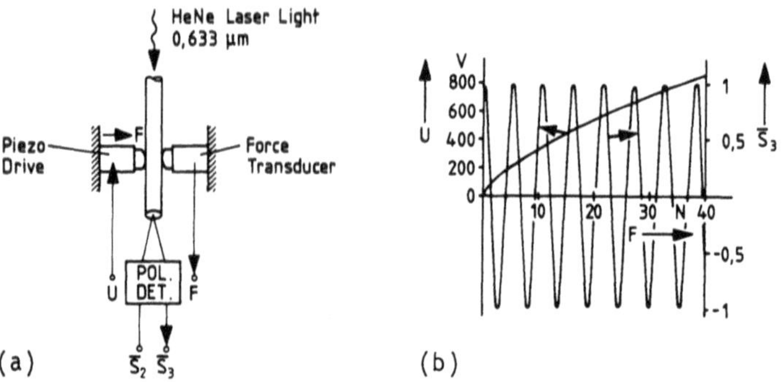

Fig. 3. Elastooptic polarization modulation in monomode fibers. (a) Test-set-up, (b) measurement results

Figure 3b shows the results. With the type of fiber used (ITT T-110, outer diameter 125 µm) approximately 4 N change of the applied force was sufficientt to change the linear birefringence by an amount of 2π rad or, equivalently, the Stokes-parameter $\bar{S}_3$ by one period. The required voltage change depended somewhat on the operating point but was always in the order of 180 V.

Figure 4 once more illustrates the polarization problems with the Sagnac interferometer. Each beam, clockwise as well as counterclockwise, passes through a polarizer, the time-varying birefringent fiber, and a second polarizer before being recombined with the complementary beam.

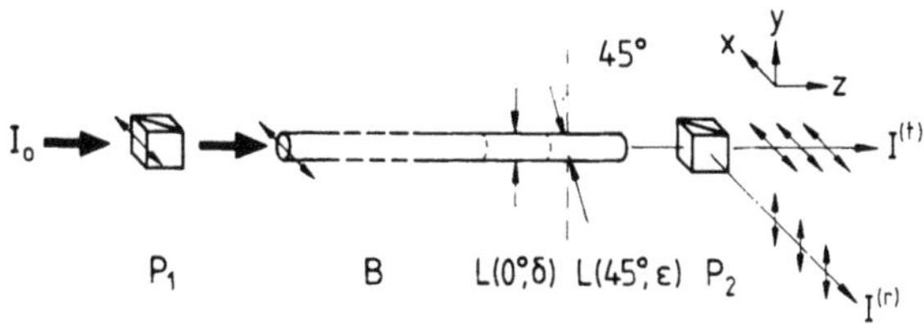

Fig. 4. In-fiber elastooptic polarization control with low insertion loss

Using Jones matrix algebra we can describe the set-up by its matrices P_1 (first polarizer), B (birefringent fiber with time-varying eigenstates and retardation), and P_2 (second polarizer). P_2 in our case is a polarizing beam splitter the transmitted and rejected polarization directions which define the x- and y-axes as shown in Fig.4. One can easily show that two linear retarders preceding P_2 with orientations 0° and 45° and retardations δ and ε, respectively, constitute a compensator suitable to transform any given input SOP to the desired output SOP (namely, linear polarization with orientation parallel to the transmittance axis of the polarizer).

In terms of Jones matrices, the reflected field $\vec{E}^{(r)}$ from P_2 reads

$$\vec{E}^{(r)} = P_2^{(r)} \cdot L(45^{\circ},\varepsilon) \cdot L(0^{\circ},\delta) \cdot B \cdot P_1 \cdot \vec{E}_0 \qquad (1)$$

where $\vec{E}_0$ is the input electrical field vector. Polarization is assumed to be parallel to the transmitted polarization direction of P_2.

Irrespective of the kinds of eigenmodes and coupling mechanisms in the fiber B can as well be imagined as a fiber-equivalent compensator composed of two linearly birefringent sections with noisy retardations δ_N and ε_N but fixed orientations 0° and 45°, respectively. Substitution of the well-known Jones-matrices for these retarders into (1) yields $\vec{E}^{(r)}$ (2) and the reflected intensity $I^{(r)}$ (3):

$$\vec{E}^{(r)} = P_2^{(r)} \cdot L(45^{\circ},\varepsilon) \cdot L(0^{\circ},\delta) \cdot L(0^{\circ},\delta_N) \cdot L(45^{\circ},\varepsilon_N) \cdot P_1 \cdot \vec{E}_0 \; (2),$$

$$I^{(r)} = \left[\cos^2\frac{\delta+\delta_N}{2} \cdot \cos^2\frac{\varepsilon+\varepsilon_N}{2} + \sin^2\frac{\delta+\delta_N}{2} \cdot \cos^2\frac{\varepsilon-\varepsilon_N}{2}\right] \cdot \frac{I_0}{2} \; (3).$$

Maximum transmittance demands that $I^{(r)}$ vanishes which further implies

$$\delta + \delta_N = \varepsilon + \varepsilon_N = 0 \, . \qquad (4)$$

So far, we have completed only one loop, namely the ε-loop, but we think completion of the δ-loop should be a mere duplication

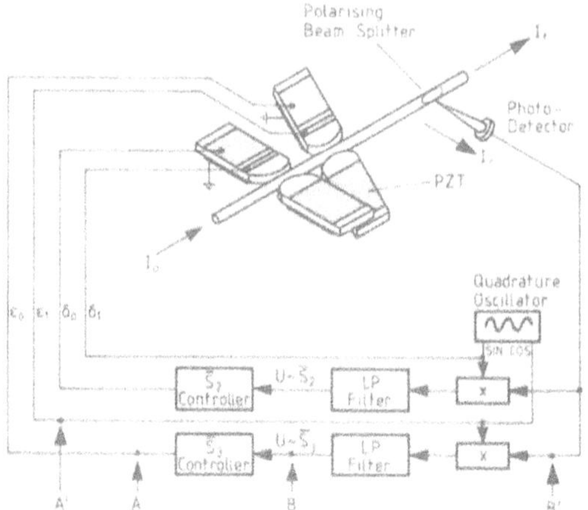

Fig. 5. Schematic representation of polarization control with modulated polarization detector

Figure 6 shows the frequency response of the ε modulator section as measured between A' and B' (see Fig.5). The ε_1 electrode is driven by 14 V_{PP} signals as it is in the closed-loop operation. Excluding the resonance peaks mainly in the 100 kHz region the response is flat up to about 1 MHz, so we can adapt the modulation frequency nearly without constraints to the requirements dictated by the interferometer (e.g. circulation time of the light).

Figure 7 shows the overall frequency response of the retarder section, i.e. including controller electrode and modulated polarization detection (as measured between A and B, Fig.5). The 3 dB corner frequency at 10 kHz is caused by the upper limit of the high-voltage driver while the fall-off above 40 kHz results from the fundamental mechanical resonance of the electrode pair.

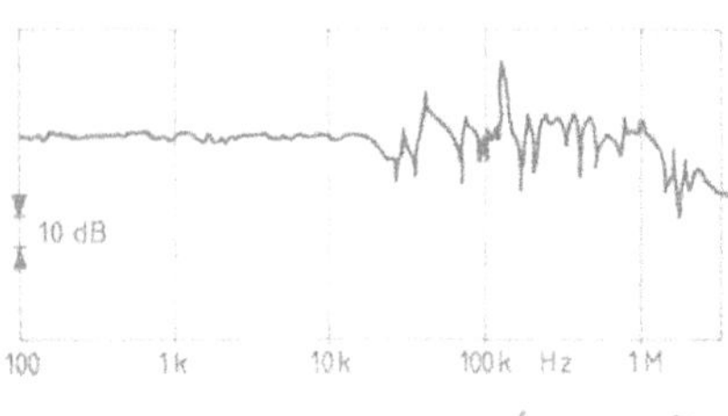

Fig. 6. Small electrode (ε_1) electrooptical transfer characteristic as measured between A' and B'

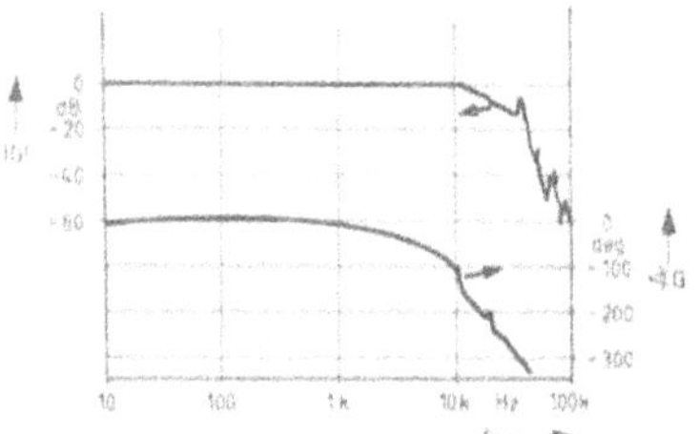

Fig. 7. Open-loop response as measured between A and B. Bias 400 V, small electrode driven by 180 kHz signal

In other words, the elastooptic compensator (δ ,ε) at the fiber end has to compensate for the noisy retardations of the fiber-equivalent compensator (δ_N, ε_N).

Generation of error signals proportional to $\delta + \delta_N$ and $\varepsilon + \varepsilon_N$ is possible by modulation of the retardations δ ,ε with high-frequency, small amplitude signals,

$$\delta = \delta_0 + \delta_1(t) \ , \quad \varepsilon = \varepsilon_0 + \varepsilon_1(t). \tag{5}$$

Substitution of (5) into (3) gives ($\delta + \delta_N$, $\varepsilon + \varepsilon_N$ small is assumed)

$$I^{(r)} = \Big\{ 1 - \cos\varepsilon_0 \cdot \cos\varepsilon_N + \sin\varepsilon_0 \cdot \sin\varepsilon_N \cdot \cos(\delta_0 + \delta_N) - \sin\varepsilon_0 \cdot \sin\varepsilon_N \cdot \sin(\delta_0 + \delta_N) \cdot \delta_1(t) + \sin(\varepsilon_0 + \varepsilon_N) \cdot \varepsilon_1(t) \Big\} \cdot \frac{I_0}{2} \quad (6)$$

Phase-coherent detection with reference signals $\delta_1(t)$ and $\varepsilon_1(t)$ generates signals proportional to $\sin(\delta_0 + \delta_N)$ and $\sin(\varepsilon_0 + \varepsilon_N)$, respectively. If ε_N vanishes, $(\delta_0 + \delta_N)$ cannot be detected but one can show that the more ε_N decreases the more the transmittance becomes independent of $(\delta_0 + \delta_N)$. The output signals of the phase-sensitive detectors can be used to control δ_0 and ε_0 via PI (proportional-integral) controller circuits.

In our experimental set-up (Fig.5), instead of the previously mentioned bulky piezo drives, we used transversely operated piezoceramics of our own fabrication (type PPK 21) with dimensions 12 x 10 x 1.2 mm. Both piezos of each pair were equipped with large electrodes for controlling δ_0 and ε_0 (controller sections), and one piezo of each pair was additionally provided with a small electrode (modulator section) used for superposing the AC-modulations $\delta_1(t)$ and $\varepsilon_1(t)$. In Fig.5 modulation with quadrature signals is assumed, the two pairs of blocks below the oscillator represent the phase-coherent detectors, the controllers consisting of PI-circuits and high-voltage amplifiers for driving the squeezers' controller sections.

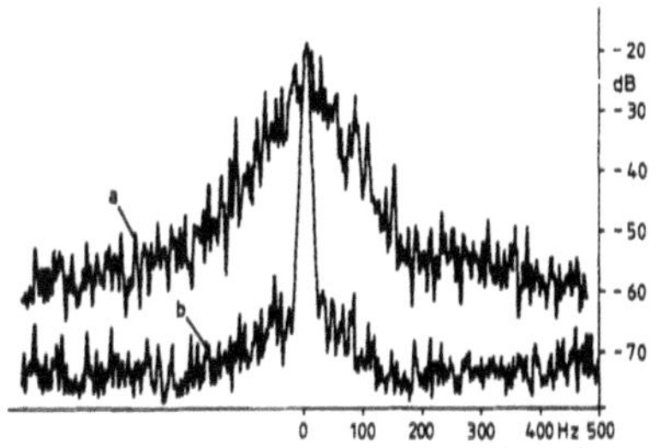

Fig. 8. Polarization noise spectrum at fiber end (S_3). Fiber length L = 1000 m, a without polarization control, b with active polarization control

The closed-loop performance of the controller section (or, in other words, the section controlling the circularity of the polarization ellipse at the output polarizer P_2) is shown in Fig.8. Once more, the polarization noise spectrum without control is shown (curve a), together with the closed-loop polarization noise spectrum (curve b). The measurement bandwidth was 10 Hz. The noise suppression ratio at the lower frequency end is 40 dB (electrical), the lower limit of 70 dB is determined by noise of the electronic circuitry. As the controller bandwidth was designed to be 10 kHz (which when considering the noise spectrum is at least a factor of 10 too high), considerable improvement can be expected by adaptation of the bandwidth to the noise upper frequency limit of several 100 Hz.

Conclusion: We propose a new type of in-fiber polarization control with low insertion loss. We used two pairs of piezoceramic squeezers one of which (ε) controls the circularity and the other (δ) the elevation angle of the polarization ellipse. The piezos are operated in transverse mode, their dimensions are 10 x 12 x 1.2 mm^3 . The system requires 200 V per 2 π rad phase retardation. If fibers with smaller overall diameter are used the required voltage can be reduced by at least a factor of two, further reduction is possible by modification of the squeezer geometry. The polarization error is detected by an AC-modulation scheme; the modulation frequency can be chosen between 40 kHz and 1 MHz. With a controller bandwidth of 10 kHz the polarization noise suppression ratio was better than 40 dB (electrical).

This work was sponsored by the Bundesministerium für Forschung und Technologie. The authors alone are responsible for the contents.

References

1 R. Ulrich and M.Johnson: Opt.Lett.4, 152 (1979)

2 J.L.Davis and S.Ezechiel: Opt.Lett.6, 505 (1981)

3 R.Ulrich: Opt.Lett.5, 173 (1980)

4 M.Johnson: Appl.Opt.18,1288(1979)

5 Y.Kidoh,Y.Suematsu,S.Tejima,J.Toda,K.Furuya:
European Conference on Optical Communication,Copenhagen,1981,paper P 10

3.3 Single-Mode Fibers

3.3.1 Theory

Polarization-Maintaining Fibers*

I.P. Kaminow

Bell Laboratories, Crawford Hill Laboratory, Holmdel, NJ 07733, USA

The parameters that determine the polarization-maintaining ability and the polarization-dispersion of a birefringent fiber are discussed in a tutorial fashion. Based on promising theoretical and experimental results, I conclude that fibers with adequate polarization-maintaining properties for sophisticated heterodyne and homodyne applications are physically realizable. It is also concluded that polarization-dispersion in nominally circular single-mode fibers is not likely to be a limitation for next-generation systems.

INTRODUCTION

Nominally circular optical fibers support two sets of modes corresponding to two orthogonal polarizations. A so-called "single-mode" fiber propagates two nearly-degenerate fundamental modes with electric fields that are orthogonally polarized along the principal x- or y-axes on the fiber z-axis. Although each mode has nonvanishing components along all three axes, in the usual paraxial ray approximation, the electric polarization is predominantly linear and transverse.

Because of the near-degeneracy of propagation constants, β_x and β_y, small perturbations along the fiber readily couple the two modes. Thus, it is found experimentally that linearly polarized light injected into a circular fiber becomes elliptically polarized after a short distance. The state of polarization is sensitive to fluctuations in temperature T, wavelength λ or mechanical pressure or position.

Present fiber applications usually rely upon detection of pulses of optical power in a photodetector that is independent of optical polarization or phase. However, some new and possibly more sophisticated applications do require that light remain polarized over long fiber lengths ranging from ~100 meters for certain sensor applications to ~100 km for a heterodyne communication system.

These applications call for new fibers that maintain the state of polarization. Some progress in this direction has been realized by designing fibers in which the degeneracy is raised by a combination of geometrical and strain anisotropy in the transverse xy-plane. The modal birefringence

$$B = (\beta_x - \beta_y)/(2\pi/\lambda) \tag{1}$$

is a measure of this anisotropy. Reviews of birefringent fibers have been given elsewhere.[1,2,3] It is our purpose here to examine some tutorial aspects of birefringent fibers and to up-date the earlier reviews in connection with (a) measurement of the spatial power spectrum of mode-coupling and (b) the dispersion of birefringent delay (the delay difference for x- and y-polarizations).

* Based on talks given at Int. Conf. on Fiberoptic Rotation Sensors, MIT, Cambridge, Massachusetts, November 1981 and NATO Advanced Study Institute, Cargese, Corsica July 1982.

PERTURBATION POWER SPECTRUM

Fig. 1(A) depicts the cross-section of an idealized anisotropic fiber with rectangular core and cladding. The principal axes are x and y. A shear strain γ, as in Fig. 1(B), rotates the principal axes so that a normal mode with x-polarization in the unperturbed fiber will be coupled to the orthogonal mode with y-polarization. The larger the anisotropy B, the larger γ must be to introduce a given cross-coupled component at a given point z along the fiber length. However, to achieve a cumulative coupling, $\gamma(z)$ must contain a periodic component with period

$$\Lambda = 2\pi/K \tag{2}$$

where the spatial frequency

$$K = \beta_x - \beta_y = (2\pi/\lambda)B. \tag{3}$$

For large birefringence B, Λ must be very small for significant coherent coupling.

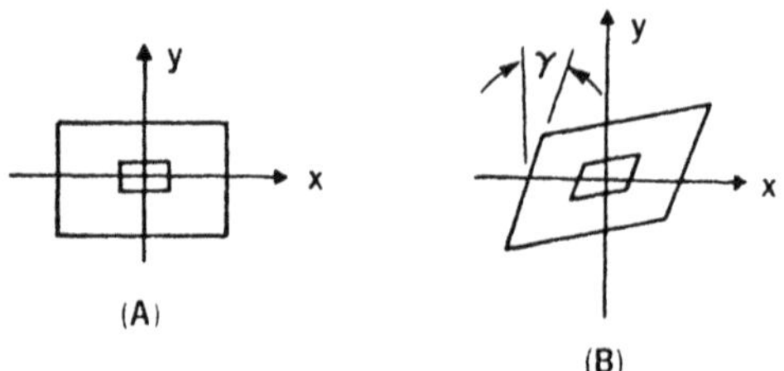

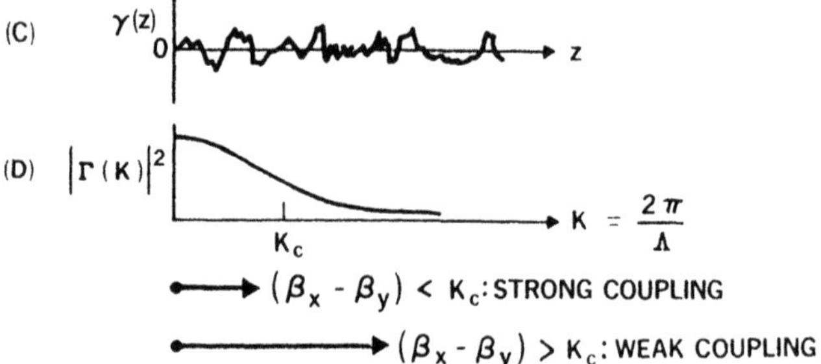

Figure 1 (A) Cross-section of an anisotropic fiber
(B) A shear strain γ that rotates the principal axes and couples the orthogonal normal modes
(C) Random variation of γ along the fiber length z
(D) Spectral power density $|\Gamma(K)|^2$ of γ in the spatial frequency domain K

The origins of $\gamma(z)$ are not known, but one may speculate that they are introduced during fabrication by variable eccentricity in the preform, the drawing apparatus and the outer jacket, or are produced afterward by external effects such as twists or lateral pressure on the finished fiber. Thus, we represent $\gamma(z)$ by a random function, Fig. 1(C), and characterize the function by its spectral power density $|\Gamma(K)|^2$ as in Fig. 1(D), where we assume a low-pass filter characteristic.[1] Thermodynamically, we know that it is impossible to sustain perturbations with $\Lambda \to 0$, therefore

$$|\Gamma(K)|^2 \to 0 \quad \text{as} \quad K \to \infty . \tag{4}$$

An example of low-pass behavior is provided by the spectral distribution $|D(K)|^2$ of the diameter variation of a standard fiber; $|D(K)|^2$ can be measured directly[4] for a given length of fiber as in Fig. 2. We see that the power density decreases by about two decades for each decade of K. The power density is extremely small for $\Lambda < 1$ mm, which must be true for any low loss fiber since the difference in propagation constants $\beta_g - \beta_r$ between guided and radiated modes is approximately 2π mm^{-1}, so that perturbations with $\Lambda \approx 1$ mm would scatter the guided mode into the cladding.

The low pass behavior means that the coupling between x- and y-polarizations can be reduced by introducing a large birefringence so that

$$|\beta_x - \beta_y| >> K_c \tag{5}$$

or

$$B >> \lambda/\Lambda_c \tag{6}$$

where K_c and Λ_c are measures of the perturbation spatial bandwidth.

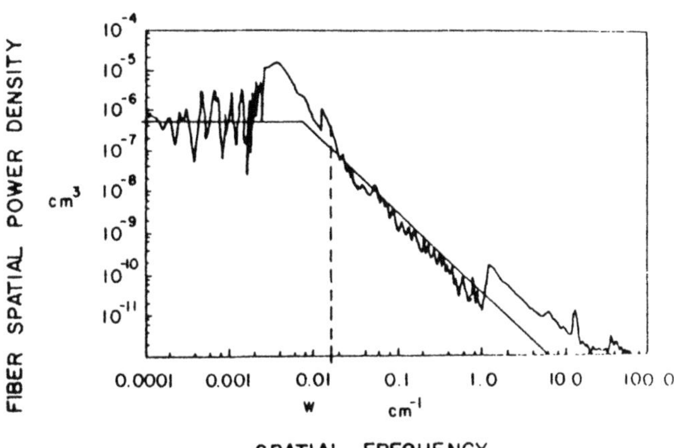

Figure 2 Spectral power density of diameter fluctuations for a standard induction furnace-drawn telecommunication fiber [4]

The modal birefringence can be expressed in terms of a "beat length" L such that

$$|\beta_x - \beta_y| L = 2\pi \tag{7}$$

or

$$L = \lambda/B\,. \tag{8}$$

The beat length can be measured directly by a variety of methods. Thus, the required birefringence to avoid polarization coupling can also be stated as

$$L << \Lambda_c\,,$$

i.e., the beat length should be much less than the critical period of the perturbations.

Alternatively, one could try to alter the shape of $|\Gamma(K)|^2$ to reduce its magnitude and bandwidth. For this purpose one would have to take pains to reduce the perturbations, the sources of which are not fully understood. Therefore, recent effort has been devoted to increasing birefringence by increasing geometrical anisotropy B_g and/or by introducing stress birefringence B_s to produce a large

$$B = B_g + B_s. \tag{9}$$

Future efforts should be devoted to controlling $|\Gamma(K)|^2$.

It can be shown[1] that a random coupling coefficient $\gamma(z)$ between normal modes gives rise to the following dependences of power P_x and P_y in the x- and y-polarized modes when all the power is in P_x at $z = 0$:

$$\langle P_x \rangle = e^{-hz} \cosh hz \,, \tag{10}$$

$$\langle P_y \rangle = e^{-hz} \sinh hz \,, \tag{11}$$

where $\langle\,\rangle$ indicates and average over an ensemble of fibers or lengths of fiber, and h is an effective coupling coefficient. The extinction coefficient η is defined as

$$\langle \eta \rangle = \langle P_y \rangle / \langle P_x \rangle = \tanh hz \tag{12}$$

$$\rightarrow 1 \quad \text{for} \quad hz \text{ large}$$

$$\rightarrow hz \quad \text{for } hz \text{ small} \,.$$

After a sufficiently long length, half the power will remain in each polarization on average. However, if h is sufficiently small we can expect to realize a small η even after long propagation length. Thus, the objective is to reduce h in order to obtain small η and good polarization maintenance over a long length, which is $\sim h^{-1}$. Note that statistically η can differ substantially from $<\eta>$, depending upon the standard deviation of η.

The theory shows that [1]

$$h = (k^2/4) \left\langle |\Gamma(Bk)|^2 \right\rangle , \tag{13}$$

where

$$k = 2\pi/\lambda \quad , \tag{14}$$

which is in agreement with the intuitive arguments given above, i.e., we want to minimize $|\Gamma(K)|^2$ for $K = Bk$. We note also that h varies as k^2 or λ^{-2} so that, all other things being equal, η is less at longer wavelengths.

Unfortunately, there is no way at present to measure $|\Gamma(K)|^2$ directly over long lengths and with the required spatial resolution $K > 1\ mm^{-1}$. Even if we could measure such a length of fiber, it would represent only one of a large number of lengths of a statistical ensemble.

An alternative approach is to measure one length of fiber using an incoherent light source rather than the monochromatic source implied in the previous discussion.[5,6] If the source has a spectral width $\Delta\lambda$, the coherence time T_c is $\lambda^2/c\Delta\lambda$. If such a source were injected into an interferometer, no interference fringes would be observed if the delay difference between the two paths of the interferometer exceeded T_c. Similarly, in a birefringent fiber, the x- and y-polarized modes of the incoherent source could not couple coherently, as assumed in the preceding coupling theory for monochromatic waves, if there is a birefringent delay difference τ such that $\tau > T_c$. The birefringent delay is defined as the group delay difference for propagation with x- or y-polarization:

$$\tau = \left(\frac{\partial\beta_x}{\partial k} - \frac{\partial\beta_y}{\partial k}\right)\left(\frac{z}{c}\right) \tag{15}$$

$$= \frac{\partial Bk}{\partial k}\left|\frac{z}{c}\right|$$

$$= \left(B + k\frac{\partial B}{\partial k}\right)\left|\frac{z}{c}\right| \,.$$

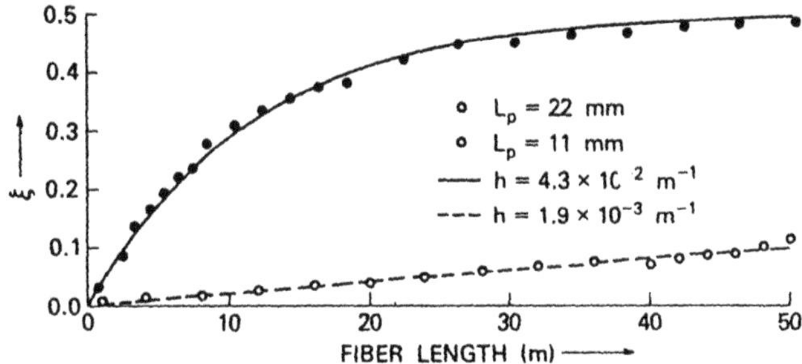

Figure 3 Length dependence of extinction coefficient ξ for two elliptical-core fibers with different birefringences [5]

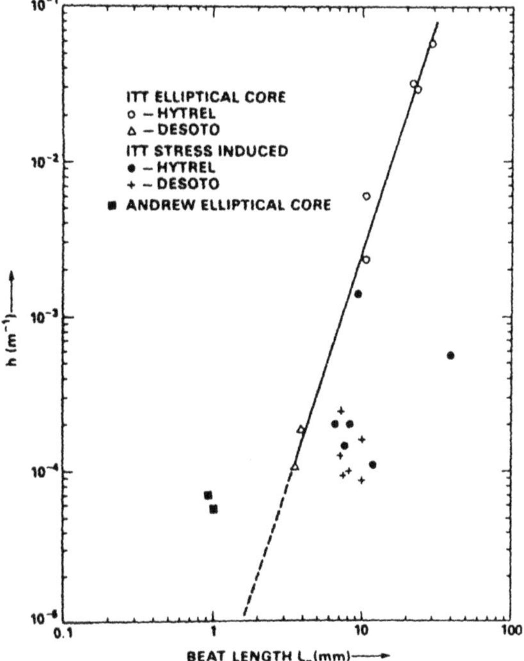

Figure 4 Coupling coefficient h versus beat length for a variety of fibers with differing birefringences [6]

Thus, each length

$$\ell_c = \lambda^2/\Delta\lambda \left[\frac{\partial Bk}{\partial k}\right]^{-1}, \tag{16}$$

corresponding to $\tau = T_c$, may be regarded as an independent member of the ensemble.

A fiber with total length ℓ will contain $N = \ell/\ell_c$ members. The total power $\langle P_y \rangle$, containing the sum of powers added incoherently from each ℓ_c, represents the ensemble average. The standard deviation of the P_y measurement is $\sim \langle P_y \rangle N^{-1/2}$. Similarly, the measured $\langle \eta \rangle$ and h will have a relative standard deviation of $N^{-1/2}$.

Measurements using this method[5] are shown in Fig. 3. A modified extinction coefficient $\xi = P_y/(P_x+P_y)$ is plotted against fiber length. It can be seen that the experimental points fit the theoretical curve.

$$\langle \xi \rangle = \frac{1}{2}\left[1 - \exp(-2h(\lambda)z)\right] , \qquad (17)$$

which follows from (10) and (11), when a suitable value of h is used. The measurements were obtained by cutting off successive lengths of fiber. The spectral width of the incoherent source covers the range 700 to 900 nm.

Similar measurements[6] on a variety of fibers with differing birefringences yield the plot of h versus beat length L shown in Fig. 4. We see that h decreases by about three decades for a decade reduction in L. Since h is proportional to $\langle |\Gamma(Bk)|^2 \rangle$ and $L = \lambda/B$, we confirm the low pass behavior of the power spectrum as proposed above and illustrated in Fig. 1(D).

BIREFRINGENCE AND DISPERSION

Various approaches have been taken to calculate the propagation constants for the orthogonally polarized modes in fibers with elliptical or rectangular cross-sections.[2] Our object here is to provide an intuitive picture of the variation in birefringence and polarization dispersion with normalized optical frequency as given by the V-value:

$$V = \frac{2\pi}{\lambda} W_i \left(n_1^2 - n_2^2\right)^{1/2} \qquad (18)$$

where n_1 is the index of the core, n_2 the index of the cladding and W_i the width of the core along the $i = x$ or y axis.

For either polarization all the energy will be in the core for $V \to \infty$ so that $\beta_x = \beta_y \to \frac{2\pi}{\lambda} n_2$ and all the energy will be in the cladding for $V = 0$ so that $\beta_x = \beta_y \to \frac{2\pi}{\lambda} n_1$. For intermediate values of V,

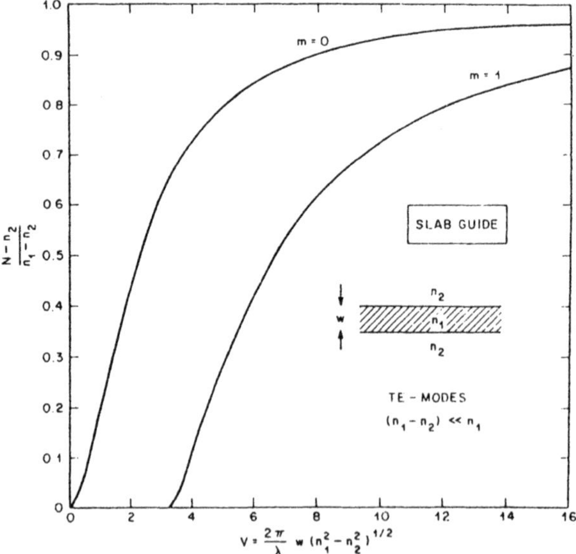

Figure 5 Dispersion of the effective index N for the TE_0 and TE_1-modes of a slab guide [7]

we define an effective index N_i such that $\beta_i = \frac{2\pi}{\lambda} N_i$, where i denotes the polarization direction. In the simple case of a planar slab waveguide of thickness W and index n_1 surrounded by a cladding with index n_2 such that $(n_1-n_2) << n_1$, N(V) has the form shown in Fig. 5 for the fundamental and first-order TE modes.[7] We use the normalized index $(N-n_2)/(n_1-n_2)$ as ordinate in Fig. 5.

Now consider two separate slabs having thicknesses W_x and W_y with $W_x/W_y = 2$ as shown in Fig. 6. The TE-mode with y-polarization corresponds to the slab W_x and V-value V_x. The slab W_y, corresponding to the x-polarization, has half the V-value at the same wavelength. The ordinate at given V for the latter curve therefore has the value of the former curve at 2V as indicated in Fig. 6. The birefringence of the two slabs is simply the difference of these two curves as shown by the dashed curve. The birefringence of a fiber with rectangular core having dimensions W_x and W_y might be expected to exhibit similar behavior. The birefringence $B = N_x - N_y$ is positive and reaches a maximum of about $0.3(n_1-n_2)$ at $V_x \approx 1.5$.

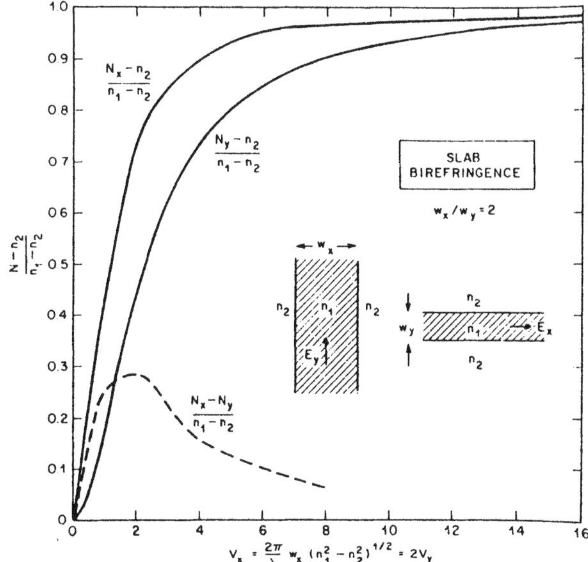

Figure 6 Dispersion curves for the TE_{ox} and TE_{oy}-modes in two orthogonal slabs. The birefringence $B = N_x - N_y$ is given by the dashed curve

We have just estimated the geometrical birefringence B_g assuming that n_1 and n_2 are isotropic. If in fact n_1 and n_2 are birefringent due to mechanical strains in the fiber core and/or cladding, we expect an added strain birefringence B_s. Far from cut-off, where most of the energy is in the core, we expect a modal birefringence

$$B_s = n_{1x} - n_{1y} \equiv b_1 \tag{19}$$

with n_{1x} and n_{1y} the core indices for the respective polarizations, and their difference the material birefringence b_1. The strain may be induced by differential thermal expansion with respect to the cladding, in which case $(n_{2x}-n_{2y})$ and $(n_{1x}-n_{1y})$ will have opposite signs, or by differential expansion with respect to a thick outer jacket, in which case the material birefringences b_1 and b_2 will have the same sign. The dispersion in the material birefringence is expected to be much less than that in the indices themselves and introduces only a very small variation in B_s with V. However, as V changes, the distribution of energy between n_1 and n_2 will vary and, if the material birefringences have opposite sign, $|B_s|$ will decrease as V decreases.

The relative signs of B_g and B_s depend upon the structure of the guide and the two components may either add or subtract. In any real fiber, both B_g and B_s will be present although one or the other may be dominant.

The polarization dispersion or delay difference between the two polarizations as given in (15) can be written

$$\tau = \left\{\frac{\partial \beta_x}{\partial k} - \frac{\partial \beta_y}{\partial k}\right\}(z/c) \tag{20}$$

$$= \left\{B + V\frac{\partial B}{\partial V}\right\}(z/c) \ .$$

As we indicated above, $\partial B_s/\partial V$, which is caused by both the variation in energy distribution between core and cladding and the dispersion in b, is usually small so that $\tau \approx B_s z/c$ for strain birefringence alone.

For geometrical birefringence alone, on the other hand, the $\partial B_y/\partial V$ term is significant and causes τ to reach a maximum value and then decrease with increasing V, eventually passing through zero. Such behavior could be observed by applying (20) to the curves in Fig. 6. Instead, however, we show the corresponding curves calculated[8] for an elliptical core fiber with ellipticity $q << 1$ as shown in Fig. 7. We see that $\tau = 0$ near $V = 2.5$, which is the cut-off value for the first higher-order mode.

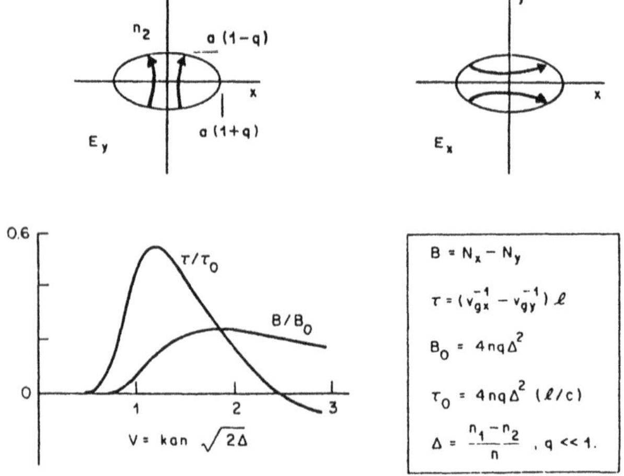

Figure 7 The x- and y-polarized modes in an elliptical-core fiber. Birefringence and polarization dispersion [8]

Since polarization dispersion can be the bandwidth-limiting factor when material dispersion has been eliminated by operation at 1.3 μm, there may be an advantage to operating near $V = 2.5$. Then the strain birefringence would be the limiting factor. Nevertheless, in nominally circular telecommunication fibers, for which $B \approx 10^{-7}$ and $\tau/z \approx B/c = 0.3$ ps/km (3×10^{12}b/s−km), polarization dispersion is not a practical limitation even for systems with 1 Gb/s pulse rate and 100 km repeater spacing. A direct measurement of τ/z gives a similar result.[9]

CONCLUSIONS

The polarization-maintaining ability of a fiber, as measured by its extinction coefficient η, can be characterized by a coupling parameter h which in turn is proportional to the coupling power-spectral-density $|\Gamma(2\pi/L)|^2$ at the spatial frequency corresponding to the beat length L. With measured values[6] of h approaching $10^{-5}m^{-1}$, we can expect that η will be less than 0.1 for fiber lengths approaching 100 km. The realization of this limiting performance will require both small L and some reduction in $|\Gamma(2\pi/L)|^2$ by improved fabrication and handling of the fiber. Several birefringent fiber structures with small L and low-loss, approaching that of nominally circular fibers, have been reported recently.[10,11] Thus, it appears that polarization-maintaining fibers for a number of sophisticated heterodyne and homodyne applications are physically realizable.

Polarization dispersion is not likely to be a severe limitation for next-generation systems utilizing nominally circular fibers.

The fibers we have been discussing employ a birefringence for linearly polarized waves. For a twisted fiber, the normal modes are circularly polarized and a large circular birefringence will also serve to reduce mode coupling.[12] In addition, the circular symmetry may make splicing simpler since the rectangular principal axes need not be aligned. Nevertheless, a pitch of approximately 14 turns/mm is required to realize a best length of 1 mm.[13] This large twist rate may be difficult to introduce in a cabled fiber.

REFERENCES

[1] I. P. Kaminow, "Polarization in Optical Fibers", IEEE J. Quant. Electr. *QE-17*, 15-22 (1981)

[2] T. Okoshi, "Single-Polarization Single-Mode Optical Fibers", IEEE J. Quant. Electr. *QE-17*, 879-884 (1981)

[3] D. N. Payne, A. J. Barlow and J. J. Ramskov Hansen, "Development of Low- and High-Birefringence Optical Fibers", IEEE J. Quant. Electr. *QE-18*, 477-488 (1982)

[4] P. H. Krawarik and L. S. Watkins "Fiber Geometry Specifications and its Relation to Measured Fiber Statistics", Appl. Opt. *17*, 3984-3989 (1978)

[5] S. C. Rashleigh, W. K. Burns, R. P. Moeller and R. Ulrich, "Polarization Holding in Birefringent Single-Mode Fibers", Optics Letters 7, 40-42 (1982)

[6] S. C. Rashleigh, W. K. Burns, M. J. Marrone and R. Ulrich, "The Characterization of Polarization-Holding in Birefringent Single-Mode Fibers", Proceedings of International Conference on Fiberoptic Rotation Sensors, MIT, Cambridge, Mass., November 1981; S. C. Rashleigh and M. J. Marrone, "Polarization Holding in Randomly Perturbed Fibers", Optical Fiber Communication Meeting, Phoenix, April 1982 - paper THCC 7

[7] H. Kogelnik and V. Ramaswamy, "Scaling Rules for Thin-Film Optical Waveguides", Appl. Optics *13* 1857-1862 (1974)

[8] D. L. A. Tjaden, "Birefringence in Single-Mode Optical Fiber due to Core Ellipticity", Philips J. Res. *33*, 254-263 (1978)

[9] N. Imoto and M. Ikeda, "Polarization Dispersion Measurement in Long Single-Mode Fibers with Zero Dispersion Wavelength at 1.5 μm", IEEE J. Quant. Electr. *QE-17*, 542-545 (1981)

[10] H. Matsumura, T. Katsuyama and T. Suganuma, "Single Polarization Fibers" in *Optical Devices and Fibers*, Chap. 3.6, Y. Suematsu, ed.(North-Holland, New York 1982)

[11] T. Hosaha, K. Okamoto, T. Miya, Y. Sasaki and T. Edahiro, "Low-loss Single Polarization Fibers with Asymmetrical Strain Birefringence", Electr. Letters *17* 530-531 (1981)

[12] L. Jeunhomme and M. Monerie, "Polarization-Maintaining Single-Mode Fibre Cable Design", Electr. Letters *16*, 921-922 (1980)

[13] R. Ulrich and A. Simon, "Polarization Optics of Twisted Single-Mode Fibers", Appl. Optics *18*, 2241-2251 (1979)

3.3.2 Fabrication

Elliptically Cored Polarization Holding Fiber

R.B. Dyott
Andrew Corporation, Orland Park, IL 60462, USA

When considering polarization in optical fiber, a distinction has to be made between the preservation of the integrity and of the direction of the polarization of the guided wave. For instance, a fiber with a perfectly circular core (1) could preserve integrity but not necessarily direction whilst a bi-refringent fiber might do both even if it were bent or twisted to some extent. A fiber which preserves integrity and direction should ideally support only a single polarized mode. This would be a simple matter with microwaves in a metal waveguide, but fibers, as dielectric guides, have no lower cut-off for the fundamental modes so that it is not possible to use the shape of the waveguide section to cut-off one fundamental mode and leave the other. See Figure (1).

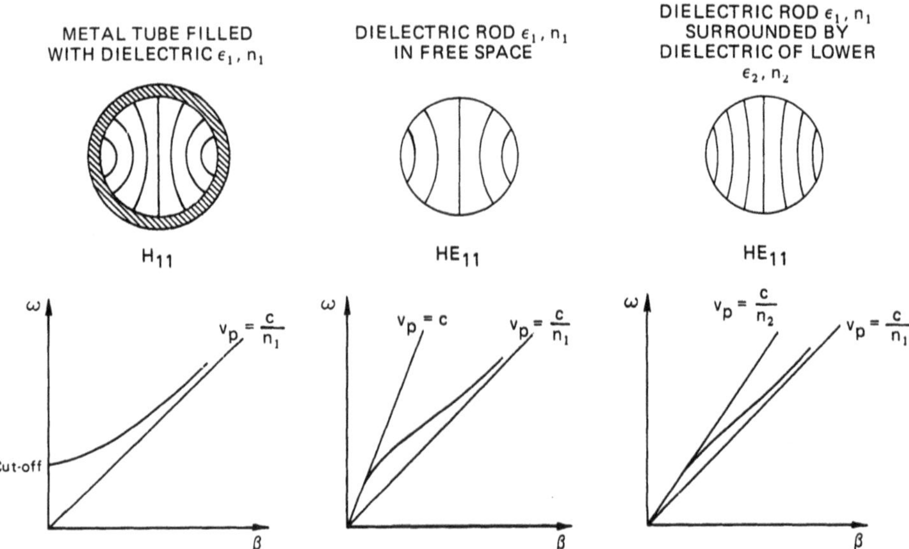

Fig. 1 Metal and dielectric waveguide

It is possible to ensure propagation in only one polarization by having a core with an anisotropic index (e.g. a crystal) with one transverse direction having an index larger than that of the cladding whilst the index in the orthogonal direction is less than that of the cladding so that light polarized in this direction is not guided[2]. See Figure (2).

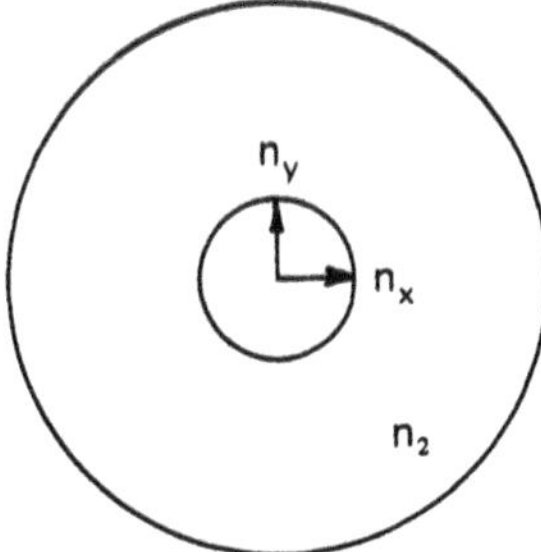

Fig. 2 Crystal-cored fiber

Unfortunately, such a crystal-cored guide is not practicable in long enough lengths to be useful.

Present day polarization preserving fibers are not, strictly speaking, single moded but have two fundamental modes whose propagation constants have been separated either by an applied stress to create an anisotropic index [3][4] or else by means of an azimuthally asymmetric geometry[5]. An example of the latter type is the elliptically-cored fiber.

Elliptically Cored Fiber

The elliptical dielectric waveguide has been investigated in detail over a number of years. Most treatments have been associated with microwave applications and deal with large differences in dielectric constant/refractive index [6][7], but there have been both analytical and numerical studies of the low index-difference approximation applicable to optical fibers [8][9][10]. An

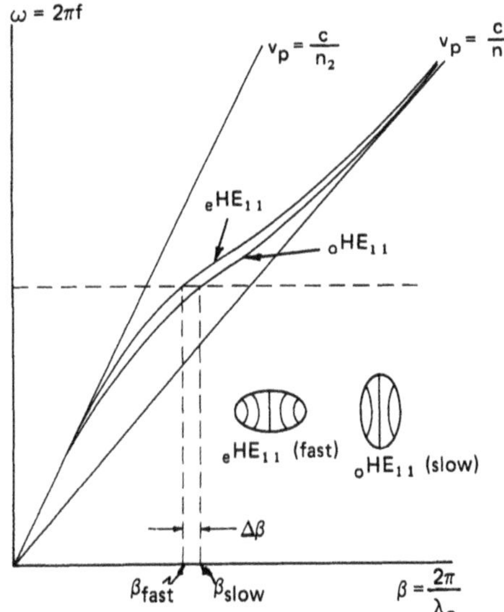

Fig. 3 Elliptical fiber waveguide

analysis produces a transcendental equation for the propagation characteristics which has exactly the same form as that for the circular guide, but with the corresponding Mathieu functions replacing the Bessel functions. The fundamental HE_{11} mode of the circular guide is split into two orthogonal modes whose transverse electrical fields lie in the direction of the major and minor axes of the ellipse. See Figure (3).

The $_{o}HE_{11}$ has the larger propagation constant so that it is more tightly bound and has a phase velocity nearer to that of the core than the $_{e}HE_{11}$ has. It is therefore the preferred mode of propagation. An alternative designation for the modes might be the $_{s}HE_{11}$ and the $_{f}HE_{11}$ with s for slow and f for fast replacing o for odd and e for even which are derived from the kinds of Mathieu function associated with the particular mode. 'Slow' and 'Fast' here refer to the phase and not the group velocities. The transcendental equation is solved numerically to give the propagation constants $_{s}\beta$, $_{f}\beta$ and so the difference between them $\Delta\beta = {}_{s}\beta - {}_{f}\beta$. $\Delta\beta$ is, to a very good approximation, proportional to $(\Delta n)^2$, where Δn is the difference in index between core and cladding.

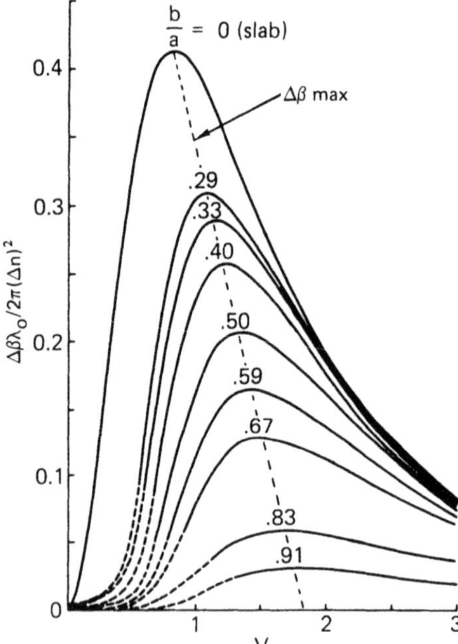

Fig. 4 Difference in propagation constants of the fundamental modes of an elliptically cored fiber

Figure (4) shows a plot of

$$\frac{\Delta\beta}{(\Delta n)^2}\,\frac{\lambda_o}{2\pi} = \frac{\Delta\beta}{\beta_o\,(\Delta n)^2}$$

against normalized frequency

$$V = \frac{2\pi b}{\lambda_o}\,[n_1{}^2 - n_2{}^2]^{1/2} \simeq \frac{2\pi b}{\lambda_o}\,[2n\,\Delta n]^{1/2}\,, \qquad n = 1{\cdot}47$$

for different values of minor axis/major axis = 2b/2a of the ellipse which goes from a circle, b/a = 1, to a slab, b/a = 0. There is an optimum value of V for maximum $\Delta\beta$ which, with ellipticities b/a $\geqslant$ approx. 0.4, is below the higher mode cut-off. As an example of the separation in propagation constants that can be achieved with an elliptical core, taking

$$\frac{b}{a} = {\cdot}4$$

then from Figure (4) ($n = 1{\cdot}47$) there is maximum separation at $V = 1{\cdot}2$ with

$$\frac{\Delta\beta}{\beta_o(\Delta n)^2} = {\cdot}26 \ .$$

With a high but achievable $\Delta n = {\cdot}062$, then

$$B = \frac{\Delta\beta}{\beta_o} = 10^{-3}$$

and a normalized beat length between modes

$$\frac{L}{\lambda_o} = \frac{1}{B} = 10^3 \ .$$

Group Delay

The group velocity

$$v_g = \frac{d\omega}{d\beta}$$

is the slope of the mode line on the ω/β diagram. Since for the elliptical guide (see Figure 3) both lines start at $\omega = 0$ and converge at $\omega = \infty$, there must be some intermediate value of ω, and therfore of

$$V \propto \frac{\omega}{c}\,[n_1^2 - n_2^2]^{1/2}$$

where the lines are parallel and the slopes/group velocities are equal.

Figure 5 shows the variation with V of group velocity expressed as a group index

$$n_g = \frac{c}{v_g}$$

for the fundamental TE and TM modes of the dielectric slab ($n_1 = 1.58$) in free space ($n_2 = 1$).

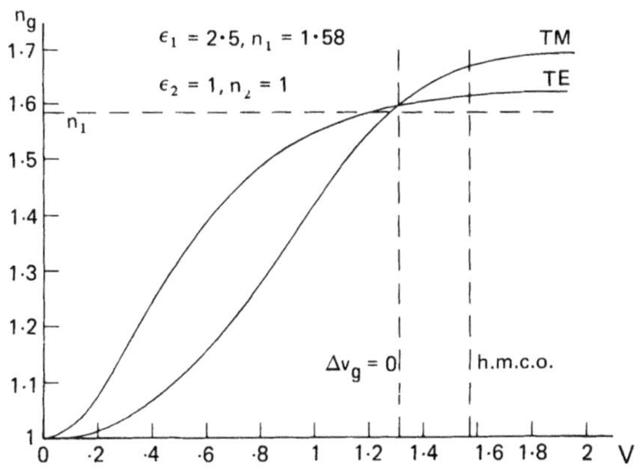

Fig. 5 Group index in slab waveguide

The cross-over point, where the group indices/velocites are equal, is at the V value which is below the higher mode cut-off

$$V = \frac{\pi}{2} \; .$$

In Figure 6 taken from ref. 9, the difference in group index ${}_f n_g - {}_s n_g$ is plotted against V for different ellipticities.

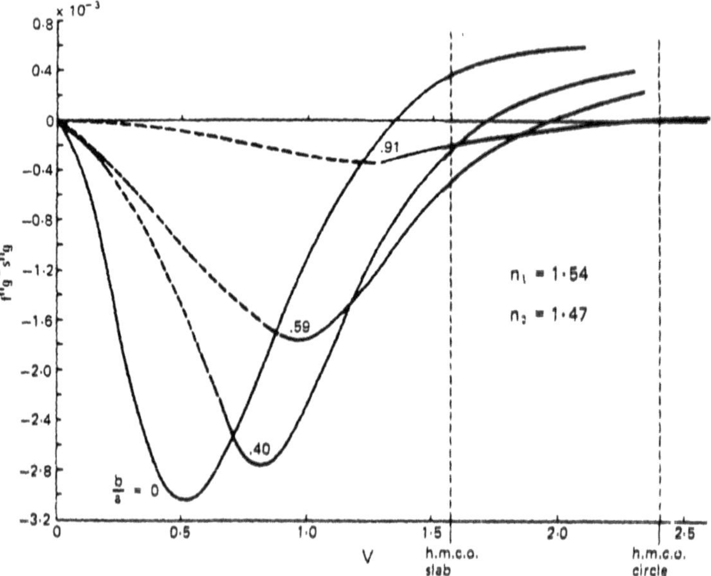

Fig. 6 Difference in group indices of the fundamental modes

For a nearly circular core the point of zero difference in group index is slightly above higher mode cut-off at $V = 2{\cdot}48$.

For an approximate range of ellipticities

$$\cdot 45 < \frac{b}{a} < \cdot 63$$

there should be a value of V below higher mode cut-off where the difference in group velocities between the fundamental modes is zero but where there is still a substantial (although not a maximum) difference in phase velocity.

This property may be useful perhaps for long distance polarization-preserving fibers of the future.

Practical Fiber

Since $\Delta\beta$ increases as $(\Delta n)^2$ there is a lot to be gained by making Δn as large as possible. It can be a great disadvantage if light is launched and then trapped in the cladding, even for a short distance. The effect can be avoided completely by making the index of the cladding less than that of the outside silica support tube. The present fiber therefore has a very highly doped germania-silica core, $\Delta n \triangleq \cdot 62$, surrounded by a non-guiding cladding which is made by depositing silica in the presence of fluorine. The fluorine serves a double purpose, decreasing the index and lowering the softening temperature sufficiently to enable the cladding to be glassed on smoothly. Photographs of preform and fiber sections are shown in Figure (7).

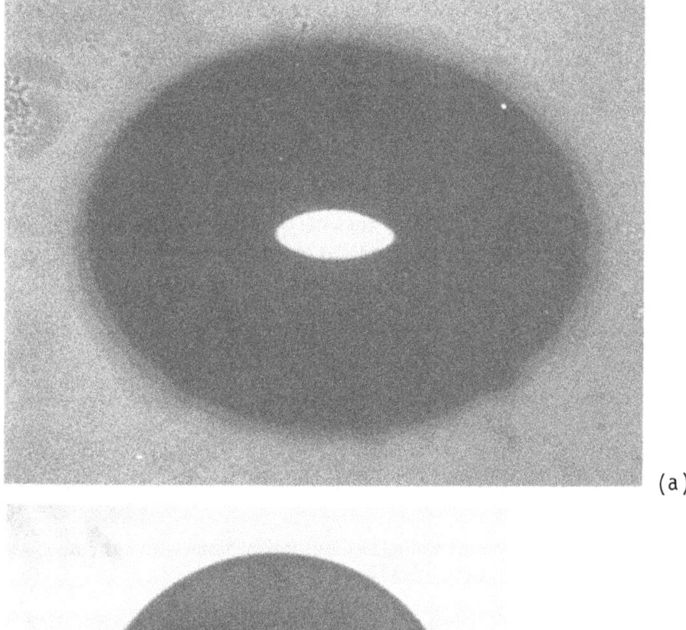

(a)

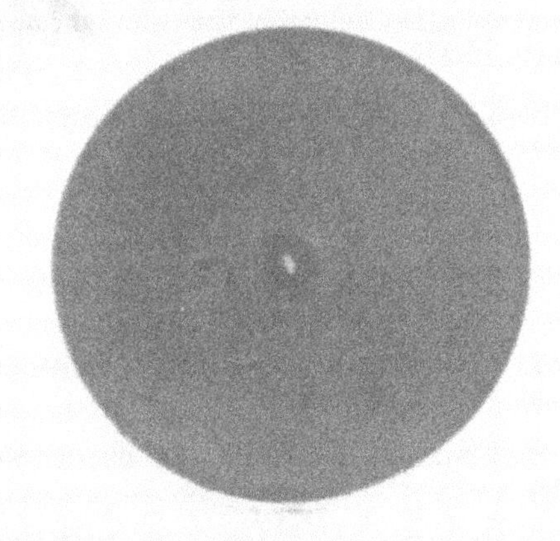

(b)

Fig. 7 Sections of (a) preform, and (b) fiber

The outside diameter of the fiber is 70 μm and the major and minor axes of the elliptical core about 2 μm and 1μm.

Independent measurements made on the fiber [11] give a beat of length of about 1 mm and a polarization preserving factor h [12] of 6 x 10^{-5} at 850 nm. The scattering loss of these particular samples was very high (80 dB at 840 nm) and must have caused some intermodal coupling thus degrading the h factor. Recently pulled fibers have a loss of 35 dB/km at 850 nm. They have yet to be tested using the method developed by Rashleigh et al. [11] but the h factor should be improved with the reduced amount of scattering.

Operating at longer wavelengths would greatly reduce the loss: for instance, the 35 dB/km at 840 nm becomes 6.5 dB/km at 1.3 μm and 3.17 dB/km at 1.55 μm. There is a discernible trend towards the longer wavelengths for some sensors such as the fiber gyro.

Applications

As well as the usual applications involving the preservation of polarization, there is the possibility that the shape of the elliptical core may in itself be an advantage as, for instance, with coupling to rectangular waveguides in integrated optics or to laser diodes. The small core dimensions, normally considered a nuisance, may be useful where a high power density is needed e.g. for non-linear effects.

References

[1] A. J. Barlow, D. N. Payne, M. R. Hadley, and R. J. Mansfield, 'Production of single-mode fibers with negligible intrinsic birefringence and polarization mode dispersion', Electron. Lett. 17, (20) 725 (1981)

[2] J. L. Stevenson and R. B. Dyott, 'Optical fiber waveguide with a single crystal core', Electron. Lett. 10, (22) 449 (1974)

[3] I. P, Kaminow and V. Ramaswamy, 'Single-Polarization optical fibers: slab model', App. Phys. Lett. 34, (4) 268 (1979)

[4] T. Katsuyama, H. Matsumura, and T. Suganuma, 'Low-loss single-polarization fibers', Electron. Lett. 17, (13) 474 (1981)

[5] T. Okoshi and K. Oyamada, 'Single-polarization single-mode optical fiber with refractive – index pits on both sides of core', Electron. Lett. 16, (18) 712 (1980)

[6] L.A. Lyubimov, G.I. Veselov, and N.A. Bei, 'Dielectric waveguide with elliptical cross section', Radio Eng. and Electron USSR. 6, 1668 (1961)

[7] C. Yeh, 'Elliptical dielectric waveguides', Jour. Appl. Phys. 33, 3235 (1962)

[8] M. J. Adams, D. N. Payne and C. M. Ragdale, 'Birefringence in optical fibers with elliptical cross-section', Electron. Lett. 15, (10) 298 (1979)

[9] R. B. Dyott, J. R. Cozens and D. G. Morris, 'Preservation of polarization in optical-fiber waveguides with elliptical cores', Electron. Lett. 15, (13) 380 (1979)

[10] S. R. Rengarajan and J. E. Lewis, 'Single mode propagation in multi-layer elliptical fiber waveguides', Radio Science 16, (4) 541 (1981)

[11] S. C. Rashleigh, W. K. Burns, N. J. Marone, and R. Ulrich 'Characterization of polarization holding in birefringent single-mode fibers', First Int. Conf. Fiberoptics rotation sensors, MIT, Nov. 1981

[12] I. P. Kaminow, 'Polarization in optical fibers', IEEE Jour. Quantum Electronics QE 17 (1) 15 (1981)

Fabrication and Properties of Low Birefringence Spun Fibers

D.N. Payne, A.J. Barlow, J.J. Ramskov-Hansen, M.R. Hadley, and
R.J. Mansfield

Department of Electronics, The University
Southampton, Hampshire, Great Britain

1. Introduction

Single-mode optical fibers are birefringent as a result of an (often unintentional) lack of circular symmetry in the core cross section. This and an associated stress anisotropy allow the fiber to support two nearly degenerate orthogonally polarised modes with a small phase-velocity difference. When only one of the modes is excited, the state of polarisation remains constant along the length of the fiber, whereas if both are present, the polarisation state evolves cyclically as the modes beat together in phase relationship. In a typical fiber the magnitude of this intrinsic birefringence is relatively small and it is found that it can be severely modified by environmental factors such as pressure, twists, and bends,which in a practical installation, vary in an unpredictable manner. The overall fiber birefringence, and thus the output polarisation state,cannot be predetermined,and, moreover, vary with time and temperature.

The unpredictability of the polarisation state presents a problem in fiber interferometers where colinearity of the interfering beams is required [1] - [4], in coupling to polarisation-sensitive integrated optics [5] and in phase-coherent detection schemes for communications [6]. A well-established solution is the "polarisation-maintaining" fiber [7] in which the internal birefringence is increased to a level well above that likely to be caused by environmental effects. The waveguide then appears essentially length invariant, since external perturbations are swamped by the high level of internal birefringence. Thus, a single linearly polarised mode can be selected and sustained without coupling energy to its orthogonally polarised partner.

By contrast, in a polarimetric sensor the variation in the output polarisation state is exploited as a measure of the magnitude of an external effect. Thus, in the Faraday-effect current monitor [8],[9], the magnetic field surrounding a conductor causes a rotation of the plane of polarisation in the fiber, while in an acoustic or magnetic-field sensor, vibration or magnetostriction of a former upon which the fiber is wound applies pressure and induces linear birefringence [10]. In these cases the internal fiber birefringence can interfere with the induced birefringence to reduce the sensitivity [8], [11], [12], and a low-birefringence fiber is an advantage.

Several in-line fiber devices have been constructed [13]-[15] which utilise the birefringence induced by controlled bends and twists to provide the fiber analog of discrete birefringent spectral filters and compensators. Internal fiber birefringence unpredictably modifies the required birefringence and, since it is temperature sensitive, detracts from device stability. Again, a low-birefringence fiber is required. A particular example of this is the tuned-coil fiber isolator [16] which utilises the Faraday effect. In this device the bending birefringence in a fiber coil is exactly matched to the coil circumferential length for up to 50 turns. Residual fiber birefringence contributes to the bend birefringence and leads to unpredictable coil tuning and poor temperature stability.

The presence of two modes in the fiber can also be a disadvantage in telecommunications systems where it leads to a reduction in bandwidth as a result of a difference in the mode group delays (polarisation dispersion). The magnitude of the effect has been estimated at 10-40 psec/km [17,18], although this would be considerably greater for a polarisation-maintaining fiber with its high birefringence. Polarisation dispersion could thus become significant in future links of 100 km or more, as presently envisaged for undersea use.

In this paper, we report a method whereby fibers can be reproducibly manufactured with negligible intrinsic birefringence or polarisation mode dispersion. Such fibers will find application in optical fiber devices, sensors and for high-bandwidth telecommunications. The fibers are produced by rotating the preform during drawing to impart a permanent twist and are referred to as 'spun fibers' to distinguish them from fibers twisted after drawing. Results are presented which illustrate the dramatic reduction in both birefringence and polarisation dispersion which may be obtained by this technique.

2. Low-Birefringence Fiber Production - Conventional Approach

"Single-mode" fibers with an asymmetric cross-section propagate two orthogonal linearly polarised modes with axes aligned with those of fiber symmetry. These modes propagate unchanged in the absence of length-dependent variations in the fiber, with a difference $\delta\beta$ (the birefringence) in their propagation constants. Two commonly used parameters which are useful in describing fibers with length-invariant properties are the normalised linear birefringence B and the mode beat length L_p at which the polarisation state is periodically repeated:

$$B = \frac{\lambda}{2\pi}\,\delta\beta\,, \tag{1}$$

$$L_p = \frac{2\pi}{\delta\beta} = \frac{\lambda}{B}\,, \tag{2}$$

where λ is the free-space wavelength of light. In general, B comprises both a waveguide shape component B_G and a stress-anisotropy component B_S. The former is a waveguiding effect in fibers with asymmetric cross-sections, whereby orthogonally polarised modes experience different guide indices.

The latter is an associated effect and is produced by an asymmetry in the internal stress distribution which is present in all fibers, since they are composed of materials having different thermal-expansion coefficients. This photo-elastic effect dominates the shape birefringence in all but fibers designed specifically to exploit the shape component [19]. Since the magnitude of the stress anisotropy is temperature dependent, the fibers exhibit a significant change in birefringence with temperature (~ 0.1 per cent/^{o}C).

Early attempts at making low-birefringence fibers [20] were aimed at reducing 1) the waveguide shape birefringence B_G by maintaining a highly circular core and 2) the elastooptic anisotropy B_S by decreasing the residual core stress. These objectives were achieved by careful attention to fabrication techniques.

The calculated retardation in degrees/m due to core ellipticity alone is plotted in Figure 1 for operation at the higher-mode cutoff (normalised frequency V = 2.4) for near-circular step-index fibers .

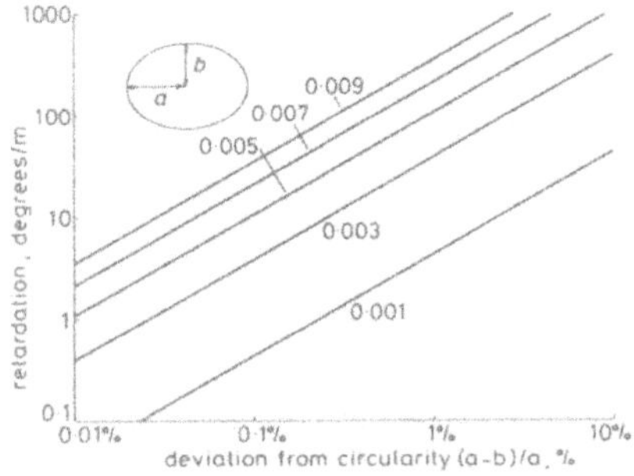

Fig.1. Retardation as a function of core circularity in a fiber having V = 2.4 at 0.633µm.
Curves are for the values of relative index difference Δ shown

From the figure it can be seen that for low values of relative index difference Δ, a circularity deviation of around 0.1% is required to obtain a retardation of < 1°/m. By careful control of circularity and attention to stress levels in the preform it was possible to manufacture a fiber with a beat length L_p = 140m at λ = 0.633µm [20]. The cross-section of the pre-

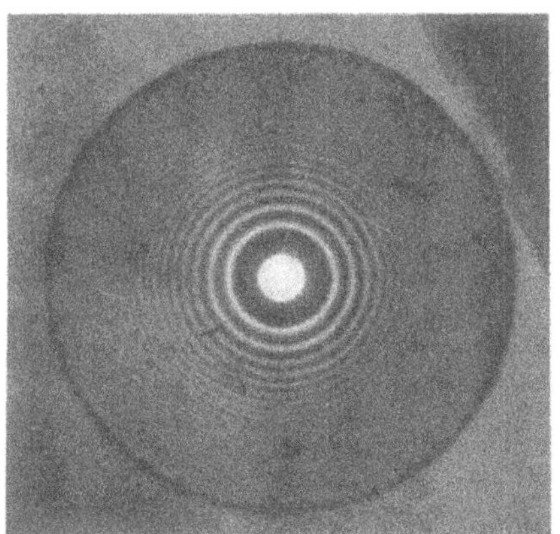

Fig.2. Cross-section of deposited region of low-birefringence preform. Rings are B_2O_3-doped layers with depressed index and the central bright region is the GeO_2-doped core

form from which this fiber was drawn is shown in Fig. 2 and the measured axial stress distribution in Fig. 3a [21]. The fiber is of the depressed cladding type and has a measured [22] refractive-index profile as shown in Fig. 3b.

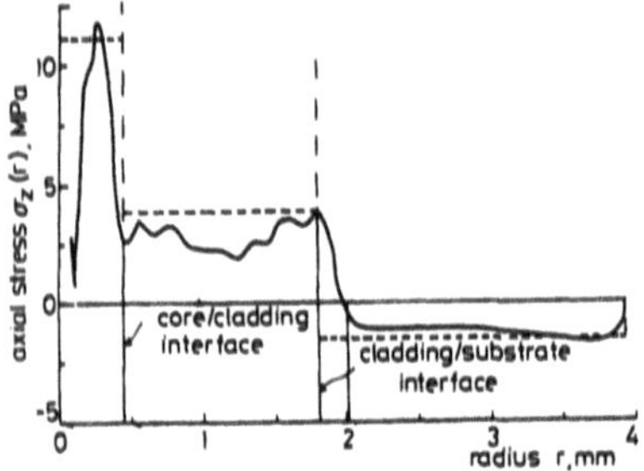

Fig.3a. Measured axial stress profile in a low-birefringence fiber preform

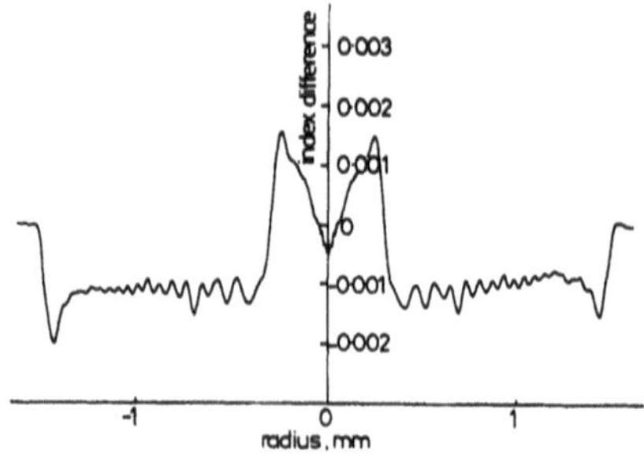

Fig.3b. Measured refractive-index profile in preform of Fig. 3a

Experience with a number of similar fibers gave low but variable birefringence results, as illustrated in the upper section of Fig. 4, and this was attributed to the inability to control the internal stress anisotropy B_S. The residual stress results from differential thermal contraction between core, cladding, and substrate materials on cooling to room temperature, and can be relatively large. Thus, a small asymmetry in the fiber cross-section produces a large stress imbalance and leads to substantial anisotropy in the core material. This conclusion was emphasized by fibers which exhibited considerably lower birefringence than allowed by their waveguide component B_G, calculated from the measured core ellipticity. Normally, B_G and B_S act in unison, provided the expansion coefficient of the core exceeds that of the cladding. If in the three-layer structure used for these fibers the core expansion coefficient is substantially less than that of the B_2O_3-doped cladding, it is possible for B_S to oppose B_G. Evidence that this occurred in some fibers was provided by gently heating the fibers to reduce the internal stress, when it was observed that the birefringence increased. This leads us to speculate that a fiber could be designed with $B_S - B_G = 0$, which would be insensitive to a production spread in ellipticity, since both B_S and B_G are approximately proportional to ellipticity.

3. Spun Fibers

3.1 Theory

When a linearly birefringent fiber is twisted [23], at a rate ξ rads/m, the azimuth of the fiber elliptic cross-section and hence the local principal axes precess. The fiber can be considered as composed of individual local sections having a length of a quarter twist period and linear birefringence which alternates in sign. Thus, although each local section may have a relatively high linear birefringence, its effect is compensated by the next rotated section. The outcome is to produce a net retardance R(z) which oscillates between a small positive and negative value along the fiber length. Thus, twist can be used to reduce the fiber linear birefringence [24] although it simultaneously introduces an elastooptic rotation of the plane of polarisation. The interaction between the local linear and circular birefringence leads to net values of the retardance $R(z)$ and rotation $\Omega(z)$ given by [25]

$$R(z) = 2\sin^{-1}\left[\frac{1}{(1+q^2)^{\frac{1}{2}}}\sin\gamma z\right] , \tag{3}$$

$$\Omega(z) = \xi z + \tan^{-1}\left[-\frac{q}{(1+q^2)^{\frac{1}{2}}}\tan\gamma z\right] , \tag{4}$$

where

$$q = \frac{2(\xi - \alpha)}{\delta\beta} , \tag{5}$$

$$\gamma = \tfrac{1}{2}(\delta\beta^2 + 4(\xi - \alpha)^2)^{\frac{1}{2}} . \tag{6}$$

Here $\delta\beta$ is the linear birefringence of the untwisted fiber and comprises both shape and stress anisotropy.

If instead of twisting the fiber, we spin the preform during drawing [26], no circular birefringence will be present ($\alpha = 0$) because the fiber is in a viscous state at the high fiber-forming temperature and cannot support significant shear stress. If the spin rate $\xi >> \delta\beta$ then the magnitude of the retardance oscillations becomes negligibly small, i.e.

$$R(z) = \frac{\delta\beta}{\xi}\sin\xi z \tag{7}$$

and the rotation $\Omega(z) \to 0$. The internal shape and stress birefringences B_G and B_S are thus reduced by the ratio of the spin pitch $P = 2\pi/\xi$ to the unspun mode beat length $L_p = 2\pi/\xi$. For small ratios the fiber appears virtually isotropic. Thus, provided $P << L_B$, any fiber, even a high-birefringence fiber, can be transformed into a low-birefringence fiber by preform spinning.

3.2 Fabrication of Spun Fibers

A preform made by the MCVD process is attached to the shaft of a gimbal-mounted speed-controlled DC motor which is fitted to the drawing tower in place of the normal preform chuck. A spring-loaded iris diaphragm is used to center the lower end of the preform as it enters the graphite-resistance furnace and also serves as a furnace gas seal. Fiber drawing is commenced in the normal way, and once stable conditions are established, the motor is run up to the desired speed. Rotation rates as high as 2000 RPM have been achieved with an accurately centered, straight preform, giving a spin pitch of 1.5cm for a drawing speed of 0.5 m/s. Shorter spin pitches (~1mm) have been achieved at lower drawing speeds.

It has been found that primary-coating process and diameter feedback control system are unaffected provided the spin rate is sufficiently high. The spin speed is linked to the fiber drawing speed to provide a constant spin pitch despite variations in drawing speed caused by automatic diameter control.

3.3 Results

The marked effect of spinning on the birefringence of a series of fibers is shown in Fig. 4. The preforms from which the fibers were drawn were all of the depressed-cladding type illustrated in Figs. 2 and 3. The upper section of Fig. 4 shows results obtained for conventional low-birefringence fibers and indicates the variability of the results obtained. The lower section demonstrates the large improvement produced by preform spinning. In some cases values for the linear birefringence are given for two adjacent sections of the same fiber, one unspun (solid bars) and the other with a spin period of a few cm (hatched bars). The polarisation properties were measured using crossed polarisers and a Soleil compensator, with the fibre suspended vertically in order to reduce the effects of external stresses. For the spun fibers a reduction in linear birefringence approaching two orders of magnitude has been achieved. The technique has been found to consistently reduce the retardation and circular rotation to levels at or below the measurement limit ($\sim 1^o$/m) and has not resulted in a measurable increase in fiber attenuation. Residual optical rotation is also found to be $<1^o$/m.

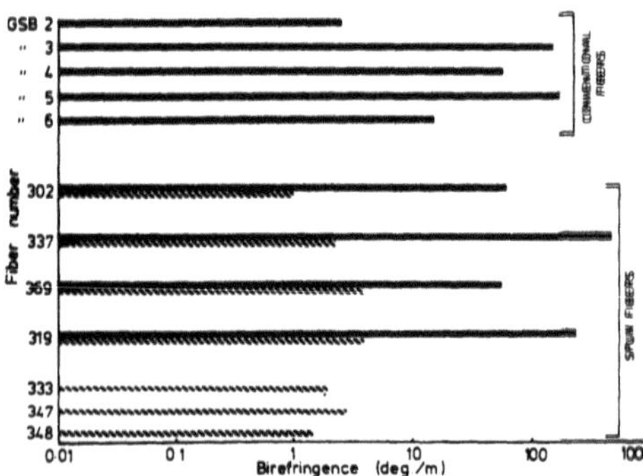

Fig.4. Comparison of birefringence results for unspun (solid bars) and spun fibers (hatched bars)

The ability of the preform spinning technique to transform any fiber into a low-birefringence fiber, provided the spin pitch P is short compared to the beat length L_p, is demonstrated in Fig. 5. A high-birefringence fiber ($L_p(0.633\mu m) \cong 19mm$) with an elliptical cladding (major/minor axis ratio of 2.5) and circular substrate was spun with a spin pitch P = 1.2mm. The spiralling cladding can be clearly seen in Fig. 5. The resulting fiber had $P/L_p = 0.06$ and, as predicted by (7), exhibited negligibly small linear and circular retardance.

It might be thought that the high-temperature spinning process helps to circularize the fiber geometry and so reduces the birefringence. However, the fiber still possesses a similar local birefringence to that in the unspun fiber, as can be demonstrated by twisting the fiber in a direction to unwind the frozen-in twist. The linear birefringence should reappear, together with an optical rotation due to the untwisting. The results of such an experiment are shown in Fig. 6. A spun fiber with P = 5cm which initially exhibited negligible retardance or rotation was untwisted and the measured linear component of birefringence compared with calculation according to (3). At the point when the fiber was completely untwisted (i.e. when the reverse twist was $\xi_r = 2\pi/P(1 - g')$, where g' is the stress-optic polarisation rotation coefficient), a linear birefringence $\delta\beta = 14.4^o$/m

Fig. 5. Transverse view of high-birefringence fiber with elliptical cladding (upper) and after spinning (lower)

reappeared, which is similar to the value $\delta\beta = 32^{\circ}/m$ measured in an unspun section 100m down the fiber. The reappearance of the birefringence with applied twist confirms that a spun fiber retains a large local anisotropy with a value approximately equal to the intrinsic fiber birefringence $\delta\beta$ before spinning. Thus the spinning process would not appear to circularise the geometry or reduce the stress anisotropy significantly.

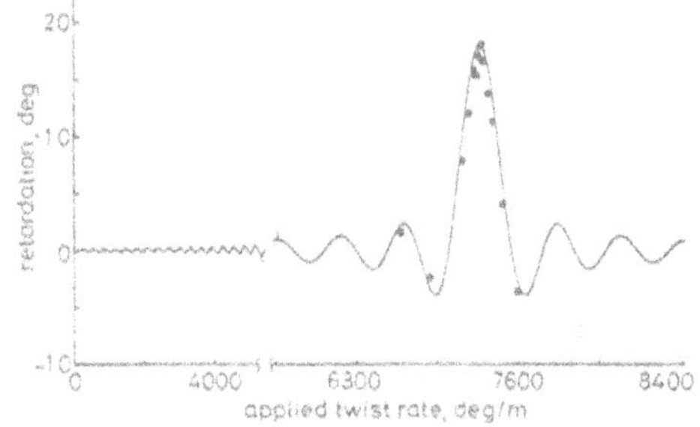

Fig. 6. Retardance in a fiber spun at a nominal P = 5 cm when untwisted after drawing. Solid line is the theoretical prediction according to (3) and is fitted to the data. Points are experimental values. Note the abscissa has a scale change

As stated earlier, the fiber is in a viscous state within the furnace during the drawing process and cannot therefore support significant shear stress. Consequently no residual torsional stresses (i.e. twist) should be present in the fiber after drawing and no photo-elastic polarisation rotation should be observed. However, no confirmatory evidence for this assumption can be expected from the foregoing experiments, since the fiber was hung vertically during the measurements and would therefore untwist to relieve any residual stress. It is of some interest to know whether it is possible to impart torsional stress during the spinning process, since it has been proposed [27] that post-draw twisted fibers be used as circular polarisation-maintaining fibers. Such fibers would transmit circularly polarised light unchanged. Preform spinning would be a convenient means of applying a large twist to the fiber, thus making it highly circularly birefringent.

The mechanism whereby torsional stress could be imparted to the fiber during drawing is through viscous drag in the molten hot zone. As the preform is rotated at high speed, shear forces will be established in the fiber neck-down region and in the length of solidified fiber between the furnace and take-up drum. However, calculations show [28] that at the viscosity of fiber drawing (~ 10^2 poise) these forces are all but completely relieved by viscous flow in the hot zone, leaving negligible shear forces (elastic twist) in the fiber wound on the take-up drum.

To confirm the above, an attempt was made to induce residual elastic twist in a fiber by using high spin rates (P = 3cm) and pulling at as low a temperature as possible (~ 1850^{0}C) to ensure a high viscosity in the hot zone. The residual twist in the fiber was measured by attaching a small weight and pointer to the fiber end and slowly unwinding up to 15m of fiber from the drum while suspended vertically. Residual twist was indicated by a permanent rotation of the pointer as the fiber was unwound. Only at the extremes of high spin rates and low temperature drawing was a significant twist measured. Then it was found that less than 0.2% of the total spin rate was incorporated as elastic twisting of the fiber. It would appear therefore that preform spinning cannot be used as a means of imparting large twist to the fiber.

4. Sensor Applications of Low-Birefringence Fibers

Low-birefringence fibers have already found application in polarimetric sensors such as the magneto-optic current monitor [8] and in fiber devices, for example, the fiber isolator [16]. However, it is clear that the low intrinsic birefringence and, perhaps more important, the lack of sensitivity of the birefringence to temperature changes, could be attractive in other devices where a stable or slowly varying output polarisation state is an advantage.

Since the fiber itself appears virtually isotropic, its birefringence properties are dominated by external effects, such as bends, twists and side pressure. With careful design cf fiber packaging it should therefore be possible to obtain a fiber length whose output polarisation state varies very slowly with time as the ambient temperature changes. Furthermore, a relatively small range of variation can be expected and the fiber would therefore be well suited to active stabilization of the output polarisation state.

A low-birefringence fiber wound in a fixed, stable configuration has principal axes and birefringence fixed by the geometry of the former, and not by an interaction between the axes of intrinsic birefringence and those induced by bending, as in the case for a conventional fiber. For example, a gyro coil of low-birefringence fiber wound on a drum has a fast axis in the plane of the bend. Whereas variations in temperature result in a change in induced birefringence due to changes in the radial pressure exerted on the fiber by the drum, the orientation of the principal axes remains invariant. This has been confirmed experimentally [29] in a prototype gyro coil. Consequently, if linearly polarised light is launched into the fiber aligned with one of the principal axes, no coupling to the orthogonally polarised mode will result, and the output remains stable and linearly polarised. Such a mode of operation may be an alternative to using a 'polarisation-maintaining' fiber.

5. Low-Birefringence Fibers in Telecommunications

Birefringence in a fiber leads to a group-delay difference between the two orthogonally polarised modes which may significantly reduce the telecommunication bandwidth. One solution is to select only one mode and prevent energy transfer to the other by using a high-birefringence polarisation-maintaining fiber. However, in the event of some power transfer occurring, the pulse dispersion will be large. The alternative solution is to use a low-birefringence fiber, preferably a spun fiber, in which the group-velocity difference is negligible.

The time-delay difference $\Delta\tau_o = z/c \cdot d(\Delta\beta)/dk$ between the linearly polarised normal modes in a length z of a conventional fiber is reduced in a spun fiber by a factor which depends on the spin rate (c is the velocity of light, k the free space wave number). In the limit of large spin, the normal modes of the fiber become circularly polarised and the delay difference τ_s is [25]

$$\tau_s = \frac{\delta\beta}{2\xi} \Delta\tau_o \quad . \tag{8}$$

Thus, spinning the preform to produce a low-birefringence fiber has the considerable additional advantage that the polarisation mode dispersion is reduced in inverse proportion to the spin rate and can be made negligibly small. Spin rates of up to 2000 r/min have been achieved [26], which at a pulling speed of 1 m/s gives a spin pitch of P = 3cm. Taking the criterion that we require L_p > 50m for ultrahigh bandwidth transmission systems, we see from (8) that this can be achieved provided the fiber has an unspun L_p >0.9m. The latter figure can be met without particular care in manufacturing.

Experimental confirmation of the considerable reduction in polarisation mode dispersion obtainable by fiber spinning is given in Fig. 7. Here the variation of birefringence with wavelength, measured using two polarisers

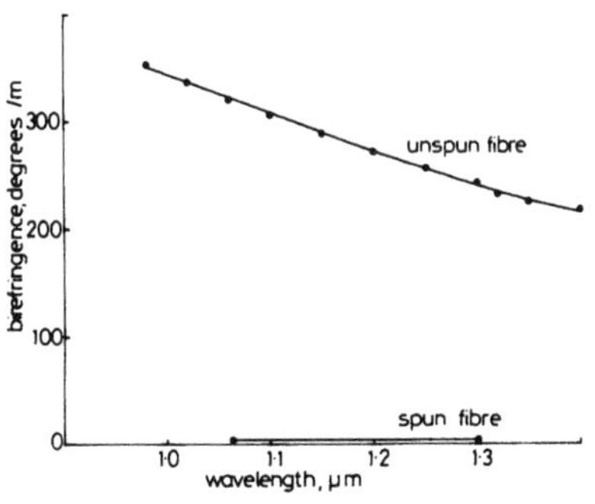

Fig. 7. Measured variation of birefringence with wavelength in spun and unspun sections of a fiber with λ_{cutoff} = 0.96 m and equivalent-step Δ = 0.5 per cent

and a Soleil compensator, is plotted for both a spun and unspun section of the same fiber.

The polarisation mode dispersion $\Delta\tau = z/c.d(\Delta\beta)/dk$ is estimated by fitting a curve to the data and taking the derivative with respect to wavelength. In the case of the spun fiber, both the retardation and rotation variations were at the limits of detection and consequently only two points are shown (Fig. 1, lower curve). The mode dispersion in this case was determined from the derivative of the variation of rotation with wavelength, since the normal modes are now circularly polarised.

For the fiber shown in Fig. 1, the intrinsic polarisation mode dispersion at a wavelength of 1.2μm was calculated to be 4.6 ps/km. When spun, this was reduced to less than 0.02 ps/km, thus illustrating the large reduction possible with the technique.

6. Conclusions

Spinning single-mode fibers provides a solution to the problems of polarisation birefringence and dispersion by simultaneously reducing both phase- and group-delay differences. Analysis has shown that, provided the spin rate exceeds the intrinsic birefringence beat length by a factor of 10 or more, residual polarisation anisotropy is reduced to a negligible value. Spinning restores the average circular symmetry of the fiber, and it thus behaves as a near-perfect isotropic medium.

Spun fibers are expected to find application in optical fiber sensors, fiber devices and in long-distance telecommunications.

7. Acknowledgements

We wish to thank E. J. Tarbox and R. D. Birch for providing preforms from which the fibers were made, and R. Calligaro, who contributed to the fabrication of low birefringence preforms and preform stress measurements.

8. References

1. R. Ulrich and M. Johnson, Opt. Lett. 4, 152 (1979)
2. G. Schiffner, W. R. Leeb, H. Krammer and J. Wittman, Appl. Opt. 18, 2096 (1979)
3. S. K. Sheem and T. G. Giallorenzi, Appl. Phys. Lett. 35, 914 (1979)
4. R. Ulrich, Opt. Lett. 5, 173 (1980)
5. T. A. Steinberg and T. G. Giallorenzi, Appl. Opt. 15, 2440 (1976)
6. Y. Yamamoto and T. Kimura, IEEE J. Quantum Electron. QE-17, 919 (1981); and F. Fabre, L. Jeunhomme, I. Joindot, M. Monerie and J. C. Simon, IEEE J. Quantum Electron. QE-17, 897 (1981)
7. V. Ramaswamy, W. G. French and R. D. Standley, Appl. Opt. 17, 3014 (1978)
8. A. M. Smith, Appl. Opt. 17, 52, (1978)
9. A. Papp and H. Harms, J. Magnetism Magnetic Materials 2, 287 (1976); and Appl. Opt. 3729 (1980)

10. S. C. Rashleigh, Opt. Lett. 6, 19 (1981); and Opt. Lett. 5, 392 (1980)
11. H. Harms, A. Papp and K. Kempter, Appl. Opt. 15, 799 (1976)
12. R.H. Stolen and E. H. Turner, Appl. Opt. 19, 842 (1980)
13. Y. Yen and R. Ulrich, Opt. Lett. 6, 278 (1981)
14. H. C. Lefevre, Electron. Lett. 16, 778 (1980)
15. M. Johnson, Opt. Lett. 5, 142 (1980)
16. G. W. Day, D. N. Payne, A. J. Barlow and J. J. Ramskov-Hansen, Opt. Lett. 7, 238 (1982)
17. S. R. Norman, D. N. Payne, M. J. Adams and A. M. Smith, Technical Digest, Second International Conference IOOC, 10.1 (1979)
18. S. C. Rashleigh and R. Ulrich, Opt. Lett. 3, 60 (1978)
19. R. B. Dyott, J. R. Cozens and D. G. Morris, Electron. Lett. 15, 380 (1979)
20. S. R. Norman, D. N. Payne, M. J. Adams and A. M. Smith, Electron. Lett. 15, 309 (1979)
21. R. B. Calligaro, D. N. Payne, R. S. Anderssen and B. A. Ellem, Electron. Lett. 18, 474 (1982)
22. York Technology P101 Preform Profiler
23. R. Ulrich and A. Simon, Appl. Opt. 18, 2241 (1979)
24. S. C. Rashleigh and R. Ulrich, Appl. Phys. Lett. 34, 768 (1979)
25. A. J. Barlow, J. J. Ramskov-Hansen and D. N. Payne, Appl. Opt. 20, 2962 (1981)
26. A. J. Barlow, D. N. Payne, M. R. Hadley and R. J. Mansfield, Electron. Lett. 17, 725 (1981)
27. L. Jeunhomme and M. Monerie, Electron. Lett. 16, 921 (1980)
28. R. J. Mansfield - Private Communication
29. M. Varnham - Private Communication

Attempt to Draw a Circular Polarization Conserving Fiber

F. Gauthier, J. Dubos, S. Blaison, Ph. Graindorge, and H.J. Arditty

Thomson CSF, Laboratoire Central de Recherche, BP 10, Domaine de Corbeville
F-91401 Orsay, France

Single mode fiber interferometers such as rotation sensors or current sensors rely on the fiber to ensure the reciprocity of the device : as the fiber is a "single mode" fiber, the propagation conditions are identical for two waves travelling in opposite directions. Unfortunately, this is not true for real fibers, since the two degenerate modes of a perfect fiber become two different modes in a real fiber. This means that two waves propagating in opposite directions in such a fiber can follow two different optical paths, and destroy the reciprocity of the device. This is why various methods have been investigated to cancel the coupling between the two modes present in the fiber. Those methods can be divided into two classes : for the first one, linear birefringence is introduced in the fiber [1], whereas for the second class of methods, circular birefringence is introduced [2]. We will present an attempt to induce circular birefringence while drawing a single mode fiber.

I - EFFECT OF BIREFRINGENCE ON THE RECIPROCITY OF A FIBER

When a single mode fiber is twisted, circular birefringence appears, which results from the elastooptic effect due to the strain on the core of the fiber [3], and also because of the combination of a "shape" effect with linear birefringence. The twist induced circular birefringence cancels a random linear birefringence that may be present in the fiber, as well as external linear birefringence (due to bending for example).

Such a birefringent fiber exhibits reduced coupling between the two circularly polarized modes of the fiber, so that a wave launched with a right-hand circular polarization will exchange no energy with the left-hand circular polarization mode ; if the same circularly polarized wave is launched at the other end of the fiber, it will also stay in the same state of polarization. Thus the fiber can be considered as a single mode fiber as far as circular polarizations are concerned, and it becomes a reciprocal device under those conditions. This method is the one used in the electrical current [4], because the Faraday effect is usually measured with circularly polarized light. Another way of using such a fiber is to launch linear polarization. As the fiber is circularly birefringent, the plane of the linear polarization will rotate while propagating through the fiber. This means that at the output of the fiber, the polarization of the light will still be linear, thus allowing mode filtering with polarizers. This can be used in the gyroscope because of the easy filtering of linear polarization. In this case, the fiber is reciprocal for this pair of linear polarizations.

II - EFFECT OF TWIST ON A SINGLE MODE FIBER

When a single mode fiber is twisted, two effects may appear, whose respective amplitudes depend on the conditions of twisting. The first, and often the main one is the elastooptic birefringence. The second is the result of the combined geometrical twist and intrinsic linear birefringence of the fiber, which we shall call the "shape" effect. The stress induced circular birefringence can be calculated to be equal to $g.\tau$ where g is related to the elastooptic coefficients of the fiber ($g = .16$) and τ is the twist rate. This twist and the associated circular birefringence almost completely "hide" a small amount of intrinsic linear birefringence of the fiber : in Fig. 1, the calculated evolution of the state of polarization, in a twisted fiber with intrinsic linear birefringence is represented on the Poincare sphere. Another phenomenon inducing circular birefringence is a "plastic" or shape twist in the fiber. This is a twist of the geometry of the fiber, but with no associated stress. When there is linear birefringence in the fiber, the result for a propagating beam is similar to the effect of circular birefringence. If the linear birefringence is small with respect to the twist rate, the resulting birefringence is approximately equal to

$$-\alpha^2/4\tau \qquad (1)$$

(where α is the linear birefringence and τ the twist rate) [5].

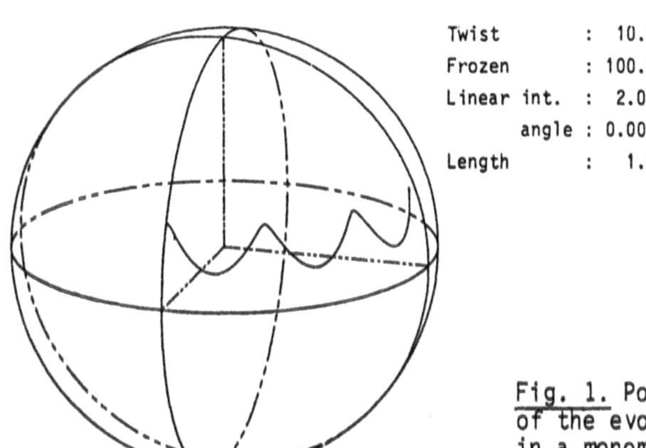

Fig. 1. Poincaré sphere representation of the evolution of the polarization in a monomode fiber

As there is usually some linear birefringence in classically drawn fibers, this effect could be of some importance. It is also important to notice that the resulting circular birefringence is of opposite direction to that due to an elastic twist.

We will now see what the characteristics of such a fiber can be.

Fig. 2 represents the evolution of the polarization of a beam propagating in a fiber with 6.28 rd/m linear birefringence and 31.4 rd/m non-elastic twist. The beam is launched with a linear polarization parallel to one of the axes of the linear birefringence. The result is an effective circular birefringence of -0.3 rd/m (this birefringence is negative with respect to the twist). If part of the twist (6.25 % in this case) is frozen into an elastic twist (see chapter III) the circular birefringence can become zero (Fig. 3) and if still more is frozen, there now is a positive circular birefringence (Fig. 4). One must note that there also is a small periodic ellipticity variation due to the linear birefringence. However, the amplitude of this variation is independent of the direction of the input polarization.

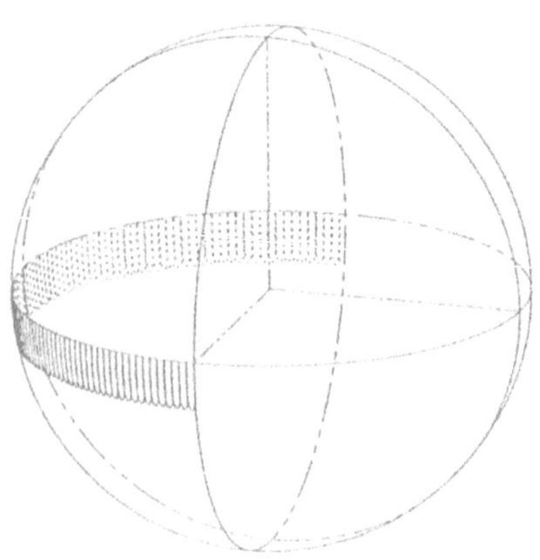

Twist : 31.4 rd/m
Frozen : 0.0 %
Linear int. : 6.28 rd/m
angle : 0.000 rd
Length : 10.0 m

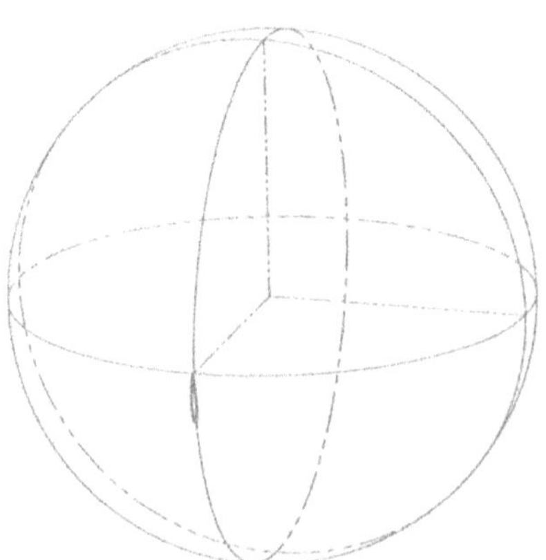

Twist : 31,4 rd/m
Frozen : 6.2 %
Linear int. : 6.28 rd/m
angle : 0.000 rd
Length : 10.0 m

Figs. 2 and 3. Evolution of the polarization with a non-frozen and partially frozen twist

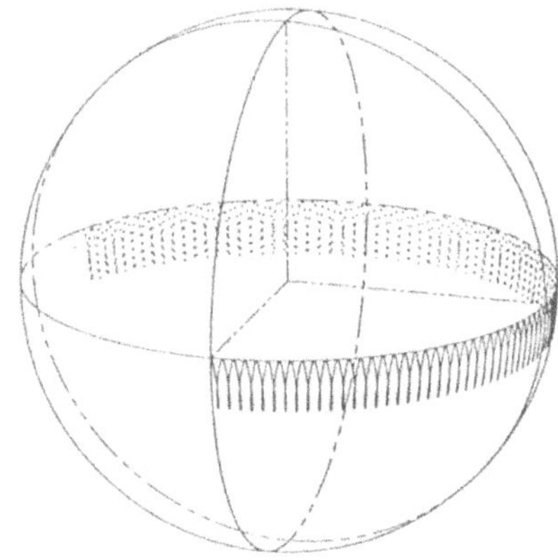

Twist : 31.4 rd/m
Frozen : 15.0 %
Linear int. : 6.28 rd/m
angle : 0.000 rd
Length : 10.0 m

Fig. 4. Evolution of the polarization with 15% frozen twist

III - METHOD OF DRAWING TWISTED SINGLE MODE FIBERS AND RESULTS

A usual way to achieve circular birefringence on a fiber is to twist it after it is drawn, and to clamp it inside a tubing to hold the elastic twist. However, this fragilizes the fiber, and may not be very reliable on the long run. We attempted to freeze the twist in line while drawing the fiber, so that it would not fragilize the fiber.

The device we used is represented in Fig. 5. The fiber itself is rotated by the machine in Fig. 6 while being drawn from a preform. Once the fiber is cooled, an epoxy-acrylate coating is deposited, both to protect the fiber, and to hold the elastic twist.

When a fiber is twisted while it is drawn, the fiber's properties are the result of the following effects. First of all, the twist is transmitted to the molten cone of glass by the elasticity of the fiber. The viscosity

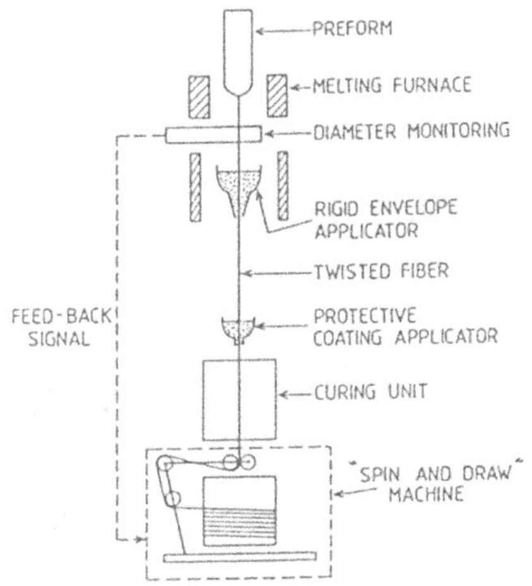

Fig. 5. "Spin and draw machine"

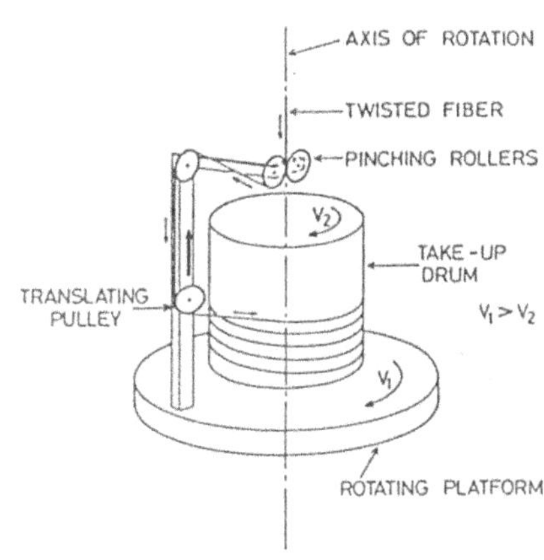

Fig. 6. Rotating machine

of the molten region causes some drag which impresses a twist to the already drawn fiber. This twist is then frozen by the application of a rigid envelope. The envelope we used was a layer of epoxy-acrylate resin applied on to the fiber. We also intend to use a more rigid material, such as glass. The second effect is the freezing of the elastic twist by the fiber itself : as the core of the fiber is composed of a less fusible material (pure silica) than the jacket (germanium or fluorine doped silica), the core freezes before the jacket. This causes the core to bear an elastic twist before the jacket freezes. The jacket itself can then "hold" the twist of the core. Finally the rotation of the molten region builds into the fiber a helix shaped core, which induces no stress on to the fiber but still causes some effective circular birefringence if associated to linear birefringence. One of the fibers we drew had a beat length of 2 meters due to circular birefringence. However, we have not determined which effect caused this birefringence. The attenuation of the fiber was between 15 and 20 dB/km. This fairly high value is the result of the special conditions of the drawing, but there was no fragilization due to twist, as the mechanical strength was measured to be identical to that of normally drawn fibers. More experiments are underway to achieve lower beat lengths, by optimizing the various parameters of the system.

Although the results are as yet not as good as expected, this method should be an efficient way to draw circular birefringent fibers, which have many applications in the field of optical sensors.

References

[1] I.P. Kaminow, J.R. Simpson, H.M. Presby, Elect. Lett., Vol. 15 (79), p. 679

[2] P. Graindorge, K. Thyagarajan, H.J. Arditty, M. Papuchon, Opt. Comm., Vol. 41 (82), p. 164

[3] R. Ulrich, A. Simon, Appl. Opt., Vol. 18 (79), p. 2241

[4] H.J. Arditty, Y. Bourbin, M. Papuchon, C. Puech, IOOC 81, Tech. Digest, paper WL3, San Francisco 81

[5] A.J. Barlow, J.J. Ramskov-Hansen, D.N. Payne, Appl. Opt., Vol. 20 (81), p. 2962

3.3.3 Characterization

The Characterization of Polarization-Holding in Birefringent Single-Mode Fibers

S.C. Rashleigh, W.K. Burns, and M.J. Marrone
Naval Research Laboratory, Washington, DC 20375, USA

R. Ulrich
Technische Universität Hamburg-Harburg
D-2100 Hamburg 90, Fed. Rep. of Germany

In single-mode fibers, irregular mechanical stresses from bending, twisting, clamping, non-uniform coating, cabling, etc. introduce perturbing birefringences. These birefringences vary sensitively and erratically with environmental conditions and thus, cause unpredictable changes in the state of polarization (SOP) along the fiber and at the fiber output. However, in several applications of single-mode fibers like interferometric sensors and coherent communications, polarization instabilities lead to noise and signal fading and must be suppressed to a very high degree if highly sensitive devices are to be realized. One approach to overcoming these instabilities has led to the development of various types of fibers specifically designed to hold, or propagate, a particular, well-defined, SOP stably along the entire fiber length, even in the presence of irregular and varying stresses [1]. These fibers exhibit a very high internal birefringence, far exceeding the mentioned perturbing birefringences.

These so-called "single-polarization" fibers are typically characterized by the strength of their birefringence $\beta_i = k_x - k_y$, or equivalently, their beat length $L_p = 2\pi/\beta_i$ which can easily be measured. Here, k_x and k_y are the propagation constants of the slow and fast polarization eigenmodes respectively. However, a more important criterion for characterizing these fibers is a measure of how well they can hold polarization that is, a measure of the average rate at which the perturbing birefringences couple power out of the desired SOP into the orthogonal one. A reliable evaluation of the polarization-holding ability of highly-birefringent fibers has proven experimentally difficult and only few, and widely differing, data has appeared in the literature. This may be due, in part, to a conceptual difficulty caused by the fact that, with a highly monochromatic source (HeNe laser), it is always possible in any given single-mode fiber to identify two linearly-polarized input states for which the output SOP's are also perfectly linear. These pairs of "linear in/out" polarizations have accidental orientations which, in general, bear no relationship to the fiber axes. Moreover, the orientations of these linear in/out states may change rapidly when the environmental conditions of the fiber vary, making it difficult to accurately locate the fiber axes. The idea of polarization-holding requires a stable SOP along the entire fiber under all conditions. A further difficulty is the random nature of the birefringence perturbations. A certain cross-polarization transfer-rate, measured at one individual fiber section, need not be representative of other sections of the same fiber.

In developing "single-polarization" fibers, the usual approach has been to increase β_1 (reduce L_p) as much as possible. The rationale behind this has been that small cross-polarization coupling will be expected if the fiber beat length is much smaller than the beat length of the perturbing birefringences. To date, fiber beat lengths as short as ≈ 1 mm have been fabricated but little effort has been devoted to determining the strength and spatial distribution of the perturbing birefringences [1].

In this paper, we show that the use of a broad-bandwidth light source allows a direct and statistically significant determination of the polarization-holding quality of high-birefringence single-mode fibers [2]. In particular, the azimuths providing best polarization holding can be simply and accurately located. Using this approach, we present a direct experimental confirmation of the theory [1] of random polarization coupling. Further, we then use this technique to characterize the polarization-holding ability of several stress-induced and elliptical-core birefringent fibers fabricated by ITT [3].

The strength and distribution of the perturbing birefringences are usually not known. Nevertheless, an estimate of the coupling caused by perturbations can be made if the perturbations are considered to be randomly distributed along the fiber. A general description of this random coupling has recently been given by Kaminow [1]. By considering a group of birefringent fibers with statistically equivalent perturbations, and the injection of monochromatic linearly polarized light into one polarization eigenmode (x), the analytical solution for the ensemble averages of the relative powers in the two polarization modes as a function of fiber length L, is given by

$$\langle P_y \rangle / \langle P \rangle = \frac{1}{2} [1 - \exp(-2hL)] \qquad (1)$$

where

$$h = (k_o^2/4) \langle |\Gamma(\beta_1)|^2 \rangle . \qquad (2)$$

Here, $\langle\ \rangle$ denotes the ensemble average, P_y the power in the cross-polarization mode, $P = P_x + P_y$ the total power, k_o the vacuum propagation constant, and $|\Gamma(\kappa)|^2$ is the power spectrum of the birefringence perturbations at spatial frequency κ.

According to Eq. (1), the parameter h describes the rate of power transfer to the cross-polarization state. The inverse, 1/h, is the characteristic distance for this polarization transfer and, therefore, is the appropriate measure for the polarization-holding quality of the perturbed fibers. Eq. (2) shows that the transfer is effected only by those components of the perturbations whose spatial frequency is $\kappa = \beta_1$, that is, their periodicity equals the fiber beat length L_p.

Eq. (1) has been formulated for monochromatic light and for a large number of statistically equivalent fibers. According to this formulation, an experimental evaluation of the polarization-holding parameter h would be difficult as it would require measurements on a large number of equivalent fibers. To overcome this problem, we have extended the theory of random polarization coupling to incorporate a finite range of wavenumbers $\Delta\nu$. For optical injection of linearly-polarized light parallel to one polarization eigenmode (x), the average power coupled to the cross-polarized eigenmode (y) as a function of fiber length L is given by

$$\xi = \{p_y\}/\{p\} = \frac{1}{2}\,[1 - \exp(-2h(\nu_o)L)] \tag{3}$$

where $h(\nu_o) = (k_o^2/4)\{|\Gamma(\beta_1)|^2\}$. Here, $\{\ \}$ denotes the spectral average over $\Delta\nu$ and p_y and $p = p_x + p_y$ are the spectral power densities. The importance of this formulation is that the parameter $h = h(\nu_o)$ appearing here and evaluated at the mean wavenumber ν_o of the interval $\Delta\nu$ is the same h as in Eq. (1). This then allows h to be determined from a single measurement of the broad-band power $\{p_y\}\Delta\nu$ in the cross-polarization state of one fiber.

The equivalence of the ensemble average and the spectral average can be explained in the time domain. Light with a uniform spectral power density over a bandwidth $\Delta\nu$ can be represented as a random sequence of quasi-periodic wavetrains, each of which has a duration equal to the coherence time $T_c = 1/c\Delta\nu$. In a highly birefringent fiber, a perturbation whose axes are not aligned with those of the fiber will couple some power out of the excited polarization (x) into the orthogonal polarization (y). Then, as this light propagates along the fiber, the wavetrain in the orthogonal polarization becomes increasingly dephased with respect to the original wavetrain because the two polarization eigenmodes have different group velocities. Eventually, after a length $\Delta\ell$, the group delay time $\tau\Delta\ell$ increases sufficiently so that the two waves do not overlap, that is $\tau\Delta\ell > T_c$, and the light is depolarized. Here, $\tau = (1/c)(d\beta_1/dk_o)$ is the group delay time difference per unit fiber length caused by the birefringence β_1 of the fiber. For fiber lengths greater than $\Delta\ell$, the wavetrain coupled to the cross-polarization state travels without interfering with the wavetrain in the originally excited polarization. At the fiber output, the cross-polarized power from each of the sections $\Delta\ell$ long can be simply detected. The fiber section $\Delta\ell$ long represents just one member of an ensemble of fibers and is equivalent to one member of the ensemble discussed by Kaminow [1]. When the contributions from all the sections $\Delta\ell$ along the fiber are added at the fiber output, it is equivalent to forming an ensemble average. The number of members in the ensemble is $N \simeq L/\Delta\ell \simeq \tau c L \Delta\nu$. For a statistically significant measurement on a single fiber, we require $N>>1$ and then, the spectral average can be characterized by its relative standard deviation which is of the order of $1/\sqrt{N}$. This fiber length $\Delta\ell (\simeq (\lambda/\Delta\lambda)L_p$ for fibers in which β_1 is due to a large internal stress) is the fiber length beyond which the polarization coupling is uncorrelated.

The simple experimental arrangement is shown in Fig. 1.

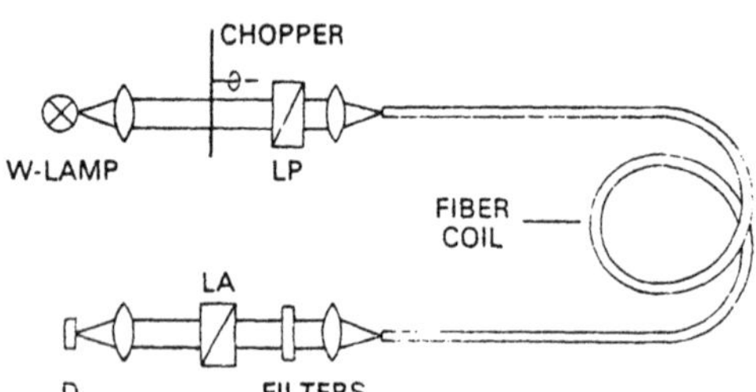

Fig. 1. Experimental setup for the characterization of highly-birefringent single-mode fibers

White light from a tungsten lamp is chopped and focused through a linear polarizer into the highly-birefringent single-mode fiber. The bandwidth $\Delta\nu$ of the light in the fiber is made as broad as possible. At low wavenumbers, it is limited by the rapidly increasing fiber bending losses and, at higher wavenumbers, by the filter at the output to be just short of the onset of the higher-order modes. Typically, this $\Delta\nu$ corresponds to the wavelength interval 700-900 nm. At the fiber output, the light is focused through a linear analyzer onto a Si photodiode and detected synchronously. The fiber axes are located by rotating the polarizer and analyzer to obtain minimum detected power. At this condition, the polarizer is oriented parallel to one of the fiber axes and the analyzer is crossed to this axis. The transmitted power is the light coupled out of the excited polarization eigenmode. Rotation of the analyzer through 90° aligns it with the excited mode. The rationale behind this determination of the axes is that, only when the input is polarized parallel to one of the polarization eigenmodes, will the output light have its highest degree of polarization. If the input polarizer excites both polarization eigenmodes, the output light will have a reduced degree of polarization.

This technique has been applied to characterize several fibers. The two major types of highly-birefringent fibers have been investigated, namely elliptical-core fibers and fibers with a large internal-stress asymmetry [1,3]. To illustrate the structure of these fibers, typical cross-sections are shown in Fig. 2. After cleaving the fiber, the fiber end is lightly etched. Regions of different refractive index etch at different rates, making the different regions clearly visible. Fig. 2(a) shows the cross-section of a standard, high NA, ITT fiber [3]. The core, cladding and support tube regions are clearly visible and the core is approximately 3 μm in diameter. An elliptical-core cross-section is shown in Fig. 2(b). For these fibers, the birefringence is porportional to the square of the index difference Δ between the core and the cladding ($\beta_1 \propto \Delta^2$) and so, to make β_1 large and still have the fiber single-moded, the core dimensions must be small. In this fiber, the core dimensions are approximately 1 μm x 1.5 μm and its beat length $L_p \simeq 2.5$ mm at $\lambda = 0.63$ μm. An example of a stress-induced fiber is shown in Fig. 2(c). In these fibers the large stress asymmetry is produced by the different thermal contractions of the various regions of the fiber. Here, two cladding regions of material with high thermal contractions are used to increase the stress anisotropy in the core. As lower index differences Δ are used, larger core dimensions are possible, in this case, being approximately 3 μm in diameter. This fiber had a beat length $L_p \simeq 6$ mm at $\lambda = 0.63$ μm.

Fig. 2. Typical cross-sections of (a) high NA nominally low-birefringence, (b) elliptical-core and, (c) stress-induced ITT fibers. A small division corresponds to 2.84 μm

The length dependence of the polarization-holding performance of two ITT elliptical-core fibers is shown in Fig. 3. The fibers were loosely coiled on 30 cm diameter drums and, for a 50 m length, the fiber axes were located. As the fiber was successively cut back, the relative power in the cross-polarized mode was measured. The experimental results are shown in Fig. 3 together with the best-fit theoretical curves according to Eq. (3). The relatively smooth string of experimental points indicates that the perturbations were fairly uniformly distributed along the fibers. As expected, the fiber with the shorter beat length holds polarization better but the very rapid fall-off of the power spectrum of the birefringence

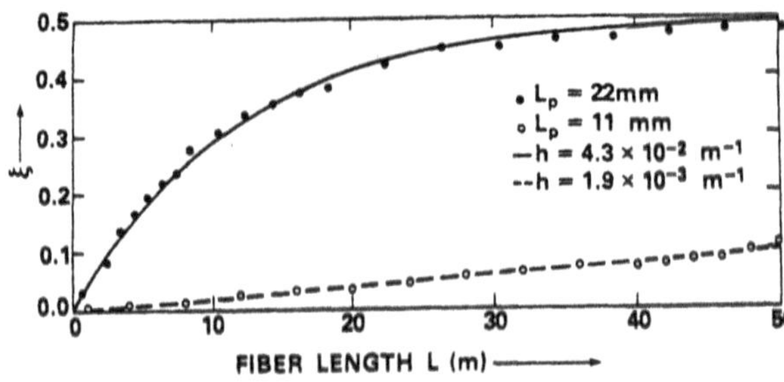

Fig. 3. Length dependence of the polarization-holding ability of two elliptical-core ITT fibers

perturbations (as indicated by h) is a little surprising. Presumably the two fibers had comparable perturbations. We consider that the good agreement between the experimental points and the best-fit theoretical curves confirms the theory of random mode coupling between the two polarizations of highly-birefringent single-mode fibers.

Finally, we have measured the polarization-holding parameter h for several highly-birefringent fibers and plotted it as a function of fiber beat length L_p (at λ = 0.83 μm) in Fig. 4. In each case, a 50 m length of fiber was loosely coiled on a 30 cm diameter drum. For the elliptical-core fibers, a very rapid decrease in h accompanies a decrease in beat length ($h \sim L_p^3$), indicating a rapid decrease in the power spectrum of the perturbing birefringences for the spatial frequencies investigated. The two different fiber jacket materials, namely the hard plastic (Hytrel [3]) jacket and the softer acrylate (Desoto [3]) jacket, appear to have little effect on the polarization holding ability of the fibers. Also shown in Fig. 4 is the h parameter for several stress-induced fibers. For the same beat length, the stress-induced fibers hold polarization much better than the elliptical-core fibers (≃ an order of magnitude) but no convincing relationship between h and L_p was evident for the stress-induced fibers. For both types of ITT fiber, the lowest value of h measured was $h \simeq 10^{-4}$ m^{-1}, indicating that, on average, 90% of the power remained in the originally-excited polarization mode after 1 km. Also shown in Fig. 4 are the polarization-holding parameters for two elliptical-core fibers fabricated by Andrew Corporation [4]. These fibers had shorter beat lengths $L_p \simeq 1$ mm and smaller h values, $h \simeq 6 \times 10^{-5}$. All the elliptical-core fibers and many of the stress-induced fibers investigated were very lossy with attenuations in the 50-250 dB/km range. It is quite possible that the mechanism which causes the very large loss is also responsible for the polarization coupling. Perhaps as the technology progresses and the fiber losses drop to acceptable levels, the polarization holding ability of these fibers will improve. It is clear that, at present, birefringence perturbations in currently available highly-birefringent fibers are still too large to allow the often-quoted performance goal of "30 dB polarization holding over a kilometer" to be realized. This would require $h \leq 10^{-6}$ m^{-1}. The rapid fall-off in the power spectrum of the perturbations in the elliptical-core fibers with increasing spatial frequency, indicates that some improvement may be gained through a further reduction in fiber beat length.

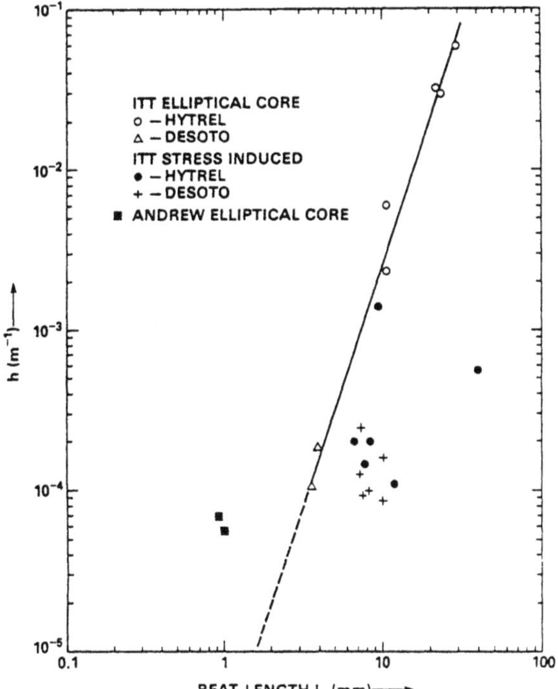

Fig. 4. Polarization-holding parameter h as a function of beat length L_p for several elliptical-core and stress-induced high-birefringence fibers

In conclusion, we have demonstrated that the implementation of a broad-band source to achieve spectral averaging is very useful in characterizing the polarization-holding ability of highly-birefringent single-mode fibers. This technique allows the location of the fiber axes to be determined simply and independent of environmental conditions. We have used this technique to confirm experimentally the theory of random mode coupling between the two polarization modes and with it, can obtain a statistically significant determination of the polarization holding parameter h from a measurement on a single fiber. We have used the technique to evaluate the polarization holding quality of several highly-birefringent fibers and believe that, through this, a measure of the power spectrum of the perturbing birefringences can be obtained.

The authors thank F. Akers [3] and R. B. Dyott [4] for supplying the fibers.

References

1. I. P. Kaminow, IEEE Journal of Quantum Electronics QE-17, 15 (1981)
2. S. C. Rashleigh, W. K. Burns, R. P. Moeller and R. Ulrich, Optics Letters 7, January 1982
3. ITT, Electro-Optical Products Division, Roanoke, VA 24019 Fabricated under contract to the Office of Naval Research.
4. Andrew Corporation, Orland Park, IL 60462

Polarization Properties of Monomode Optical Fibres: The Use of P.O.T.D.R. to Determine Spatial Distributions

A.J. Rogers*

Central Electricity Research Laboratories
Leatherhead, Surrey, Great Britain

1. INTRODUCTION

As a result of the important role which polarization effects play in various measurement sensors which utilize monomode fibres [1, 2] it is essential to acquire a detailed understanding of the polarization behaviour of these fibres. Such understanding is especially important in the case of the optical-fibre gyroscope.

Polarization-Optical Time Domain Reflectometry (POTDR) [3] is a valuable analytical tool in this respect since it provides the means whereby the spatial distribution of some of the fibre's polarization characteristics may be determined along its length. It also provides the means whereby one may determine the spatial distribution of any external field which may be capable of modifying the fibre's polarization characteristics (e.g. electric field, magnetic field, pressure, temperature, etc).

The basic arrangement for POTDR is shown in fig. 1. Narrow light pulses from a convenient laser are launched into a monomode fibre via a beamsplitter whose polarization properties must be accurately known. As each pulse propagates down the fibre it suffers Rayleigh scattering and some of the back-scattered signal is returned to the launch point, where it is polarization-analysed and time resolved. It follows that, at time 2t after launch, the polarization state of the back-scattered light re-emerging from the fibre will be determined, for a given launch state-of-polarization (SOP), by the polarization effect of the go-and-return passage of the light through a length ct of the fibre (c being the mean group velocity of the light in the fibre medium).

The task in hand is thus to determine what information may be extracted from this returned signal time-function with regard to the spatial distribution of the fibre's polarization characteristics.

*Presently Royal Society Research Fellow at King's College, London

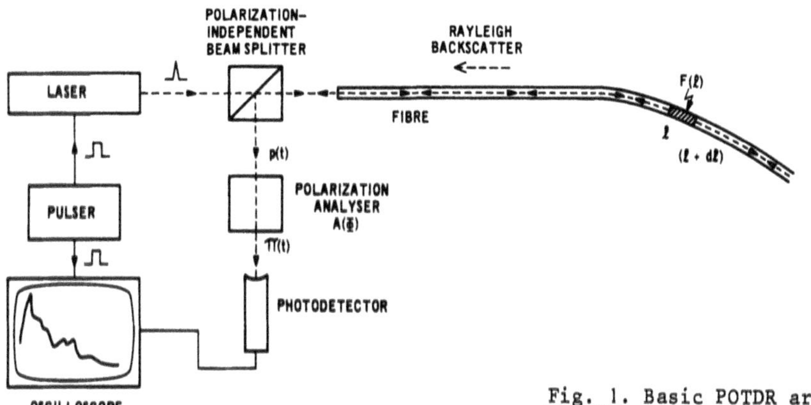

Fig. 1. Basic POTDR arrangement

2. POINCARÉ SPHERE

The properties of the Poincaré sphere are fairly well known [4] so that they will be only briefly summarized here. The Poincaré sphere is a Cartesian representation, in three dimensions, of the Stokes parameters (S_1, S_2, S_3: See fig. 2). The Stokes parameters define completely the SOP of any polarized light beam. To each elliptical polarization state there corresponds a point on the sphere; all linearly polarized states lie on the equator and the two circularly polarized states lie, respectively, at the North and South Poles.

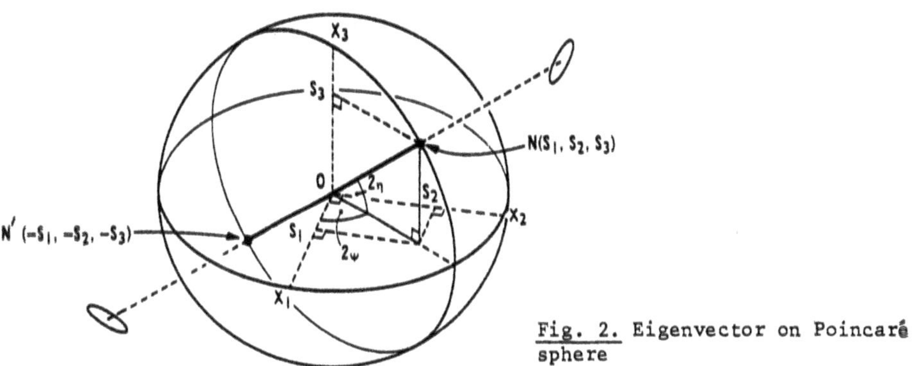

Fig. 2. Eigenvector on Poincaré sphere

The prime importance of the Poincaré sphere in the analysis of polarization phenomena is that it reduces most problems in the analysis to considerations in three-dimensional geometry, thus providing both an elegant visual representation and also a convenient method of solution.

The action of a lossless polarization element is represented in terms of its polarization eigenmodes - the SOPs which are unchanged in form by the action of the element. The two eigenmodes (N, N') are joined on the sphere by a diameter, and the action of the element is to rotate the sphere about this diameter through an angle equal to the phase delay introduced between the eigenmodes by the element. Thus any given input SOP will be converted into a unique output SOP.

3. POTDR ACTION

Consider the situation depicted in fig. 3. Light with a given input SOP is passed up to point ℓ in the fibre and back; its final SOP is measured. Time resolution of the returning signal allows any value of ℓ to be chosen up to the total fibre length. If we are able to determine, for any length ℓ, the eigenmodes together with their relative phase delay, we shall be able to determine, by differentiation, the complete distribution of the fibre's polarization characteristics.

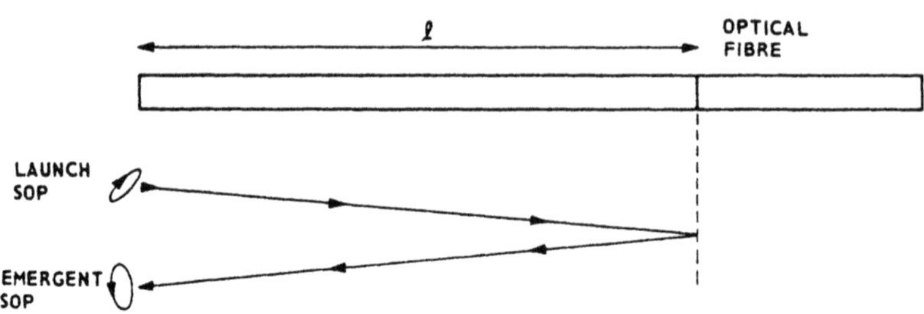

FIG. 3: POTDR POLARIZATION ELEMENT

Now for any given lossless polarization element we may use the Poincaré sphere to see how we may, in general, determine its eigenmodes and their relative phase delay. If we measure one pair of input and output SOPs, then the rotation which connects them must take place about a diameter lying in the plane which bisects the two points on the sphere [3]. It is clear that by measuring another such pair of points (i.e. by changing the input SOP) we shall similarly define another such plane, and the line along which these two planes intersect must be the diameter connecting the eigenmodes. Having determined the eigenmode diameter it is a simple geometrical matter to determine the rotation necessary to connect either pair of SOPs.

It is clear that the method just described relies on access to both ends of the element. For POTDR action, however, we have access only to one end of the fibre and we must now examine the implications of this.

It is evident, from the Poincaré sphere, that rotation about any eigenmode diameter may be resolved into one about an equatorial diameter and one about a polar diameter. This is the geometrical representation of the well known fact that any lossless polarization element is equivalent to a retarder-rotator combination, where the axes of the retarder, the retardation of the retarder and the rotation of the rotator are all uniquely defined (within the range $0-\pi$).

Let us, firstly, assume that the length ℓ of fibre has reciprocal polarization properties, i.e. that the characterisation of these properties is independent of the propagation direction of the light. If this is so then the length may be represented as a retarder-rotator combination as shown in fig. 4 a . (This combination will, in all polarization respects, be equivalent to the length ℓ of fibre, even though the properties of the fibre will not, in general, be homogeneous). We see immediately that, on go-and-return passage through this combination, the effect of the rotator must be cancelled.

Thus, of the three pieces of information we require for complete knowledge of the element:

(i) Orientation of retarder's axes (ψ)
(ii) Retardation of retarder (δ)
(iii) Rotation of rotator (ρ).

The last (ρ) can never be available. Hence we see that our element will always behave, on double passage, as a pure retarder whose axes and retardation we may indeed measure directly (fig. 4 b). Its eigenmodes are, of course, two linearly polarized orthogonal states.

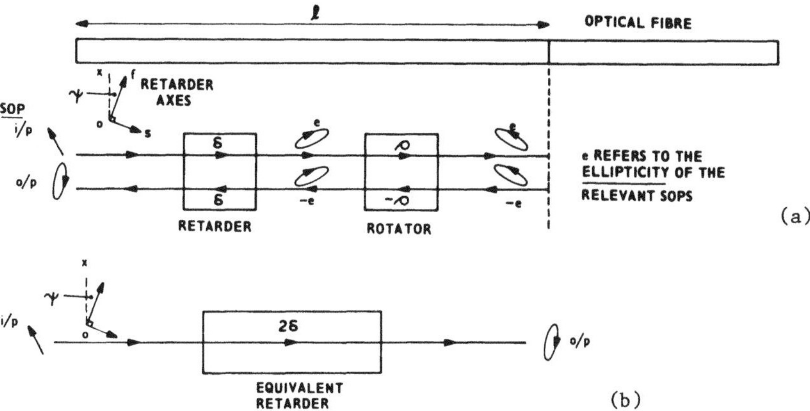

Fig. 4. Discrete equivalents for POTDR polarization element

Another interesting point emerges. Having specified the retarder we know what will be the SOP resulting from its action for any given input SOP. The action of the rotator obviously does not alter the ellipticity of this SOP but only its orientation (fig. 4 a). Consequently our ignorance of the value of the rotation can be expressed alternatively by stating that, with use of POTDR, we may determine the ellipticity of the SOP at any point within the reciprocal fibre, but we may never determine its orientation. *

We may use the results of the above discussion to define more specifically how our knowledge of the fibre's polarization properties is limited in practice.

4. DETERMINATION OF POLARIZATION DISTRIBUTIONS

The spatial resolution of any POTDR determination of the spatial distribution of the fibre's polarization properties will be limited by the laser pulse width or the response time of the photodetection circuit, whichever is the larger. It is important that the polarization properties do not change significantly within the resolution length, for this will lead to an effective depolarization of the received signal (there being, in this case, a spread of SOPs within the detection period). The signal may thus be analyzed effec-

* In coming to this conclusion I have benefited from discussions with Dr D.N. Payne and Dr E. Brinkmeyer.

tively only at times t and t + dt, where dt corresponds to the resolution length. It follows that the ellipticities of the SOPs may be determined at the corresponding points ℓ and $\ell + d\ell$.

The following statements can be made concerning the fibre's polarization properties:

(i) If the element $d\ell$ is a pure rotator this will be known, for the two ellipticities will always be the same regardless of the initial SOP at the launch end: the magnitude of the rotation, however, cannot be determined.

(ii) If the ellipticities at ℓ and $\ell + d\ell$ are not always the same the element is not a pure rotator. If we know it to be a pure retarder we may determine the magnitude of the retardation but not the orientation of the axes. This may be deduced from the Poincaré sphere, for we know that rotation must occur about an equatorial diameter. If we ensure, for example, that the ellipticity at ℓ is zero (i.e. circular polarization) then the required retardation is the angular distance between that pole and the latitude of the observed ellipticity at $\ell + d\ell$. There is nothing in the situation, however, which allows the direction of the diameter to be determined within the equatorial plane.

(iii) If the orientation of the retarder's axes is known then a knowledge either of the retardation or of the rotation allows the other of these to be determined, for, in a given azimuthal plane, there is only one diameter which has the required ellipticity and arc length to connect the two observed ellipticities ℓ_ℓ and $\ell_\ell + d\ell$.

5. APPLICATION TO OPTICAL-FIBRE GYROSCOPE

For effective operation of the optical-fibre gyroscope it is required that, in the absence of the Sagnac effect, the phase delay experienced by a given SOP (usually linear) should be the same in each direction around the fibre loop. Any non-reciprocal coupling between the given SOP and any other SOP will violate this condition.

The most common cause of such coupling is non-reciprocal rotation caused by, for example, local magnetic fields.

If non-reciprocal rotation is present in a length ℓ of monomode fibre, POTDR will immediately reveal it for, in this case, a different retarder-rotator combination is required for each direction of propagation, and thus the go-and-return passage will no longer be equivalent to a single retarder, but to a retarder-rotator combination itself. The fact that the eigenmode diameter no longer lies in the equatorial plane is a direct statement of such non-reciprocity (fig. 5).

The double passage may more conveniently be represented as a retarder-rotator-retarder triplet for, in this case, the value of the rotation caused by the rotator should be a direct measure of the rotational non-reciprocity, provided only that this is small compared with the linear retardation. Since POTDR allows the rotation to be measured as a function of position in the fibre, it follows that it will allow the spatial distribution of the non-reciprocity to be measured.

Non-reciprocal linear retardation, apart from any such caused by the Sagnac effect, is rare. If it exists in a monomode fibre, it may, however, also

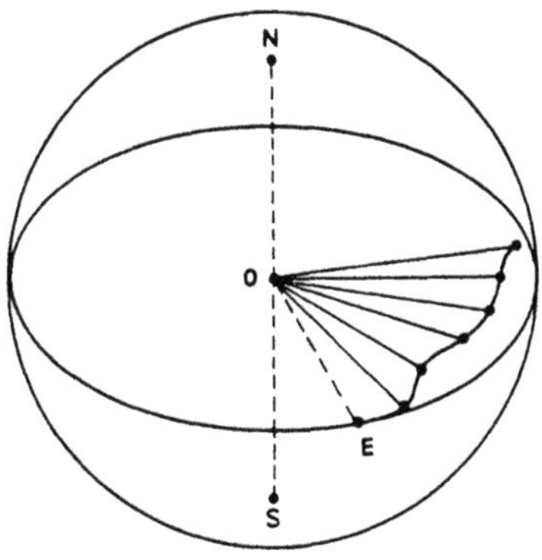

Fig. 5. Mapping of non-reciprocity on Poincaré sphere

be revealed by POTDR, for it will again result in the rotator of the equivalent retarder-rotator pair possessing non-zero rotation. The exception to this is the special case where the fibre length is, in fact, comprised of a pure retarder with axes of fixed orientation; in this case the eigenmodes will remain on the equator even though the phase delay between them will not be the same for the two directions. This singularity may, however, be removed by a slight twisting of the fibre. Such twisting will result in each propagation direction again being equivalent to a retarder-rotator pair, so that any non-reciprocity will show up as a non-zero rotation, as before.

This section may be summarized thus:

(i) If POTDR indicates that rotation is non-zero for the equivalent retarder-rotator pair appropriate to any given length of fibre, then non-reciprocity is present in that length.

(ii) A given length of fibre should only be considered reciprocal if the rotation remains zero even after slight twisting of the fibre.

(iii) If non-reciprocity is present and the rotation is small compared with the retardation, then the magnitude of the rotation is a direct measure of the non-reciprocity.

(iv) The POTDR technique will thus allow the spatial distribution of the fibre's non-reciprocity to be determined, though other knowledge will be required for a complete statement of the spatial distribution of the non-reciprocal polarization properties.

6. CONCLUSIONS

POTDR is a valuable technique for determining the spatial distribution of the polarization properties of monomode optical fibres, even though it cannot determine the exact distribution in the absence of other knowledge.

POTDR will always allow detection of non-reciprocity in monomode fibres and will provide sound indication of its spatial distribution. This latter capability is of especial importance to the effective operation of the optical fibre gyroscope for it will allow identification of regions of fibre which exhibit non-reciprocal properties, and will thus aid identification of the mechanisms responsible for non-reciprocal effects in optical fibres.

7. ACKNOWLEDGEMENTS

This paper is published by permission of the Central Electricity Generating Board. The author is grateful to the Royal Society and the Science and Engineering Research Council for the award of an Industrial Fellowship, presently held in the Electrical Engineering Department of King's College London, and during the tenure of which this paper was written.

8. REFERENCES

1. Rogers A.J., 'Optical Measurement of Current and Voltage on Power Systems', IEE J. Electric Power Applications 1979, 2 (4) pp 120-124

2. Smith A.M., 'Polarization and Magneto-Optic Properties of Single Mode Optical Fibre', Appl. Opt. 1978, 17, pp 52-56

3. Rogers A.J., 'Polarization-Optical Time Domain Reflectometry', Elec. Lett. June 1980, 16 (13) pp 489-490

4. Jerrard H.G., 'Transmission of Light through Birefringent and Optically Active Media', J.O.S.A. 1954, 44 (8) pp 634-640

Induced Circular Birefringence and Ellipticity Measurement in a Faraday Effect Fiber Ring Interferometer

P. Ferdinand and J.-L. Lesne

Direction des Etudes et Recherches, Electricité de France
F-78400 Chatou, France

1. Abstract

In order to build a ring interferometer able to detect the Faraday effect, a new measurement method of the induced circular birefringence using a Stabilized Transverse Zeeman Laser is described.

Conservation of the Circular State Of Polarization in the fiber is also controlled by the ellipticity measurement. A good agreement with theory has been achieved.

This experimental set-up gives a simple and new method of obtaining a fiber which preserves the State of Circular Polarization in our optical current sensor.

2. Introduction

Our purpose is to build an optical fiber ring interferometer to detect the Faraday effect induced by a high current line encircled by the fiber [1], Fig.1.

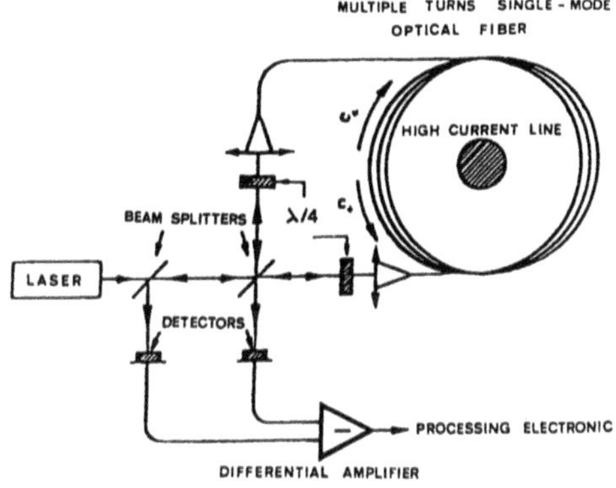

Fig.1 Principle diagram of the optical ring interferometer current sensor

It is of prime interest in this interferometric fiber sensor to maintain the State Of Polarization (S.O.P.) of a light beam during its propagation.

With our optical scheme the current measurement requires the detection of a small difference of phase between two circularly polarized beams propagating clockwise and counter clockwise in the fiber coil. This method will be effective if the circular polarization is preserved during the propagation in the fiber. We report a new simple way to measure the circular birefringence induced by twisting the fiber and to control the ellipticity of the beam. To perform this measurement we use a Stabilized Transverse Zeeman Laser (S.T.Z.L.) [2].

The induced circular birefringence is then the difference of phase between the light beating of the S.T.Z.L. detected at the input and output of the twisted fiber.

3. Measurement of the induced circular birefringence

The experimental set-up (Fig.2) used for measuring the induced cirular birefringence by twisting a fiber includes :

A home made S.T.Z. Laser followed by a quarter wave plate, the two emerging light waves being cross-circular polarized.

A glass plate reflecting a small fraction of the incident light to a reference photodiode.

A single mode optical fiber which it is possible to twist by means of a rotatable graduated support.

A second photodiode which detects the light emerging from the twisted fiber.

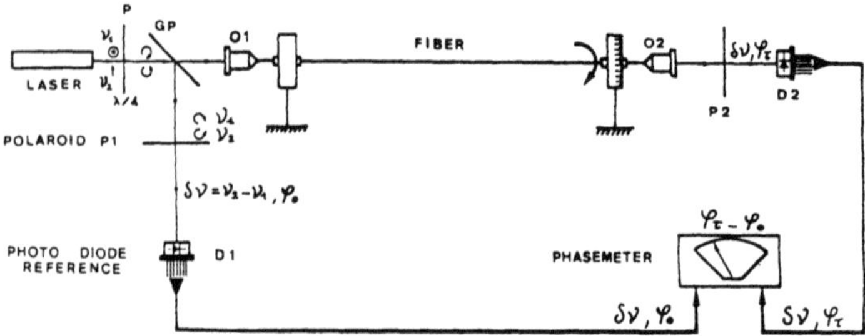

Fig.2 Measurement of circular birefringence induced by twist

The measurement of the phase difference between the beat signals ($\Delta f \simeq 130$ kHz) is performed by a phasemeter. This measurement is proportional to the twist rate applied to the fiber.

Let Δn be the difference of refractive indices for two crossed circular polarized waves. Neglecting the fiber dispersion we can write the difference of phase

$$\Delta\phi = Ln\,\Delta\omega/C + L\omega\,\Delta n/C = \Delta\phi_1 + \Delta\phi_2 \ . \qquad (1)$$

$\Delta\phi_1$ accounts for the delay due to the propagation into a fiber of length L for a signal at frequency equal to the intramode beat $\Delta\omega/2\pi$, n is the refractive index of the fiber : $\Delta\phi_1 \simeq 0.23°/m$. C is the light speed.

$\Delta\phi_2$ is the phase shift due to the induced circular birefringence.

As far as the fiber length is less than a few meters we can neglect $\Delta\phi_1$ in comparison with $\Delta\phi_2$. Then the measured phase shift $\Delta\phi$ is directly proportional to the circular birefringence induced by twisting the fiber,

$$\Delta\phi \simeq (L\omega/C)\Delta n = (2\pi L/\lambda)\Delta n \sim \Delta n \quad . \tag{2}$$

With some turns the eigenstates of polarization are more and more circular.

On the POINCARE sphere (Fig.3) the $\vec{\Gamma}$ vector (equal to the composition of the linear $\vec{g}_0$ and the induced circular birefringence $\vec{k}$) becomes about parallel to the polar axis.

The evolution of $\Delta\phi$ is linear with the twist τ [3] [5].

The slope provides the proportionality coefficient g between the strain-induced circular birefringence and the physical twist τ,

$$\Delta\phi = g\,\tau \sim \text{twist} \quad . \tag{3}$$

Using (1),

$$g = (\omega\,\Delta n) / (C.\tau), \qquad \tau \text{ [turns by meters]} \quad . \tag{4}$$

Figure 4 shows the evolution of $\Delta\phi$ with the twist rate τ in the single mode fiber.

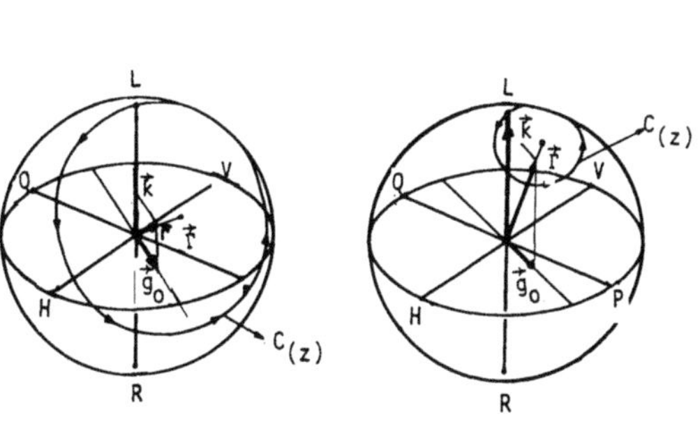

Fig. 3. Evolution of the S.O.P. during the propagation. Rotation around $\vec{\Gamma}$. Initial circular polarization. Fiber referential

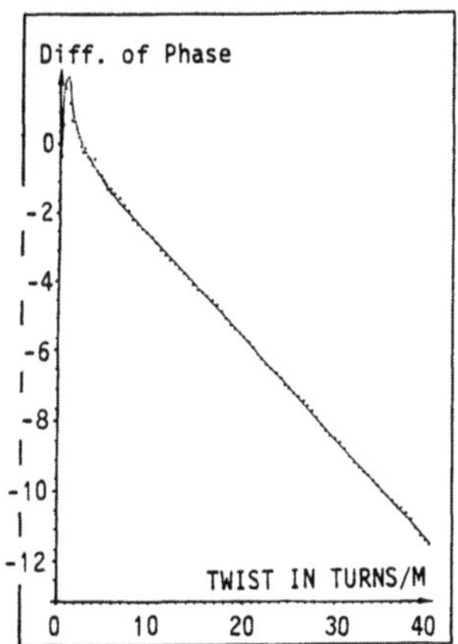

Fig. 4. Difference of phase between inverse circular waves

Our experimental curves (g = 0.148, 0.149) for two different fibers in silica are in good agreement with the results given in [3] (g = 0.13, 0.16).

4. Ellipticity measurement

For an optical fiber interferometer sensor it is necessary to maintain a particular S.O.P. during the propagation. The more effective the preservation of the circular S.O.P., the more effective the Faraday effect in the fiber coil. The experimental set-up used for the ellipticity measurement is shown in Fig.5.

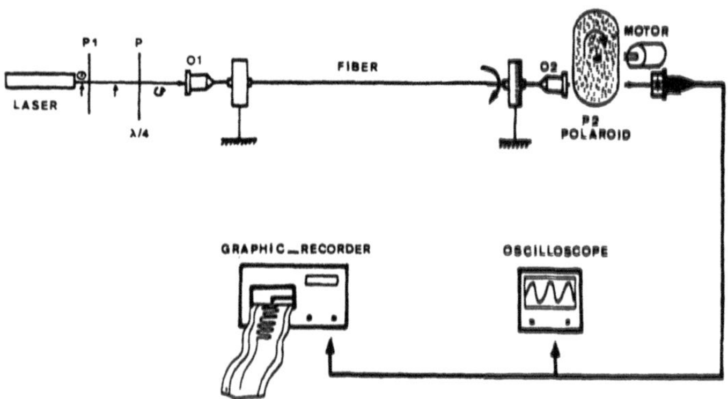

Fig.5 Measurement of the ellipticity

One of the two polarizations of the laser is maintained and converted into circular polarization. The beam is focused into the fiber by a microscope objective O_1. The supports and the second objective O_2 are the same as in the previous experiment. The light passes through an analyser P_2 rotated by a motor. The intensity of the beam detected by a photodiode is recorded on a graphic paper. Quantitative analysis allows the measurement of the ellipticity of the S.O.P. of the output beam.

Using JONES's matrix representation [4], the polarization evolution of the twisted fiber is specified by

$$\begin{pmatrix} E_x(Z) \\ E_y(Z) \end{pmatrix} = [M] \begin{pmatrix} E_x(o) \\ E_y(o) \end{pmatrix} = \begin{bmatrix} A & B \\ -B & A^* \end{bmatrix} \begin{pmatrix} 1 \\ i \end{pmatrix} \tag{5}$$

in the fiber system, where

$$A = \cos(\Gamma Z) + i(g_o/\Gamma)\sin(\Gamma Z), \tag{6}$$

$$B = ((k - \omega_b)/\Gamma)\sin(\Gamma Z), \tag{7}$$

$$\Gamma = (g_o^2 + (\omega_b - k)^2)^{1/2}, \tag{8}$$

ω_b = the circular intrinsic birefringence which can be neglected,
g_o = the linear intrinsic birefringence of the fiber,
k = the circular induced birefringence,
Z = the fiber length.

For e, the ellipticity of the output light beam, analytic calculation gives :

$$e = \frac{(g_0^{\,2} + k^2) - 2\, g_0\, |\sin(\Gamma Z)|\, [k^2 + g_0^{\,2}\cos^2(\Gamma Z)]^{1/2}}{k^2 + g_0^{\,2}\cos(2\Gamma Z)} \quad . \tag{9}$$

The experimental ellipticity is defined as the ratio of the two electric components, then as the square root of the ratio of the extreme values of the intensity,

$$e_{exp} = (I_{min}/I_{max})^{1/2} \quad . \tag{10}$$

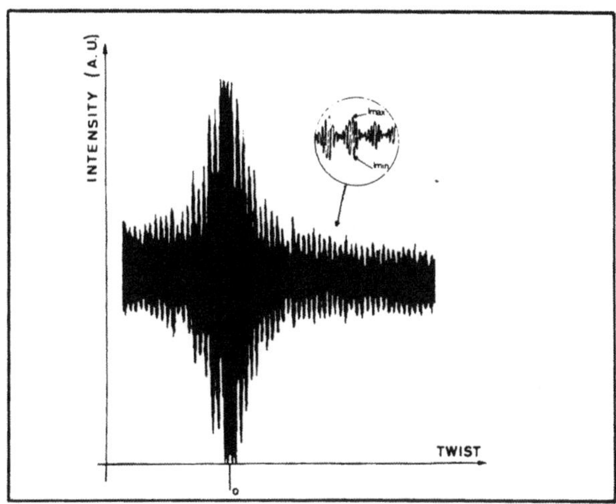

Fig. 6. Intensity as a function of twist

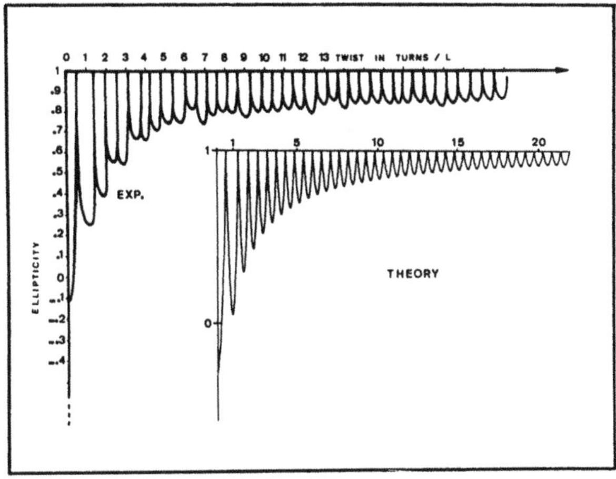

Fig. 7. Experimental and theoretical ellipticity

Figure 6 gives the output light beam intensity during the twist and modulated by the rotating analyser.

Figure 7 shows a good agreement between experiment and theory : with a circular input polarization the ellipticity during the propagation oscillates between circular, 1, and a lower value depending of the rate of linear intrinsic birefringence.

For the considered fiber a twist of $\sim$ 20 turns per meter is needed to keep the ellipticity close to .85 (L = 102 cm).

The maximum value, 1, is not obtained due to the very high linear birefringence of the fiber.

From Fig.8 and with (11) the maximum ellipticity for different values of linear birefringence can be determined in function of twist. These curves are derived from the theoretical ellipticity (9),

$$e_{min} = (k - g_0) / (k + g_0) \quad . \qquad (11)$$

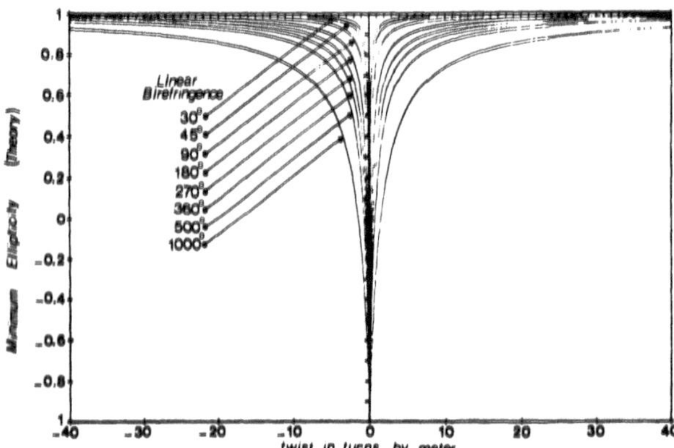

Fig. 8. Abacus: Twist necessary to obtain a given minimum ellipticity

Figure 8 allows the definition of the theoretical twist required to obtain a given minimum ellipticity and thus to preserve the quasi circular S.O.P.
The experimental set-up of the ellipticity measurement allows one to verify the good conservation of the S.O.P. and to optimize the twist rate.

5. Conclusion

The experimental results are in good agreement with theory.

The use of a S.T.Z.L. provides an easy method of measuring the induced circular birefringence in a single-mode fiber.

The circular birefringence induced by twist is directly proportional to the twist rate.

The twist rate per meter necessary to obtain an ellipticity close to 1 is only a function of the intrinsic linear birefringence.

This experimentation gives a simple method of obtaining a fiber which preserves the State Of Circular Polarization in a fiber ring interferometer our optical current sensor.

Acknowledgments. The authors are grateful to C. Cahen and B. Dessus for helpful discussions. They would like to thank H. Arditty and P. Graindorge for their help and C. Raffaeli for typewriting.

6. References

1. B. Dessus, J-L. Lesne, "Conférence Européenne d'Optique 1980", Horizons de l'Optique, Pont-à-Mousson, p.V-15 (22-25 Avril 1980)

2. V. Umeda, M. Tsukiji, H. Takasaki, Appl.Opt. 19, 442 (1980)

3. R. Ulrich, A. Simon, Appl.Opt. 18, 2241 (1979)

4. C. Jones, J.Opt.Soc.Am. 38, 671 (1948)

5. M. Monnerie, P. Lamouler, Electron.Lett. 17, 252 (1981)

3.4 Light Sources

Single Longitudinal Mode Modified CSP Injection Laser for Single-Mode Optical Fiber

A.D.Ceruzzi, T.E. Stockton, and J.B. McNeely

Laser Diode Laboratories, A M/ACOM Company
New Brunswick, NJ 08901, USA

1. Abstract

Stable single transverse and longitudinal mode cw operation has been demonstrated using a newly developed gaussian channel substrate (GCSP) laser structure. These devices operate in excess of 14mW and are predominately single longitudinal mode over their entire operating range. Highly reliable performance is also expected due to low threshold current (16mA) and very high differential quantum efficiency (52% for single face). Using a tapered polished fiber technique, coupling efficiencies as high as 35% using 3 μm core single mode fiber were attained for cw lasers in the 840nm range. These devices appear ideally suited for use as optical sources for fiber optic interferometric applications.

2. Introduction

This paper describes injection lasers now available as miniature high quality optical sources for fiber optic rotation sensors. The recently-developed gaussian channel substrate planar (GCSP) laser is ideally suited as the optical source in single mode fiber interferometric sensor systems and in fiber optic communication applications. This GCSP structure provides transverse mode stability, linear light output-current characteristics, and low spontaneous noise by incorporating a built-in effective index guide that varies across the lateral junction plane. This device operates in a fundamental transverse and single longitudinal mode up to 14mW of output power per facet.

Fabrication of this device requires only single step liquid phase epitaxy (LPE) and is described in Section 3. A first order approximation of the waveguide is described in Section 4 as well as the mechanism of operation. The excellent spatial, spectral and performance characteristics will be discussed in Section 5.

The method used to increase the effective numerical aperture of single mode fiber is a tapered polished fiber (TPF) process and is described in Section 6. Launch efficiencies as high as 35% have been obtained using a taper ratio of 1.76.

Finally, Section 7 demonstrates the coupling of GCSP lasers to single mode fiber as a function of taper ratio.

3. Liquid Phase Epitaxial Growth and Fabrication

Gaussian channel substrate planar (GCSP) [1] devices are fabricated by single-step liquid-phase epitaxial growth over channeled substrates. This allows the formation of a built-in lateral waveguide in the junction plane and, as a consequence, provides lasers which operate single mode [2,3].

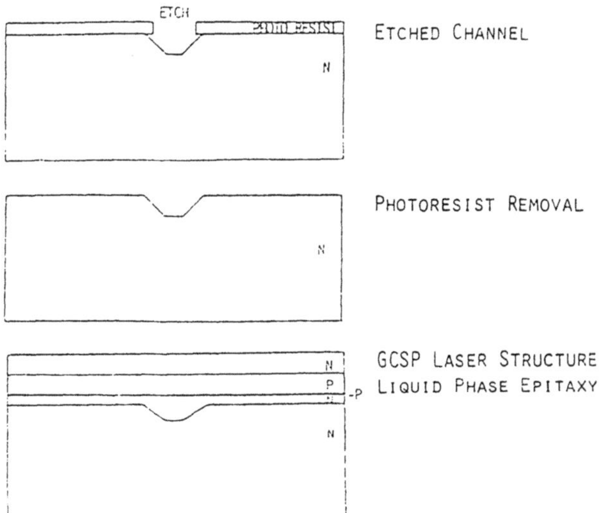

Fig. 1 Illustration of Processing Steps for the GCSP Laser Structure

Shown in Figure 1 are the processing steps which lead to the GCSP laser structure. The GaAs substrate is Si doped ($3\times10^{18}/cm^3$) and has a controlled amount of misorientation, α, in the parallel direction, which plays an important role in determining both the channel shape and the lasing cavity size. The top surface of the wafer is misoriented with respect to the (100) planes in the direction parallel to the channel and is shown in Figure 2. It is important to control α to within $0.5° \pm 0.1°$ in order to obtain the proper channel cross-section during LPE growth.

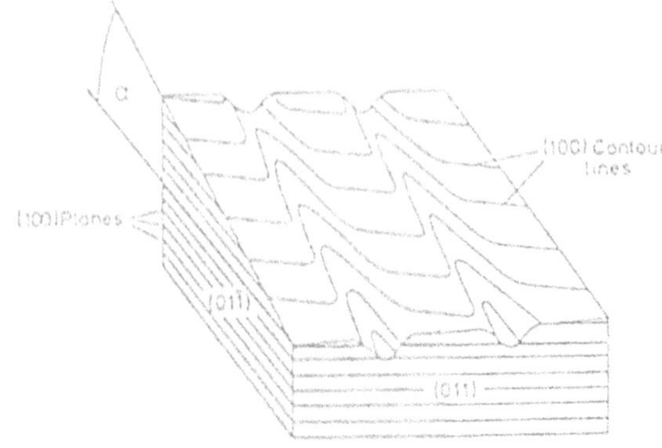

Fig. 2. Top surface of a channeled substrate wafer illustrating the (100) contour lines formed by the intersection of successive equispaced (100) planes and the top surface of the wafer. The top surface of the wafer is misoriented with respect to the (100) plane by an angle   in the direction parallel to the channels. Ref.: G.H.B. Thompson, Appl. Phys. Lett. (1976)

The channel etching is performed by using a photoresist mask and a 3:1:1 (H_3PO_4:H_2O_2:H_2O) at 25°C. Five micron stripes are formed using standard photoresist techniques. The substrates are dipped into the etchant for ∿30sec. to form 1 μm deep channels and 5.5 μm wide top widths. The resist is then removed with hot acetone and plasma etched. The channel substrates are cleaned and loaded into the LPE boat. Figure 3 is an SEM photomicrograph depicting the cross section of the trapezoidal shaped channel.

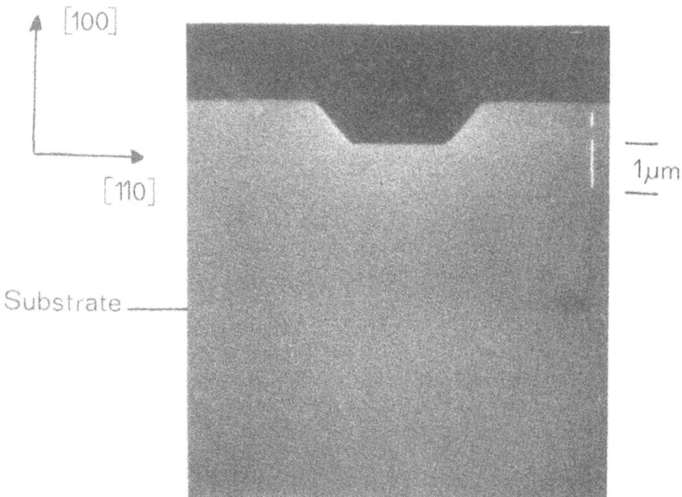

Fig. 3. Cross section of a trapezoidal channel before growth

Liquid-phase epitaxy is performed by the thin solution technique [4] and allows reproducibility and uniformity in the deposition of the thin layers (0.15μm). Using a modified low temperature LPE method a gaussian shaped channel is formed via controlled meltback prior to initiating growth of the waveguide. Four layers, n-$Al_xGa_{1-x}As$ passive, p-$Al_xGa_{1-x}As$ active, p-$Al_xGa_{1-x}As$ passive and n-GaAs cap layer are successively grown over the guassian shaped channel to form the double heterostructure waveguide. The resultant structure is shown in Figure 4. In order to avoid heavy melt-etch of the channel's shoulders, a controlled amount of super cooling (10°) is provided before the introduction of the substrate into the first solution.

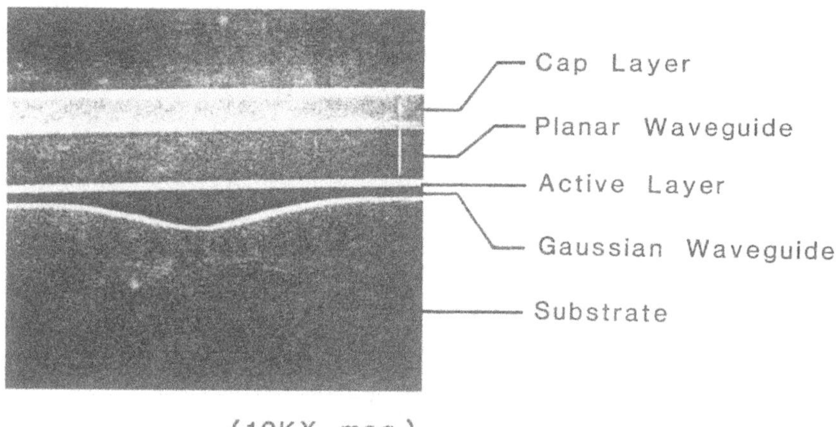

Fig 4. SEM photomicrograph of GCSP laser structure

Figure 5 is a schematic cross-section of a selective zinc diffusion through a 5.0 μm mask which penetrates one-third into the p-passive layer. The process utilized is a 700°C vapor phase semi-sealed ampoule diffusion with an encapsulation layer to minimize arsenic dissociation from the epi wafer. The zinc diffusion is made 1.0 μm wider than the channel in order to provide a uniform spatial gain distribution over the waveguide region.

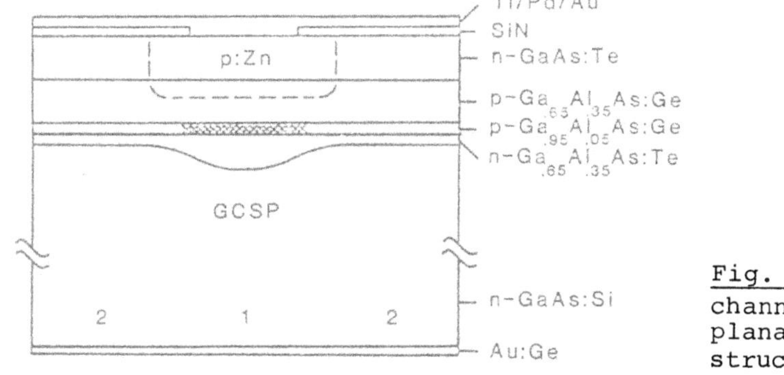

Fig. 5. Gaussian channel substrate planar laser structure

Ohmic contacts are provided by thermal evaporation of Ti/Pd/Au on the cap layer and Ni/Au/Ge/Au on the substrate surface. The wafer is cleaved into 200 µm wide bars and Al_2O_3 passivation coatings are applied by electron beam evaporation. The bars are then scribed into discrete laser chips, probed and mounted junction down on Pt/Au/In evaporated copper heat sinks using an automated eutectic die bonder.

4. Mechanism of Operation and Waveguide

The GCSP laser obtains its mode selection as a result of variable radiation losses across the lateral junction plane. This is accomplished by the channel geometry and refractive index profile as shown in Figure 6. As a result, the higher order transverse modes are absorbed by the leaky guide layer and fundamental mode operation occurs. In addition, the low loss at the channel center is responsible for low threshold current as well as extremely low spontaneous emission intensity observed in this laser.

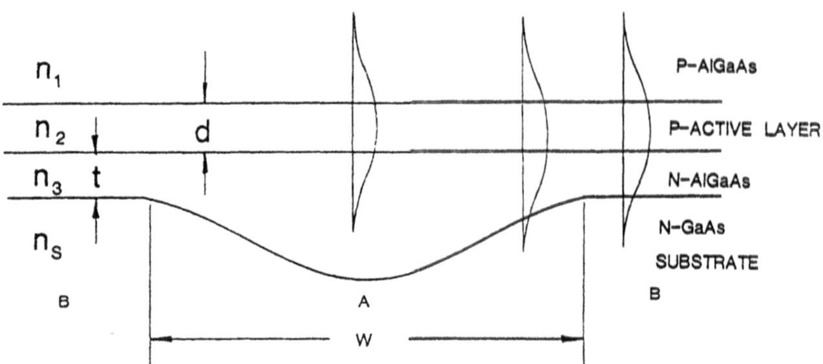

Fig. 6. Schematic cross section of a GCSP laser structure

As a 1st order approximation to evaluate the optical distribution of the guided wave, we describe a $\cosh^{-2}$ index profile [5] by

$$n^2(x)=n_s^2 + 2n_s\Delta n/\cosh^2(2x/h) . \qquad (1)$$

For the analysis we take the x axis as the transverse direction in the plane of the junction and the z axis as the propagation direction. Shown in Figure 7 is a plot of the effective index, n^2, as a function the transverse direction, x. The channel width is defined by h and n_s is the substrate refractive index. The wave confinement is obtained by an effective index that varies gradually over the guide cross section and confines the propagated wave. In addition, light is absorbed at the edges of the distribution and is shown by the index profile. This type of positive waveguiding, demonstrated by GCSP lasers, favors the zero order transverse mode.

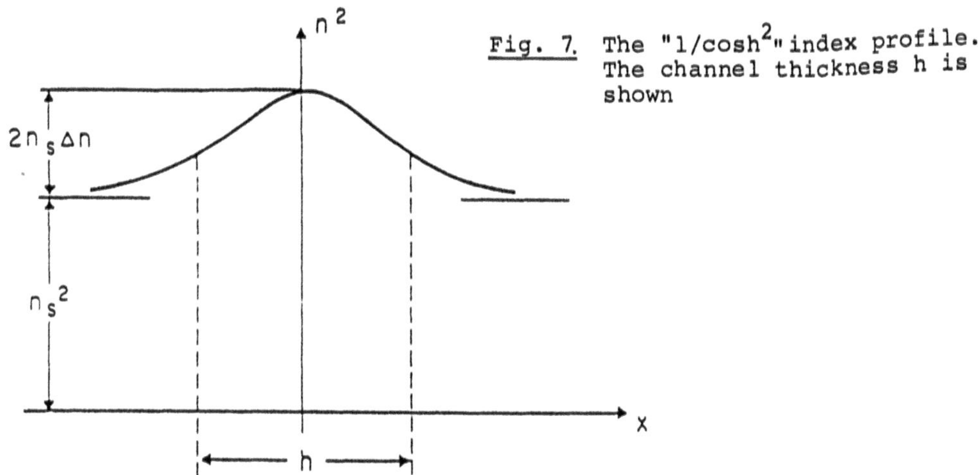

Fig. 7. The "$1/\cosh^2$" index profile. The channel thickness h is shown

5. Electro-optical Characteristics

By using the fabrication procedure shown in Figure 1, single tranverse and longitudinal mode cw lasers have been obtained [1]. The performance characteristics of such lasers are shown in Figures 8

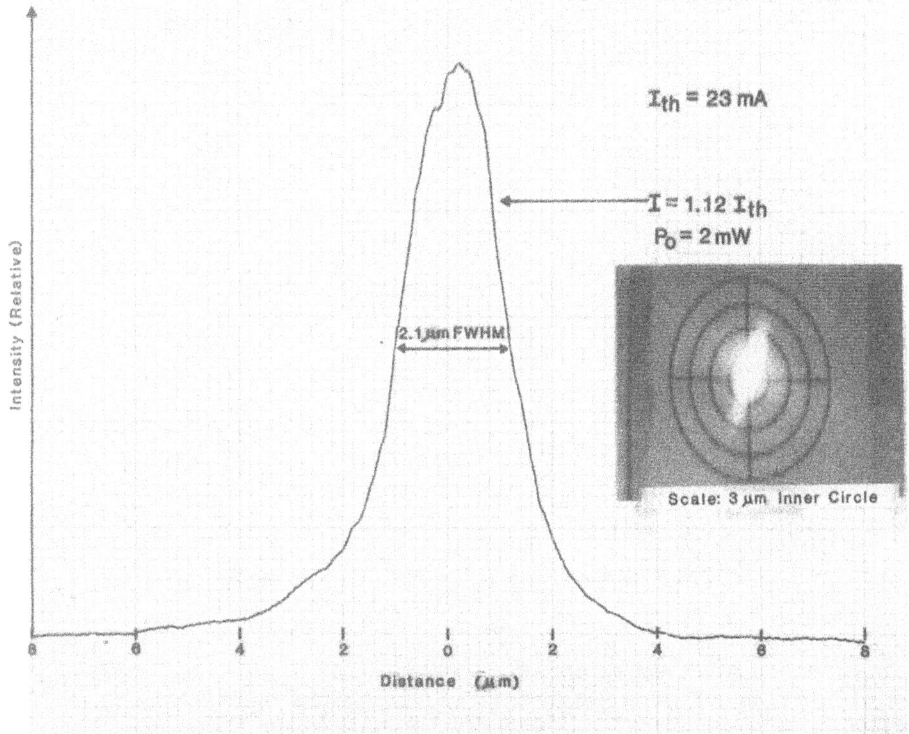

Fig. 8. Near-field intensity distribution of lasing emission as a function of current

through 14. The channel was 5.5 μm wide, 1.0 μm in depth and the active region thickness was 0.15 μm as shown in Figure 4. A near field lasing spot of 2.1 μm total width (measured at $1/e^2$ intensity points) was obtained and is shown in Figure 8.

The reason for such a small spot relative to the channel width (5.5 μm) is because of variable radiation loss achieved by the GCSP refractive index profile across the junction plane. As shown in Figure 6, the tails of the evanescent wave are absorbed by the higher refractive index of the substrate. As the wave propagates, there is a minimal loss at the center of the channel and, as a consequence, the lasing spot size is smaller than the channel width.

A block diagram of the fast scanning spectrometer used for spectral characterization of cw diodes is shown in Figure 9. This system permits real-time display of the entire lasing spectrum and is especially useful for studying the dynamic behavior of single longitudinal mode lasers. The spectrometer is equipped with a custom grating and detector suitable for wavelengths ranging from 0.7μm to 2.0μm.

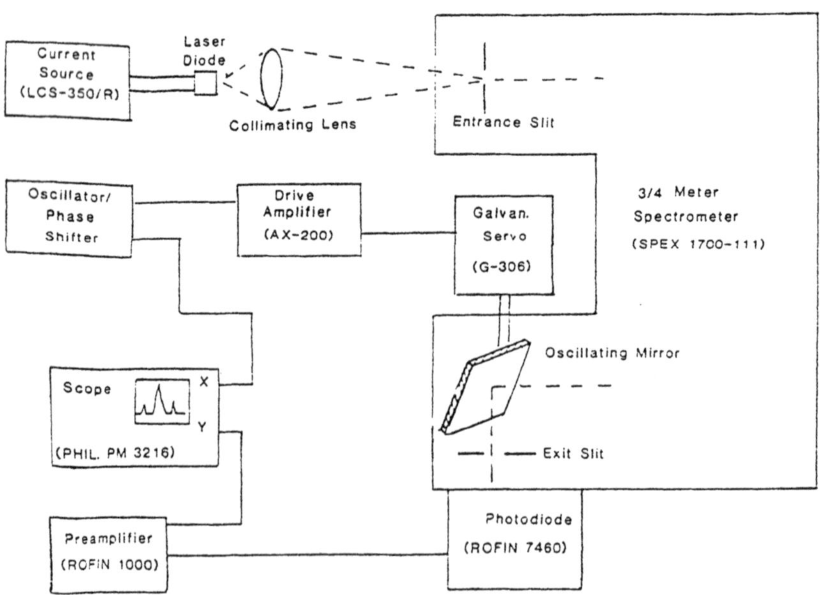

Fig. 9. Block diagram for rapid scanning spectrometer

Figure 10 demonstrates the spectral behavior of a SLM GCSP laser as a function of drive current. Photographs were taken from just below threshold (spontaneous) to a power level of 14mW. The spectrum starts out multi-longitudinal mode below threshold and becomes SLM from 1mW to 14mW having a stable linewidth of less

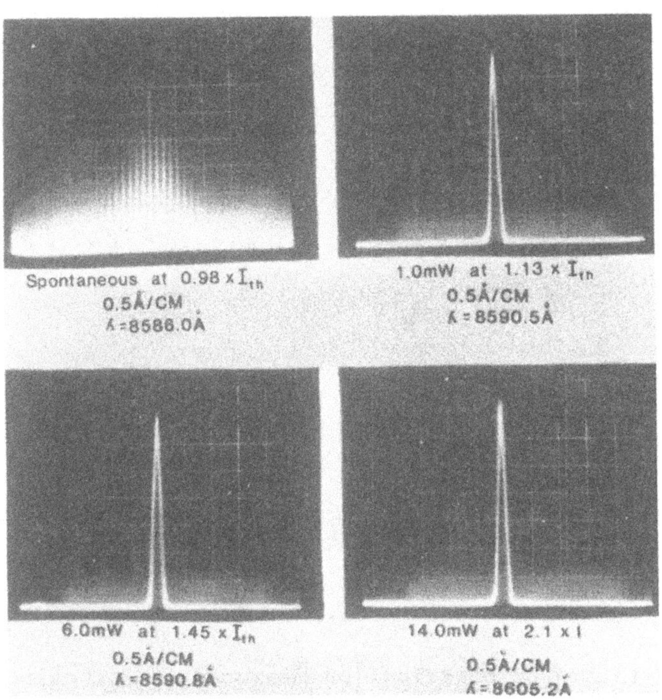

Fig. 10. GCSP-7#30 spectral measurements using rapid scanning spectrometer

than 0.2Å. In addition, there is no mode hopping to longer wavelengths from 1mW to 6mW. It is suspected that the reason for this behavior is due to small temperature sensitivity, extremely low threshold, extremely high slope efficiency and the relatively wide longitudinal mode spacing (4.5Å) associated with the short cavity length.

Figures 11 and 12 show the farfield intensity patterns taken parallel and perpendicular to the junction plane. In both cases, the curves have a gaussian distribution and the peaks remain central. The total absence of beam wander and satellite modes over a large power range demonstrates the effectiveness of the GCSP waveguiding mechanism. The full width at half maximum intensity (FWHM) for the perpendicular ($\Theta_{\perp}$) and parallel ($\Theta_{\parallel}$) beam divergence is 37° and 12°, respectively.

A cw light-current characteristic for a typical GCSP laser is shown in Figure 13 using a 200 μm cavity length, for threshold to 14mW output power. The very low spontaneous emission intensity below threshold (<0.2mW) is due to enhanced current confinement within the diffused region [6] and the effective index guide generated by the GCSP waveguide mechanism. The external differential quantum efficiency (η_{ext}) is 52% for a single face (rear facets contain 85% reflective coating).

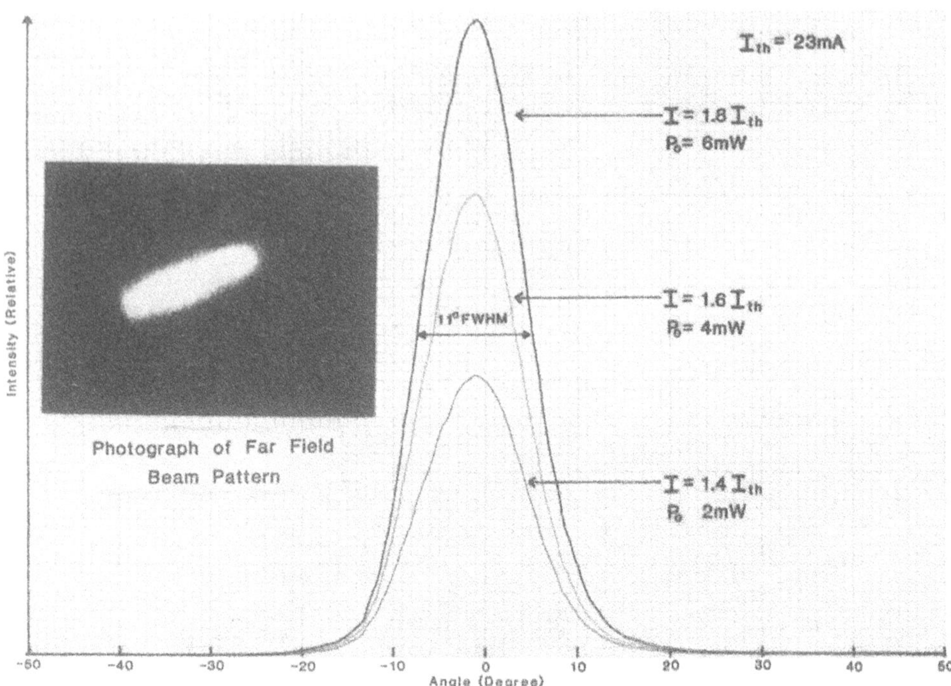

Fig. 11. Far-field intensity distribution measured in direction parallel to junction plane

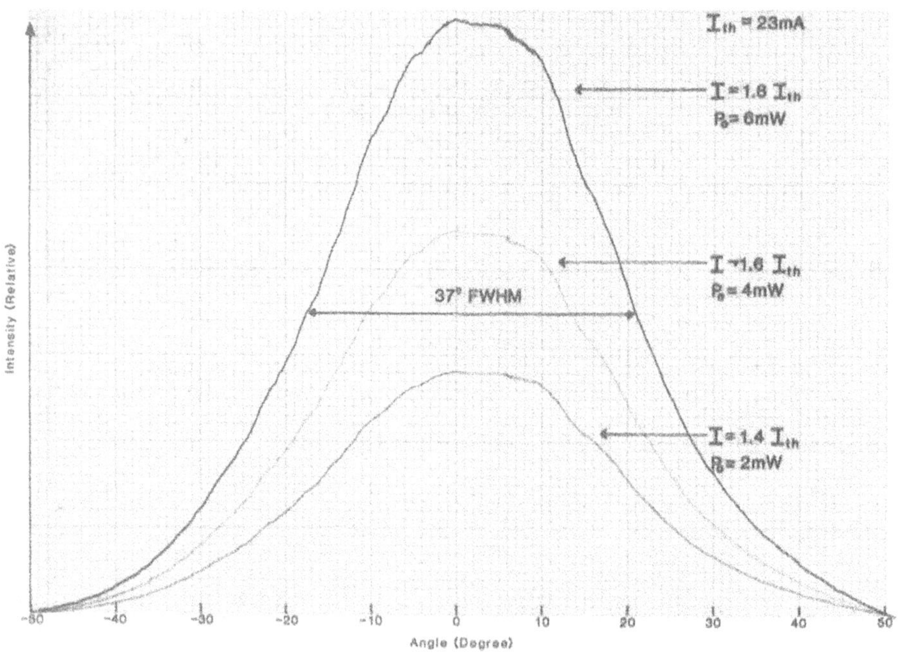

Fig. 12. Far-field intensity distribution measured in direction perpendicular to junction plane

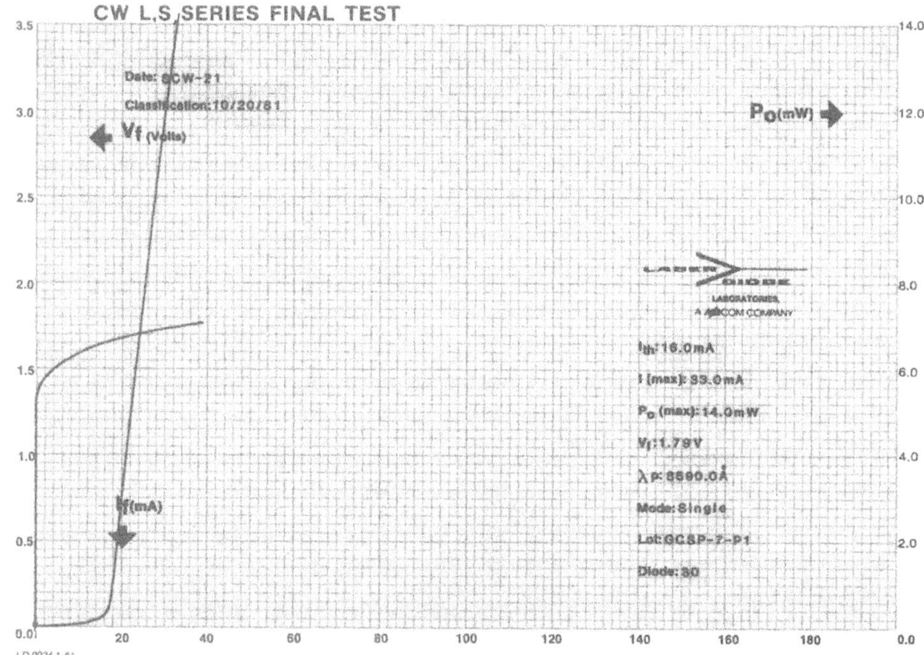

Fig. 13. Typical P-I curve of a GCSP laser diode

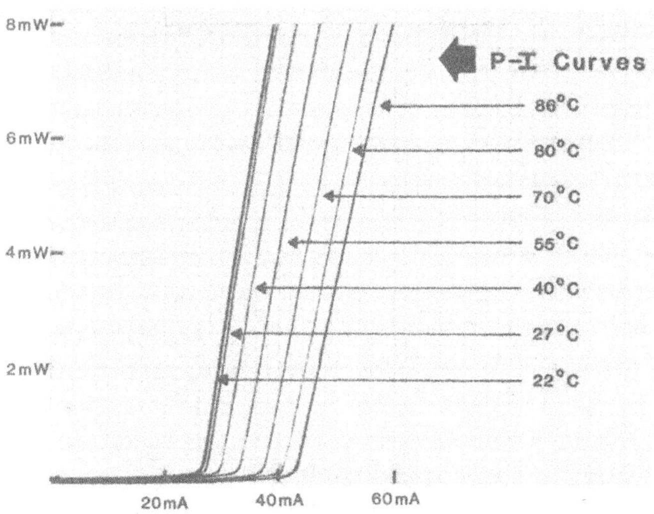

Fig. 14. P-I curves of GCSP as a function of heat sink temperature

6. Taper Polished Fiber (TPF) Process

Several techniques have been investigated to optimize the coupling efficiency between single mode optical fibers and injection laser diodes (ILD) such as HF etching, epoxy lensing [7] and quandrangular pyramid shaped hemiellipsoidal ends [8]. The method developed at LDL was a tapered polished fiber technique which has demonstrated excellent launch efficiency and requires only a simple drawing apparatus, oxyacetylene microtorch and an Al_2O_3 cleaving tool. Shown in Figure 15 is a bare fiber vertically suspended with a small weight attached to the fiber end. The torch is then passed slowly across the fiber in the horizontal plane and stretching of the glass takes place. Proper adjustment of the flame, weight and fiber position are needed to provide highly reproducible stretching of the fiber. By cleaving the fiber at the appropriate point and fire polishing the fiber end, very high launch efficiencies can be obtained reproducibly. Taper ratio, R, one of the important parameters in determining coupling efficiency, is defined as a_2/a_1 where a_2 is the diameter of the fiber end after the stretching, cleaving and fire polishing is complete. Figure 16 is a photomicrograph showing the tip of a TPF magnified 100x. The taper length and taper ratio is 850 µm and 1.76 respectively. In the tapered region, the core radius grows smaller in proportion to the cladding radius. High coupling efficiencies have been obtained with this technique because the TPF end increases the effective numerical aperture of the fiber. Also, the taper ratio can be accurately adjusted to match the source size.

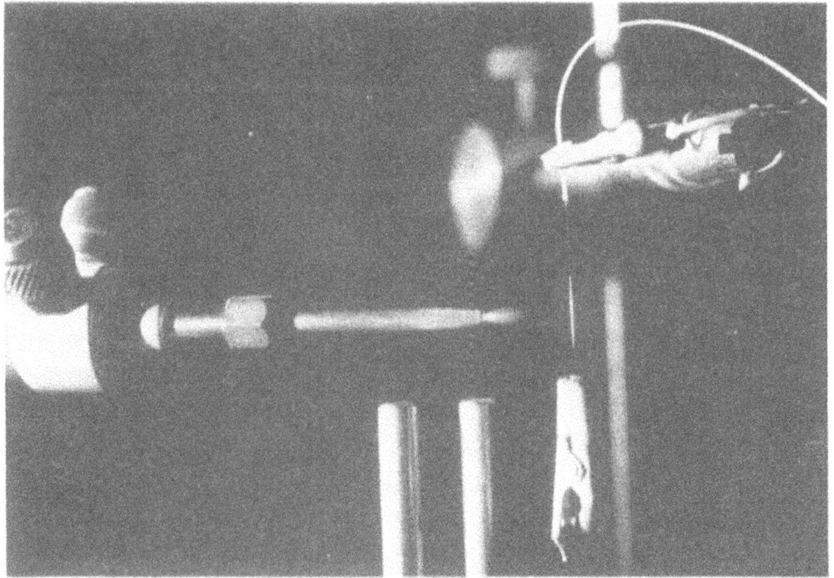

Fig. 15. Fiber drawing apparatus used to stretch and polish single mode fiber

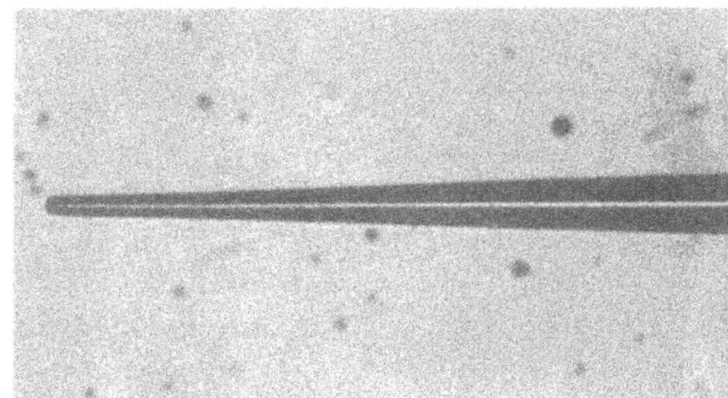

Fig. 16. Tip of a tapered polished fiber with a taper ratio of 1.76

7. Fiber/Laser Assembly and Coupling Efficiency

High efficiency power coupling has been demonstrated by using the TPF technique and the GCSP injection laser diode. The single mode fiber has a 3 μm core diameter, N.A. of 0.09 and a cladding diameter of 65 μm. The coupling between the GCSP laser and tapered polished fiber was accomplished by aligning the input end of the tapered fiber to the laser diode emitting area. Figure 17 is a photograph of the three-dimensional micromanipulator used to precisely align the fiber end to the laser emitting area to obtain maximum coupling efficiency.

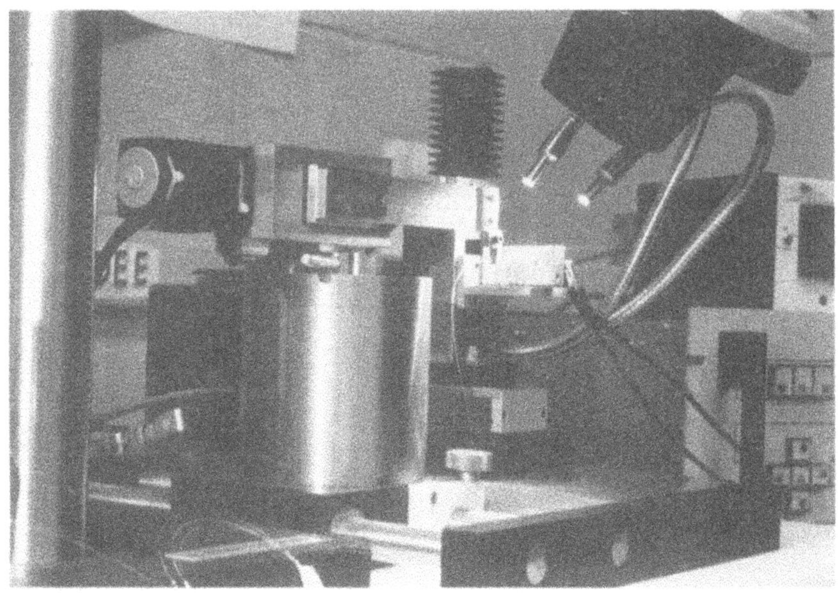

Fig. 17. Three-D photograph of micromanipulator used to align fiber to laser

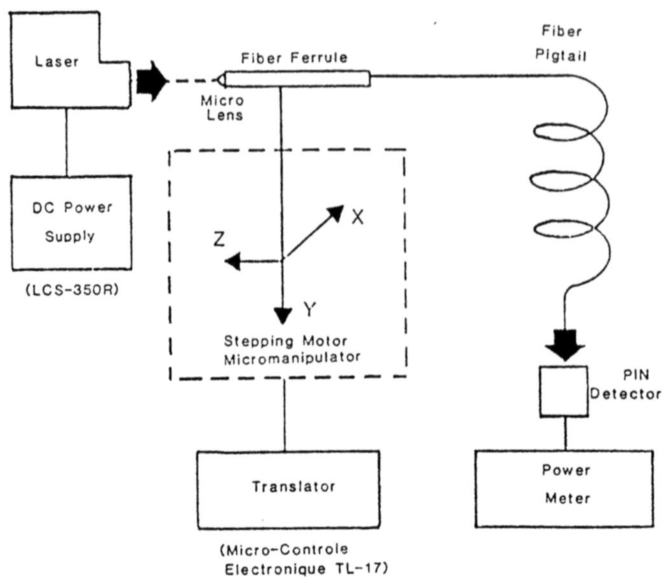

Fig. 18. Block diagram of fiber alignment apparatus

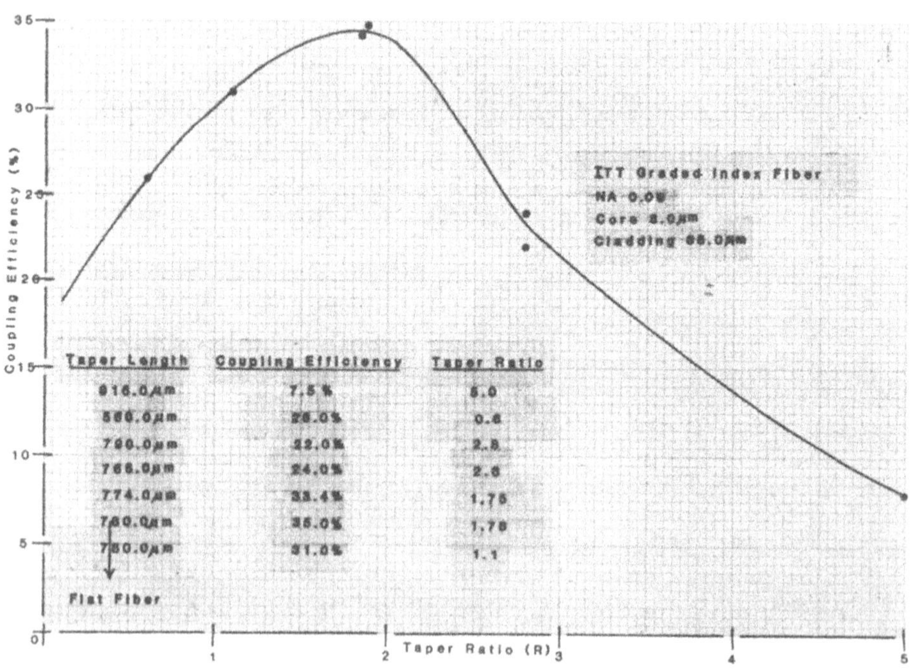

Fig. 19. Calculated coupling efficiency between TPF and laser diode as a function of R

Figure 18 is a block diagram illustrating the key components of the alignment system. The laser is mounted on a stepping motor controlled x-y stage which has a 0.1 μm step resolution. The tapered fiber pigtail is embedded in a rigid brass ferrule with Sn/Pb (60%/40%) solder and is held in the z-axis manipulator. The laser is then operated and the output end of the fiber pigtail is coupled to an EG&G silicon photodetector. Both the drive current and photodiode current are displayed simultaneously. The x, y and z stepping motors are moved until the photodetector output is peaked on the power meter.

Shown in Figure 19 is the calculated coupling efficiency of the GCSP laser versus the taper ratio of the TPF. A summary of the fiber parameters and laser characteristics for this experiment are shown in Table 1. The same diode and drive current conditions were used in all the data taken in Figure 19. The maximum coupling efficiency attained here was 35% using a 1.76 taper ratio. In addition, the coupling efficiency ranged from 22% to 35% using a taper ratio between 0.6 to 2.8 giving this technique good flexibility. Using the above fibering method and (60%/40%) SnPb solder, hermetically sealed fiber-pigtail units were fabricated for Naval Research Laboratories under contract # N00014-81-C-2259.

Table 1. Summary of fiber parameters and diode characteristics

Fiber Type	Single Mode, Graded Index
Inner Cladding Diameter	62 [μm]
Core Diameter	3 [μm]
Attenuation	2.5 [db/Km]
Numerical Aperture, N.A.	0.09
Laser Type	SCW-21, GCSP
Threshold Current, I_{th}	16 [mA]
Forward Current, I_f	22 [mA]
Forward Voltage at I_f, V_f	1.72 [v]
CW Radiant Flux at I_f, P_o	5.0 [mW]
Peak Wavelength, λp	859.0 [nm]
Spectral Width at 3 db, $\Delta\lambda$	0.02[nm]
Beam Aspect Ratio, Θ_n/Θ_p	3.2

8. Conclusion

A GCSP laser which has a smoothly varying refractive index profile in the lateral plane and operates in the fundamental transverse and single longitudinal mode has been demonstrated. The device also exhibits very low threshold current (16mA), high external differential quantum efficiency (52%) and linear light/current characteristics in excess of 14mW per facet. It was also shown at a 35% coupling efficiency can be obtained between a GCSP laser and single mode fiber using a tapered polished fiber technique.

9. Acknowledgements

This work was supported by Naval Research Laboratories. The authors would like to express their thanks to Art Mantie for liquid phase epitaxial growth.

References

1. A. Ceruzzi, T. E. Stockton, Fiber Optic Conference Proceedings, East, Boston, pg. 91, (1980)

2. R. D. Burnham and D. R. Scifres, Applied Physics Letters, Vol. 27, pg. 510 (1975)

3. P. A. Kirkby and G. H. B. Thompson, Journal Applied Physics, Vol. 47, pg. 4578 (1976)

4. H. F. Lockwood and M. Ettenberg, Journal Crystal Growth, Vol. 15, pg. 81 (1972)

5. H. Kogelnik, "Topics in Applied Physics", Vol. 7 (Springer-Verlag, Berlin, Heidelberg) pg. 55 (1975)

6. G. H. B. Thompson, "Physics of Semiconductor Laser Devices", (John Wiley & Sons) pg. 441 (1980)

7. C. C. Timmermann, Applied Optics, Vol. 15, pg. 2432 (1976)

8. H. Sakaguch et al., Electronic Letters, Vol. 17, pg. 425 (1981)

The Temporal Coherence of Various Semiconductor Light Sources Used in Optical Fibre Sensors

R.E. Epworth

Standard Telecommunication Laboratories Ltd.
London Road, Harlow, Essex, Great Britain

Many kinds of coherent sensors have been devised which use optical fibres. The choice of light source is dictated by many requirements including: high brightness (hence high launched power), ruggedness, insensitivity to reflections, reliability, low AM and FM noise, cost and power consumption. Most early experimental sensors used HeNe lasers, one benefit being direct visibility to the eye. Such lasers are also highly coherent and this was thought by many to be an essential requirement. However, high coherence is only required if the sensing interferometer has a large path difference, ie. if $(T_2 - T_1) \gg 2\pi/\omega$ in Fig. 1.

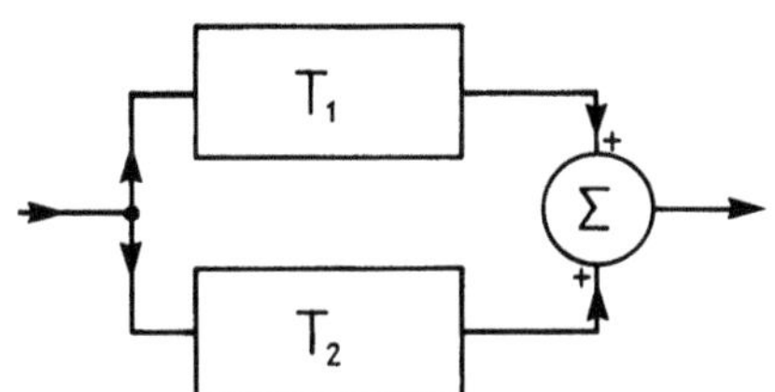

Fig. 1. 2-path interferometer with path delays T_1 and T_2

Furthermore, the source need only have high coherence at the specific time delay equal to the difference in delays of the interfering paths. (Note that an interferometer is an FM discriminator with sensitivity proportional to path difference so there is good reason to closely match the path lengths or even to phase lock to the zero order fringe.) If the interferometer is nominally balanced (e.g. the Sagnac fibre gyro) then a low coherence source may be used. This is the basis of the Michelson stellar interferometer 1 which was used to measure the angular dimensions of stars despite their optical incoherence.

Fig. 2 illustrates how the degree of coherence of a finite coherence source varies with path difference when measured using a 2-path interferometer. Fig. 3 shows Moiré patterns formed when two identical ruled patterns are superimposed at a small angle to each other. Fig. 3a shows a pair of highly coherent, regularly spaced gratings, whilst in Fig. 3b the line spacings are randomized, i.e. have low coherence. Note that in (b) only the zero order fringe appears.

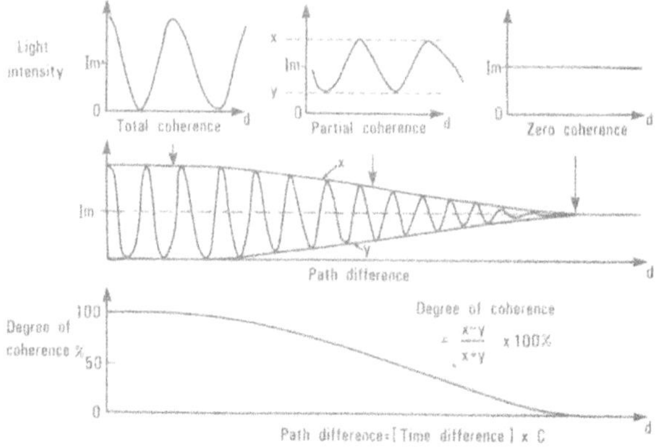

Fig. 2. Fringe pattern and degree of coherence vs. path difference for a source of finite coherence

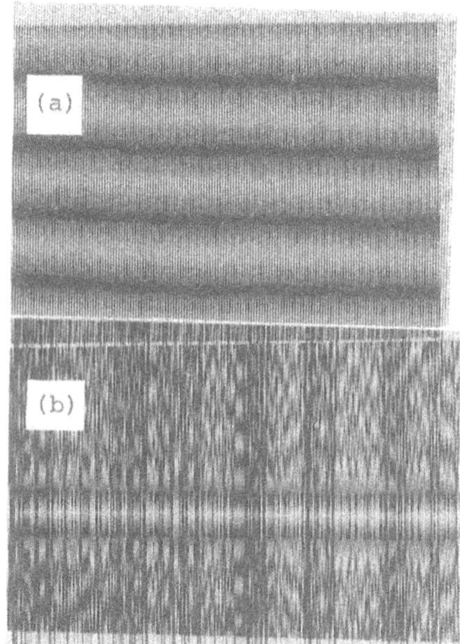

Fig. 3. Moiré patterns generated by overlaid pairs of gratings - (a) coherent, (b) incoherent

High source coherence, therefore, is not an essential requirement, indeed it can be a severe disadvantage[2]. In a Sagnac fibre interferometer, the counter-propagating backward scattered waves scattered from a region about the midpoint of the interferometer are mutually coherent, and this region extends either side to a distance equivalent to the source coherence length/time. Thus this fraction of the scattered power adds coherently and causes an error signal. This may be made negligible by using a low coherence source[3]. Alternatively, the coherence of the propagating light may be sufficiently reduced by phase modulating the primary wave[4]

Polarization fading can be a problem in fibre sensors. However, if the delay between two orthogonally polarized modes exceeds the source coherence time, they become mutually incoherent. Fig. 4 shows how linearly polarized light launched at 45° to the fast axis of a birefringent element is depolarized if the polarization dispersion $\Delta T_1 > T_c$, the source coherence time. However, complete depolarization only occurs for launch at 45° to the principal axes. This problem is overcome in the arrangement shown in Fig. 5. The output from the first element is arranged to be at exactly 45° to the second element, thus ensuring depolarization for all input states. Any polarization-selective receiver will then always see half the power, so no fading occurs.

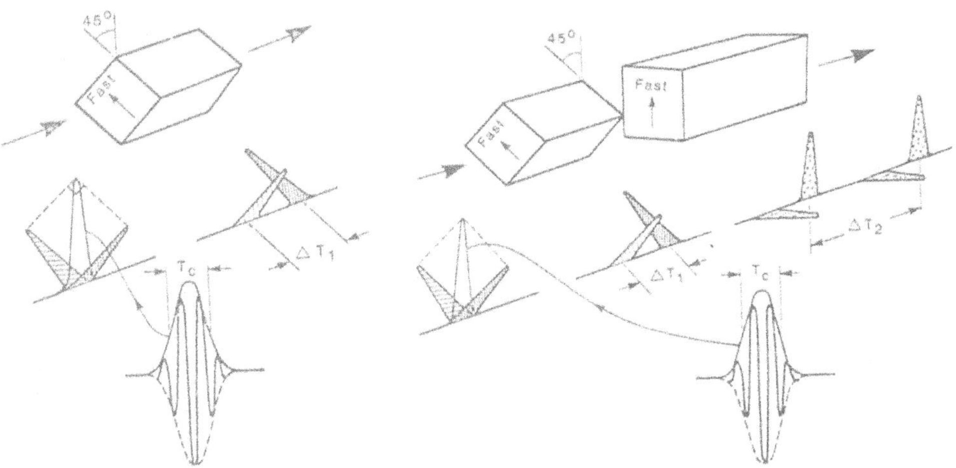

Fig. 4. Depolarization of light with short coherence time, T_c, by polarization of a birefringent element

Fig. 5. Lyot depolarizer ensures depolarization for all input orientations

In the Lyot depolarizer, ΔT_2 and ΔT_1 are in the ratio 2:1, which is the optimum for a given total birefringence. Similarly, low coherence may be used to prevent unwanted interference between modes of different order in multimode Sagnac fibre sensors, or to reduce the effects of insufficiently stripped cladding modes. This interference appears as speckle which can be suppressed if the source coherence time is shorter than the various intermode delays [5].

Several kinds of semiconductor light sources are available and these have widely different coherent characteristics. Conventional surface emitting LEDs have low temporal coherence, but their low spatial coherence and low brightness make them generally unsuitable for single mode sensors. Stripe geometry devices offer high spatial coherence and brightness; these include edge emitting super radiant LEDs, low temporal coherence multimode lasers and single longitudinal mode lasers.

Several semiconductor sources were measured using a Michelson interferometer to determine their temporal coherence characteristics [5]. Fig. 6 shows the degree of coherence versus path difference for an ELED; its coherence falls off after a few wavelengths' path difference.

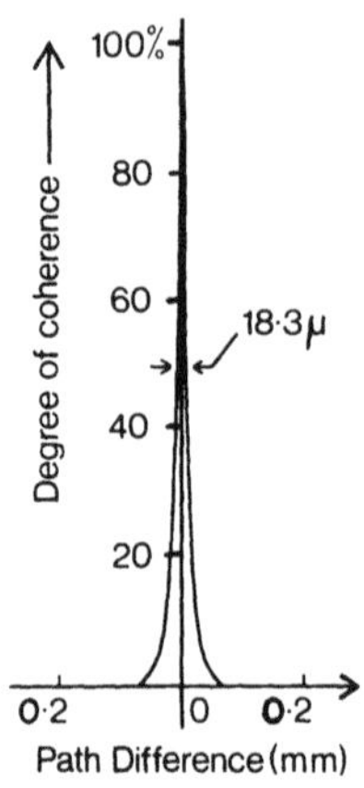

Fig. 6. Measured coherence function of an edge emitting LED

The coherence functions of multi-longitudinal mode lasers are more complex; they exhibit peaks, spaced at multiples of the effective laser cavity length/round trip time, separated by regions of low coherence. This is explained in Fig. 7, which shows the superposition of the interference fringes of different numbers of modes. The result is a reduction in fringe contrast except at multiples of the laser cavity length, where all the separate coherence functions become in-phase.

These peaks are clearly seen in the measured coherence function of a stable low coherence multimode laser (Fig.8). The envelope of these peaks represents the coherence function of a typical individual mode - note that it falls off over a few tens of millimetres. The multimode spectrum is shown inset. The measured degree of coherence often varies with measurement bandwidth. A laser which is jumping between modes at a few MHz will appear multimode when measured with a slow response system, but may be quite coherent in the shorter term. Fig. 9 shows the zero and first coherence peaks of the laser in Fig. 8. Almost identical curves were obtained with wide and narrow measurement bandwidth, indicating good mode stability (low partition noise). With this type of laser the coherence varies with operating power level.

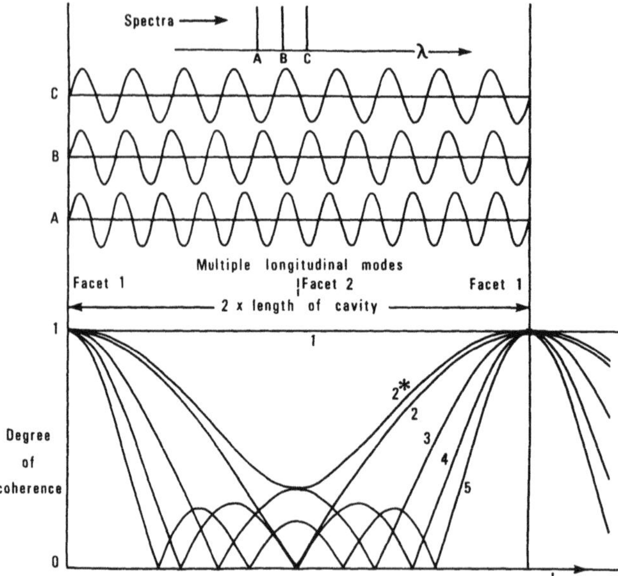

Curve 2 shows that the addition of two adjacent modes of equal intensity results in zero coherence at the midpoint. If the intensities of the two modes are in the ratio 2 : 1 an incomplete null appears, as in curve 2*.

Fig. 7. The superposition of the coherence function of various numbers of longitudinal modes results in coherence peaks at multiples of the laser cavity length

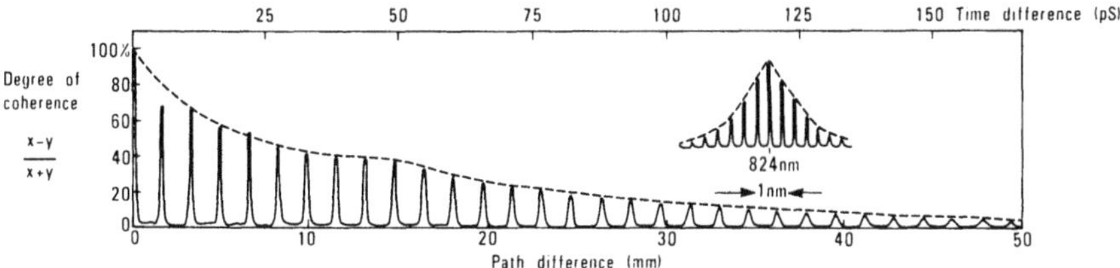

Fig. 8. Measured coherence function and spectrum of gain guided laser (STL 3 μm stripe width)

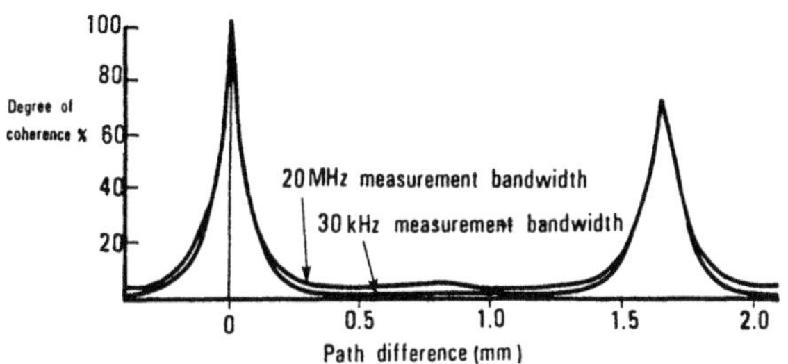

Fig. 9. First part of coherence function of previous type of laser measured in wide and narrow bandwidth

The majority of available semiconductor lasers are not of the type just described. They are much more coherent, as can be seen in Fig. 10, which shows the coherence functions of several nominally single mode lasers measured up to a path difference of one metre. They all show significant coherence over several nanoseconds. Curves A and B are thought to be limited by the experimental configuration. The curves for lasers C and D exhibit

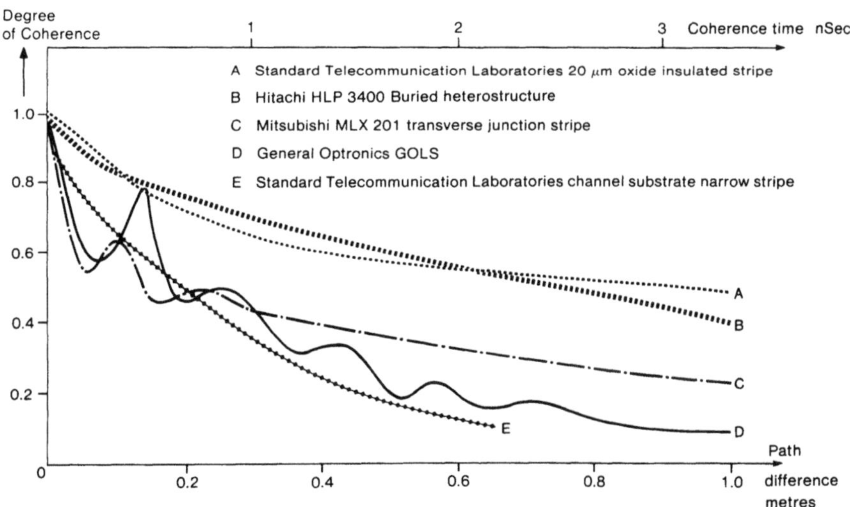

Fig. 10. Coherence functions of several "single mode" type lasers, measured up to 1 metre path difference

ripples; these may be caused by residual longitudinal modes of unequal mode spacing. Laser E had a short cavity; samples with longer cavities gave results similar to A and B. A word of caution: even low levels of reflection back into coherent lasers are sufficient to drastically reduce the coherence and increase the FM and AM noise. Thus care should be taken to suppress reflections in practical sensors and whilst making measurements such as these.

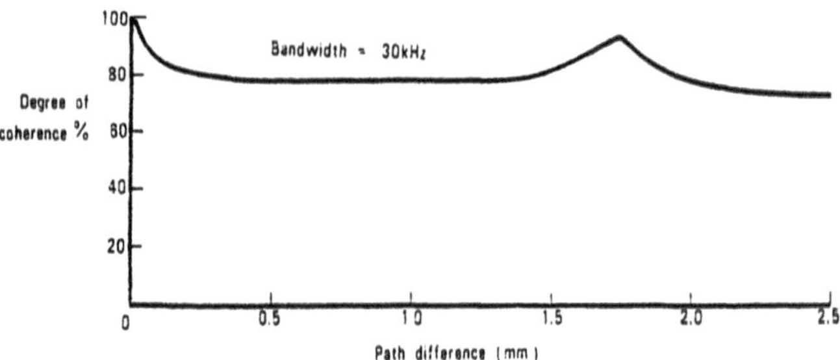

Fig. 11. Measured short length coherence function of stable single mode laser

The first few millimetre of the measured coherence function of a Transverse Junction Stripe laser is shown in Fig. 11. This device showed high coherence even in a narrow measurement bandwidth. Note the small coherence peaks caused by the low level auxiliary modes (c.f. Fig. 9). By comparison laser D mode jumped at a few MHz and hence showed wide differences in coherence function with measurement bandwidth (Fig. 12). Note the high degree of coherence at the first coherence peak (c.f. Fig. 9).

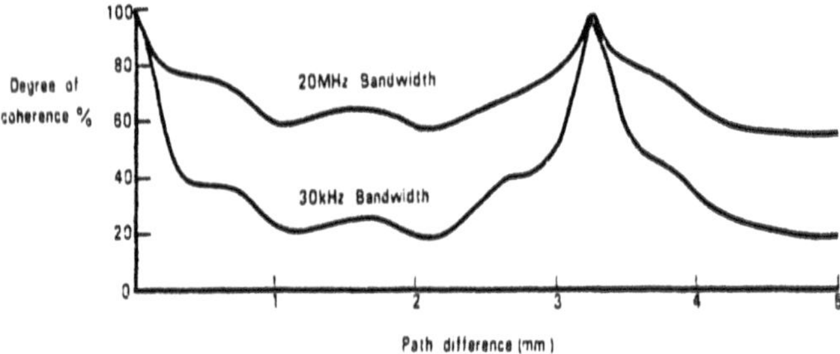

Fig. 12. Measured short length coherence functions of unstable mode jumping single mode laser

It is possible to reduce the coherence of the more coherent types of semiconductor lasers by various means [5,6]. Fig. 13 shows schematically how the spectrum of a laser DC biassed above threshold can vary with operating point. Virtually all so-called single mode lasers will mode jump at some conditions of power and temperature (and external reflections). Also, both the mean centre frequency and the frequency of the individual longitudinal modes change with optical power level. The latter is seen as a modulation of the effective laser cavity length.

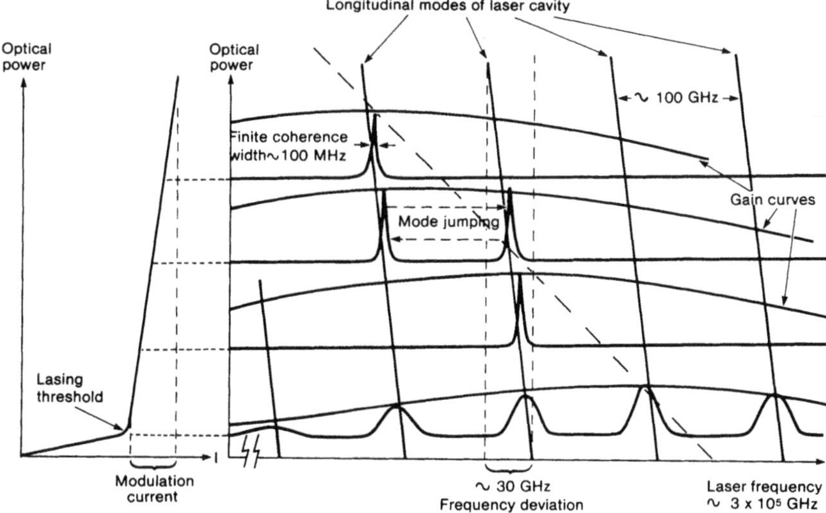

Fig. 13. Illustration of spectral variations caused by direct modulation, showing FM and jumping of longitudinal modes

Thus, modulating the laser current not only causes intensity modulation, but also causes frequency modulation. Hence, if the applied modulation is high frequency compared with the measurement bandwidth, the effective coherence may be reduced. The magnitude of the frequency deviation varies with the modulating frequency and with laser type, as shown in Fig. 14. Below 1 MHz the deviation is dominated by thermal effects, whilst above 10 MHz it is caused by carrier density variations.

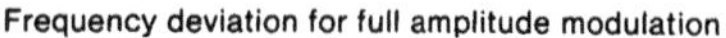

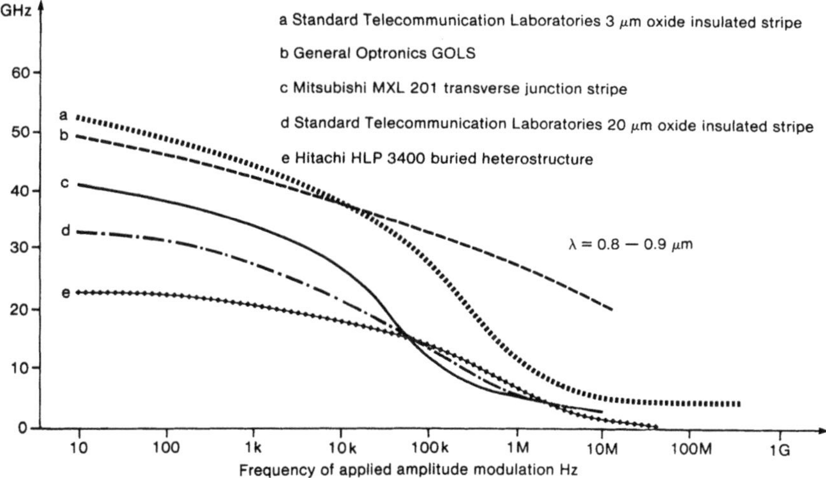

Fig. 14. Frequency deviation of longitudinal modes of various lasers, measured over a wide modulation frequency range

Another method of reducing the coherence is to modulate the laser from below its threshold with short pulses of only a few nanoseconds duration. The laser has insufficient time to settle into a single coherent mode of oscillation and the resulting spectrum and coherence function are similar to those of a low coherence multimode laser. Similarly the coherence may be reduced by modulating the laser with a microwave signal.

In conclusion, the need or otherwise for temporal coherence in fibre sensors has been discussed. The measurements of various sources are presented, together with suggested methods of reducing coherence.

The author would like to thank I. Hardcastle and M. J. PETTITT of STL, who assisted in the coherence measurements.

References

1. Michelson Stellar Interferometer, "Principles of Optics", M. Born & E. Wolf, 7.3.6

2. "Limitation of rotation sensing by scattering", C. Cutler, S. Newton, H. Shaw, Opt. Lett. 5, 488, Nov. 1980

3. "Low drift fibre gyro using a superluminescent diode", K. Bohm, P. Marten, K. Petermann, E. Weidel, R. Ulrich, Elect. Lett. Vol 17, No. 10, May 1981

4. "Reduction of backscattering error in an optical fibre gyro", S. Newton, C. Cutler, H. Shaw, CLEO 81, ThM6

5. "Modal noise causes and cures", R. E. Epworth, Laser Focus, Sept. 1981, pp. 109 - 115

6. "Measurements of static and dynamic coherence phenomena using a Michelson interferometer", R. E. Epworth, Proc. 5th European Conference on Optical Communication, Amsterdam, 1979

Noise in Diode Lasers

A. Dandridge

Naval Research Laboratory, Washington, DC 20375, USA

The small size, high efficiency and high output power of solid state GaAlAs single-mode diode lasers make them a convenient source for fiber optic sensor systems. However, the transition from laboratory setups to packaged devices has led to a number of problems primarily caused by the source. The problems of the output of these single-mode devices, which result in a loss of sensor sensitivity may be split into three basic groups:

1. Coherence length
2. Amplitude noise
3. Phase noise.

The lasers' spectral and noise properties are also strongly dependent on light fed back into the laser cavity. The amplitude and phase of the light fed back into the laser cavity as well as the external cavity length are all important in determining the lasers' properties. Although many different structures of single-mode GaAlAs lasers have been fabricated, they usually show generally similar behavior when their optical output is considered.

In the Fiber Optic Sensor System Program at NRL, the following seven types of laser have been investigated: 1) Hitachi HLP 1400, channel substrate planar (CSP), 2) Hitachi HLP 2400U, buried heterostructure, 3) Hitachi HLP 3400, buried heterostructure, 4) Mitsubishi ML 4307, transverse junction stripe, 5) General Optronics TB47 proton bombarded buried heterostructure, 6) Laser Diode Labs CSP-7 channel substrate planar, 7) Laser Diode Labs CS234 channel substrate planar. The last two devices are both Gaussian CSP structures; the CS234 device has a high threshold current ($\sim$ 40 mA), whereas the CSP-7 device has a lower threshold ($\sim$ 20 mA). The lasers were chosen for their differing structures; HLP 2400U and 3400 were solely index guided and the others were a combination of both gain and index guiding. In the LDL CSP-7 device, the active planar region contains a slight dip which provides a stronger index guidance to the mode than the conventional CSP structure. Also two other devices, a Laser Diode Labs and an RCA anti-reflection coated laser have been investigated. These devices with their short coherence length were chosen as possible candidates for fiber gyro systems.

1. Coherence Length

Above threshold (> 1.15 I_{thres}) all seven lasers were observed to run on a single longitudinal mode. The relative energy in the dominant mode compared to that in the dominant mode plus the twenty adjacent modes was found to be between 0.95 and 0.98 for the lasers investigated. Linewidth measurements were made with a piezoelectrically scanned Fabry-Perot interferometer; standard deconvolution procedures were employed to obtain the true linewidth. Great care was taken to reduce feedback into the laser cavity. Feedback $>10^{-7}$ has been shown to produce line narrowing in GaAlAs lasers, reducing linewidths by greater than a factor of 100 [1,2], wheras , feedback $>10^{-4}$ has been shown to produce satellite mode generation, multimode behavior and line broadening [3]. Typically, the free-running linewidth of the lasers investigated lay between 10 MHz and 150 MHz. Lasers of identical structure also showed large linewidth variations presumably dependent on sample variations and age.

In the free-running mode all the lasers investigated had a coherence length >1 m. Consequently they were suitable for use in interferometric sensors with balanced arms. However, backscattering from long lengths (>100 m) of fiber can induce satellite mode generation, line broadening and multimode behavior [4]. Consequently, isolation techniques may be necessary for use in sensors which will employ long sensor lengths (e.g., magnetic field sensor) [5]. For these sensors, longer wavelength lasers may provide a solution owing to the lower transmission losses and reduced backscattering.

In the fiber gyro, short coherence length lasers that have the output facet antireflection coated and consequently have a broad band output (multimode), have been used to reduce the backscatter-induced phase noise [6]. However, feedback from the external cavities to such a device can induce single longitudinal operation with a narrow linewidth, thereby producing a highly coherent source [7]. Consequently, to maintain broadband operation, isolation between the laser and gyro may be necessary.

2. Amplitude Noise

Amplitude noise measurements were obtained by detecting the laser output with a large area Si photodiode run photoconductively with a load resistance of $10^4\ \Omega$ and biased at 9V. The large area detector was used to insure that almost all the radiation from the front facet of the laser was collected by the photodiode. For laser outputs in excess of 1 mW a neutral density filter was placed between the laser and photodetector so as to keep the photodiode response in the linear region. Care was taken to insure that optical feedback into the laser cavity was less than 10^{-7}. The output of the photodetector was analyzed using a Tektronics 7L5 and a Hewlett Packard 3582A spectrum analyzer. Noise values agreed to within $\pm$ 1 dBV using both spectrum analyzers in the region where their frequency range overlapped.

Both spectrum analyzers gave their output in terms of a measured value of dBV (dBV_{mea}) which was of course, dependent on the output voltage of the detector V. In these measurements the parameter of interest is the normalized laser noise (i.e., $\frac{dI}{I}$ where I is the laser's intensity and dI the r.m.s. fluctuation at frequency f). However, it is the convention in noise studies to use the electrical noise power generated (in this case by the laser light) in the detector (ie proportional to $(dI)^2$). The following equation was used to calculate the normalized noise output in dB:

$$dB = 20 \log \left|\frac{dV}{V}\right| = dBV_{mea} - 20 \log V$$

where dV is the rms voltage fluctuation (noise) which is proportional to dI. Thus, a 10^{-5} fluctuation corresponds to -100 dB. All results were subsequently normalized to a 1 Hz bandwidth.

Measurements on a number of lasers were made to insure that the observed noise was not a function of the laser's environment, the lasers being run in vacuuo, in air and with different heat sinks. The results were identical to within 1 dB. Measurements of the noise of the current supply (Ni-Cd cells) indicated that the resultant noise was typically 30 to 40 dB below that which would produce the observed amplitude noise.

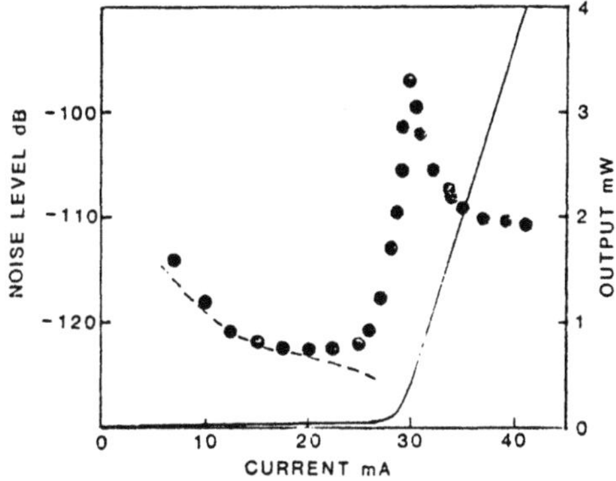

Fig. 1. Laser amplitude noise (●) and output power of a Mitsubishi TJS laser as a function of driving current. Noise data taken at 1 kHz. Dotted line indicates the shot noise limit.

Figure 1 shows the output characteristics of the Mitsubishi TJS laser, one of the lowest noise lasers investigated. The calculated value of the shot noise limit of the detector (i.e., indicating the quantum noise limit of the emitted light) is also shown in Fig. 1. It can be seen that below threshold the device is shot noise limited; at threshold the relative noise increases by 25 dB. As the laser becomes single-moded, the relative noise drops by 10 dB, above threshold the decrease in $\frac{dI}{I}$ is caused by the increase in output power I, dI remaining constant (± 1 dBV). All the lasers investigated showed behavior similar to the TJS laser, but in general the noise levels were 5 to 20 dB higher. All the lasers investigated showed a $f^{-1/2}$ frequency dependence of $\frac{dI}{I}$ (i.e., 1/f when electrical noise power is considered), examples of which are shown in Fig. 2. Typically the observed spread of amplitude noise values for lasers of the same type was ± 2 dB and

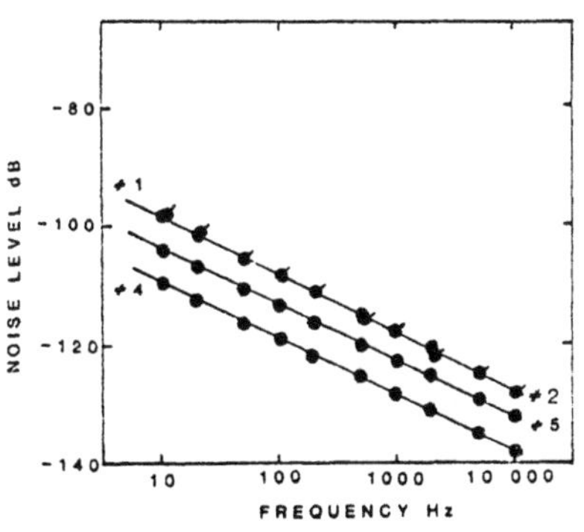

Fig. 2. Free-running low frequency amplitude noise, data normalized to a 1Hz bandwidth; lasers numbered in text

the observed distribution of amplitude noise in Fig.2 is unlikely to be caused by sample variations. Consequently, care should be taken in choosing low noise laser structure for fiber interferometers.

Figure 3 shows the frequency dependence of the amplitude noise for two of the antireflection coated devices. The behavior is close to the shot noise limit. These devices appear to be suitable for fiber gyro applications.

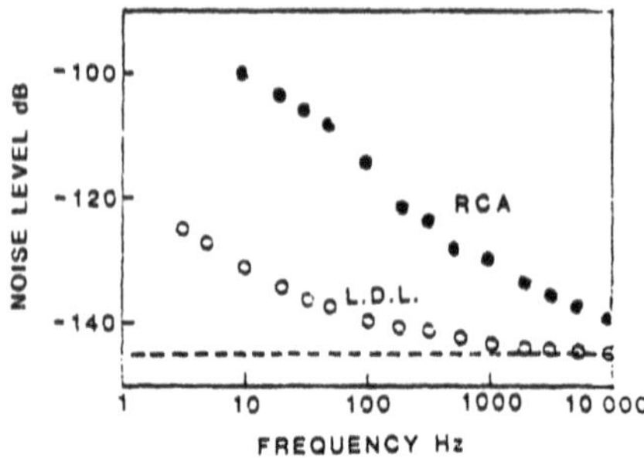

Fig. 3. Amplitude noise as a function of frequency of an RCA and a Laser Diode Labs antireflection laser. Dotted line indicates the shot noise limit

3. Phase Noise

Phase noise in an interferometer system is due to a change in the wavelength of the interferometer's light source and is in general strongly dependent on the optical path difference between the arms of the interferometer. The experimental arrangement used to measure the phase noise was that of an unbalanced Michelson interferometer.

The variation of the interferometer (phase) noise with optical path difference (OPD) is shown in Fig. 4, the slope of ~ 1.0 indicating a linear relationship between the noise and the OPD. Typically, the noise contribution increases a thousandfold from zero to 1 m OPD. Consequently, the phase noise requires accurate path matching in fiber interferometer systems. Two methods of phase noise reduction have been attempted: a) feedback-induced line narrowing [2], b) electronic compensation of the phase noise [8]. These methods will be discussed in detail in the presentation.

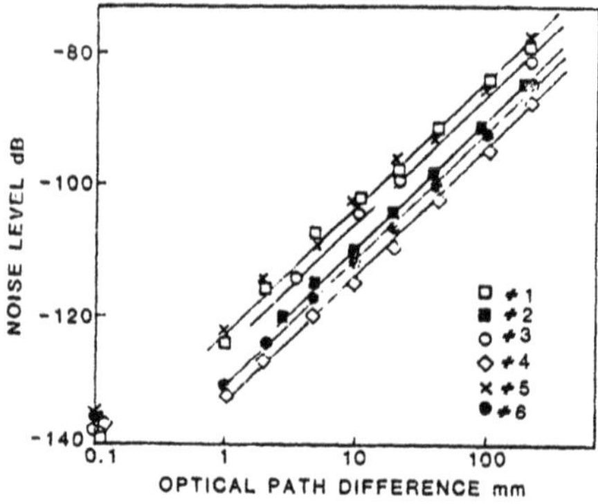

Fig. 4. Variation of the noise output of the interferometer (phase noise) in dB as a function of path length difference at 2 kHz with a 1 Hz bandwidth; lasers numbered in text

The author would like to thank his coworkers A. B. Tveten, R. O. Miles, Lew Goldberg, S.J. Petuchowski and T. G. Giallorenzi for their help.

References

1 S. Saito and Y. Yamamoto, Electron. Lett. 17, 325, (1981)

2 L. Goldberg, A. Dandridge, R. O. Miles, T. G. Giallorenzi, and J. F. Weller, Electron. Lett. 17, 677, (1981)

3 R. O. Miles, A. Dandridge, A. B. Tveten, H. F. Taylor and T. G. Giallorenzi, Appl. Phys. Lett. 37, 990, (1980)

4 A. Dandridge and R. O. Miles, Electron. Lett. 17, 273, (1981)

5 A. Dandridge, A. B. Tveten, G. H. Sigel, Jr., E. J. West and T. G. Giallorenzi, Electron. Lett. 16, 408, (1980)

6 K. Bohm, P. Martin, K. Petermann, E. Weichel and R. Ulrich, Electron. Lett. 17, 352 (1981)

7 S. J. Petuchowski, R. O. Miles, A. Dandridge and T. G. Giallorenzi, submitted to Appl. Phys. Lett.

8 A. Dandridge and A. B. Tveten, submitted to Electron. Lett.

Part 4

Fiber Optic Rotation Sensor Systems

4.1 Open-Loop Operation

All Single Mode Fiber Optic Gyroscope

R.A. Bergh, H.C. Lefèvre, and H.J. Shaw

Edward L. Ginzton Laboratory, Stanford University
Stanford, CA 94305, USA

The intrinsic qualities of fiber optic components have improved the sensitivity of single mode fiber gyroscopes reported to date [1]. We investigate the residual sources of short term noise and long term drift and discuss the solutions adopted to reduce their effects. We have used the reciprocal configuration [2-4] of the fiber Sagnac interferometer to avoid phase shifts due to changes in birefringence of the fiber coil. Bulk optic beamsplitters, polarizers and birefringent plates have been replaced by fiber optic directional couplers [5], polarizers [6] and polarization controllers [7], fabricated on a single continuous strand of fiber (Fig.1). The biased rotation signal is obtained with a delayed reciprocal phase modulation technique [3,4] which, theoretically, allows one a very good bias stability.

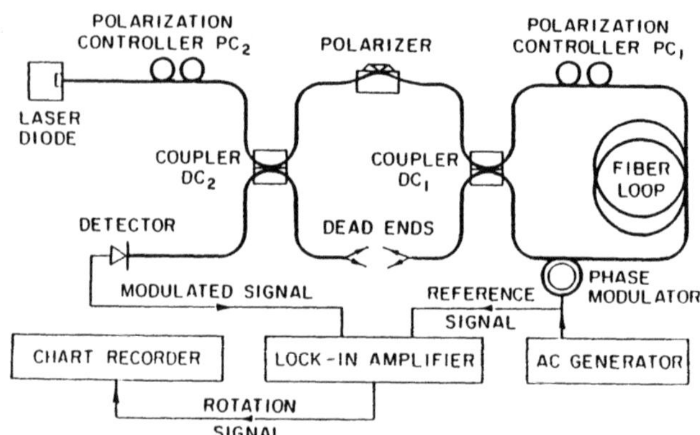

Fig.1 Fiber gyro schematic

When a temporally coherent source is used, the main short term noise source is parasitic interference due to coherent Rayleigh backscattering [8]. Although it is possible to reduce this with additional phase modulation [1,3], we obtained the best results using a solid state GaAs laser which decreases the coherence of the backscattered waves with respect to the primary waves. Fig.2 shows the short term response of the system with integration time of 1 sec, using a fiber length of 580 m, coil radius of 7 cm, and wavelength of 820 nm. This short term result compares well with active laser gyros. The averaged value of the noise spectral density of the rotation signal is 0.02 $(deg/h)^2/Hz$. A recent report on Honeywell laser gyro tests [9] gives a typical range of 0.016 to 0.04 $(deg/h)^2/Hz$.

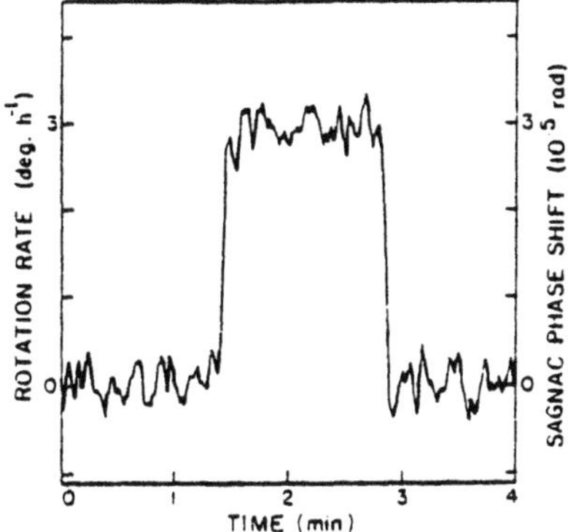

Fig.2 Short term response of fiber gyro (1 sec integration time)

Further reduction in noise level may be possible because our system is not photon-noise limited. The all-fiber components have very low loss: less than 2 dB for the set, three of which (directional couplers and polarizer) are each passed through twice. As a result, even with the low input power injected into the single mode fiber (400 μW), we have 30 μW incident on the detector. The photon noise spectral density is 0.001 $(deg/h)^2/Hz$. The excess short-term noise observed may be due to acoustic effects acting on the fiber in the sensing loop. This is possible because, although the loop is optically reciprocal at zero modulation frequency, and thus self-compensates well against reciprocally introduced low frequency noise, this compensation degrades as the modulation frequency increases and disappears at the loop synchronous frequency, where the period of the modulation is twice the loop transit time. In our present system, the Hytrel jacket on the fiber is known to be very acoustically sensitive. The short term noise might possibly be improved by using another coating on the fiber, such as glass or metal [10].

To develop this further we look briefly at the way in which the noise-compensation property of a reciprocal sensing loop modifies the spectral density of noise introduced into the loop. If we suppose that we have a phase noise generator at a certain point in the fiber, the phase of each counterpropagating wave will suffer a change $\phi(t)$. This random function $\phi(t)$ has a correlation function $\Gamma_\phi(\tau) = \langle \phi(t) \cdot \phi(t-\tau) \rangle$ and a spectral density $\gamma_\phi(\omega) = \int \Gamma_\phi(\tau) \exp(i\omega\tau)\, d\tau$. $\phi(t)$ has rms value σ_ϕ given by

$$\sigma_\phi^2 = \Gamma_\phi(0) = \int \gamma_\phi(\omega)\, d\omega \quad .$$

When the two waves interfere, their phase difference is:

$$\Delta\phi(t) = \phi\,(t - t_p) - \phi\,(t - t_p')$$

where t_p and t_p' are the propagation times in the two opposite directions between the phase noise generator and the coupler. $\Delta\phi(t)$ is also a random function. Its correlation function is:

$$\Gamma_{\Delta\phi}(\tau) = \langle \Delta\phi(t) \cdot \Delta\phi(t - \tau) \rangle ,$$

$$\Gamma_{\Delta\phi}(\tau) = 2\Gamma_{\phi}(\tau) - \Gamma_{\phi}(\tau + \Delta t_p) - \Gamma_{\phi}(\tau - \Delta t_p)$$

where $\Delta t_p = t_p - t_p'$. The linearity of the Fourier transform and the shift theorem leads to a spectral density of $\Delta\phi$

$$\gamma_{\Delta\phi}(\omega) = 2\, \gamma_{\phi}(\omega) \, [1 - \cos\, (\omega\, \Delta t_p)] \quad .$$

For low frequencies $\omega << \omega_0 = \frac{\pi}{t_\ell}$, where t_ℓ is the group transit time of light around the loop,

$$\gamma_{\Delta\phi}(\omega) \approx \omega^2\, \Delta t_p^2 \cdot \gamma_{\phi}(\omega) \quad .$$

Thus there is a compensation proportional to ω^2 between the noise of $\Delta\phi(t)$ and the noise of $\phi(t)$. For a real gyroscope in a noisy (e.g., vibrating) environment phase noise is generated continuously along the fiber but the same kind of compensation applies.

The effect of the modification of the spectral density of the noise by this compensation will be felt in the way the noise level in the output signal varies with output integration time T . For example, white noise in the signal will vary as $T^{-1/2}$. On the other hand white noise generators located within the loop will lead to noise at the output having spectral density proportional to ω^2 when ω is small compared to ω_0 . In this case the noise level in the output signal will vary as $T^{-3/2}$. In practice we can expect the dependence on T to be between the above two rates. Experimentally we observed an rms value of noise on the rotation signal of 0.02 deg/h for T = 10 sec, which is 7 times smaller than for T = 1 sec, which is consistent with the above prediction.

Further motivation for desensitizing the fiber to acoustic noise results from the desire to reduce the response time below 1 sec, which increases the noise level approximately proportional to the bandwidth.

The use of gyroscopes in inertial navigation systems requires a very good long term stability. There are several parasitic effects which produce signals of the order of 10^{-4} rad (i.e., 10 deg h^{-1}), drifting slowly over periods of hours. A simple temporal integration of the output signal gives, as we have seen an important noise reduction over several minutes but does not remove these long term drifts. We have previously described three such effects. [1] As was shown recently [11], an imperfect polarizer could induce nonreciprocity but the very high quality of the present single mode fiber polarizer (90 dB extinction) avoids this problem. The high second harmonic component in the output signal produced by the phase modulation process, can also be a source unless it is filtered out before reaching the lock-in amplifier used in the detection system. Polarization modulation and harmonic generation within the piezoelectric ceramic phase modulator have to be cancelled out using the proper frequency $f_p = 1/2t_\ell$, where t_ℓ is the time of group propagation through the loop.[p]

We have found that, when the above steps are taken, another process leading to long term drift becomes observable at the low level of phase shifts being measured. The Earth's magnetic field creates a residual non-reciprocal Faraday effect, even if the line integral of this field around the closed optical path is zero. The net effect of the Faraday mechanism on the relative phases of the counterpropagating waves depends on their state of polarization: maximum for circular, zero for linear. The residual integration of the Faraday effect will not be proportional to the integral of the magnetic field alone. Thus even in the presence of a curl free magnetic field, the integration can have a finite result, and its value will be environmentally dependent through changes in the light polarization.

Slow changes in the birefringence modify the amount of residual Faraday effect and give a bias drift. This can be suppressed by putting the sensing coil in a field-free environment. With the sensing coil located in a magnetic shield, the bias stability over a few hours is a few tenths of a degree per hour.

In conclusion, it appears that the short term noise of the bias of our all fiber gyroscope compares well with laser gyros. It can probably be improved further by reducing residual acoustic effects. However, the result for the long term stability is still an order of magnitude too high.

References

1 R. A. Bergh, H. C. Lefevre and H. J. Shaw, Optics Letters 6, 502 (1981)

2 H. Arditty, H. J. Shaw, M. Chodorow and R. Kompfner, Proc. Soc. Photo-Opt. Instrum. Eng. 157, 138 (1978)

3 H. Arditty, M. Papuchon, C. Puech and K. Thyagaran, Technical Digest of Topical Meeting on Integrated and Guided-Wave Optics, Incline Village, Nevada, January 28-30, 1980

4 R. Ulrich, Optics Letters 5, 173 (1980)

5 R. A. Bergh, G. Kotler and H. J. Shaw, Electronics Letters 16, 260 (1980)

6 R. A. Bergh, H. C. Lefevre and H. J. Shaw, Optics Letters 5, 479 (1980)

7 H. C. Lefevre, Electronics Letters 16, 778 (1980)

8 C. C. Cutler, S. A. Newton and H. J. Shaw, Optics Letters 5, 488 (1980)

9 J. Feldman and S. Helfant, p. 451, presented at NAECON 77, Dayton, Ohio, May 17-19, 1977

10 N. Lagakos and J. A. Bucaro, Applied Optics 20, 2716 (1981)

11 E. C. Kintner, Optics Letters 6, 154 (1981)

This work was supported by the Atlantic Richfield Co.

Digital Fiber Optic Rate Sensor Development

J.M. Martin, W.S. Brockett, R.L. Selleck, and M.G. Croteau

Martin Marietta Corporation, Orlando Division, P.O. Box 5837
Orlando, FL 32855, USA

Introduction

One of the most important applications of the fiber optic rate sensor (FORS) will be in systems requiring a very wide dynamic response ($\sim 10^8$) operating in a highly dynamic environment. This level of performance requires a rate sensor which 1) is insensitive to acceleration (g), shock, and vibration; 2) has low drift; 3) has the capability for linear response over a wide rate range; and 4) has a fast dynamic response. Other challenging requirements such a rate sensor must meet include rapid warmup, bias stability over long storage periods, and low cost. Presently available sensors include the mechanical spinning mass types and the ring laser gyro (RLG).

The spinning mass gyro is inaccurate in high dynamic environments since its bias has both g and g^2 sensitivities. Further, the spinning mass gyro has limited dynamic range capability ($\sim 10^3$). This is because of an analog readout and scale factor nonlinearity due to the bearings and mass imbalance. The spinning mass inertia required for sensitivity results in this gyro having an inherently slow response time. The principal cost drivers in the spinning mass gyro are the precision machining tolerances on parts and assembly tolerances, rotor balancing precision, and cleanliness of the assembly.

The RLG overcomes some of the problems associated with the spinning mass gyro, but adds new ones. The RLG does not have a spinning mass; hence, there are no direct g and g^2 sensitivities. However, to avoid frequency locking at low rates, RLGs require a bias modulation supplied by either mechanically oscillating the gyro axis or inserting a nonreciprocal phase shifter within the cavity. Further, the HeNe gas ring laser, which is the actual sensor, introduces additional problem requirements, including high-voltage, extremely low scatter optics, a precision-balanced gas discharge, and a highly clean optical assembly.

The FORS overcomes the problems of spinning mass gyros because it has no moving parts. It overcomes the problems of RLGs because it is nonresonant and has all-solid-state optics and electronics, requiring only low voltages

(<25V). Further, assembly tolerance and cleanliness requirements are believed to be substantially reduced with the FORS, although alignment tolerances will be comparable to those of the RLG.

This paper presents progress made toward development of a digital readout fiber optic rate sensor which has demonstrated a wide dynamic range (10 deg/h to 1000 deg/s) and a highly linear scale factor (0.03 percent) over this range. An experimental model, the FORS Model II, was built. It has an optical sensor and all electronics necessary to output a 16-bit digital word at a 2.5 kHz rate. (Over 400 hours of rate testing were performed.)

Principles of Operation

The fiber optic rate sensor is designed to employ the Sagnac effect for measuring rotation rate [1]. The Sagnac effect yields a phase change between two optical beams, one traveling clockwise (cw), and the other counterclockwise (ccw) on a fiber coil.

The phase change, ϕ, is proportional to the angular rate according to

$$\phi = \frac{4\pi RL}{c_o \lambda} \Omega, \qquad (1)$$

where R is the fiber coil radius, L is fiber length, c is the speed of light, λ_o is the freespace wavelength of light, and Ω is the rotation rate.

The counter-propagating light beams are combined on a photodetector. The detector current is proportional to $|\vec{E}_{cw} + \vec{E}_{ccw}|^2$, where $\vec{E}_{cw}$ and $\vec{E}_{ccw}$ are the complex amplitudes of the cw and ccw beams, respectively. In typical implementations [2-4], a modulation is applied which produces a frequency difference between the two beams such that the photodetector current, i(t), will have the form

$$i(t) = P_o R_D \left[\sigma_1 + \sigma_2 - 2\varepsilon\sqrt{\sigma_1 \sigma_2} \cos \; (\omega_1 - \omega_2)(t - nL/c) + \phi_1 - \phi_2\right] \qquad (2)$$

where P_o is total optical power of the source, R_D is detector responsivity, σ_1 and σ_2 are the transmittances of each beam through the optical system, ε is the efficiency of the detection scheme, ω_1 and ω_2 are the instantaneous frequencies of the beams, n is the refractive index of the fiber, and ϕ_1 and ϕ_2 are the phase angles of the beams, which include the Sagnac effect phase shift as well as error terms previously described [5,6].

In our model, the modulation is produced by an acousto-optic (AO) modulator such that the frequency difference, $\omega_1 - \omega_2$, is constant and is equal to 20.5 MHz. This approach permits direct measurement of the Sagnac phase shift as a time delay on an intermediate frequency. Advantages of this approach are a very wide dynamic range and a stable, linear scale factor. Analog approaches to measuring the phase shift in which the signal amplitude is measured are more limited in dynamic range. Additionally, indirect measurement of the phase shift results in reduced scale factor stability.

In the FORS Model II, the frequency difference is produced with a $PbMO_4$ AO modulator which is driven at two phase-locked frequencies (71.750 and 92.250 MHz), as shown in Fig. 1. A net deflection efficiency of 67 percent was achieved over a 2 mm aperture with 2W of RF power. The two Bragg reflected beams are focused one on each fiber-loop end, and exit from the opposite ends of the fiber. The beams are detected as shown in Fig. 2. The beamsplitter shown in the figure samples the two beams which exit from the fiber ends. The beams propagate at a relative angle of 5 mrads which gives an interference fringe spacing of 200 μm (about ten fringes across the 2 mm diameter beam). To enhance the detected signal, Ronchi rulings with the same spacing are placed over the detectors. The resultant signals are 63 percent of the signal available were the beams' angle zero. Thus, ϵ in (2) is equal to 0.63 $(2/\pi)$. A polarizer is included between the fiber and beamsplitter to minimize phase drift due to fiber polarization effects [5]. Also shown in Fig. 2 is an optical-reference detector which produces a reference signal similar to the data signal but without a fiber. The purpose of the optical reference signal is to compensate phase drift introduced by the modulator and its drive electronics.

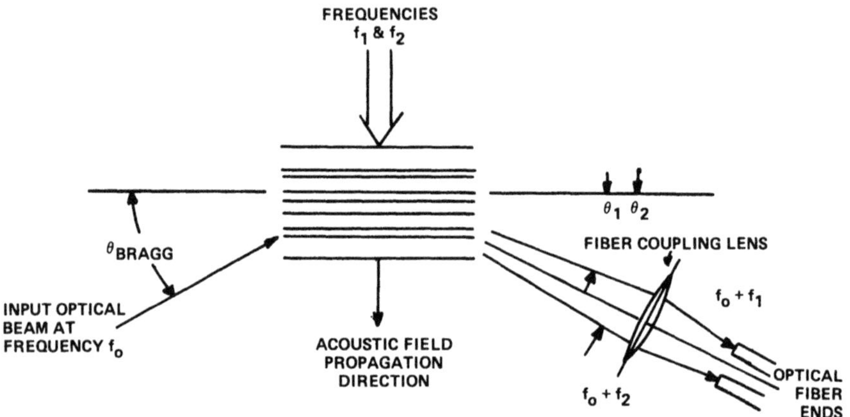

Fig. 1. Digital fiber optic rate sensor dual frequency acousto-optic modulator operation

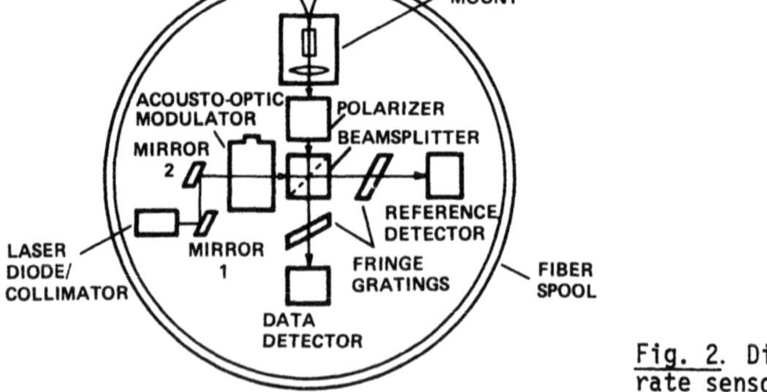

Fig. 2. Digital fiber optic rate sensor optics

A schematic of the processing electronics used to produce a digital output is shown in Fig. 3. The photodetector signals at 20.5 MHz are amplified and mixed with a local oscillator down to 80 kHz. The two channels are balanced to minimize environmental effects and phase stability effects of the local oscillator. Mixing increases the time delay between the two channels by a factor of 256 to 1. The mixer output signals are filtered by an 8 kHz bandwidth bandpass filter, converted to a square wave by a zero crossing detector, and counted down by a synchronous binary counter. The last step allows the original 80 kHz signal to be tracked over 16 cycles of delay. The counter output is fed to a standard digital-phase detector whose output is decoded into phase sign and a gate on an 82 MHz counter. This technique produces a 16-bit output with a least significant bit value of 0.36 degrees of phase shift and a dynamic range of 1000 deg/s. The output logic includes parallel-to-serial conversion for transmission over the rate table slip rings.

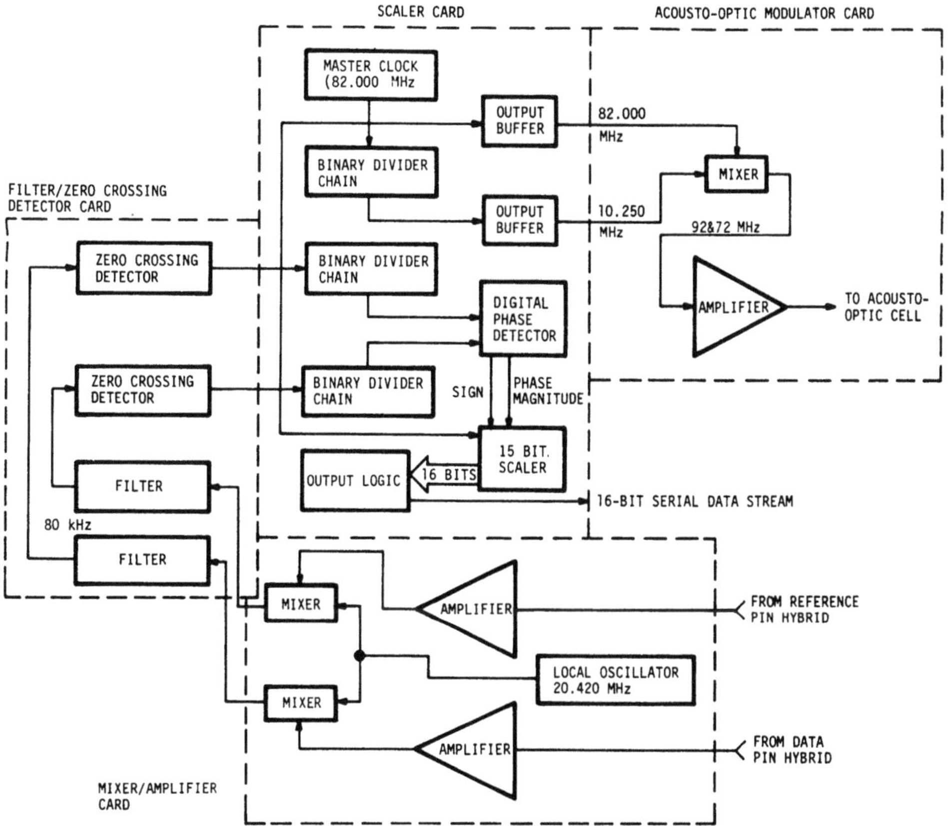

Fig. 3. Digital fiber optic rate sensor optics

Photographs of the FORS Model II on the Genisco Series 1100 rate table in the inertial guidance laboratory are shown in Figs. 4 and 5. Figure 4 shows all the electronics and the rate sensor with one half of the thermal housing removed. Heat was applied for elevated-temperature tests. Using a thermistor located at the inside center of the FORS optical baseplate, top and bottom heater elements were controlled by a proportional controller. The rate sensor with all covers removed is shown in Fig. 5. The baseplate,

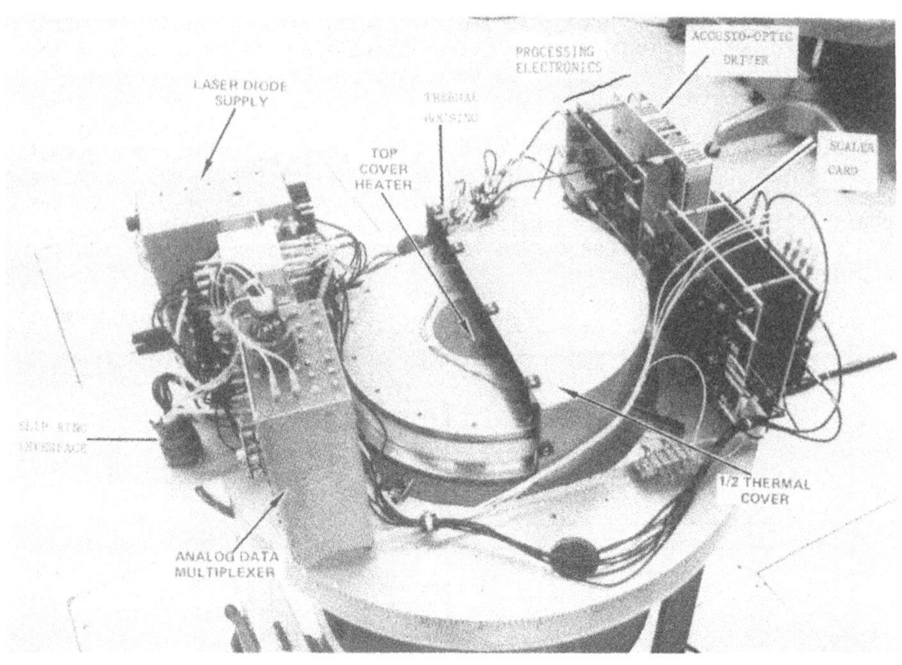

Fig. 4. Fiber optic rate sensor (one-half cover removed)

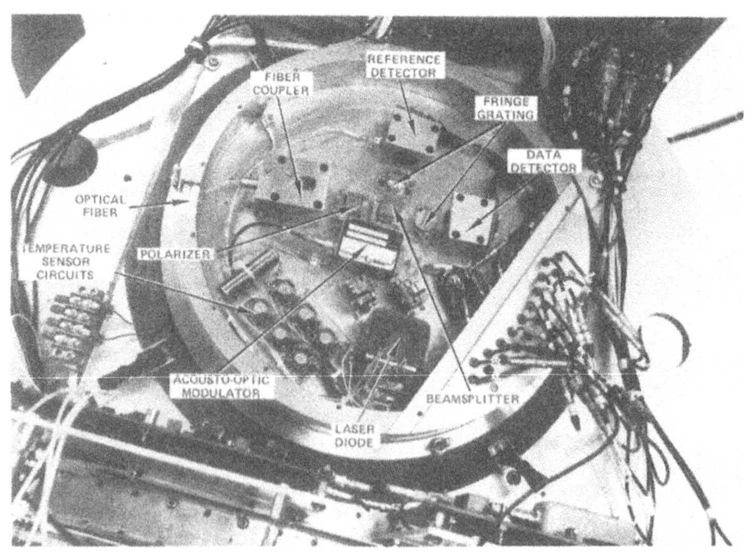

Fig. 5. Fiber optic rate sensor (covers removed)

spool, and optical mounts were made from stainless steel to minimize thermal and mechanical effects on drift and scale factor stability. The fiber spool is 30 cm in diameter and contains 1200m of ITT type T-110 single-mode optical fiber wrapped in 17 layers totaling 1300 turns. The excess spooling loss was 0.5 dB. GaAlAs laser diodes of 830 nm wavelength were used. The lasers were operated using a commercially available power supply. The flux coupled through the fiber was typically 5 to 7 μW. The detectors are silicon PIN photo diodes. Also shown in Fig. 5 are six temperature monitor circuits. These monitored the fiber, spool, baseplate, AO modulator, and laser temperatures throughout testing.

Rate Sensor Tests

The FORS Model II was subjected to rate sensor performance testing according to standard practice [7]. Figure 6 shows the 24 hour drift performance uncorrected rate sensor output versus time, and fiber monitor temperature versus time. Figure 7 shows the data of Fig. 6 after normalization to a

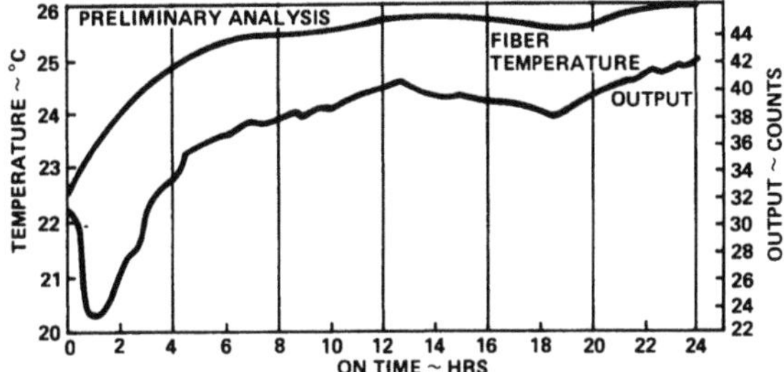

Fig. 6. Preliminary 24-hour initial performance drift test uncorrected - output vs time and fiber temperature vs time

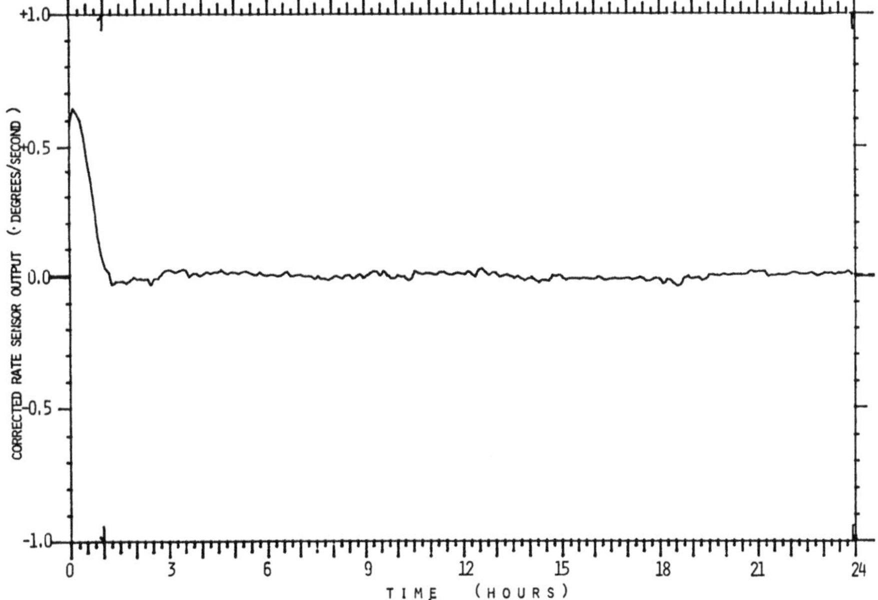

Fig. 7. Corrected rate sensor output vs time-initial drift performance

fixed set of conditions through a multiple linear regression analysis. In general, the output was normalized to fiber temperature and AO modulator drive voltage. The normalization coefficients are those produced by the multiple linear regression analysis. Typical values were 0.29 deg/s/°C temperature coefficient and 0.416 deg/s/v AO modulator voltage coefficient. The rate sensor produced a 16-bit output (phase measurement) every 394 μs which was summed for 0.2 second (508 16-bit words). This sum was stored in the test computer random access memory (RAM). The process was continued for 1.28 minutes, at which time the RAM was full. Finally, data collection was stopped and the RAM contents were transferred to disc storage. The drift test plots are of data averaged over 30 samples, which is 6 seconds of data.

Results of a short off-cycle 1 hour drift test after a 24 hour warmup are shown in Fig. 8. The off time was 1 minute. The drift over the 3 hours was within ±0.008 deg/s. Figure 9 shows a portion of the 1000 deg/s rate sweep test which began at 5 deg/s and increased in 1 deg/s step increments every table revolution. The scale factor linearity was determined by taking data for 100 revolutions at selected rates over the ±1000 deg/s range. The RMS deviation was 0.03 percent of nominal. The nonlinearity is believed to be due to drift error and not to sensor loop thermal and mechanical sensitivity. The noise and response to a step input is shown in Fig. 10. This is a raw data plot on the 0.2 second sample interval.

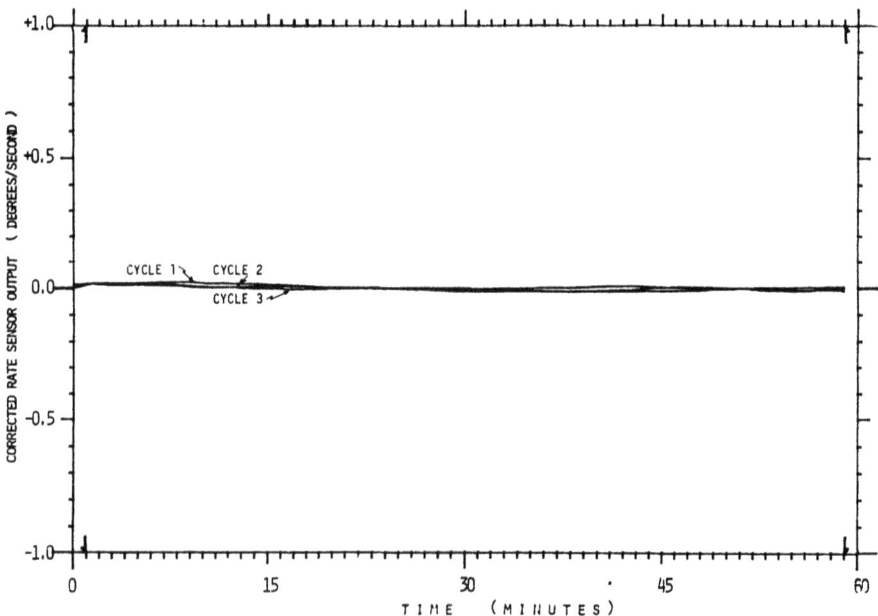

Fig. 8. Corrected rate sensor output vs time - 8 hour drift turn-on transient - ambient temperature

Table I summarizes test results and compares them with predictions. The drift random noise is obtained from short-term noise measurements to avoid drift thermal sensitivity. The drift stability is an average of the RMS drift over all drift runs. the scale factor is given in arcseconds and

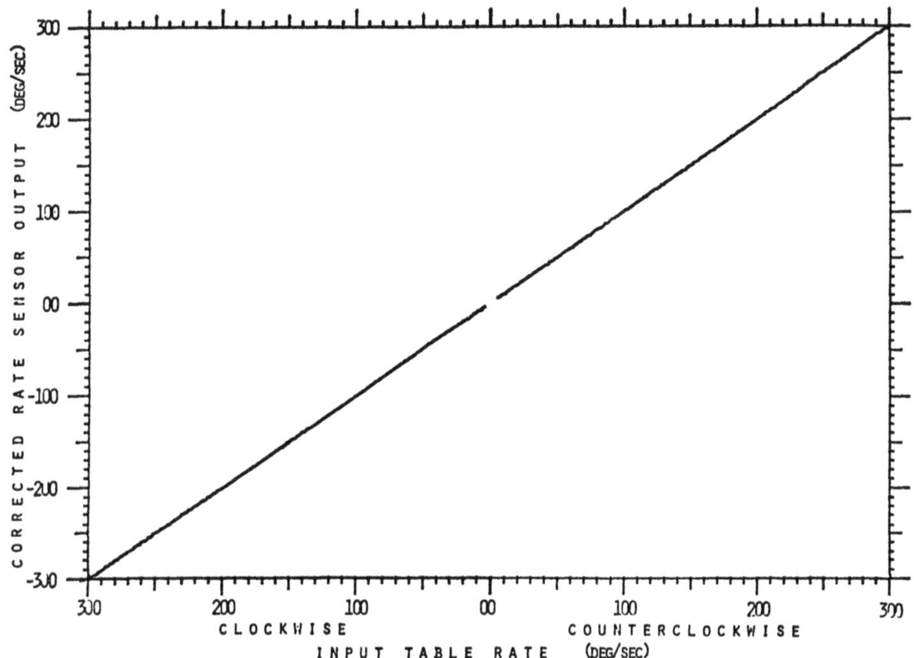

Fig. 9. Rate sweep - corrected rate sensor output vs input rate

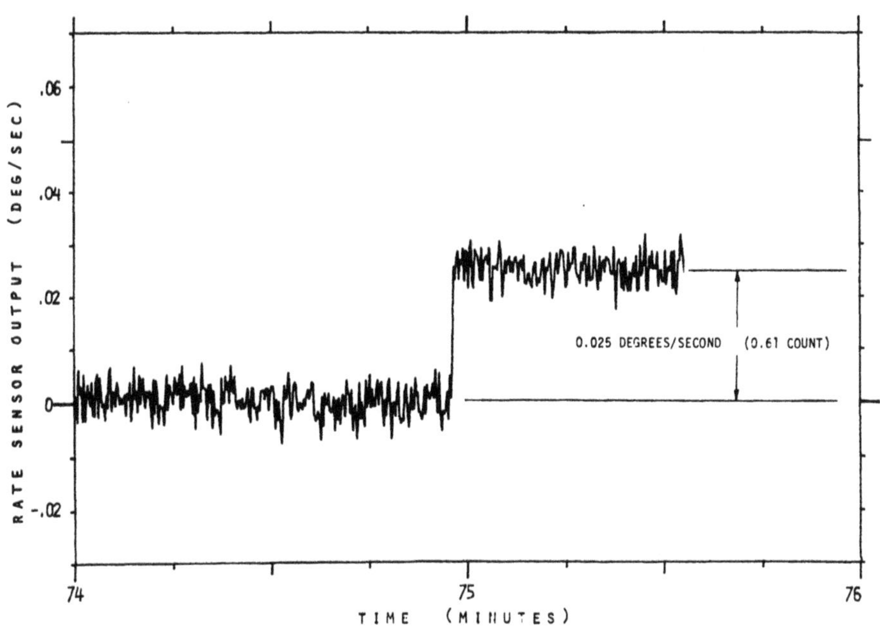

Fig. 10. Corrected rate sensor output vs time for input rate sweep change

deg/s to illustrate that all results may be interpreted in terms of a rate integrating gyro which produces 22.2 million counts per revolution, independent of rotation rate. The scale factor thermal sensitivity listed is an upper bound because the drift-thermal sensitivity limited the measurement.

The drift thermal sensitivity prediction is based upon the stability of the bias term in (2), $(\omega_1-\omega_2)\, nL/c$. The observed sensitivity was 5 times greater. The discrepancy is not likely to be due to polarization effects nor due to thermal gradients along the fiber. Thermal gradients across the fiber spool were measured. During tests, the temperature difference across the spool varied at the rate of 0.2°C/h. This rate of change is 10^3 times too small to account for the observed sensitivity resulting from the thermally induced nonreciprocity effect predicted by SHUPE [6].

Table I. Comparison of measured and predicted performance

Performance Parameter	Model II (Analytical)	Model II (Experimental)
Drift		
Magnitude (deg/s)	-	<0.012
Random noise (deg/$\sqrt{h}$)	0.055	0.062
Stability (deg/s)	-	0.012
Sensitivity (temperature) (deg/s/°C)	0.06	0.288
Scale factor (rate)	0.0403	0.0412
(deg/s/count)		
(arcseconds/count)		0.05837
Stability (ppm)	230	<360
Linearity (percent)	0	0.03
Sensitivity (temperature)(ppm/°C)	35	<200
Threshold		
Dynamic (deg/s)	0.05*	<0.012
Static (deg/s)	0.05*	<0.012
Lifetime		
Storage (y)	-	-
Operating (y)	1	1
g sensitivity (deg/s/g)	0	0
Dynamic range (deg/s/)	±1,000	±1,000
Lock-in rate (deg/s)	0	0
Sample interval (ms)	200	200

* Digital resolution

Drift tests were performed using polarizers with extinctions of $2X10^{-5}$ and 10^{-3} and then using no polarizer. The thermal sensitivity measured without using a polarizer was decreased by 16 times when employing the poorer polarizer. There was an additional improvement of 1.4 when the better polarizer was used. However, the expected improvement is by the square root of the extinction [5]. Therefore, an improvement of 50 should have been observed, were there no other factors contributing to the sensitivity. The small improvement observed indicates that a factor other than polarization effects in the fiber is contributing to the thermal sensitivity. Work is continuing on analyzing this sensitivity. It is believed that 1) the bias term is the significant source of the observed sensitivity, 2) thermal expansion of the jacket may have a greater effect than predicted, and 3) the thermal sensitivity of the fiber refractive index may be higher than the value assumed ($7X10^{-6}$/°C).

Conclusions

We have demonstrated a direct phase measurement, digital, all-solid-state fiber optic rate sensor with a $\sim 10^5$ dynamic range and a linear scale factor over this range. The device tested included all electronics and optics necessary to output a 16-bit digital word, including sign. Predictions of random noise coefficient and scale factor agreed well with the experimental values. Temperature sensitivities predicted were within a factor of 10 or better of measured values. The most significant sensitivity predicted and observed was due to the large phase bias term $(\omega_1 - \omega_2)$ nL/c. No component failures occurred over the 400 hours of tests.

Work is continuing on the digital FORS. Techniques for canceling or eliminating the phase bias term have been developed and are undergoing testing. These techniques and other improvements in the optics and electronics are being incorporated into a sensor design which will be 9 cm in diameter and 3 cm in height.

References

1. E.J. Post, Review of Modern Physics 39, 475 (1967)

2. W.S. Brockett, J.M. Martin, J.M. Hoimes, T.L. Johnson in "Proceedings of the International Conference on Lasers '79", Society for Optical and Quantum Electronics, Orlando, FL (1979)

3. J.L. Davis and S. Ezekiel, Optics Letters 6, 505 (1981) including references, numbers 1-12

4. R.A. Bergh, H.C. Leferre, and H.J. Shaw, Optics Letters 6, 502 (1981)

5. E.C. Kinter, Optics Letters 6, 154 (1981)

6. D.M. Shupe, Applied Optics 19, 654 (1980)

7. J. Feldman and S. Helfant, "A Laser Gyro Evaluation Plan and Test Results," SPIE Proceedings "Laser Inertial Rotation Sensors", 157, 196 (1976) and "Laser Gyro Evaluation Test Plan", Charles Stark Draper Laboratory Report, No. R-988 (July 1977)

Fiberoptic Rotation Sensor: Analysis of Effects Limiting Sensitivity and Accuracy

G. Schiffner, B. Nottbeck, and G. Schöner

Research Laboratories of Siemens AG
D-8000 München 83, Fed. Rep. of Germany

1. Introduction and Experimental Results

In Fig.1 the setup is shown which is under consideration here. We are investigating a conventional Sagnac-type fiber ring-interferometer [1-6]. In our experiment the fiber coil has 12.3 cm diameter and we use about 2880 m single-mode (SM) fiber made in our laboratories. The effective area of the coil is 88.6 m^2. For ease in alignment and to have more opportunities to perform experiments we have designed a symmetrical interferometer arrangement containing two lasers. However, we use only one laser at a time. The lasers are semiconductor laser diodes made in our laboratories, oscillating at about 880 nm and with a spectral bandwidth of 2 to 4 nm.

To obtain an analog signal proportional to the angular velocity a phase modulation is introduced with a piezoceramic element controlling the length of a short fiber section [7]. The modulation frequency is about 38 kHz. Only the optical signal travelling back to the source is used which is picked up by a silicon pin photodiode and amplified in a FET low noise amplifier. This signal is analyzed by means of two lock-in amplifiers which are tuned to the fundamental frequency and to twice the fundamental frequency. The ratio of the output signals is obtained by means of a divider and this ratio is plotted versus time. This method delivers a signal independent of laser power and fiber attenuation. The setup was presented at the 1981 Hannover Industrial Fair and at the Laser 81 Exhibition in Munich.

In Fig.2 the sensitivity of our setup is demonstrated by detecting the earth's rotation rate. For this purpose the fiber coil was mounted in such a way that the axis was oriented parallel to the base plate. This platform could be rotated about a vertical axis. In intervals the axis of the fiber

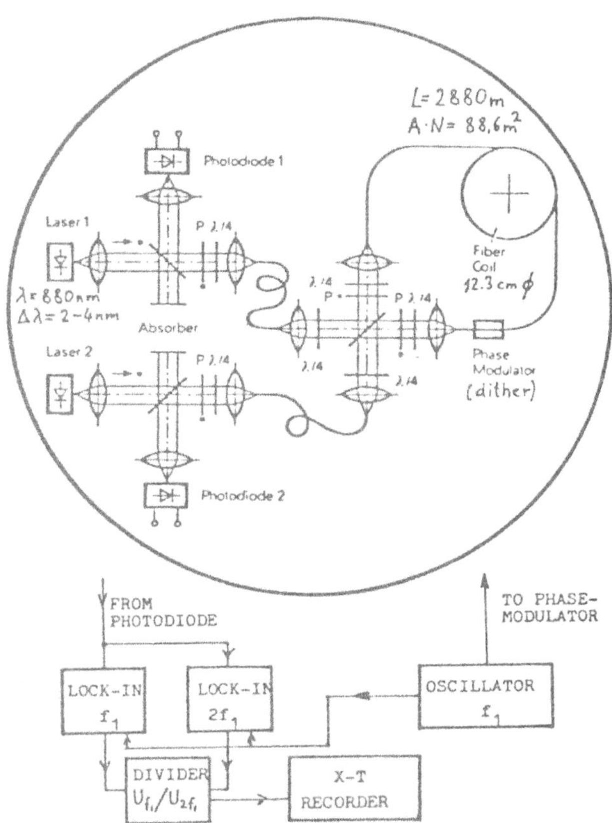

Fig.1. Scheme of the fiber-optic rotation sensor as tested in our experiments. The optical parts are mounted on a platform about 60 cm in diameter.

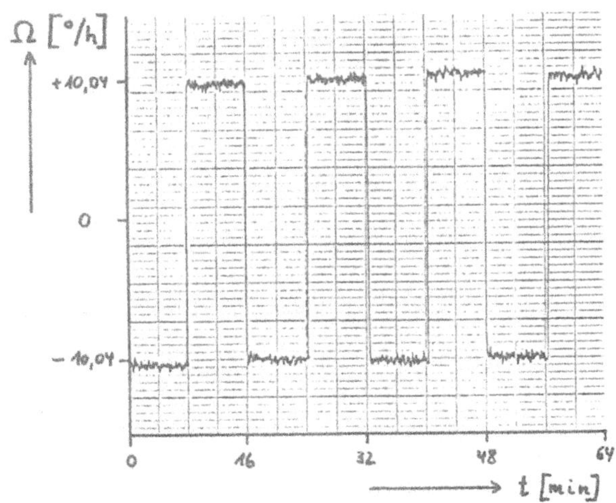

Fig.2. Detection of the earth's rotation rate. The output signals of the lock-in amplifiers are plotted versus time. The horizontally oriented axis of the fiber coil was alternatively turned to north and to south.

coil was aimed to north and to south, respectively. The horizontal component of the earth's rotation at 48° latitude is 10°/h and, therefore, the rotation rate difference between both directions is 20°/h. The ratio of the output signals of the lock-in amplifiers have been plotted versus time using a time constant of three seconds at both lock-in amplifiers. We found that the noise corresponds to about 0.3°/h and the long term drift was about 0.5°/h.

During the experimental work we noticed a number of effects causing drift, scale factor changes and noise. In the following we shall discuss some effects which, in our opinion, are most severe.

2. Drift Effects

Table 1 gives an overview of effects causing drift and means are discussed of reducing these effects.

Table 1. Effects causing drift

Effect	Means to reduce the effect
1 Beamsplitter with free beam propagation: light beams propagating to the fiber ends and beams coming from the fiber ends do not have the same axis	. Insert a good spatial filter (SM fiber filter) and use only the optical signal running back to the source . Use directional couplers, avoid freely propagating beams
2 Losses in the beamsplitter or directional coupler	. Use only the optical signal running back to the source
3 Birefringence of fiber	. Insert polarizers . Stabilization of state of polarization, use polarization preserving fibers
4 Exposure of fiber to thermal radiation	. Shield fiber coil
5 Magnetic field (earth's field)	. Shield fiber coil magnetically
6 Polarization modulation (amplitude modulation) by the phase modulator	. Good design of phase modulator (low PM) . Choose fiber length L according to $L = \frac{c}{2nf_1}$ n group refractive index of fiber, f_1 modulation frequency

Item 1 can be observed if freely propagating light beams are used between the fiber ends and the beamsplitter. By using a piece of SM-fiber between the laser and the beam-splitter acting as spatial filter the effect can be considerably reduced for the otpcali signal travelling back to the laser source. The effect is one reason for the fact that the other optical signal cannot be used. In case the beamsplitter is replaced by a low loss directional coupler and all optical waves are guided in single-mode waveguides, the effect should disappear.

Item 2: In the case that the beamsplitter or the directional coupler shows losses the phase angle between reflected and transmitted beam may deviate from 90 degrees. But this affects only the optical signal not returning to the source.

Item 3: Without polarizers the birefringence of the fiber may cause a very large drift. It has been shown that placing polarizers in front of the fiber ends has the effect that the fiber becomes a reciprocal two-port network. Since the birefringence in connection with polarizers may cause large attenuation a scheme to stabilize the polarization was proposed. Also special SM-fibers maintaining the state of polarization are available and it might be advantageous to use these fibers. In our experiments we used only polarizers and we had no severe problems with changing attenuation.

Item 4: We observed some drift when the fiber coil was exposed to the thermal radiation of a hand, e.g., at a distance of about 30 cm. This is probably due to a change of optical path length near one end of the fiber. Shielding the coil with aluminum foil reduced this effect considerably.

Item 5: We observed some drift and signal offset due to the earth's magnetic field. Shielding the fiber coil magnetically also reduces the thermal effect.

Item 6: As mentioned before we use a phase modulation to obtain a signal proportional to the rotation rate. The elongation of the fiber may cause a polarization modulation which is converted to an amplitude modulation at the polarizers. This modulation is detected by the lock-in amplifier and produces an error signal. In early experiments we used a mechanical dithering of the fiber coil to superimpose an oscillating angular velocity. Later we replaced the mechanical resonator by a phase modulator arranged at one end of the fiber. In both cases we observed the effect.

Good design of the phase modulator such that it changes the state of polarization only very little is one means of reducing the effect. Further we found that choosing a fiber length L such that L is given by $c/2nf_1$ causes a 180 degree phase shift of the modulation signal. This cancels the resulting amplitude modulation for the optical signal going back to the source at low angular velocities. However, this is not a final solution.

3. Noise

In Table 2 the main noise sources are discussed (see also [8].In early experiments using a He-Ne laser as the source, we observed large fluctuations in the photodetector signals which had their origin in additional interferences by scattered light. We changed the light source and switched to a semiconductor laser diode. These lasers have a spectral width of a few nanometers and the coherence length is below 1 mm. Due to this small coherence length the scattered light could not interfere and the signal became free of noise as observed by means of an oscilloscope.

Table 2. Effects causing noise

Effect	Means to reduce the effect
1 Noise due to light scattering in fibers (Rayleigh) and in components	. Light source with low coherence length (multimode semiconductor diode laser), frequency modulated source
2 Photodetector noise	
2.1 1/f-noise	. Phase modulation frequency >1 kHz
2.2 amplifier noise	. Low C input circuit, large R, low noise FET
2.3 shot noise	. High power laser, low attenuation fiber, low fiber coupling losses

The electrical noise coming from the photodetector and its preamplifier can be reduced going to frequencies higher than about 1 kHz to reduce the 1/f noise component. The preamplifier noise may be reduced by well-known methods. The shot noise is always there and may only be reduced by increasing the optical signal.

4. Stability of the Scale Factor

Up to now most efforts have been directed to increasing the sensitivity and stability. However, not very much work has been performed to investigate the stability of the scale factor.

Table 3 gives the main items. The area A of the fiber coil changes slightly with temperature because of the nonzero thermal expansion coefficient of silica fibers. This effect may be taken into account by sensing the temperature and correcting for the rotation rate accordingly, for example, in a data processing unit which evaluates the sensor signals.

Table 3. Effects causing scale factor changes

Effect	Means to reduce the effect
1 Fiber coil	
1.1 Change of area A with temperature	. Sense temperature T, correct for T, stabilize T
1.2 Change of attenuation and birefringence	. Evaluate signals such, that signal amplitude has no effect on scale factor
2 Laser	
2.1 Change of laser wavelength	. Sense for λ, correct for λ
2.2 Change of spectral distribution	. Use low bandwidth source
2.3 Change of output power	. See 1.2
2.4 Effect of light returning to laser	. Insert optical isolator or attenuator, use lasers insensitive to returning light
3 Phase modulation	
3.1 Change of modulation amplitude	
3.2 Change of modulation frequency	. Sense amplitude and frequency, correct for ϕ_1 and f_1

If the attenuation of the fiber is changed due to temperature variation, vibration, and shock this can be taken into account. First, the effects mentioned may be reduced by carefully winding the coil, for example, on a silical glass core which has the same thermal expansion coefficient as the fiber.

If the attenuation of the fiber changes, the optical signal will change in amplitude. By evaluation of this signal with a single lock-in amplifier a change of the scale factor would result. However, in our setup we use a more sophisticated signal processing method which considers both the ω_1- and the $2\omega_1$-component. Both components are proportional to the attenuation factor and it is possible to determine Ω independent of the optical signal power.

If the laser wavelength changes this effect must be taken into account in the data processing unit. If λ is a function of temperature only, a temperature sensor is sufficient. Otherwise a means of measuring the wavelength must be provided.

If the source has a relatively large bandwidth, the mean wavelength is essential for the scale factor. When the spectral distribution changes this might affect the mean wavelength. Since it is not easy to determine the spectral distribution, it might be advisable not to use a source with a large bandwidth.

Some authors mentioned that light returning to the laser disturbs the emission and causes amplitude or wavelength fluctuations. We have not observed this effect. However, optical isolators, high signal attenuation, or lasers which are insensitive to this effect might help.

The stability of the phase modulation amplitude and the phase modulation frequency is important for a stable scale factor. Some stabilization might be achieved by electronic means. Further, the amplitude and the frequency might be sensed and the rotation rate should be corrected accordingly.

5. Miniaturization

Using SM-fiber directional couplers is probably the simplest way of obtaining a miniaturized optical interferometer circuit. For this purpose we have developed a SM-fiber with two cores within the cladding (refer to Fig.3, [9,10] . The core separation is about 10 μm and the core diameter about 4 μm. Light of variable wavelength was coupled directly into one core by means of a short section of an SM-fiber, and the coupling factor was measured versus wavelength. The resulting curve is given in Fig.4 for a fiber piece 24.2 mm in length. The coupling factor is close to predicted values. The coupler has nearly no losses and is independent of polarization. Therefore, it would be ideal for the fiber rotation sensor. The only disadvantage is that we have not yet managed to split the fiber ends to gain better access to individual waveguides.

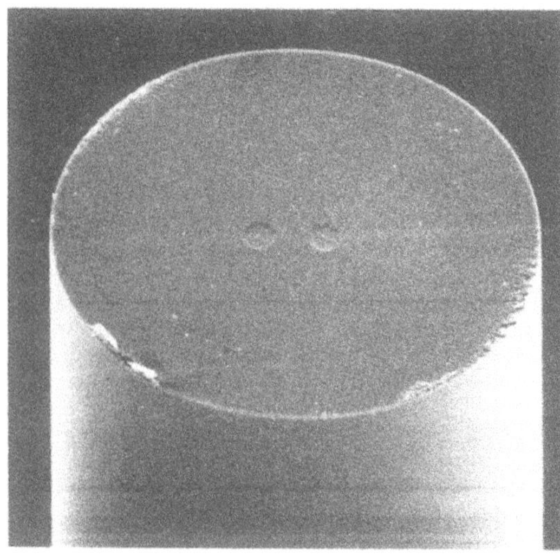

Fig.3. SEM photograph of double-core fiber endface etched with hydrofluoric acid

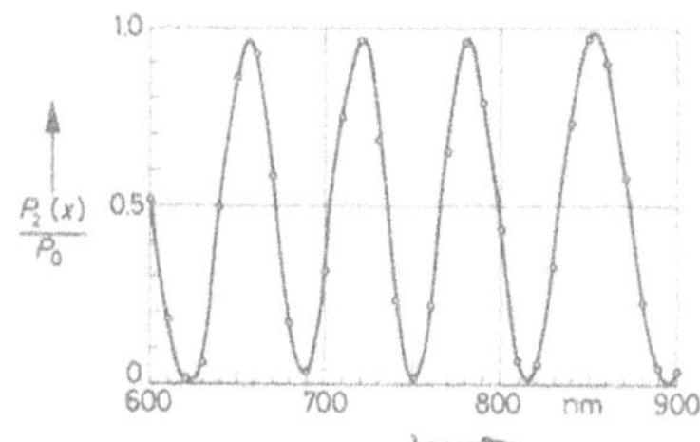

Fig.4. Experimentally determined coupling factor $P_2(x)$ of a double-core fiber versus wavelength λ

We have also made SM-fibers with a core near the surface and the outer cross section is nearly semicircular. We call this fiber eccentric core fiber [11]. Arranging two fibers as shown in Fig.5 leads to a directional coupler. The fibers are embedded in Sylgard. It has been observed that bending the fiber coupler within the plane of both cores changes the coupling properties.

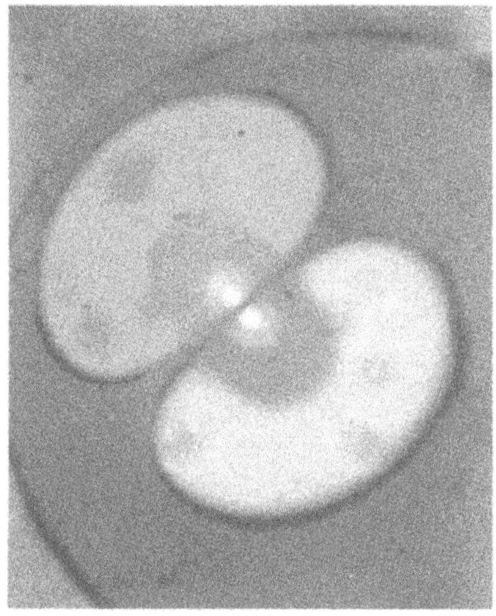

Fig.5. Microscope photograph of a polished cross section of an eccentric core fiber coupler. Light emitted from the cross section was coupled into one of the coupler tails to make the cores visible.

6. Conclusions

Some progress has been achieved during the last years. However, more work must be done to obtain a resolution, accuracy, and dynamic behavior as needed in navigational systems. Due to the fact that many parameters can be changed and due to progress in fiber and diode laser technology there is some hope of further improving the properties of the fiber gyroscope.

Another difficult task is miniaturization by using fiber directional couplers or integrated optics. It is possible that more effects exist which influence the accuracy of the sensor. By increasing the sensitivity automatically new effects might be found.

References

1. E.J. Post: Rev. Mod. Phys. 39, 475-493 (1967)
2. R.A. Bergh, H.C. Lefevre, H.J. Shaw: Opt. Lett. 6, 198-200 (1981)
3. K. Böhm, P. Marten, K. Petermann, E. Weidel, R. Ulrich: Electron. Lett. 17, 352-353 (1981)
4. W.R. Leeb, G. Schiffner, E. Scheiterer: Appl. Opt. 18, 1293-1295 (1979)

5. G. Schiffner, W.R. Leeb, H. Krammer, J. Wittmann: Appl. Opt. 18, 2096-2097 (1979)
6. G. Schiffner: Siemens Res. & Development Rep. 9, 16-25 (1980)
7. J.L. Davis, S. Ezekiel: Proc. of the Soc. of Photo-Optical Instrumentation Engineers, Laser Inertial Rotation Sensors, Aug. 30-31, 1978, San Diego, Calif. 157, 131-136 (1978)
8. S.-Ch. Lin, T.G. Giallorenzi: Appl. Opt. 18, 915-931 (1979)
9. G. Schiffner, H. Schneider, G. Schöner: Appl. Phys. 23, 41-45 (1980)
10. G. Schöner, G. Schiffner: Siemens Res. & Development Rep. 10, 172-178 (1981)
11. G. Schöner, E. Klement, G. Schiffner, N. Douklias: to be published

Investigations on a Fiber Gyro for Heading Reference Applications

K.U. Baron and K. Wickert
TELDIX GmbH
D-6900 Heidelberg, Fed. Rep. of Germany

Summary

For future vehicle navigation and vehicle location systems active as well as passive laser gyros are taken into account as heading sensors. After a short presentation of system considerations on frequently updated vehicle location systems, investigations on a fiber gyro setup are reported. They were aimed at clarifying the physical and engineering problems coming up with the work to be accomplished. In comparison to the requirements derived from the POLA-vehicle location system the results obtained show that the application of fiber optic sensors will be possible in the near future.

1. Introduction

For a long time, TELDIX has been engaged in the research, development and production of navigation systems based on the dead reckoning principle. High accuracy systems, e.g. the TELDIX FNA System[1] use a north seeking gyro for providing the initial heading reference [1] and a directional gyro for measuring changes in heading. Using a gyro with a drift rate of the 0.3°/h class, autonomous navigation over a period of two hours is possible without updating.

For medium accuracy systems, e.g. the TELDIX FOA System[2] a stored north direction is used as an initial reference which, as well as the computed present position, is assumed to be updatable by position fixes between defined time intervals. Therefore, one can apply a directional gyro with a drift rate of the 5°/h class [2] to obtain a navigation accuracy in the

---

[1] Fahrzeugnavigationsanlage, i.e. military land vehicle navigation system.

[2] Fahrzeugorientierungsanlage, i.e. vehicle orientation system for military land vehicles.

order of 1 to 2 % of the distance travelled. Finally, the TELDIX POLA System[3] aimed at the location of special purpose land vehicles within urban areas represents a low accuracy navigation system for which the heading information can in principle be provided by a mechanical gyro or a magnetic sensor.

The POLA System philosophy is based on the fact that the continuously computed vehicle locations can be compared with the city map data stored in a central computer and updated by plausibility considerations.

Location errors caused by a modest drift behaviour of the heading reference may thereby be eliminated. Taking into account the various anticipated mission profiles and characteristics, which are not to be discussed here, we can allow for a maximum heading angle error of up to 8° within two minutes.

Due to various advantages widely discussed in previous publications the utilization of active as well as passive laser gyros is taken into consideration for future navigation systems. TELDIX has therefore started its own developmental programs in the field of active and passive laser gyros. In the following, investigations on a fiber optic gyro are discussed. Such a gyro might be an attractive alternative for getting the heading information within the POLA System since mechanical gyros do not meet the low cost requirement while a pair of probes for measuring the Earth's magnetic field is affected by external fields, e.g. from street cars.

2. Experimental Investigations

The aim of the investigations performed was to get basic information about the performance behaviour of a fiber optic gyro, especially about the sensitivity limiting effects, or, to say it in other words, what physical, engineering and technological efforts have to be made to reach a specific performance behaviour.

2.1 Fiber Gyro Setup

The experimental setup used by TELDIX is shown in Fig. 1. A commercial monomode fiber (154 m) is wound on a plexiglas ring of 0.3 m in diameter resulting in an enclosed optical area of 11.4 m². The light is coupled into the fiber by the lens systems L_1L_4 and L_1L_5. The imaging properties of this coupling system reduce any lateral displacement relative to the beam by more than seven times depending on the beam's divergence angle. The two polarizers P3 and P4 are adjusted in such a way that, by activating the polarization adjustment tool (PAT), the oppositely travelling waves are of the same plane of polarization when reimpinging on the beam splitter. Without a change in the

[3] Positionsbestimmung von Landfahrzeugen, i.e. land vehicle location system.

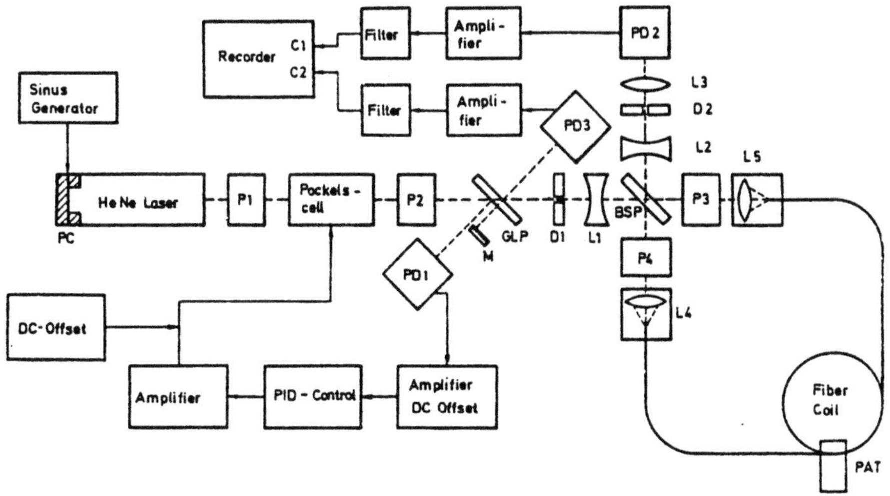

Fig.1. Fiber Gyro Experimental Setup

induced birefringence (see for example [3] to [5]) by external disturbances, this polarization plane remains constant, for which, in practice, special means have to be provided. The intensity stabilization of the incoming light is achieved by a Pockels cell and two polarizers P1 and P2 operating within a closed servo loop. The intensity changes due to inertial rotation are detected by the photodetector PD2 and recorded on one channel of a two-channel chart recorder. On the other channel, the stabilized intensity of the incoming light is monitored.

2.2 Measurements

Figure 2 shows the gyro's output due to angular rates equal or below 0.2°/s and a simultaneous recording of the laser intensity which is stable to within 10^{-4} . To obtain quanti-

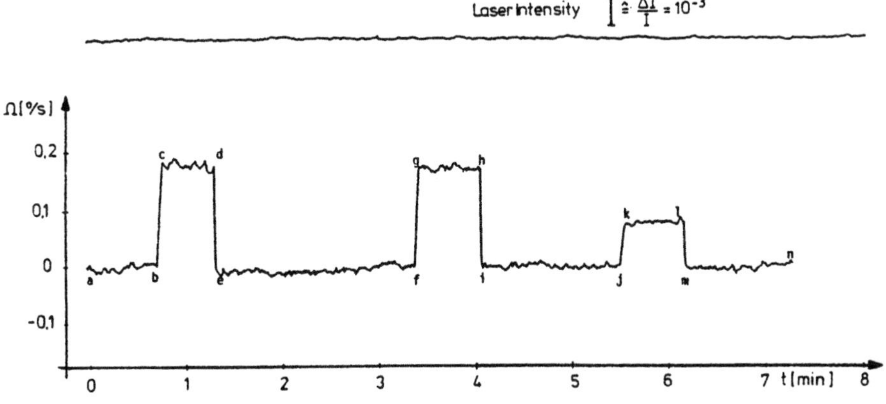

Fig.2. Fiber Gyro Output Signal

tative information about the fiber gyro's performance, the curve of the output signal was digitized and computer evaluated. Thereby, the noise equivalent rotation rate (NERR) computed from the mean value of the standard deviation of all sections a to b until m to n was found to be $\Delta\Omega = 3.8 \cdot 10^{-3}\,°/s$. In terms of gyro technology, this corresponds to a short-term random drift. The slope of the base line gained by linear regression was found to be $\Delta\Omega/\Delta t = 32 \cdot 10^{-3}\,(°/s)/h = 8.9 \cdot 10^{-6}\,°/s^2$. Again in terms of gyro technology, this quadratic drift term corresponds to a random drift ramp.

Table 1. Summary of Error Sources in the Experimental Setup

Error Source	Simul./ Assumed Err.Sig.	Effective Error Signal Quantity	Erroneous Sagnac Signal	Remarks
Laser Intensity Change	$\frac{\Delta I}{I} = 10^{-4}$		$3.8 \cdot 10^{-3}\,°/s$	Calcul.
			$3.8 \cdot 10^{-3}\,°/s$	Exper.1
		$\frac{\Delta I}{I} \approx 10^{-4}$	$3.3 \cdot 10^{-3}\,°/s$	Exper.2
Electr.				
I. Noise				
a=Intens. Stabil.		$\Delta U = 1 \cdot 10^{-2}\,V$	$4.2 \cdot 10^{-3}\,°/s$	
b=Mod. Syst.		$\Delta U = 0.13\,V$	$0.2 \cdot 10^{-3}\,°/s$	Time const. ~ 0.1 s
c=Det. Syst.		$\Delta U = 5 \cdot 10^{-5}\,V$	$1.0 \cdot 10^{-3}\,°/s$	
$\sqrt{a^2+b^2+c^2}$			$4.3 \cdot 10^{-3}\,°/s$	
II. Drift				
a=Intens. Stabil.		$\frac{\Delta U}{\Delta t} = 1.0 \cdot 10^{-2}\,V/h$	$4.4 \cdot 10^{-3}\,(°/s)/h$	
b=Mod. Syst.		$\frac{\Delta U}{\Delta t} = 0.12\,V/h$	$0.2 \cdot 10^{-3}\,(°/s)/h$	
c=Det. Syst.		$\frac{\Delta U}{\Delta T} = 1.8 \cdot 10^{-3}\,V/h$	$8.9 \cdot 10^{-3}\,(°/s)/h$	
$\sqrt{a^2+b^2+c^2}$			$10 \cdot 10^{-3}\,(°/s)/h$	Total drift p.hour
Thermal				
Piezo Expans.	$\Delta T = 0.1\,°C$		$3.8 \cdot 10^{-3}\,°/s$	
Area Change	$\Delta T = 1\,°C$		$0.15 \cdot 10^{-3}\,°/s$	Calcul.
Room Temp. Drift	$\Delta T = 0.01\,°C$		$40 \cdot 10^{-3}\,°/s$	Time const. ~ 2 h
Laser Frequen. Change	$\Delta\nu = 200\,MHz$ $d = 100 \cdot \lambda$		$8 \cdot 10^{-3}\,°/s$	Calcul.

2.3 Discussion of Error Sources

The error sources discussed in the following represent some sensitivity limitations in the experimental setup shown above. They are summarized in Table 1. Each error source listed herein will lead to an erroneous Sagnac signal. It has been distinguished whether the quantitative value of an error source was gained from a simulation or an assumption (column 2) or from a measurement (column 3).

2.3.1 Intensity Fluctuations

Referring to the basic equation representing the behaviour of the Sagnac interferometer

$$I\ (\Delta\Omega, \gamma) = 2\ I_o \cdot \cos^2\left(\frac{\Delta\Omega}{2} + \frac{\gamma}{2}\right) \tag{1}$$

($\Delta\varphi$ = Sagnac phase, γ = phase offset, I_o = incoming light intensity) it turns out that any change in I_o will result in an intensity change of the interference pattern and is not discernable from one caused by a phase shift due to inertial rotation. The figures in Table 1 show a good agreement between the apparent Sagnac signal gained from calculation and measurements.

2.3.2 Electronic Error Sources

According to Fig.1, there are three electronic systems which account for noise and signal drift. These are the intensity stabilization system, the modulation system, and the detection system.

Intensity Stabilization System

Due to the voltage dependent transmission of the Pockels cell any voltage instability at its input leads to an erroneous Sagnac signal which in this setup amounted to $4.2 \cdot 10^{-3}\ °/s$.

Modulation System

Any change in the piezo crystal's length will lead to a laser frequency shift which, in turn, simulates an input rotation rate. The measured noise voltage of $\Delta U = 0.13$ V will therefore cause a rotation rate of $\Delta\Omega = 0.2 \cdot 10^{-3}\ °/s$. The voltage drift is $\Delta U/\Delta t = 0.12$ V/h causing a signal drift of $\Delta\Omega/\Delta t = 0.2 \cdot 10^{-3}\ (°/s)/h$.

Detection System

The detection system also accounts for noise and drift effects. With the light beam blocked off the photodetector, the noise voltage was measured at the amplifier's output to be $\Delta U = 5 \cdot 10^{-5}$ V (within 30 s) corresponding to an angular rate of $\Delta\Omega = 1 \cdot 10^{-3}\ °/s$. The electronic drift was $\Delta U/\Delta t = 1.8 \cdot 10^{-3}$ V/h corresponding to a signal drift of $\Delta\Omega/\Delta t = 8.9 \cdot 10^{-3}\ (°/s)/h$.

Taking the root of the quadratic sum of the different electronic error sources, we obtain a NERR due to the latter ones only amounting to $\Delta\Omega = 4.3 \cdot 10^{-3}$ °/s and a quadratic drift term of $\Delta\Omega/\Delta t = 1 \cdot 10^{-2}$ (°/s)/h.

2.3.3 Thermal Error Sources

Thermal error sources may affect the gyro's performance in a great variety of ways, e.g. by a thermal expansion of the interferometer's area, a thermal expansion of the piezo crystal, a displacement of the coupling units and a change in the fiber's birefringence. Recording the temperature dependance of the interferometer will therefore represent an overall temperature sensitivity, thus summarizing all different effects. Fig.3 shows two recorded temperature dependences. The first was taken during 40 hours including two night periods and one day period in the course of a workday. Within the first 15 hours, there is a decrease in temperature during the evening and night followed by a temperature rise during a workday period and finally again a smaller decrease during the following night. It turns out that a long-term temperature change of $\Delta T = 0.01$°C will lead to an erroneous rotation of $\Delta\Omega = 40 \cdot 10^{-3}$ °/s, while short-term temperature variations are nearly cancelled out. The other graph shows a Sagnac signal and temperature recording over 13 hours in which the temperature variations are within $\Delta T \sim 0.1$°C. The measured drift of the Sagnac signal within this

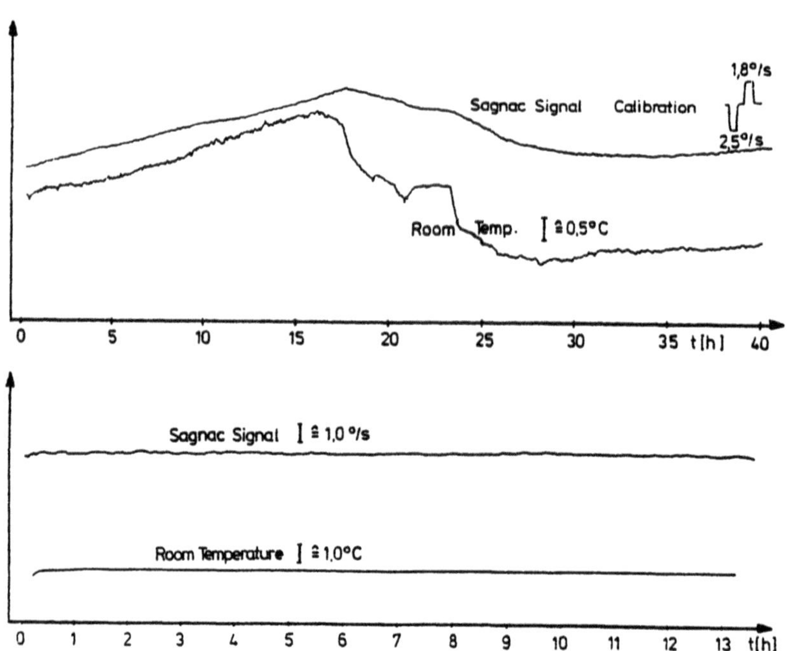

Fig.3. Temperature Dependence of the Fiber Gyro Output Signal

period is in the order of $\Delta\Omega/\Delta t = 3.3 \cdot 10^{-6}$ °/s². The above diagrams show that the relatively simple test setup which, of course, was by no means designed for low temperature sensitivity is severely affected by thermal effects. Thus, for practical applications particularly under stringent environmental conditions special consideration must be paid to avoid these effects.

2.3.4 Phase Fluctuations

Referring to the relationship between the relative phase γ and the optical path difference Δd, which is governed by
$\gamma = 2\pi/\lambda \cdot \Delta d = \nu \cdot 2\pi/c \cdot \Delta d$,
we can see that any change in one of these quantities will alter the phase. The optical path difference Δd is affected by rotation, changes in the birefringence, mechanical disturbances, etc. Even if Δd is constant but unequal to zero, there will be a phase change if the laser frequency is changing. This behaviour is shown in Fig.4. If there is a frequency change of 220 MHz caused, for example, by a change in length of the piezocrystal of 0.1 μm we will obtain an erroneous Sagnac signal of $\Delta\Omega = 5.5 \cdot 10^{-3}$ °/s even if Δd is only about 50 λ.

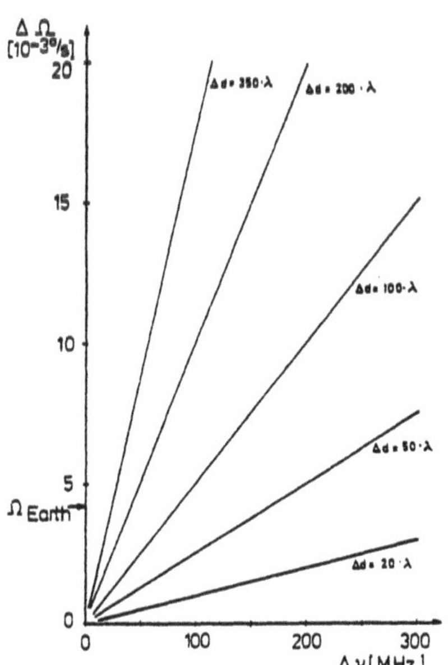

Fig.4 Apparent Rate Signal $\Delta\Omega$ as a Function of a Laser Frequency Change

2.3.5 Influence of Retroreflection

Back coupling of light intensity into the laser resonator itself seriously affects the laser output stability. Light reflections at optical components, the fiber end faces and at imperfections within the fiber core itself contribute to this overall back coupled light intensity [6], [7], [8]. We have reduced this problem by modulating the laser resonator length, thus achieving an improvement of the NERR by a factor of about 75 compared to the case without modulation.

3. Phase Offset and Signal Sensitivity

In considering the various noise sources acting in the fiber-optic gyro we will see that in the case of our DC-detection scheme the specific phase offset γ leading to an optimal signal-to-noise ratio depends on the noise sources dominating in the actual setup.

The signal S to be evaluated arises from the intensity change ΔI due to a rotation induced phase shift $\Delta\varphi$ and is equal to

$$S = \Delta I(\Delta\varphi,\gamma)=I(\Delta\varphi,\gamma)-I(\Delta\varphi=0,\gamma)=a\cdot I_0\cdot\sin(\Delta\varphi+\gamma)\cdot\Delta\Phi \ . \qquad (2)$$

The total rms noise N is assumed to be the sum of the rms contributions from different noise sources which have been mentioned before. Calculating the optimal signal-to-noise ratio with respect to the phase offset γ one will get the following relationship for γ

$$\gamma_{opt} = \text{arc cos}\left(-\frac{1}{1+b_2/b_1 I_0}\right). \qquad (3)$$

Herein the constant b_1 includes all noise sources which are proportional to the intensity and b_2 corresponds to the electronical noise. For a detailed derivation of this expression see reference [9]. A brief discussion of this formula shows that in the general case of comparable contributions from all noise sources the optimum phase bias lies between $\pi/2$ and π. For the TELDIX fiberoptic gyro, measurements have revealed a ratio of $(b_1 \cdot I_0)/b_2 \sim 3.8$ leading to a calculated optimal phase offset of $\gamma_{opt} \sim 142°$.

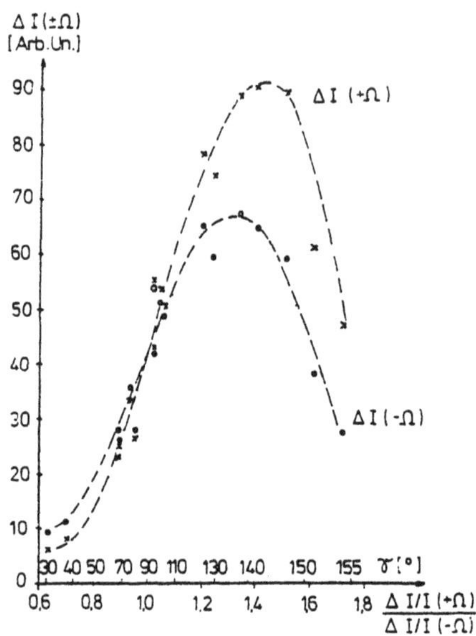

Fig.5. Measured Signal Sensitivity as a Function of the Phase Offset γ

Figure 5 shows the measured influence of the phase offset γ on the signal intensity for two angular rates of equal amount (8.33°/s) but of opposite direction. It turns out that the maximum signal gain will be obtained with a phase offset γ in the region of 140°, showing a fairly good agreement with the estimation performed above.

4. Conclusions

The investigations performed have shown that the requirements defined by the POLA System can be met, at least under laboratory conditions. The heading angle error, $\Delta\Psi$, accumulating between two position fixes will be lower than 1°, resulting from contributions which are linear and quadratic in time, and proportional to the square root of time. In addition, it has been demonstrated that reaching a sensitivity limit of the order of the earth's rotation rate is possible with a comparatively simple laboratory setup, especially without applying a phase sensitive detection technique. It is quite obvious that for field applications, particularly under stringent environmental conditions, a lot of basically physical and engineering work has to be done.

Acknowledgement

We highly appreciate the scientific cooperation with Prof. Dr. W. Demtröder from the University of Kaiserslautern.

References

1 R. Sindlinger, H. Riethmüller: North Seeking Gyro with Automatic Calibration, DGON-Symposium Gyro Technology, Stuttgart, 1980

2 M. Foitzik: KK-25 A new Low Cost Directional Gyro, DGON-Symposium Gyro Technology, Stuttgart, 1979

3 A. Simon, R. Ulrich: Evolution of Polarization Along a Single-Mode Fiber, Appl. Phys. Lett. 31, 517 (1977)

4 A. M. Smith: Polarization and Magnetooptic Properties of Single-Mode Optical Fiber, Appl. Opt. 17, 52 (1978)

5 S. C. Rashleigh, R. Ulrich: Polarization Mode Dispersion in Single-Mode Fibers, Opt. Lett. 3, 60 (1978)

6 D. E. Thompson, D. B. Anderson, S. K. Yao, B. R. Youmans: Sagnac Fiber-Ring Interferometer Gyro with Electronic Phase Sensing Using a (GaAl) As Laser, Appl. Phys. Lett. 33, 940 (1978)

7 G. Schiffner: Lichtleitfaser-Rotationssensor auf der Grundlage des Sagnac-Effekts Siemens Forsch.- u. Entwicklungs-Ber., Bd. 9 (1980) Nr. 1, S. 16

8 K. Böhm, P. Russer, E. Weidel, R. Ulrich: Low-Noise Fiber-Optic Rotation Sensing, Opt. Lett. 6, 64 (1981)

9 K. Wickert, W. Demtröder, K. Baron Experimental Investigations of Sensitivity Limiting Factors in an Optical Fiber Gyro, Opt. Comm. to be published

Dual Polarization Gyro

J.G. Hanse, G.L. Mitchell, and P.E. Bjork
Honeywell, Inc., Minneapolis, MN 55413, USA

This paper presents a new fiber optic gyro configuration that is designed to provide independence from the principal bias errors due to the phase modulator. The configuration operates with optimum $\pi/2$ phase shift for gyro sensitivity, 2x increase in scale factor and has been experimentally shown to provide a 10x improvement in immunity to modulator-induced bias errors.

Introduction

This report presents results from an investigation of dual polarization gyros. The dual polarization concept is a new fiber optic gyro configuration designed to provide independence from principal gyro bias error sources. Experimental results indicate an order of magnitude improvement in bias error performance.

Gyro Error Terms

The most significant difference between rotation sensors and other sensors such as hydrophones or magnetometers is the requirement for absolute measurements. The hydrophone, for example, only needs stability for periods of time on the order of a fraction of a second. Gyros must maintain an absolute reference from hours to years, depending on the mechanization.

This absolute reference requirement is easier to meet with the Sagnac interferometer currently used for the active ring laser gyro than, for example, a Mach-Zehnder interferometer. Both geometric paths in the Sagnac are identical. Components in the Sagnac interferometer are sensitive to environmental effects, giving rise to non-reciprocal phase shifts and, hence, a non-ideal path for light traveling in the two directions. This reduces the stability of the device, causing gyro bias (drift) errors.

It is important to note that despite the popularity of shot-noise-based error calculations, the principal problems in gyro performance have always been component-related. One of the most significant components in terms of stability is the phase modulator, which is used to achieve a $\pi/2$ offset between the two counter rotating beams. The environmental sensitivity of phase modulator components is especially important in military gyro applications, where one normally requires instant-on performance in a rough environment without temperature compensation. These requirements dictate the need for a concept that will remove or reduce the phase modulator errors.

Dual-Polarization Concept

The dual-polarization fiber optic gyro is a modification and expansion of the standard homodyne measurement. The basic principle is to simply operate two independent rate gyros simultaneously in the same fiber and optical components. Each of the gyros is a complete homodyne rate gyro; thus, the errors caused by heterodyning are avoided. If the phase modulator can be made to act oppositely for the two gyros, then it is possible to combine the outputs of the two gyros so that the modulator errors are removed. This is in principle the same as a so-called DILAG (differential laser gyro) concept. The two independent gyros can be made to coexist if their polarizations are orthogonal. The desired phase modulation will be shown to be a natural outgrowth of the orthogonal polarizations. The results of proper implementation of the concept are optimum $\pi/2$ phase shift for gyro sensitivity, 2x increase in scale factor, no first order length bias errors and cancellation of the modulator bias errors.

To aid in understanding the dual-polarization performance with polarization-preserving fibers, let us define a coordinate system convention. For this discussion, we will use a coordinate system tied to the light and aligned with the orthogonal vertical and horizontal axes of the fiber. Figure 1 shows examples of linear polarization states propagating in opposite directions in two fibers. The box diagrams alongside the fibers are how the polarization state appears to the observer who is "looking into the light" with the direction of propagation coming out of the page. If the light is circular polarized, the direction of electric vector rotation in the fiber is represented by an arrow in the diagram with clockwise rotation indicating right circular polarization. This convention will be used in following illustrations to trace the polarization state through the dual polarization fiber gyro. In some diagrams, other information, such as the orientation of wave plate fast/slow axes or accumulated phase, will be presented with the diagrams.

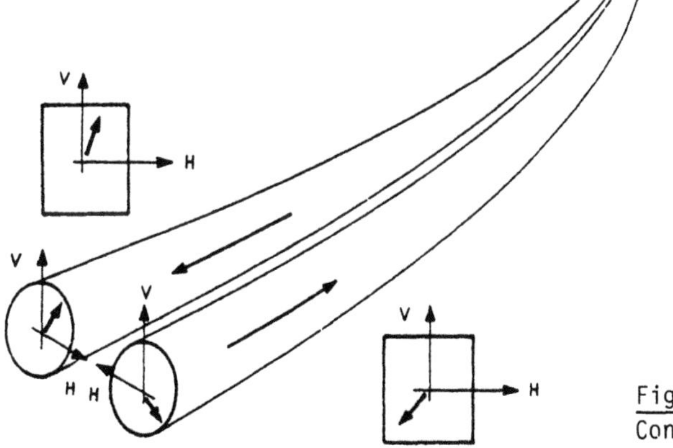

Figure 1. Fiber Polarization Convention

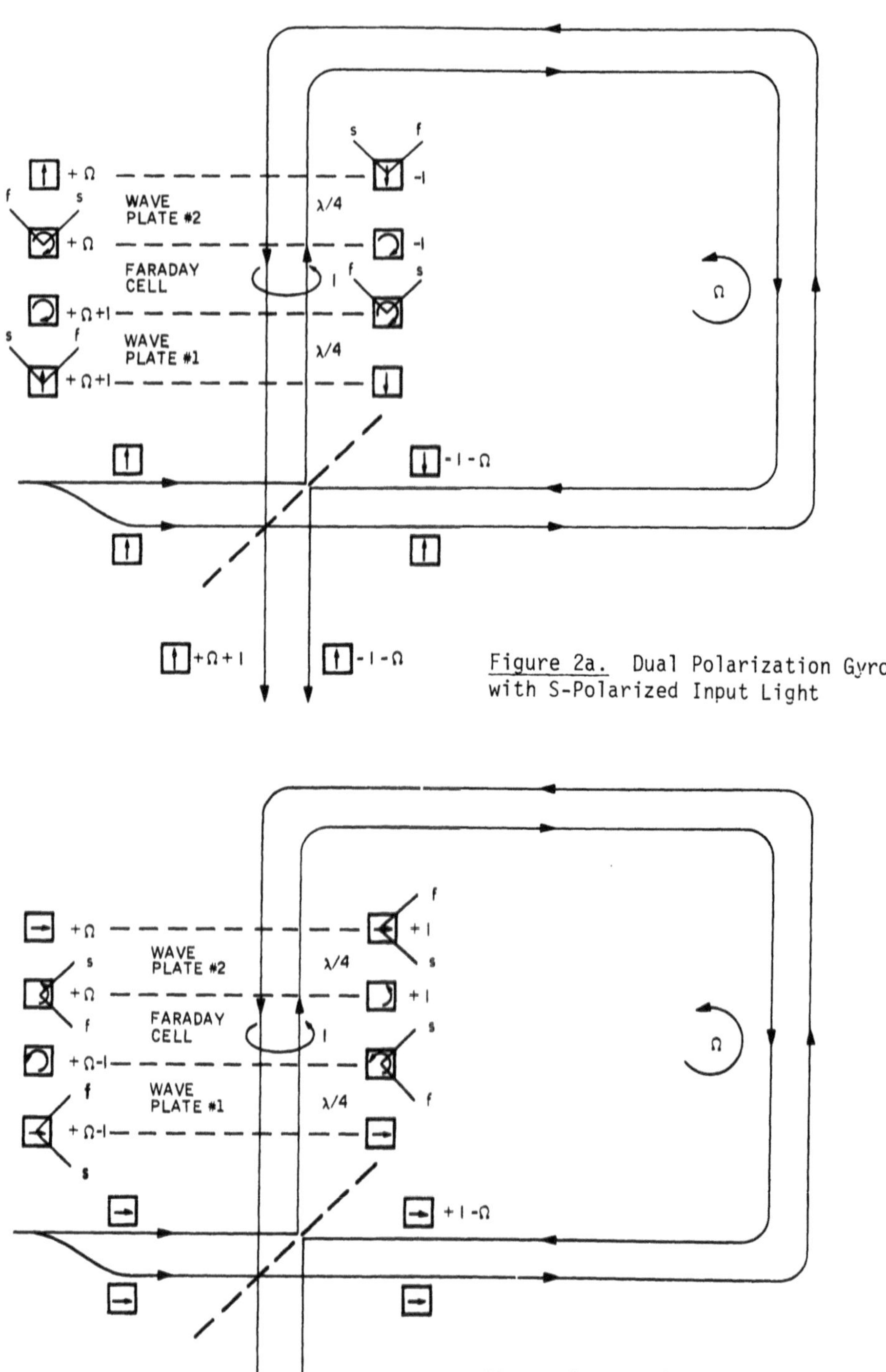

Figure 2a. Dual Polarization Gyro with S-Polarized Input Light

Figure 2b. Dual Polarization Gyro with P-Polarized Input Light

Figure 2a shows a dual polarization gyro configuration using isotropic or birefringent fiber. The source polarization is aligned with the fast axis of the fiber. This beam is divided at the beam splitter, whose axis of polarization is such that the reflected beam undergoes a π phase change. The reflected beam passes through the first $\lambda/4$ plate, becomes RCP (right circuit polarized), passes through the cell, and undergoes a phase retardation. The second $\lambda/4$ plate returns the RCP light to the same linear polarization which entered the first $\lambda/4$ plate. This is a result of having the two plates rotated 90 deg with respect to each other. The linear polarized light enters the fiber aligned with the fiber fast-axis, and propagates clockwise around the loop, causing a Sagnac phase retardation. The output from the fiber is reflected at the beam splitter, accumulates another π change in phase, and strikes the detector. A perfectly isotropic fiber or one with polarization compensation can be used as well as a polarization-preserving fiber.

The other half of the light from the source beam passes through the beam splitter and through the fiber, picking up a Sagnac phase advance. In wave plate #2 it becomes RCP, and since it is passing through the Faraday cell in the opposite direction, it now experiences a phase advance. Wave plate #1 returns the RCP light to the original linear polarization, which passes through the beam splitter and combines with the counter-rotating beam to form an interference pattern whose phase difference can be expressed as

$$\Delta\phi_1 = 2\Omega + 2I.$$

Now consider Figure 2b, which is the same interferometer with the polarization of the source rotated by 90 degrees. The reflected beam is not phase changed. Wave plate #1 creates LCP light which is phase advanced in the Faraday cell. Wave plate #2 returns the light to the initial polarization state; the rotating fiber coil generates a phase retardation, and the output is reflected onto the detector. The transmitted portion of the source light is phase advanced in the fiber, and is changed to LCP light by wave plate #2. Because the LCP light is passing through the cell in the opposite direction from the other beam, it will receive a phase retardation. Wave plate #1 converts the output of the Faraday cell back to the original polarization, which passes through the beam splitter to form an interference pattern with the counter-rotating beam. The phase separation of this interferometer can be expressed as

$$\Delta\phi_2 = 2\Omega - 2I.$$

The phase output of the two polarization situations can be shown to be

$$\Delta\phi_1 = 2\Omega + 2I + \delta\phi_1 ,$$

$$\Delta\phi_2 = 2\Omega - \delta 2I + \delta\phi_2$$

where

Ω = Sagnac phase shift due to rotation,

I = phase shift introduced by a Faraday cell,

$\delta\phi_{1,2}$ = non-reciprocal fiber effects for the fast and slow fiber polarizations.

The intensity at the detector is therefore

$$E_1 \simeq 1 - \cos\left(2\Omega + (2I + \delta\phi_1) \right) ,$$

$$E_2 \simeq 1 - \cos\left(2\Omega - (2I - \delta\phi_2) \right) .$$

The difference of these intensities can then be expressed as

$$\Delta E = (\phi_1 + \phi_2) + 4\Omega \quad 1 + 1/2 \quad I^2 + 1/4 (\phi_1^2 + \phi_2^2)$$

$$+ \frac{I}{2} (\phi_1 - \phi_2) .$$

Notice that, just as in the standard homodyne technique, the fiber non-reciprocal terms $\delta\phi_1 + \delta\phi_2$ appear directly as a bias error. However, the phase modulator instability enters as a scale factor error. The significance of this can be demonstrated in a simple example. Consider a homodyne measurement with a perfect fiber of length 1 km in a 0.1m diameter loop, and a perfect laser source at 0.8μ. The output phase shift would be

$$Z_1 = \frac{4\pi LR}{\lambda c} \Omega + \delta$$

where Ω is the input rate and δ is the phase modulator error.

For the same measurement in the dual-polarization mode,

$$Z_2 = 2 \frac{4\pi LR}{\lambda c} \Omega \left(1 + \frac{\delta^2}{2} \right) .$$

Now, if an undetected modulation shift of 1×10^{-4} rad was made in the presence of a 10 deg/hr input rate, the output phase shift would be

$$Z_1 = 2.27 \times 10^{-4} \text{ rad}$$

for the normal homodyne measurement and

$$Z_2 = 2.54 \times 10^{-4} + 1.27 \times 10^{-12} \text{ rad}$$

for the dual polarization measurement.

For this 10 deg/hr input rate, the conventional homodyne gyro would read out a rate of 17.9 deg/hr, while the dual-polarization gyro would read out 10.006 deg/hr. The result clearly shows that the modulator stability in the dual-polarization scheme is no longer the major error source that it is in the standard detection schemes.

Experimental Details

The experimental set-up used is shown in Figure 3. Here optical components are arranged on a plate which has been mounted on a rate table. Fiber is wound around the periphery of the plate. In this experiment approximately 150 meters of fiber are used on a 0.9 meter diameter circle. A Tropel 100 helium-neon light source is used for the initial experiment. This laser produced noise due to light feedback from the experiment. In later arrangements a Faraday cell isolator was used to reduce the returning light, thereby reducing oscillations in the laser. Laser output power is sampled, as

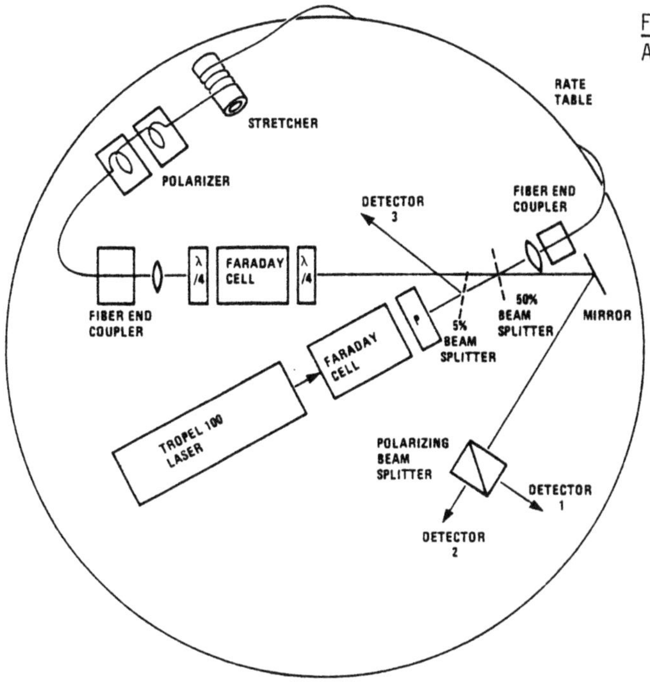

Figure 3. Gyro Component Arrangement or Rate Table

shown, by detector 3. This power measurement and others at various places on the table are connected to a dedicated computer facility used to record data, analyze performance, and plot gyro characteristics.

The 50% beamsplitter which allows the coupling of light to both fiber ends is made from quartz substrate and designed to work at near normal incidence, minimizing polarization-dependent effects. A second Faraday cell with associated quarter wave plates provides the non-reciprocal element needed for dual polarization gyro biasing. Quarter wave plates on either end of this Faraday cell are aligned with a 90° difference with respect to each other.

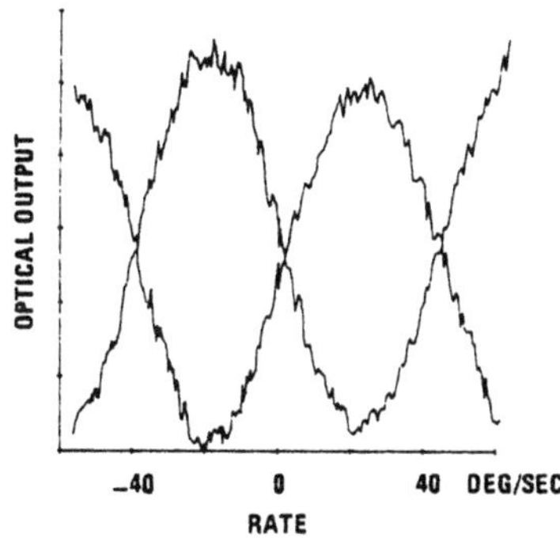

Figure 4. Gyro Output vs. Rotation Rate for Horizontal and Vertical Gyro Polarizations

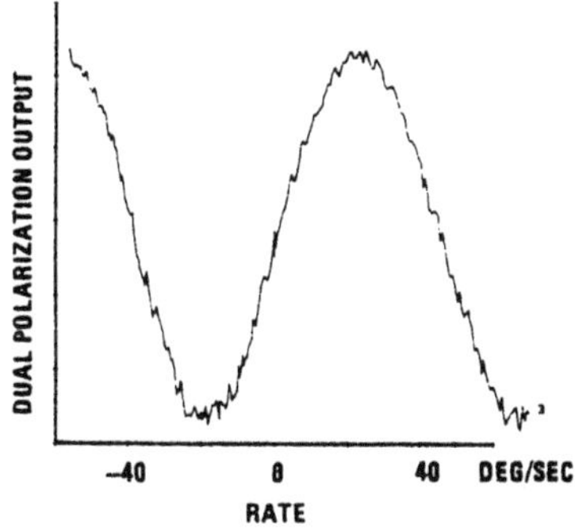

Figure 5. The Combination of Two Polarization Outputs

The optical components on the rate table are connected through a slip ring assembly to a dedicated computer data retrieval system. This is centered around a Fluke model 1720A instrument controller. This system includes an analog scanner which feeds a remotely controlled 6-1/2 digit voltmeter. Data is stored in the 1720A memory, processed, and then stored on disk file.

The dual polarization principle is verified and illustrated in the following figures. Using the breadboard set-up of Figure 3, we were able to change optical components and vary experimental conditions to optimize the gyro operation. Figure 4 illustrates a dual polarization gyro output showing the horizontal and vertical detector signals. This plot shows several of the unique features of dual polarization configurations.

At zero-rate the two outputs are observed to be at the quadrature point ($\pi/2$), and of opposite slope. This situation is obtained with the gyro aligned for maximum output. These two polarization outputs have been combined algebraically in Figure 5. Comparing Figures 4 and 5, one can note that the scale factor for the gyro has been increased by a factor of two and the noise decreased. The nature of the noise is obvious from re-examining Figure 4, in which the correlated peaks can be seen.

In order to decrease the noise in the gyro output, we analyzed the noise sources. There are two major contributors to the output noise. They are laser generated noise and backscattering from inside the fiber. The laser generated noise was primarily due to the feedback of energy from the rate sensor into the laser. This feedback was substantially removed by installing a Faraday rotator/polarizer to isolate the laser from the rate sensor. The fiber backscatter noise was removed by installing a fiber stretcher in the fiber loop. The fiber stretcher was implemented to decorrelate backscattering, allowing further improvement in the gyro performance. Both frequency swept (chirped) and continuous sine wave exertation for the PZT stretcher were evaluated. The most effective decorrelation was found to be in the 4.0 - 4.5 KHz region. A 1-centimeter diameter PZT cylinder with about 15V was used with 10 fiber turns.

The gyro output characteristics using the noise reduction techniques described above are shown in Figure 6. This output compares favorably with the previous output, Figure 5. RMS measurements of noise between these runs show a factor of 30 improvement in noise performance of the rate output.

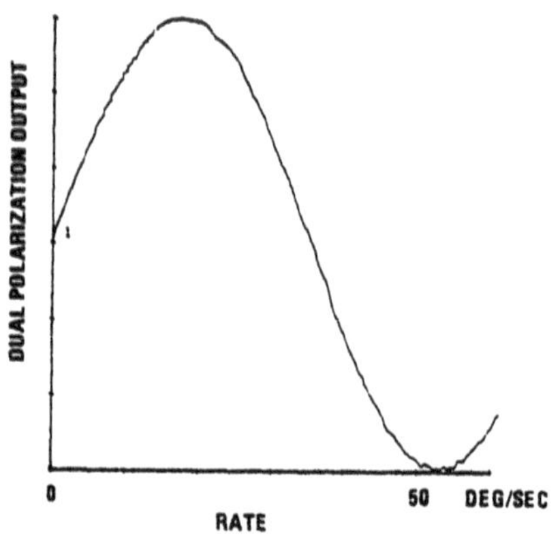

Figure 6. Dual Polarization Gyro Output After Noise Reduction

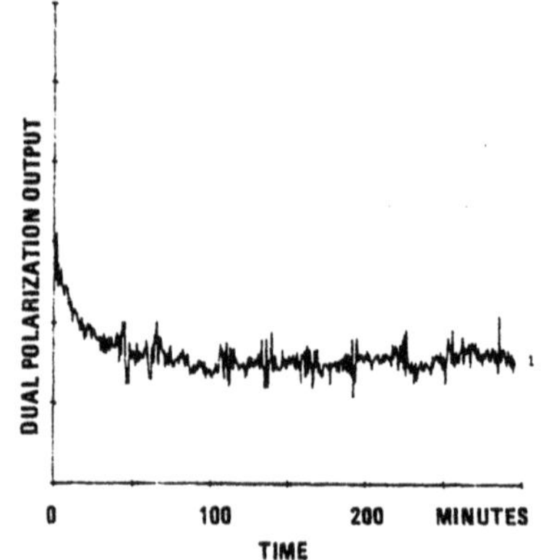

Figure 7. Long-Term Drift Run with Expanded Vertical Scale

A long-term drift run is shown in Figure 7. Here the data are plotted with an expanded vertical scale, allowing more detailed examination of the noise. Aside from a turn-on transient, the data are relatively stable. The drift indicated by the portion of the plot after the stabilization is approximately 50 deg/hr/hr. RMS noise in the best portions of the drift run were 1.6 deg/hour.

The most important objective of this experimental effort was to demonstrate dual polarization gyro independence from phase modulator errors. To accomplish this it is necessary to measure gyro bias while changing phase modulator settings. The principal factor in phase modulator stability is environmental temperature; this parameter was varied over a time period of about one hour, and the gyro drift recorded. The gyro temperature was raised $\sim 10^{0}C$ above ambient and allowed to cool. The observed results are shown in Figure 8. The conventional outputs (labeled 0 and 2) show the expected shift in output with temperature. Curve 1 shows the dual polarization performance of this gyro. It is an order of magnitude more stable than the conventional (single) polarization gyro. This confirms the expectation of better drift performance for the dual polarization gyro.

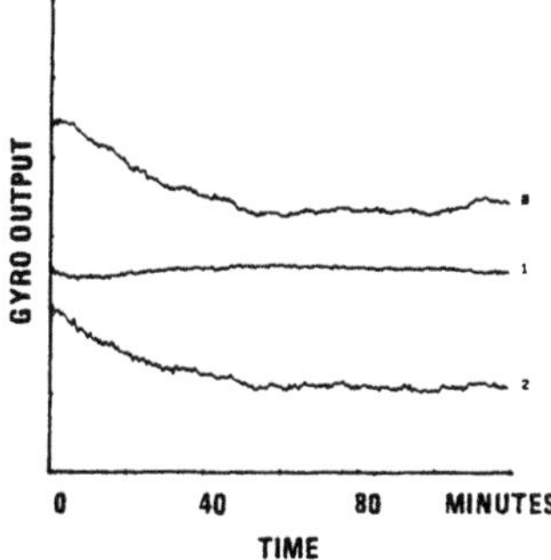

Figure 8. Single Polarization (curves 0, 2) and Dual Polarization (curve 1) Output Showing an Order of Magnitude Stability Improvement for Dual Polarization

A Fiber Gyroscope Based on a Two Frequency Zeeman Laser

S. Donati and V. Annovazzi-Lodi
Istituto di Elettronica e Com. Elettriche, Università di Pavia,
I-27100 Pavia, Italy

Abstract

In this work we present a new approach for implementing a fiber gyroscope, based on a frequency-stabilized Zeeman He-Ne laser as the source feeding the fiber coil. The two modes of the source are generated at slightly different frequencies and have orthogonal polarizations. These features make it possible to measure the inertial phase shift with its sign, and give high linearity on a wide dynamic range of rotation and inherently good sensitivity.

1. Introduction

In the development of the fiber gyroscope, some measurement problems have been encountered in processing the output signals, i.e.:

- linearity error due to the cos ϕ dependence of the photodetector beating signal, where $\phi = 8\pi AN\Omega/\lambda c$ is the inertial phase shift
- sign detection on Ω
- offset in the $\Omega = 0$ output signal
- errors in the integration operation by which the rotation angle $\psi = \int \Omega dt$ can be obtained.

Though different approaches have been proposed [1-5], the simplest way to overcome these problems appears to be the introduction of a frequency offset between the two modes, which propagate in the opposite directions of the fiber coil. Indeed, a number of papers have reported gyro schemes making use of acoustooptic modulators both internal and external to the fiber coil [6-9], but when the two modes are not inherently supplied to the source, some residual criticality still remains because of alignment and wavefront matching problems.

We used a Zeeman-split two-mode source to sense the inertial phase shift with an offset frequency, obtaining a beating signal of the form $\cos(\omega_1 - \omega_2)t + \phi$. With this, one can measure Ω with no linearity and bias errors if a reference signal $\cos(\omega_1 - \omega_2)t$ is taken from the laser beam before it is injected into the fiber coil.

In addition, the two fields are orthogonally polarized and can be easily separated and recombined by a Glan cube. The sign of Ω is detected by comparing the measure and reference sine waves, and the digital integration of Ω can also supply the angular position ψ without the usual errors due to an analog integration.

2. The Frequency-Stabilized Zeeman Source

The source was implemented by Zeeman splitting of the Ne atomic line [10]. A stack of C shaped, AlNiCo permanent magnets was applied to the capillary tube of a commercial unit (0.5 mW, internal mirrors, Spectra-Physics 144) on a 80-mm segment so as to supply a transversal magnetic field of about 300 G with good uniformity.

In the transverse Zeeman effect, the active medium exhibits a double-peaked atomic line for the mode oscillating with a polarization normal to the magnetic field, while the line pertinent to the mode parallel to the field is unperturbed. Thus, due to the difference in their pulling effects, the two orthogonal modes will oscillate with a small frequency offset, which is a function of detuning between the atomic center and the cavity resonance frequencies [10-12]. As the detuning is increased up to the limit of the single-mode regime, the frequency offset increases, e.g., from a few kHz to some 100 kHz, and thus one can adjust Δf by controlling the cavity length L. This control, which yields frequency stabilization, is, however, necessary to overcome minute fluctuations of L (e.g., due to thermal drifts and microphonics). Of the different possibilities, we selected the thermal-control technique [10] based on the expansion produced by a short coil of NiCr wire wound directly on the capillary tube (Fig. 1).

The coil is driven at a few watt level by the amplified error signal of the beating frequency Δf, converted to a voltage with respect to a reference voltage. We found for the loop gain G an optimum value $G = 10^3$ at a loop frequency cutoff of f = 3 Hz. Typical stabilities at Δf = 50 kHz were ±1 Hz (T = 1 s) and ±100 Hz (T = 1 h).

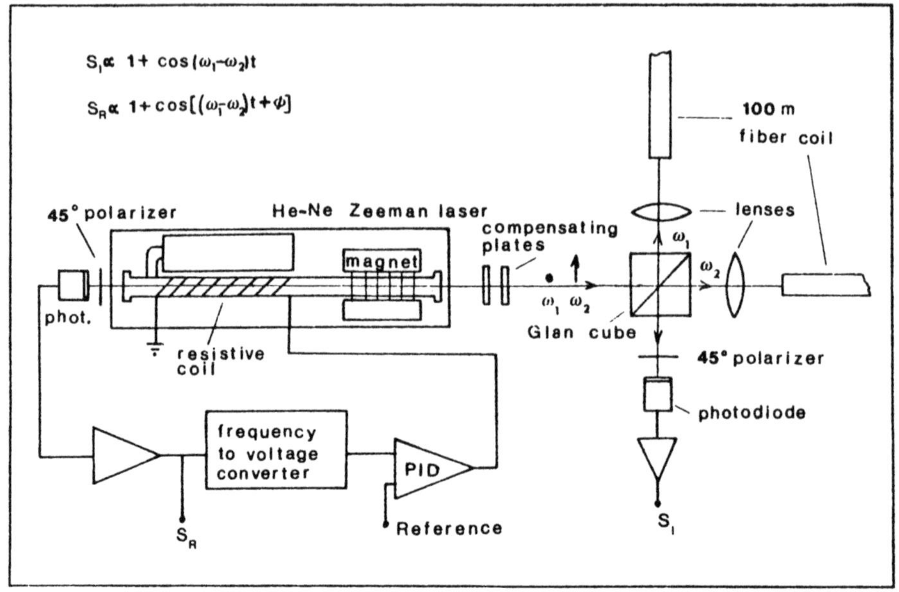

Fig. 1. Gyro setup showing the source stabilization loop

The two-mode Zeeman operation was found to be feasible in most other He-Ne tubes of the side-arm, internal-mirror type, and it was also feasible with a longitudinal field for which one obtains two modes with opposite circular polarization at the output.

3. Fiber Gyroscope Setup

As the propagation medium we used a 100-m sample of Valtec SM05 monomode (850 nm) fiber. Careful winding was performed, to avoid torsion, on a 250-mm diameter coil. Since the fiber allows two LP modes to propagate at the He-Ne wavelength, a mode filter is used at both ends of the coil. This made by winding a short piece (1 m) of the fiber ends on two 40-mm diameter rings. This filters out the LP_{11} modes while introducing negligible attenuation on the fundamental LP_{01} mode.

The basic configuration of the gyroscope is shown in Fig. 1. The two modes at frequencies ω_1 and ω_2 are separated by the polarizing cube and focussed onto the fiber ends. After propagation, the same polarizations are directed toward the photodiode, where a beating is obtained by inserting a 45°-oriented polarizer. The signal S_I is compared to S_R, taken on the back mirror of the Zeeman laser, so as to extract the inertial phase shift ϕ. S_I and S_R are passed through zero-crossing discriminators, which feed the J and K inputs of

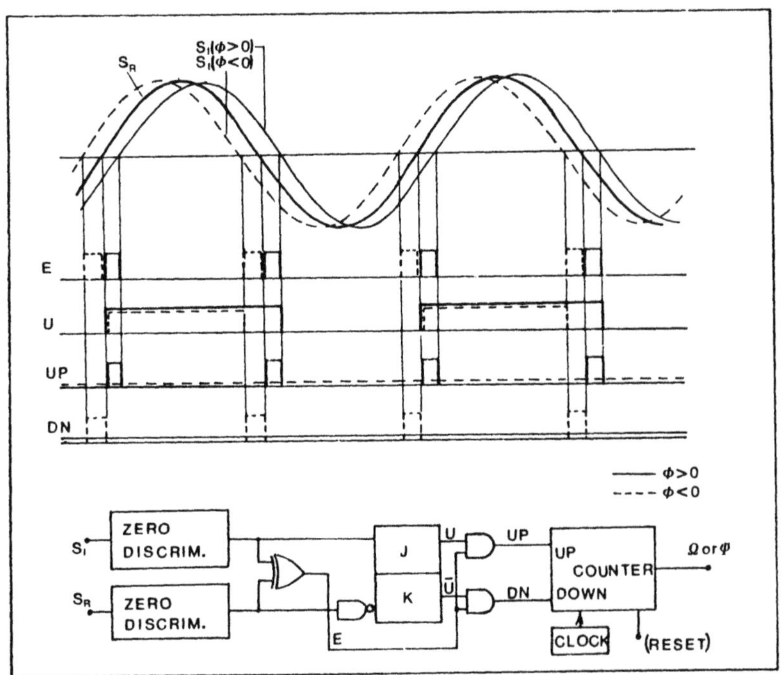

Fig. 2. Electronic handling schematics for Ω and ψ detection

a flip-flop. The output U of the flip-flop (Fig. 2), when combined with the result of either S_I or S_R, gives the sign information in the form of UP and DN logic signals as shown in the timing diagram in Fig. 2. The counter can give either the integrated phase ψ when allowed to count freely without any reset, or the Ω signal if it is periodically reset at a frequency f_r; the ratio f_{clock}/f_r is the scale factor of countings $n = (f_{clock}/f_r)\ 8A\Omega/\lambda c$. Note that in this measurement scheme no linearity error is introduced in the full range $-\pi < \phi < +\pi$. However, if the two modes are not exactly orthogonal or have a little ellipticity, a spurious term is added to S_R and S_I which cannot be cancelled out in the whole dynamic range. While the origin of this error appears to be attributable to nonuniformity of the magnetic field and residual coupling, an a posteriori compensation can be provided by a variable birefringent plate ($\lambda/2$ plus $\lambda/4$, see Fig. 1) adjusted at the proper angle. In this way we can reduce the spurious term, to which the nonlinearity error is proportional, to less than 1% (Fig. 3).

The responsivity of our setup, with 100-m fiber length, is $\phi/\Omega = 0.8$ rad/(rad/s). The best sensitivity was 200 μr (S/N = 1), but it was not always reproducible. When a couple of 45°-oriented polarizers were inserted at the

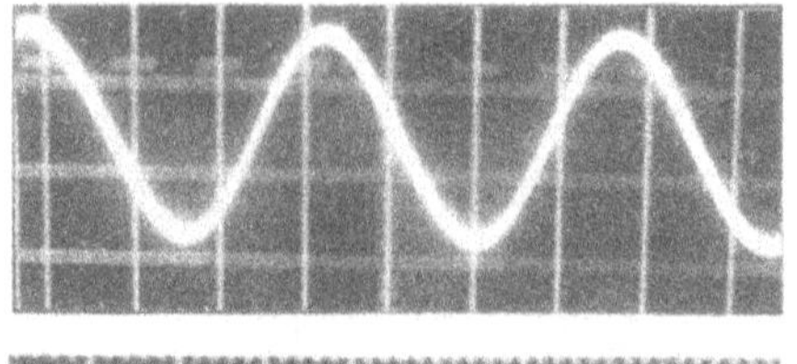

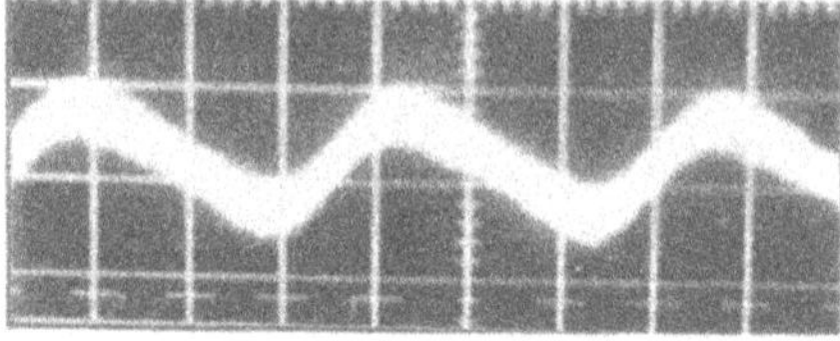

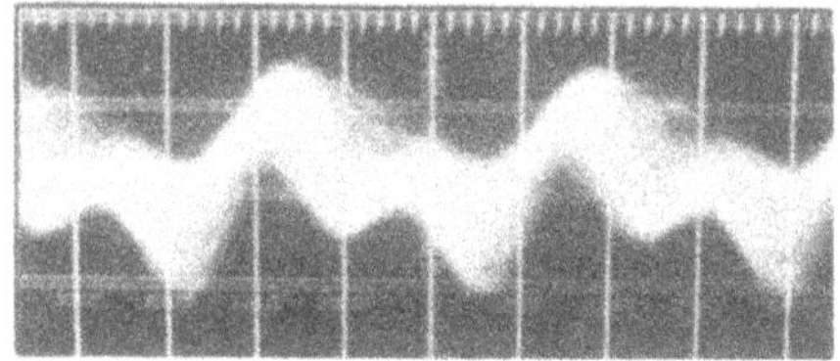

Fig. 3. Orthogonality of Zeeman laser modes: the beating signal S_R is shown with the polarizer at 45° (upper trace, 100 mV/div) and at 0° and 90° (middle) and botton traces, 1 mV/ div)

fiber inputs so as to propagate identical polarizations, i.e., ensure reciprocal propagation in the fiber, the results showed a fair improvement and became much more reproducible. Also, the stability of the zero-signal baseline dropped to typical values of 50 μr in a 1-h period.

Acknowledgements. This work was supported by the Consiglio Nazionale delle Ricerche. Valerio Annovazzi-Lodi was sponsored by a scholarship from the Soc. Selenia S.p.A., Rome.

References

1 R. Ulrich: Opt. Lett. *5*, 173 (1980)
2 S.C. Rashleigh, W.K. Burns: Opt. Lett. *5*, 482 (1980)
3 R.A. Berg, H.C. Lefevre, H.J. Shaw: Opt. Lett. *6*, 198 (1981)
4 C.H. Bulmer, R.P. Moeller: Opt. Lett. *6*, 572 (1981)
5 K. Hotate, Y. Yoshida, M. Higashiguchi, N. Niwa: Appl. Opt. *20*, 4313 (1981)
6 D. Thompson, D.B. Anderson, S.K. Yao, B.R. Youmans: Appl. Phys. Lett. *33*, 940 (1978)
7 R.F. Cahill, E. Udd: Opt. Lett. *4*, 93 (1979)
8 J.L. Davis, S. Ezekiel: Opt. Lett. *6*, 505 (1981)
9 G.A. Sanders, M.G. Prentiss, S. Ezekiel: Opt. Lett. *6*, 569 (1981)
10 S. Donati: J. Appl. Phys. *49*, 495 (1978)
11 S. Donati: *Conference Digest CPEM 78*, Conference on Precision Electromagnetic Measurements, Ottawa (IEEE Press, New York 1978) p.175
12 S. Donati, V. Speziali: Alta Freq. *47*, 172 (1978)

4.2 Closed-Loop Operation

Closed Loop, High Sensitivity Fiberoptic Gyroscope

M.G. Prentiss, J.L. Davis, and S. Ezekiel

Research Laboratory of Electronics, Massachusetts Institute of Technology
Cambridge, MA 02139, USA

The possibility of using a multiturn fiber Sagnac interferometer as a sensor of absolute rotation has aroused much interest over the past few years. Because such sensors are based on the detection of very small nonreciprocal phase shifts in kilometer long fibers, the research has been very challenging as well as very exciting. A number of different approaches have been studied [1-14] and, as these studies progressed, a number of interesting problems have been uncovered [15-19]. Methods of solving such problems have been developed enabling a rather rapid improvement in performance, particularly over the past few months [13,14].

In this paper we present the results of preliminary studies using a 200 m long fiber gyro that exhibits low drift and short term noise that is close to that predicted by photon shot noise relevant to our setup. Our approach is based on a closed loop technique that we proposed several years ago [4] which employs two acoustooptic (A/O) frequency shifters within the fiber interferometer.

Our setup, which at present uses discrete components, is shown in Figure 1. Light from a 5 mW He-Ne laser passes through an A/O isolator followed by a short single mode coupling fiber. After transmission through a vertical polarizer P, the light enters the Sagnac interferometer at beam splitter BS2 where it splits into clockwise (cw) and counterclockwise (ccw) beams that propagate through a single mode fiber coil that is 200 m long and 19 cm in diameter. The cw and ccw beams pass through A/O shifters placed at equal distances from the beam splitter (BS2). One of the A/O shifters is driven at a fixed frequency f_2 = 40 MHz and the other A/O shifter is driven at f_1 by a voltage controlled oscillator (VCO) around 40 MHz. In addition, the cw and ccw beams pass through an electrooptic phase modulator (E/O) placed near the beam splitter (BS2) and driven at 470 kHz. That portion of the light that returns through the coupling fiber is detected by a photodetector (PD) after reflection at beam splitter BS1.

As discussed elsewhere [4,14], measurement of nonreciprocal phase shift is accomplished by generating a nonreciprocal phase modulation at a rate f_m using the E/O crystal and subsequently demodulating the output of the photodetector in a phase sensitive demodulator (PSD) at f_m. Closed loop operation is achieved by feeding the output of the PSD into a servo amplifier which in turn adjusts the VCO frequency f_2 so as to maintain a net nonreciprocal phase shift of zero. In this way, the frequency difference $\Delta f_{12} = f_1 - f_2$ is directly proportional to nonreciprocal phase shift generated by a rotation rate Ω given by

$$\Delta f_{12} = \frac{4A}{\lambda_o P_n} \tag{1}$$

where A is the area enclosed by one turn of the fiber coil, λ_o is the vacuum wavelength of the laser, and P_n is the optical length of one turn of the fiber coil. It should be noted that for a fixed fiber coil geometry, Δf_{12} is independent of the total fiber length L and the number of turns N in the coil.

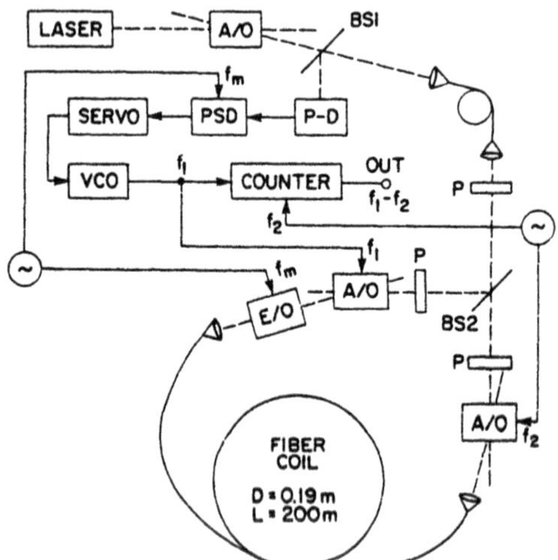

Fig.1. Experimental set-up

One of the basic limitations on the measurement of a rotation rate Ω is set by the shot noise associated with the light received by the photodetector. This photon shot noise [4,15] causes an uncertainty $\delta\Omega$ in the measurement of rotation rate given approximately by

$$\delta\Omega \approx \frac{\lambda_o c_o}{2\pi L D} \frac{1}{(N_{ph}\ \eta_D \tau)^{1/2}} \tag{2}$$

where c_o = the speed of light in vacuum, L = the total length of fiber, D = the diameter of the coil, N_{ph} = the number of photons/sec reaching the detector, η_D = quantum efficiency of the detector, and τ = the averaging time.

In practice, there are a number of noise sources that are much larger than the photon shot noise. For example, Fig. 2 (left) shows the output of the PSD (τ = 1 second, 6 dB/octave filter) with the feedback loop open and

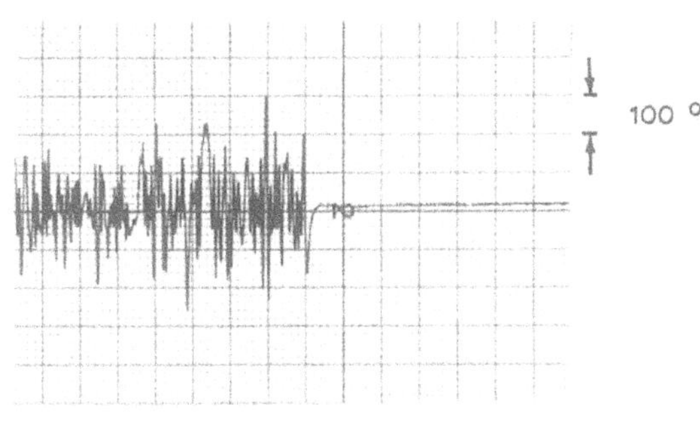

Fig.2. Gyro short term noise without laser jitter (left) and with laser jitter (right)

with $f_1 = f_2$. The peak-to-peak noise corresponds to rotation rates of several 100 degrees an hour and is caused mainly by backscattering in the fiber and at interfaces within the fiber interferometer. This type of noise can be drastically reduced by making f_1 not equal to f_2, by jittering the laser frequency, or by using a broadband light source. In a previous publication [14] we used both $f_1 \neq f_2$ and laser jitter to reduce backscattering noise. In the set-up discussed here we used laser jitter only to reduce the noise to that shown in Fig. 2 (right). A longer stretch of the short term noise data (τ = 1 sec) taken at a more sensitive scale is shown in Fig. 3 a . In comparison, Fig. 2 b is the laser intensity noise at the detector when demodulated at $f_m(\tau = 1$ sec) in the absence of the modulation voltage on the E/O crystal.

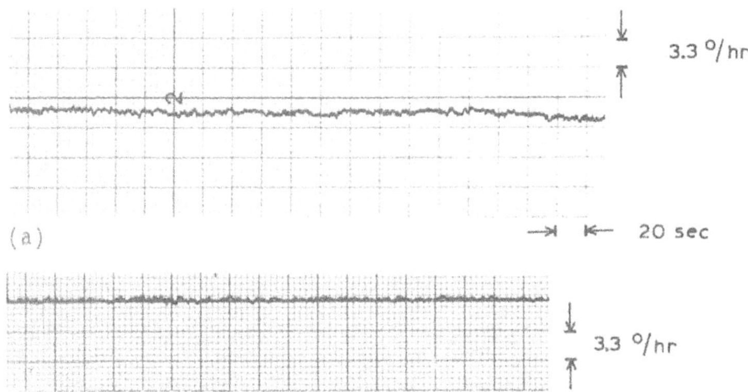

Fig.3. (a) Gyro short term noise (τ = 1 sec, 6 dB/octave filter)
(b) Laser intensity noise, conditions as in (a)

As can be seen, the noise in Fig. 3 a and Fig. 3 b are comparable. Since the rms value of the noise in Fig. 3 b is close to the calculated photon shot noise at the detector, we conclude that the short term noise of the gyro Fig. 3 a is limited primarily by photon shot noise in our present set-up.

The long term behavior was studied using closed loop operation. In this way, the Sagnac interferometer is held at the peak of the zero fringe by feedback control of f_1 as shown in Fig. 1. Data was assembled by counting $f_1 - f_2$ and averaging over 1 second intervals for several runs of about 10 minute duration. The solid circles in Fig. 4 are the rms values of $\delta\Omega$ for various counting intervals τ.

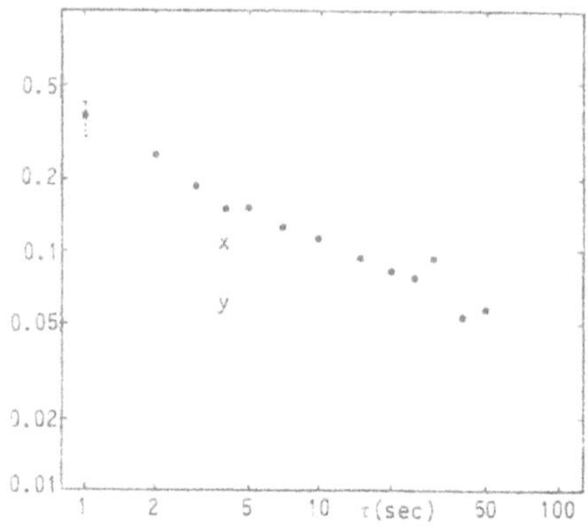

Fig.4. Plot of rms short term noise $\delta\Omega$ vs τ using closed loop system

As can be seen the variation of $\delta\Omega$ with τ approximates $\tau^{-1/2}$ as expected for shot noise behavior in (2). Also on Fig. 4 we have indicated with an x the point that corresponds to the open loop rms noise estimated from Fig. 3 a which is close to the noise in the light in Fig. 3 b . In addition we have indicated with Y the calculated shot noise limit based on the measured detector current and the simple formula in (2). The discrepancy between point x and the closed loop data may be attributed to our crude estimate of rms noise in Fig. 3 a . The discrepancy between point Y and the closed loop data shows that the shot noise formula in (2) needs refinement.

It should be remembered that the data in Fig. 4 is obtained with a 200 m long fiber. If a 2 km low loss fiber is used at 1.3μ, we would expect a reduction in $\delta\Omega$ by a factor of 5.

The major problem at present is the long term drift caused mainly by mechanical instability of our discrete component set-up. It is clear that an all-integrated system would alleviate such problems.

We have also examined the dependence of gyro output on temperature. Fig. 5 shows the open loop PSD output for a 3.5°C variation in fiber coil

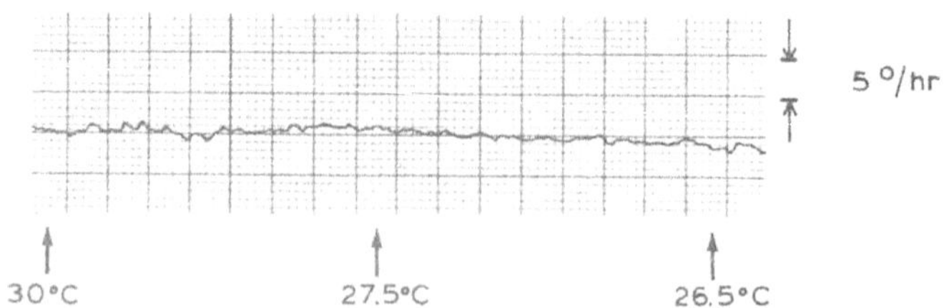

Fig.5. Dependence of gyro output on temperature (τ = 1 sec)

temperature. In principle, a change in temperature should only affect the scale factor, i.e. $2\pi LD/\lambda_0 c_0$ in the case of open loop measurement of phase, or $D/\lambda_0 n$ in the case of the closed loop scheme in our set-up (n is the refractive index of the fiber). The negligible variation of PSD output with temperature in Fig. 5 is consistent with the expectation that only the scale factor is influenced by temperature. Moreover, the data in Fig. 5 shows that there is no significant temperature-dependent bias.

Acknowledgments

This work was supported by the Joint Services Electronics Program at the Massachusetts Institute of Technology.

References

1. V. Vali and R.W. Shorthill, Appl. Opt. 15, 1099 (1976)
2. R. Goldstein and W.C. Goss, Proc. Soc. Photo-Opt. Instrum. Eng. 157, 122 (1978)
3. M.N. McLandrich and H.F. Rast, Proc. Soc. Photo-Opt. Instrum. Eng. 157, 127 (1978)
4. J.L. Davis and S. Ezekiel, Proc. Soc. Photo-Opt. Instrum. Eng. 157, 131 (1978)
5. D.E. Thompson, D.B. Anderson, S.K. Yao and B.R. Youmans, Appl. Phys. Lett. 33, 940 (1978)
6. H. Arditty, H.J. Shaw, M. Chodorow and R. Kompfner, Proc. Soc. Photo-Opt. Instrum. Eng. 157, 138 (1978)
7. R.F. Cahill and E. Udd, Opt. Lett. 4, 93 (1979)
8. R. Ulrich, Opt. Lett. 5, 173 (1980)
9. H. Arditty, M. Papuchon, C. Puech and K. Thyagarapan, in Digest of the Topical Meeting on Integrated and Guided-Wave Optics (Optical Society of America, Washington, D.C., 1980)
10. S.C. Rashleigh and W.K. Burns, Opt. Lett. 5, 482 (1980)
11. K. Böhm, P. Russer, E. Weidel and R. Ulrich, Opt. Lett. 6, 64 (1981)
12. R.A. Bergh, H.C. Lefèvre and H.J. Shaw, Opt. Lett. 6, 198 (1981)
13. R.A. Bergh, H.C. Lefèvre and H.J. Shaw, Opt. Lett. 6, 502 (1981)
14. J.L. Davis and S. Ezekiel, Opt. Lett. 6, 505 (1981)
15. S. Lin and T.G. Giallorenzi, Appl. Opt. 18, 915 (1979)
16. G. Schiffner W.R. Leeb, H. Krammer and J. Wittman, Appl. Opt. 18, 2096 (1979)
17. D.M. Shupe, Appl. Opt. 19, 654 (1980)
18. C.C. Cutler, S.A. Newton and H.J. Shaw, Opt. Lett. 5, 488 (1980)
19. E.C. Kintner, Opt. Lett. 6, 154 (1981)

Compact Fiber-Optic Gyro

E. Udd and R.F. Cahill

McDonnell Douglas Astronautics Company, 5301 Bolsa Avenue
Huntington Beach, CA 92647, USA

Abstract

A 6.4 cm diameter, 1.7 cm thick fiber-optic gyro that employs closed loop phase compensation has been fabricated and tested. The operation of this device is described as well as experimental results using a 100 m length, 6.4 cm diameter fiber-optic coil and an 1100 m length, 14.2 cm diameter fiber-optic coil. Measurements are reported for bias drift rate, linearity, and scale factor error.

Introduction

Advanced strapdown guidance and navigation systems require small, highly reliable gyroscopes of wide dynamic range. Mechanical gyros currently available are suitable for many applications, but maximum rate limitations, reliability problems, g-sensitivity, and long spin-up time have spurred the development of ring laser gyros in Government and industry. The efforts in developing ring laser gyros over the past 17 years [1-4] have resulted in high performance levels through the use of sophisticated techniques to circumvent the lock-in problem of ring laser gyros at low rotation rates.

Over the past few years, efforts have been made to bypass the inherent problems of ring laser gyros by adopting passive techniques employing fiber optics [5-7]. Most of these devices have nonlinear, analog output with limited dynamic range, and testing has often been confined to laboratory breadboards.

The phase-nulling optical gyro (PNOG) [8-11], on the other hand, is a linear rotation sensor rather than a sinusoidal sensor. It produces an inherently digital output via a frequency change proportional to rotation rate. The PNOG uses the nonreciprocal phase shift resulting from an induced frequency difference between counterpropagating beams in a fiber-optic coil to null out non-reciprocal phase shifts due to rotation. The PNOG thus has the potential for wide dynamic range, high sensitivity, linear rotation sensing, and the inherently digital output desirable for modern guidance systems. A compact PNOG packaged in a manner compatible with size constraints imposed by many applications is described.

Description of the 6.4 cm diameter PNOG

Residual error sources present in an earlier 12.7 cm diameter model [11] were investigated. Based on these results, a 6.4 cm diameter, 1.7 cm thick PNOG was designed. Fig.1 illustrates the layout of the package. Significant design improvements incorporated into this package that were not present in the earlier model include the following: (1) Reduction of component count by approximately a factor of two. This feature, in combination with greatly reduced separation between key components and rigid potting of each element to thick alignment plates, minimizes partial misalignments caused by vibration and mechanical flexing. (2) Through the appropriate placement of a single-mode spatial filter and polarizer, reciprocity conditions of the light path are guaranteed [12], reducing environmental sensitivity. (3) The acousto-optic modulator has heatsinking to prevent beam deflection due to thermal gradients. The modulator is used in a beamsplitting mode, increasing the optical efficiency of the system and greatly relaxing requirements on the modulator for optimum performance. (4) The ends of the fiber have been potted in line, allowing the acousto-optic modulator to operate in the beam-

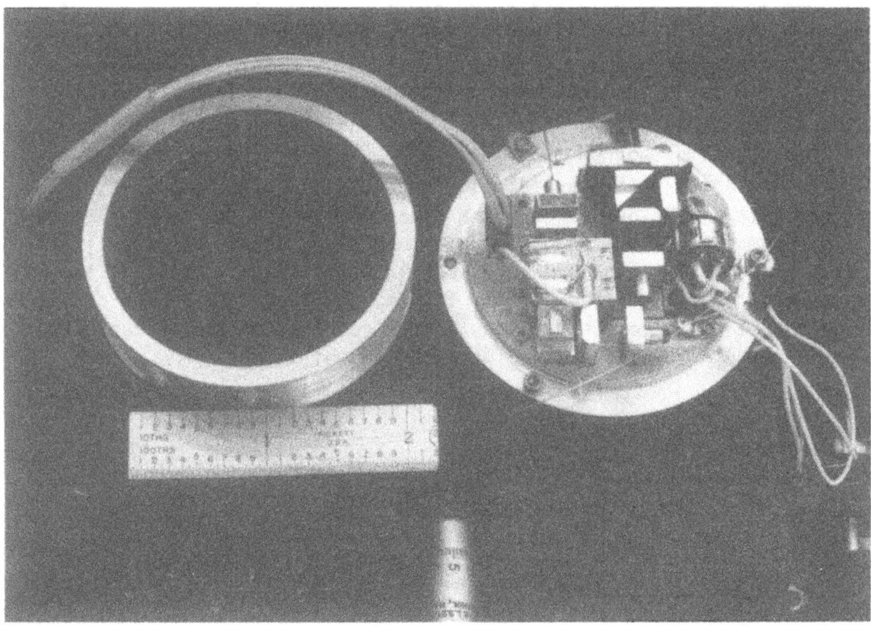

Fig.1a. Photo of the 6.4 cm diameter optical gyro

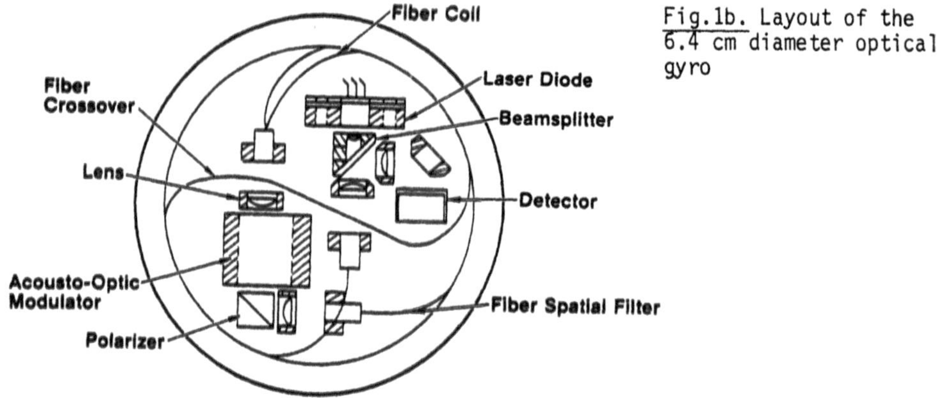

Fig.1b. Layout of the 6.4 cm diameter optical gyro

splitting mode and preventing relative motion of these important positions. (5) Extensive baffling and shielding has been incorporated into the package to prevent stray light from entering the detector.

The 6.4 cm diameter package employs a single-frequency, single-mode ML-3001 laser diode whose output is collimated by a single-element lens and passes through a beamsplitter. The light beam is then focused into the end of a fiber spatial filter that ensures the input to the system is a single-spatial mode and aids in maintaining reciprocity conditions. The output of the fiber spatial filter is collimated and reflected by a polarizing beam-splitter cube that ensures a single polarization is maintained throughout the system. The light beam enters the acousto-optic modulator, where the light beam is split into zero and first order diffracted beams which are coupled into opposite ends of the fiber coil. This procedure is shown in Fig.2. The two light beams generated by the acousto-optic modulator circulate in opposite directions through the fiber coil. The two beams exit the coil, are recombined by the acousto-optic modulator, and directed by the polarizing beamsplitter cube and lens back into the fiber spatial filter. The mixed beams exit the fiber spatial filter, are recollimated and deflected by the output beamsplitter onto a lens/mirror combination that focuses the light beams through an aperture and onto the output detector.

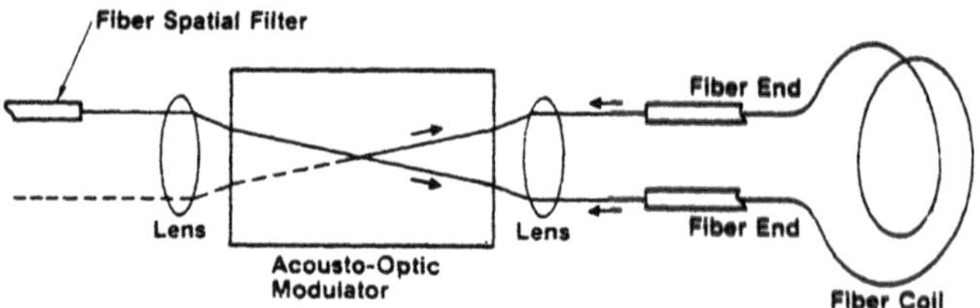

Fig.2. Top view of acousto-optic modulator configuration

Experimental Results

The gyro of Fig.1 has been the subject of extensive testing. In general, it exhibits a reduction in environmental sensitivity of more than an order of magnitude compared to an earlier model [11]. Remaining error terms due to environmental effects are largely due to wavelength drift of the laser diode. Fig.3 is a graph of frequency output of the gyro using a 100 meter, 6.4 cm diameter fiber coil during a 300 second bias drift test. Fig.4 illustrates the results of a 4000 second bias drift test by plotting the change in heading angle versus time. For this test the 6.4 cm fiber coil was replaced with

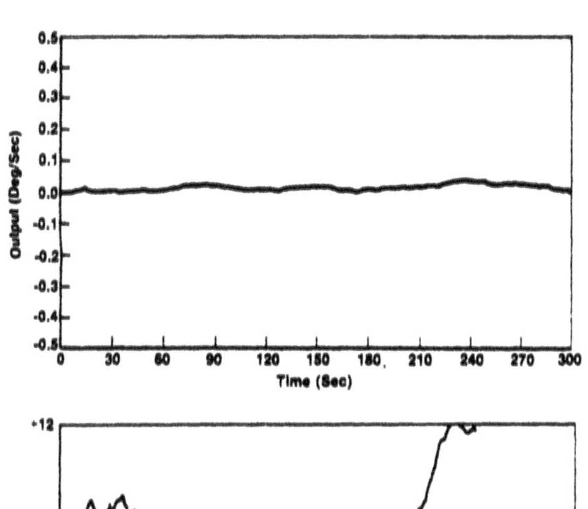

Fig.3. Bias drift test, 1 msec integration time, 6.4 cm diameter optical gyro

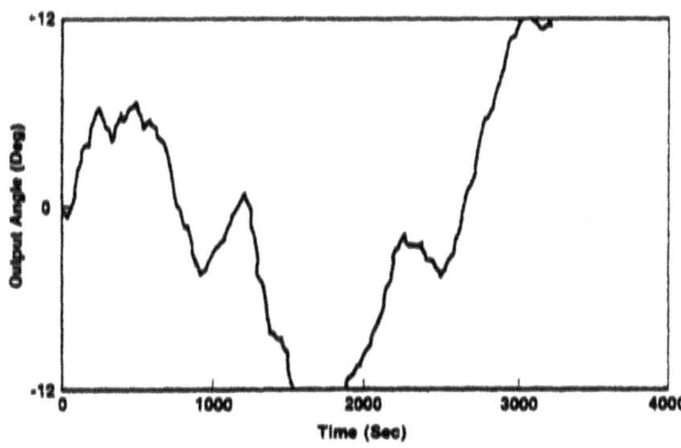

Fig.4. Bias drift test, 1 sec integration time, optical gyro using 1100 m fiber coil

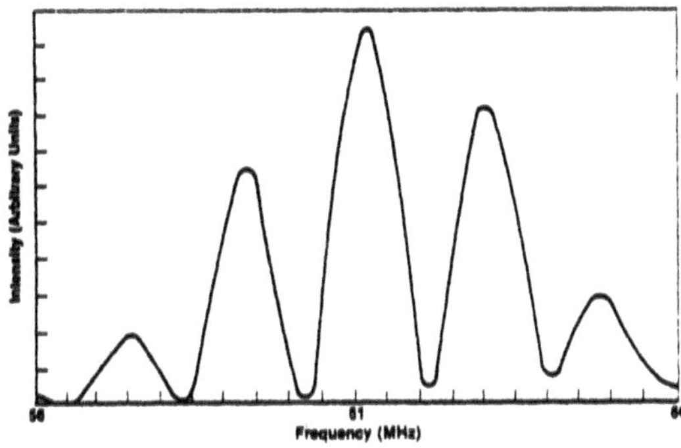

Fig.5. Output of the optical gyro as a function of frequency using a 100 m fiber optical coil

an 1100 meter, 14.2 cm diameter fiber coil to increase sensitivity. It can be seen that bias drift has been reduced to about 12 deg/hr. No attempt was made to control temperature, and most of the residual drift is attributable to wavelength change of the laser diode with temperature. Fig.5 illustrates the output of the gyro as the frequency of the acousto-optic modulator is scanned. In this case the 100 meter fiber coil was used, which in fact can be determined by noting that the separation between fringes is given by c/Ln where c is the speed of light in vacuum, L is the length of the fiber coil, and n is the index of refraction of the fiber. If the length of the fiber

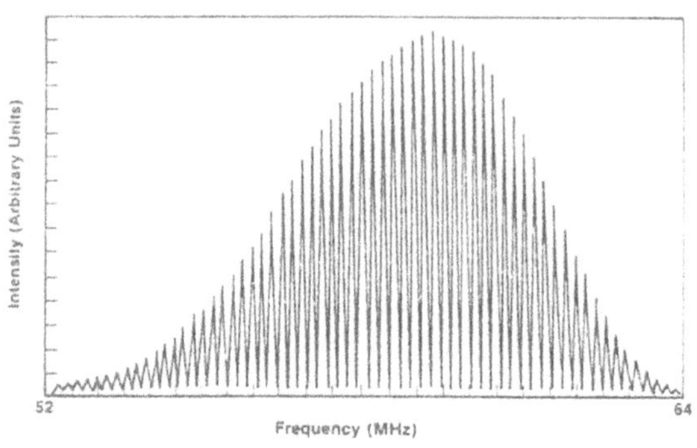

Fig.6. Output of the optical gyro versus frequency using an 1100 m fiber coil

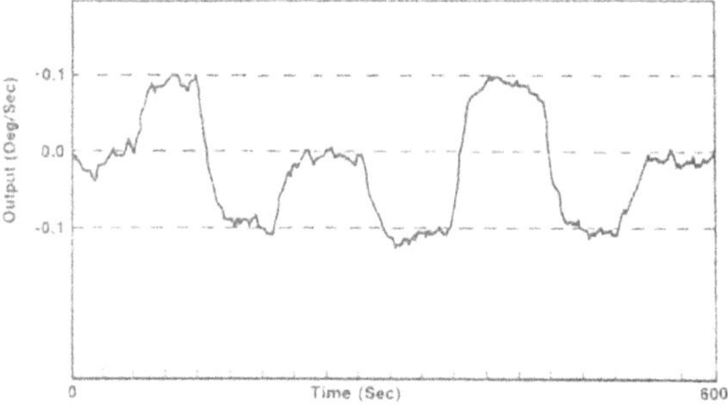

Fig. 7. Response of the gyro to a 0.1°/sec input rotation rate, 1100 m coil

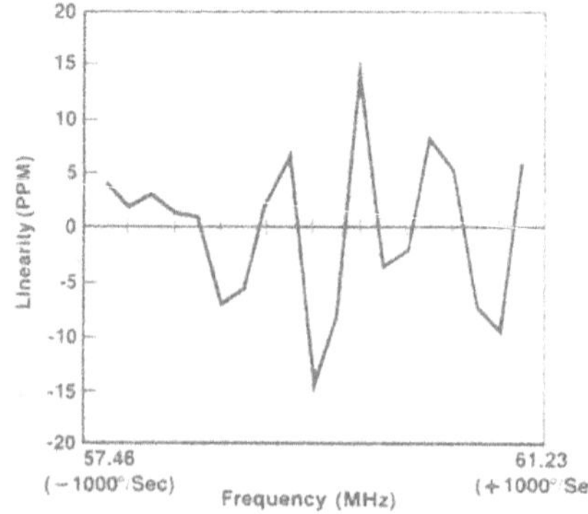

Fig.8. Linearity of the optical gyro versus frequency, 1100 m coil

coil is known, the separation between adjacent fringes can be used to monitor wavelength as n is a function of wavelength. Fig.6 illustrates the output of the gyro during a similar scan using an 1100 meter fiber coil.

Figure 7 shows the response of the gyro to a 0.1 deg/sec input rotation rate using the 1100 meter, 14.2 cm diameter fiber-optic coil. Fig.8 illustrates the linearity of the gyro over a frequency range of about four MHz. The residual nonlinearity is believed to be due to bias drift over the measurement period.

References

1. J. Kilpatrick, IEEE Spectrum 4,44 (1967)
2. F. Aronowitz and L.W. Lim, Applied Optics 13,338 (1977)
3. F. Aronowitz, Laser Applications (M. Ross, Editor, Academic Press), 131 (1971)
4. M.O. Scully, V. Sanders and M. Sargent, Optics Letters 3,43 (1978)
5. V. Vali, R.W. Shorthill and M.F. Berg, Applied Optics 16,2605 (1977)
6. V. Vali and R.W. Shorthill, Applied Optics 16,290 (1977)
7. V. Vali and R.W. Shorthill, Proceedings of SPIE 76,110 (1976)
8. R.F. Cahill and E. Udd, Optics Letters 4,93 (1979)
9. R.F. Cahill and E. Udd, Proceedings of the National Aerospace and Electronics Conference, Dayton, Ohio, 8 (1979)
10. E. Udd and R.F. Cahill, Proceedings of Lasers 1979, Orlando, Florida, 694 (1979)
11. R.F. Cahill and E. Udd, Applied Optics 19,3054 (1980)
12. R. Ulrich and M. Johnson, Optics Letters 4,152 (1979)

Fiberoptic Gyro Using Magneto-Optic Phase Nulling Feedback

W.C. Davis, W.L. Pondrom, and D.E. Thompson
Rockwell International, Anaheim, CA 92803, USA

Introduction

The potential for high sensitivity in fiberoptic ring gyroscopes has been demonstrated [1], and, as a result of considerable attention to the device, many studies have been made which have advanced the state-of-the-art. This paper describes an approach to the design of a fiberoptic gyro which has the intent of producing an operational device. The objective of the design is to implement simply and effectively a closed loop phase nulling system with a minimum of encumbrance.

Design Approach

To meet the objective the Faraday magneto-optic effect was selected to produce the offsetting phase shift to null the Sagnac phase shift. This method of producing the offsetting phase shift could be accomplished without introducing any active optical components into the otherwise passive interferometer. The inherent stability of this mechanization is evident in that in addition to maintaining a simple optical system, the field coupling is bias free, and therefore drift free, and can be anticipated to inject a minimum of noise into the readout system.

The principle of the fiberoptic gyro is to form an interferometer of two counter propagating beams in an optical fiber coil. Rotation of the coil about its axis will induce a phase difference proportional to the rotation rate which in turn will result in variation of the intensity of the fringe. It is essential that the signal output, i.e., variation in fringe intensity is a function only of the phase difference. Thus phase modulation is induced with the result that the signal measurement is of the change in the modulation form, rather than a purely intensity measurement. To produce the modulation the Faraday effect in the fiber is again used for the same reasons as cited for the feedback loop.

System Description

Figure 1 is a schematic diagram of the system. The output of the source, at the left, is passed to a collimating lens and linear polarizer to the beam

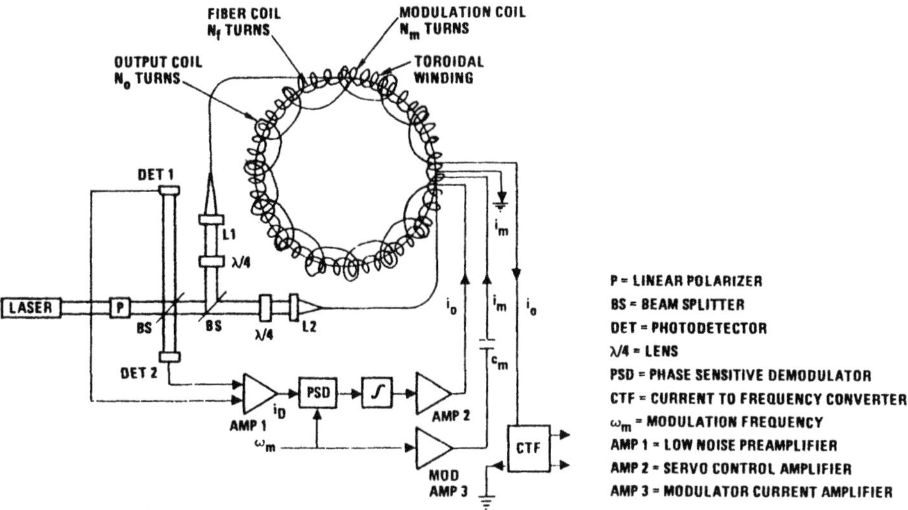

Figure 1. Fiberoptic Gyro Schematic

divider. The two output beams from the beam divider pass via quarter wave plates and focussing lenses, to the ends of the optical fiber. After traversing the fiber the beams are recombined and directed to a photodetector at the bottom of the diagram. The beam divider element is comprised of two beam dividers with the output to the detector taken from the left most divider. Thus each beam experiences an equal number of reflections and transmissions at this element. The result is that the phase shift between the beams is zero and the airy disc of the diffraction pattern has maximum intensity. Two toroidally wound coils enclose the fiber coil, one of which is the control or phase nulling coil, the other being the modulation coil. The modulation coil is driven by an oscillator via a current driver which is capacitively coupled to the coil. The oscillator is tuned to match the series frequency of the capacitor and coil (3.5KHz). The output of the detector is coupled via a low noise pre-amplifier to a synchronous demodulator, low pass filter, and current driver, to the control coil. The current driver is a high impedance (100KΩ) source for the coil; thus the control current is free of any effects of the modulation field. The current which drives the coil is then coupled to a high precision, low drift, current-to-frequency converter where the signal can be scaled to read in pulses per unit angle.

Experimental System

The experimental system was designed to meet two criteria. The first is to demonstrate the principle outlined in the design approach section. The second was to facilitate experimentation. The most significant aspect of the system, was the design of a coil assembly which would allow interchanging the fiber without unwinding the modulation coil. The control coil is only a few turns and can be readily removed and installed; however the modulation coil is of the order of 10^4 turns and could not be readily removed. Thus the gyro is comprised of two basic elements, the optical (interferometer) platform, and the coil assembly. In Fig. 2 the system is shown mounted upon a Leitz rotation table. In Fig. 3 the optical platform is shown. In this figure the source is at the left. The detector is in the foreground. In this photo two

Figure 2. Assembled Fiberoptic Gyro

Figure 3. Optical Platform

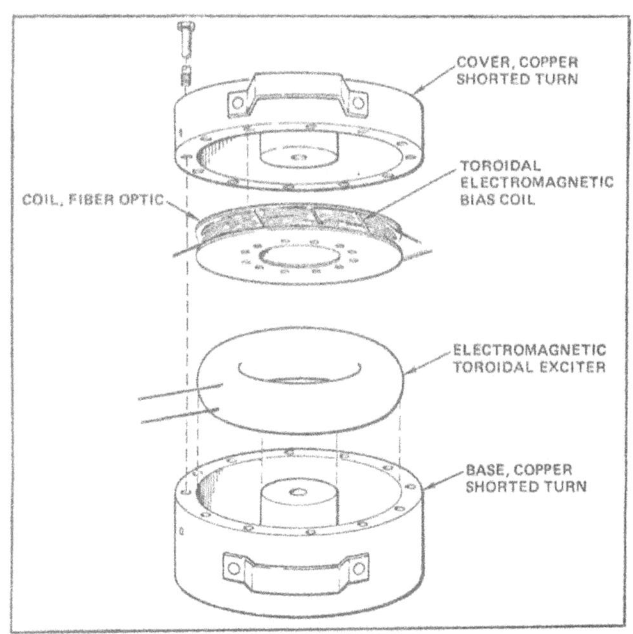

Figure 4. Coil Assembly Schematic

Figure 5. Open Coil Assembly (Without Fiber Coil)

additional components are shown, both linear polarizers, lying between the beam divider and the quarter wave plates in the path to the fiber. The fibers are held in positioners which have 5 axis freedom of motion. A sixth axis can be added by rotating the fiber about its own axis. In Fig. 4 a schematic diagram of the coil assembly is shown. The principle of operation is to couple a toroidal modulation coil to the fiber by means of a single turn secondary which forms a case for the assembly. The control coil is wound in a toroid directly around the fiber coil form. Figure 5 is a photograph of the coil assembly opened. The fiber coil is not shown.

A final element whose description is germane to the discussion of the experimental system is shown in Figure 6. This mechanism is designed to wind fiber on a variety of coil forms while simultaneously imparting a twist to the fiber. The fiber coil form used in the experiments is of the type shown in the photo in the process of being wound. With this device fibers have been wound successfully with as many as 100 twists per meter.

Figure 6. Fiber Coil Winder

Discussion of Parameters

The Faraday magneto-optic effect produces a phase shift $\emptyset$ in response to an applied field according to

$$\emptyset = H V \ell$$

where H is the field strength in oersteds, V is the Verdet constant for the medium (radians/amp-turn) and ℓ the length of the medium. For fiber of length ℓ in a coil with a toroidal current winding about it

$$\ell = N_f (2 \pi r)$$

and H can be written $H = N_i \, i/(2 \pi r)$

where r is the radius of the fiber coil, N_f the number of turns of fiber, N_i the number of the current windings, and i the current. Then the phase shift due to the field can be written

$$\emptyset = N_i N_f V i .$$

In the gyro, light is propagating bi-directionally in the fiber. If the light is circularly polarized, and of the same hand, the phase difference in the counter propagating beams is

$$\emptyset_F = 2N_i N_f V i .$$

Similarly the modulation phase shift is given by

$$\emptyset_{FM} = 2\, N_i\, N_f\, V\, i\, \cos\omega_m t$$

where ω_m is the modulation frequency.

To implement the Faraday effect to offset the Sagnac phase shift and produce phase modulation, ideally, the medium should support circularly polarized light without altering it. An ideal single mode fiber would provide the required medium. However single mode fibers are not ideal, but exhibit an inherent linear birefringence and, when wound upon a coil form, an additional stress induced linear birefringence is evident and the presence of the linear birefringence quenches the Faraday effect if the fiber length exceeds half the polarization beat length. By sufficiently twisting the fiber to induce a strong circular birefringence, the linear birefringence in the fiber can be suppressed [2]. The fiber used in the experiment exhibited a beat length of approximately 2 meters prior to being twisted. The beat length ℓ_b, is given as

$$\ell_b = 2\,\pi\,/\,\beta$$

where β is the differential propagation constant due to line birefringence in the medium between the fast and slow axes. For the fiber used (Corning single mode fiber, 6 µm core diameter) the birefringence coefficient is given by

$$\beta = 2\,\pi\,/\,2\ \text{(meters)} \cong 3\ \text{rad / meter.}$$

It can be shown that the twist induced circular birefringence coefficient, α, in terms of the twist rate ξ, is given by

$$\alpha = g\,\xi$$

where g is an experimentally determined value equal to 0.08...0.12 for silica fibers. By applying a twist rate of 43 per meter and using a nominal value we find

$$\alpha = 27\ \text{rad / meter}$$

which is nearly ten times the linear birefringence, and can be expected to support circular polarized light in the fiber.

To determine the magnitude of modulation required for maximum sensitivity to the Sagnac phase shift we write for the detector current i,

$$i = k\,P$$

where P is the incident power and k the responsivity of the detector. The total phase shift, $\emptyset$, between the two counter propagating beams is given by

$$\emptyset = \emptyset_s - \emptyset_F + \emptyset_m \cos \omega_m t \quad .$$

Thus if P_1 and P_2 are the powers falling on the detector due to each beam acting alone, the detector current, i_D, is given by

$$i_D = K\,(P_1 + P_2 + 2\,(P_1 P_2)^{1/2} \cos \emptyset)\ .$$

The maximum signal to noise ratio in the control loop occurs when $\emptyset_m$ is approximately $\pi/2$, and the apparatus is adjusted to obtain this value.

With closed loop operation a rotational input of 0.13 deg/sec caused a current of 6 ma to flow in the control coil. Using the expression for the Sagnac phase shift ϕ_s

$$\phi_s = \frac{8 \pi N_f A \Omega}{c \lambda}$$

where A is the area enclosed by the fiber coil, Ω the rotational rate, c the velocity of light and λ the wavelength of the source, and applying the values from the system parameters and measurements, $N_f = 400$, $A = 63\ cm^2$ and the source wavelength of 0.82 μm,

$$\phi_s = 5.8 \times 10^{-4}\ rad\ .$$

Using the expression for the Faraday effect offsetting phase shift ϕ_F

$$\phi_f = 2\ N_i\ N_f\ i\ V\ ,$$

where N_i = 30 turns, i = 6 ma, and assuming the Verdet constant = 3.43×10^6 rad/a-t

$$\phi_F = 5.0 \times 10^{-4}\ rad\ .$$

This would indicate that the length of the fiber is considerably less than one-half of a beat length, i.e., the twist rate of 43 twists per meter is sufficient to suppress the linear birefringence in the fiber.

Noise Sensitivity

Polarizers have been introduced between the beam dividers and the quarter wave plates in order to define a specific port to port two way transmission [3] which is reciprocal in the absence of the Faraday effect and Sagnac effect. This has the effect of reducing the noise by a factor of three in the output. The implication of this result is that we do not quite have a polarization maintaining circularly birefringent fiber. It is believed that in principle the quarter wave plates, and the twist of the fiber ends should be adjusted so that the mode of propagation is circular at the length midpoint of the fiber coil in order to obtain maximum modulator efficiency and scale factor stability, at the lowest noise level. A satisfactory way of adjusting to this condition has not been found due to instabilities in the current bulk optical set-up.

Perhaps the single largest noise in this system appears to be a result of the open air paths between the optical elements in the interferometer. Upon placing an enclosure about the entire system a dramatic (order of magnitude) reduction in noise occurs. It is expected that even further noise reduction will occur if no air path exists between components.

The system is operated with the enclosure in place and the resulting chart record is shown in Fig. 7. The rotational table is turned on and off with a step of $\pm .06^\circ$/sec, and a signal to noise ratio of 10 is obtained (τ = 0.1 sec) without further electronic signal processing. This is a noise level of less than a signal corresponding to two times earth rate.

In summary, we have designed, built, and operated a Sagnac fiber ring gyro demonstrating that the Faraday magneto-optic effect in the fiber can be used to produce the modulation and feedback necessary to perform the closed loop

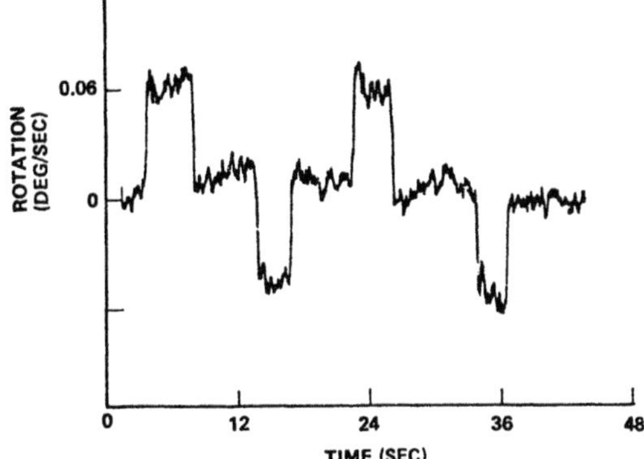

Figure 7. Gyro Performance

phase nulling operation. The noise and drift associated with the bulk optic interferometer limit the overall performance of the system. It is anticipated that these problems can be overcome while maintaining the basically simple, rugged aspect of using the Faraday effect to produce an operational device.

References

1. V. Vali, R. W. Shorthill, M. F. Berg, Applied Optics 16, 2605 (1977)

2. S. C. Rashleigh and R. Urich, Appl. Phys. Lett. 34, No. 11, 1 June 1979

3. G. Schiffner et al., Applied Optics 18, No. 13, 1 July 1979

Heterodyne Fibre Gyro With Complete Reciprocity

I.P. Giles and B. Culshaw

Department of Electronic and Electrical Engineering
University College London, Torrington Place, London WC1E 7JE, Great Britain

Abstract

This paper describes an optical fibre gyroscope which uses a completely reciprocal heterodyne detection technique. The system is based on the Mach Zehnder interferometer and provides a phase and polarisation reference for the beams transmitted through the fibre. Initial measurements have been made on the system and its feasibility demonstrated. Further developments are being made to the system to produce an instrument which will investigate the noise limitations imposed by environmental effects on the fibre loop.

1. Introduction

Development of the optical fibre rotation sensor into a practical high quality strap-down gyroscope is restricted by certain fundamental limitations. The limitations of the system can be considered in two groups, those imposed by the Sagnac interferometer and noise effects caused by the use of an optical fibre in the optical path.

A closed loop Sagnac interferometer (figure 1) which is a homodyne system detects the rotation induced phase shift as an intensity change of the interference pattern. For an ideal system with a 3dB beamsplitter and

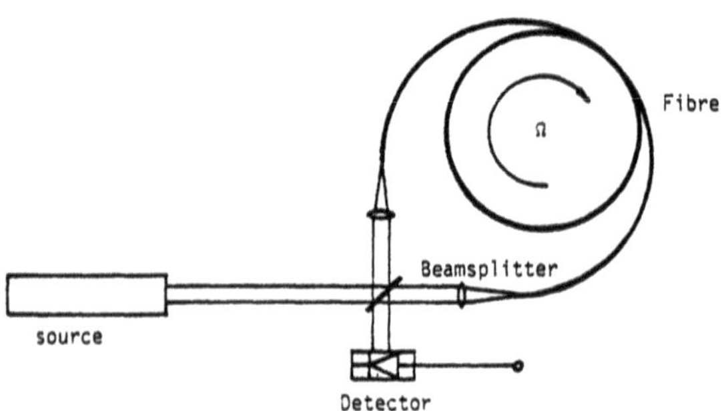

Figure 1 Basic optical fibre Sagnac Interferometer

identical propagation in each direction the detector output is given by

$$I \alpha |E|^2 \{1 + \cos(\Delta\phi + \phi_+ - \phi_-)\} \quad (1)$$

where I is the detector output,
E is the optical beam amplitude,
$\Delta\phi$ is the Sagnac phase shift,
φ_+ is the phase shift in the clockwise direction,
ϕ_- is the phase shift in the anti-clockwise direction.

From this equation several potential performance limitations can be seen.

(1) The system has minimum sensitivity at low rotation rates, i.e. $\Delta\phi \rightarrow 0$.

(2) At large rotation rates ($|\Delta\phi| > \pi$) ambiguities arise hence limiting the dynamic range of the system.

(3) ϕ_+ and ϕ_- are very large compared to $\Delta\phi$ and therefore there must be identical optical paths in each direction i.e. $\phi_+ = \phi_-$.

(4) The detector output is sensitive to intensity fluctuations which are indistinguishable from rotation induced phase shifts in the same bandwidth.

(5) Homodyne detection schemes are baseband systems and as such are susceptible to i/f noise.

The fibre loop in such an interferometer introduces several effects which reduce the signal to noise ratio in the detection.

(1) Polarisation coupling effects in the fibre are non-reciprocal and the condition $\phi_+ = \phi_-$ is not met. This can be improved by correct polarisation analysis, but intensity fluctuations are then observed [1].

(2) Coherent backscatter causes phase noise in the signal bandwidth [2]. The effect of the backscatter is dependent upon the coherence length of the optical source.

(3) Thermal gradients across the fibre induce non-reciprocal phase effects [3] which also limit the performance.

The performance limitations which have been enumerated are well known and and various forms of detection techniques have been investigated to improve rotation detection. This paper discusses a new detection scheme which uses a heterodyne system to overcome the Sagnac interferometer imposed limitations, giving a scheme which can be developed to investigate the noise effects produced by the fibre.

2. Heterodyne Techniques

The heterodyne detection system is based upon the Mach-Zehnder interferometer (figure 2). One path is a length of optical fibre and the other path contains a Bragg cell which frequency shifts the optical beam. The detected output from this is the intermediate frequency, defined by the Bragg cell drive frequency (f_0) with the phase difference between both arms. Such a system requires a long coherence length laser because of the large path imbalance.

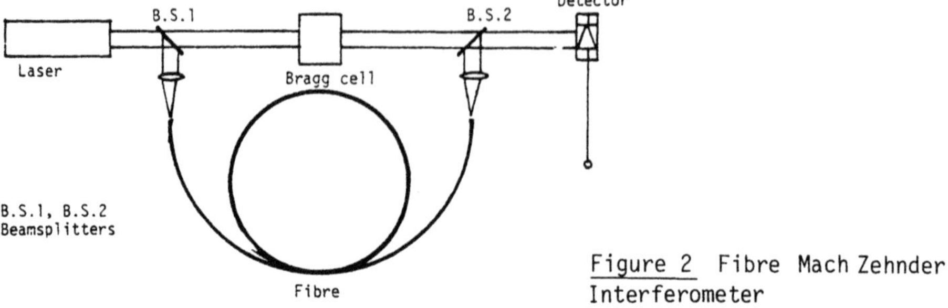

Figure 2 Fibre Mach Zehnder Interferometer

Figure 3 shows the extension of the Mach-Zehnder interferometer into a heterodyne Sagnac interferometer. B.S.1 divides the beam into two parts which are the inputs to two interconnected Mach-Zehnder interferometers. The first beam is split by B.S.2, half launched into the fibre and half passed through the Bragg cell. Both beams are recombined by B.S.3 and mixed on detector 1. The second beam is similarly split by B.S.3, recombined by B.S.2 and mixed on detector 2. Each detector gives the i.f with a phase which represents the difference between the propagation direction and the reference arm. As the reference arm is identical for both directions, comparing the i.f. in a phase sensitive detector produces an output proportional to the rotation rate of the system.

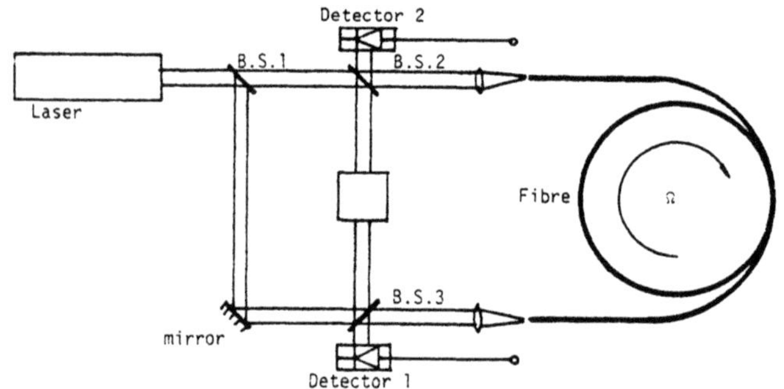

Figure 3 Heterodyne Sagnac interferometer

Heterodyne detection has the well known advantage that 1/f noise is avoided and by synchronous detection a.m. caused by intensity fluctuations can be removed. This method applied to the Sagnac interferometer offers additional benefits. An optical rotation sensor must be reciprocal and the system gives complete reciprocity if the identical frequencies are propagated through the fibre in each direction and path length changes in the optical path external to the fibre (the reference path) are identical for both directions. The direct reference path gives a high "local oscillator" power to the detector which ensures shot noise limiting and allows the use of p.i.n. diode detection.

Single mode fibre is used in all fibre gyroscope systems, to achieve identical path propagation, but environmentally induced polarisation coupling destroys this path reciprocity. By correct polarisation analysis, launch and detection in same polarisation path [1], reciprocity holds. The phase reference arm in the heterodyne interferometer also provides a polarisation reference in that the detector will only mix signals of the same polarisation orientation. It is also possible to investigate the polarisation properties of the fibre loop by simply rotating the input polarisation. This system gives access to the phase information in each direction through the fibre which can be separately processed and performs all of the signal processing electronically. In this way it is possible to investigate the polarisation and backscatter "noise" induced by birefringence caused by environmental effects. The system can then be used as an instrument to characterise the limitations of the optical fibre gyroscope.

3. Experimental System

The heterodyne system was built on a steel base with each component firmly located into the base to produce a rigid, compact device. A single mode HeNe laser provided the long coherence source, pellicle beamsplitters were used and an 80MHz Bragg cell. The fibre loop was 130m long wound in a 10cm radius coil and single mode at 0.633μm. Each detector output was amplified and limited and then both compared in a phase sensitive detector. Rotating the supporting table gave a phase shift at the output demonstrating the feasibility of the system. Rotation sensitivity in this system, however, was limited by the phase error introduced by fibre end reflections which mixed coherently with the reference beam. As discussed previously, it is necessary in this system to use a long coherence length laser, but for a Sagnac rotation sensor, short coherence sources are generally used to reduce backscatter effects. It was therefore necessary to code the input optical signal in some way to avoid backscatter effects. Two methods of coding are possible, to use a pulsed optical input and allow the backscatter to clear from the system in between the pulses or to use a continuous beam and vary the frequency using FM CW radar techniques.

The latter method was adopted for this system in preference to pulsing techniques.

4. Frequency Modulation

Varying the frequency of the input beam linearly gives an output frequency from the detector which is proportional to the time delay in the fibre path. In this way the backscatter components are separated in frequency from the transmitted signal. Figure 4 shows the principle for a triangular frequency modulation and the ideal required output spectrum. The spectrum contains a large transmitted signal separated by a frequency Δf from the centre frequency (f_0) which corresponds to the time delay set by the fibre where

$$\Delta f = \tau . \frac{fm}{T} . \quad (2)$$

τ is the fibre delay time,

fm/T is the linear frequency ramp ratio.

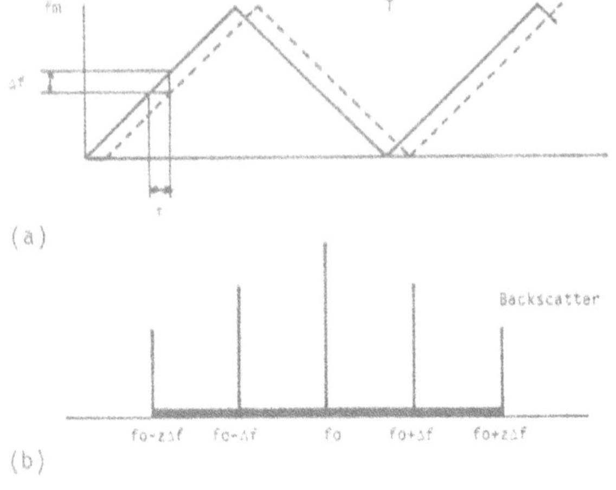

Figure 4 (a) Triangular frequency ramp applied to optical input and output delayed fibre.
(b) Ideal spectrum; near end reflections appear at f_o, transmitted signal at $f_o \pm \Delta f$, for end reflections at $f_o \pm 2\Delta f$

Reflections from either end of the fibre are detected at the centre frequency and at $f_o \pm 2\Delta f$. Between these two frequency limits there is a band of noise caused by the fibre backscatter. In practice, the ideal is not met and the spectrum detected is an envelope of frequencies separated by the modulation frequency.

Frequency modulation techniques were used in the experimental system by incorporating a second Bragg cell at the laser output. A triangular frequency ramp was applied to the Bragg cell at a rate of ∿5kHz fundamental with a modulation depth of 20MHz. The spectral output from one detector is shown in figure 5. The transmitted signals can be seen quite clearly and the near end and far end reflections can also be seen when the transmitted beam is blocked (figure 5b). The envelope of frequencies is wider than expected due to the spatial variation of the Bragg cell output causing amplitude modulation and also to the distortions on the triangular drive.

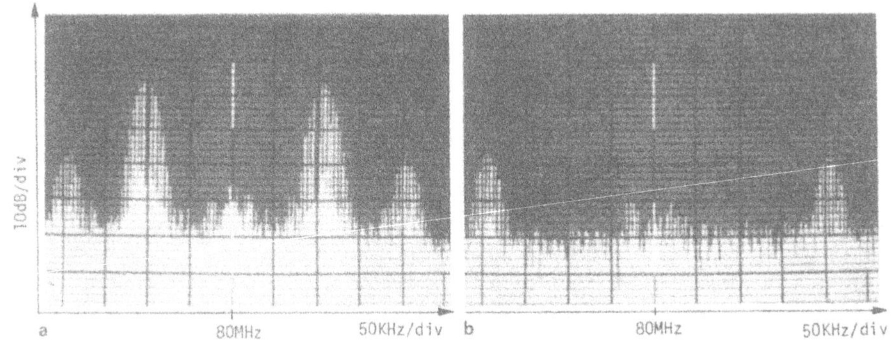

Figure 5 Detector output displayed on a spectrum analyser
(a) This shows reflections from both fibre ends and the transmitted signal
(b) With the transmitted beam blocked at launch the reflections are shown

The outputs from each detector were mixed down to baseband, filtered and the phase compared by a vector voltmeter. Sensitivity levels of the order of 100°/hr were recorded. This noise level was induced by the electronic detection introducing a differential phase variation in each channel. Present work is aimed at eliminating this source of error by matching the channels more closely and then obtaining a better characterisation of the limitations of the optical system.

5. Conclusions

A new detection scheme has been discussed which uses a heterodyne technique to overcome basic problems associated with basedband techniques. The system is able to investigate the phase modulation introduced in each direction and will be used to characterise the fibre noise effects. In order to do this the system must be completely reciprocal and operate with a single polarisation.

A prototype of this system has been built and behaves as expected with present limitations set by the electronic signal processing techniques. Development of the system is now in progress to reduce electronically induced noise and simplify the optical system by removing the Bragg cells and directly modulating the laser source.

References

1. "Fiber-ring interferometer: polarisation analysis", R. Ulrich, M. Johnson, Optics Letts. 4, 1979, pp.152-154

2. "Low-noise fibre-optic rotation sensing", K. Bohm, P. Russer, E. Weidel, R. Ulrich, Opts. Letts. 6, No.2, 1981, pp.64-66

3. "Thermally induced non reciprocity in the fiber-optic interferometer", D.M. Shupe , App. Opts. 19, No.5, 1980, pp.654-655

4. "Solid-state phase-null optical gyro", R.F. Cahill, E. Udd, App. Opts. 19, 1980, pp.3054-3056

5. "Closed-loop, low noise fiber-optic rotation sensor", J.L. Davis, S. Ezekiel, Opts. Letts. 6, No.10, 1981, pp.505-507

6. "All single mode fiber optic gyroscope with long term stability", R.A. Bergh, H.C. Lefevre. H.J. Shaw, Opts. Letts. 6, No.10, 1981, pp.502-504

Optical Fiber Laser Gyro: Homodyne and Heterodyne Detections

K. Hotate, N. Okuma, M. Higashiguchi, and N. Niwa

University of Tokyo, Institute of Interdisciplinary Research
Faculty of Engineering, 4-6-1 Kamaba, Meguro-ku, Tokyo 153, Japan

1. Introduction

The ring interferometer with a multiturn single-mode optical fiber loop has been investigated for use as a rotation rate sensor utilizing the Sagnac effect. We have proposed an optical system with homodyne detection, in which the phase-difference bias can easily be introduced, and the rotation detection experiment was performed using the preliminary test setup with a He-Ne laser [1].

Two optical systems for the gyro are described in this paper. In the first half of the paper, the theoretical considerations on homodyne detection system with a laser diode as an optical source and the result of the rotation detection experiment are described. Short-time resolution of the rotation rate better than 0.87mrad/s has been achieved with good linearity.

In the latter half, a novel optical heterodyne system is proposed. A phase nulling detection is useful for the gyro that is required to have a wide dynamic range and linearity. The optical heterodyne technique enables the phase nulling with an appropriate electronics. However, former optical heterodyne systems have a residual pathlength difference between counterpropagating waves in fiber [2], that is shown to induce the output drift when the optical source spectrum fluctuates. The optical heterodyne system proposed in this paper has only a quite small pathlength difference. Moreover, the system has no frequency difference between two waves in the fiber, that causes output drift with temperature change. Preliminary experimental results are also described.

2. Description of the Homodyne Detection System

2.1 Optical Configuration

Figure 1 shows a conceptual diagram of the homodyne detection system. The laser output, divided by beamsplitter BS_1, is launched into the single-mode optical fiber. Two waves propagate in the optical fiber loop in opposite di-

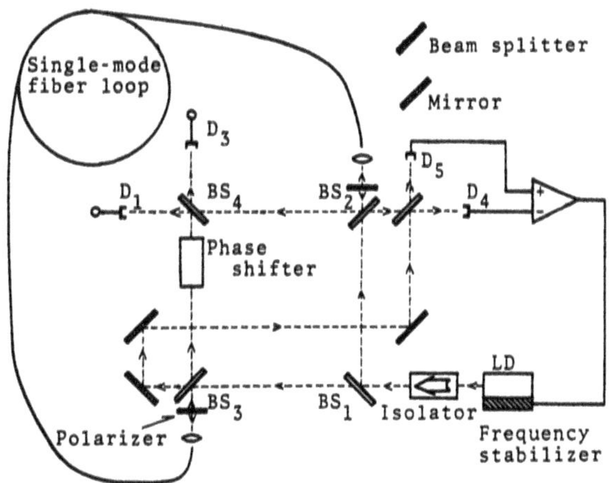

Fig.1 Conceptual diagram of homodyne detection system

rections, after which they are recombined by BS_4. The most significant feature of the proposed optical system is that the phase-difference bias between two waves can easily be adjusted by an ordinary phase shifter placed in the BS_3-BS_4 path to optimize rotation sensitivity. The reason for introducing the other components will be discussed later.

2.2 Consideration of Additional Noise Sources

In the optical system, the counterpropagating waves are separated from each other. Then additional noise sources may be introduced.

Relative motion of the optical components may cause additional noise. However, it may be successfully reduced using an integrated optical circuit or microoptics technology.

In the optical system a residual pathlength difference between counterpropagating waves may exist. In such a case, thermal fluctuation of the residual pathlength difference and/or of the laser frequency causes output drift. Denoting the residual pathlength difference as $\Delta\ell$, the corresponding phase difference φ is expressed as

$$\varphi = k\Delta\ell \tag{1}$$

where k denotes the wave number in vacuum. Thermally induced drift of φ is expressed as

$$\Delta\varphi = \frac{1}{c}\omega\,\Delta\ell\left(\frac{1}{\Delta\ell}\frac{d\Delta\ell}{dT}\Delta T_\ell + \frac{1}{\omega}\frac{d\omega}{dT}\Delta T_\omega\right) \tag{2}$$

where ΔT_ℓ and ΔT_ω denote the temperature fluctuation in the optical circuit and the laser diode, respectively.

For example, for a GaAlAs laser diode of 850nm wavelength, the oscillating frequency changes ~20 GHz/°C ($1/\omega \cdot d\omega/dT = 5.7\times10^{-5}$/°C). Therefore, the relation between $\Delta\varphi$, $\Delta\ell$, and ΔT_ω is computed as shown in Fig.2. It is indicated in Fig.2 that the reduction of the pathlength difference or the stabilization of the laser frequency[3] must be performed to realize the highly stabilized gyro.

The interferometer having detectors D_4 and D_5 can be used to measure the phase-output drift $\Delta\varphi$. That is, assuming that the thermal fluctuation coefficient of the optical circuit, excluding the fiber loop and the laser diode, is independent of the location, the phase change $\Delta\varphi_c$ detected by D_4 and D_5 is proportional to $\Delta\varphi$. It should be noted that the output drift induced by fluctuation in the optical circuit, which is expressed by the first term in parentheses in (2), can be reduced only by controlling the laser frequency[3,4] so that $\Delta\varphi_c = 0$ within the above condition.

In a ring interferometer, the laser output returns on itself. Then, for a single-mode laser diode, oscillating mode hopping and/or multiple-mode oscillation takes places [5]. The oscillating mode hopping to the adjacent longitudinal mode causes a frequency shift of ∿100GHz, which induces a significant output change, as shown in Fig.3, even for a minimum pathlength difference of $\lambda/4$. The multiple-mode oscillation degrades the visibility. Therefore, an optical isolator must be used as shown in Fig.1.

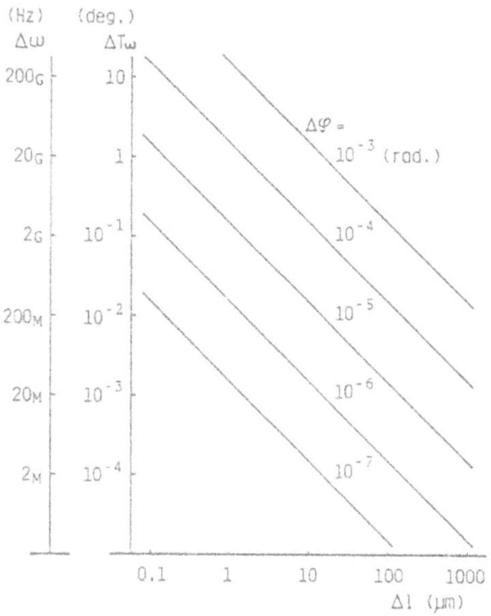

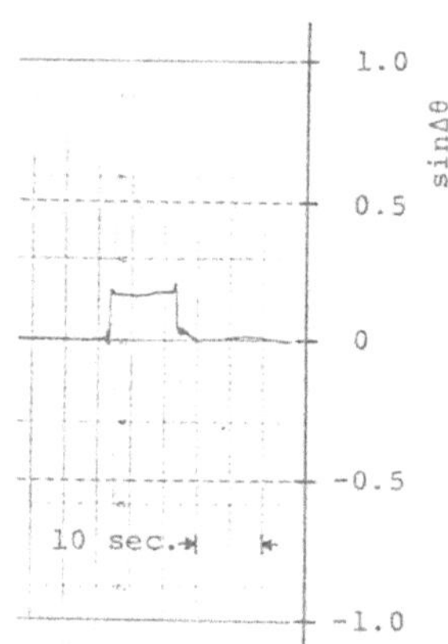

Fig.3 Output change induced by the mode hopping of LD

◁ Fig.2 Output phase drift induced by the oscillating frequency change of a GaAlAs laser diode

3. Experiment

3.1 Experimental setup

In the experimental setup, the light source is a single-mode GaAlAs laser diode with 816nm wavelength (Hitachi HLP3400), whose frequency is stabilized by controlling its temperature. The controller stabilizes the temperature of its heat sink within ±0.01°C, which corresponds to a frequency stability of ±200MHz. Moreover, an optical isolator (NEC LC-4202-02) is located between BS_1 and LD, whose isolation rate is ∿30dB. The stable interfered output is obtained with the isolator. A 150μm thick glass plate is used as a phase shifter whose insertion angle is adjusted to optimize the sensitivity. The single-mode 300m long optical fiber is wound on a 34cm radius drum. To reduce the reflected light at the fiber endfaces, index matching oil is inserted between the endfaces and lenses.

Because the beam splitter used in the setup is not ideal, the complementary relation between the outputs of D_1 and D_3 cannot be obtained. Therefore, an additional beamsplitter and detector are introduced between BS_3 and the polarizer, and D_1 output is divided by that of the detector to compensate the drift in the polarization fluctuation.

3.2 Estimation of the Residual Pathlength Difference

In a single-mode laser diode, longitudinal mode hopping takes place at a certain temperature and injection current. Using this property, the residual pathlength difference can be estimated.

Under the pathlength difference $\Delta\ell$, the laser frequency change of $\Delta\omega$ induces the output phase shift

$$\Delta\overline{\Theta} = \frac{\Delta\omega}{c}\Delta\ell \ . \tag{3}$$

In Fig.3 the output phase shift $\Delta\overline{\Theta}$ of 0.15rad is experimentally observed when the longitudinal mode is shifted to the adjacent mode by adjusting the injection current. In this case the frequency shift is 112GHz. Therefore, the residual pathlength difference of the setup is $\sim$65μm.

The method discussed in this section can be used in trimming to reduce the residual pathlength difference.

3.3 Rotation Detection

A rotation detection experiment was performed using the experimental setup. Figure 4 shows the divider output when rotation rates from ±0.25 deg./s to ±9.0 deg./s are applied to the rotation table. The ordinate indicates $\sin\Delta\Theta$. The drift shown in fig.4 is caused mainly by the Duralmin board fluctuation, which is used as a base for the optical setup. The rapid variation with a small amplitude is caused by the vibration of the components induced by the rotation. The short-time resolution limited by the rapid variation is better than 0.05 deg./s (0.87mrad/s).

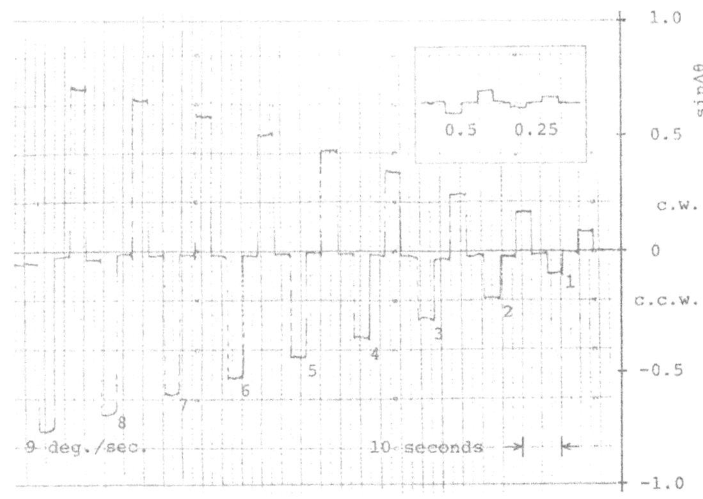

Fig.4 Rotation detection by the experimental setup

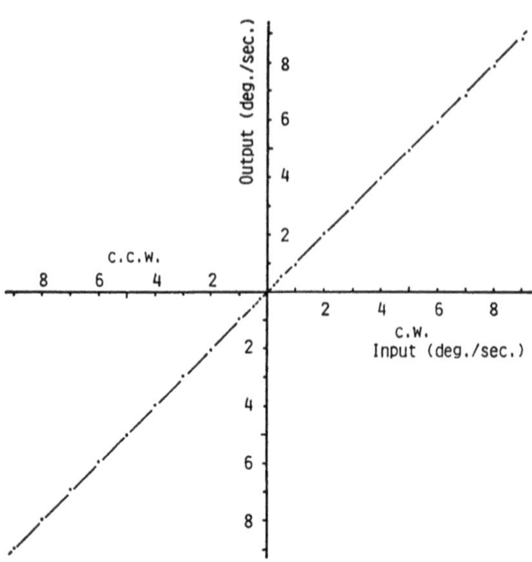

Fig.5 Linearity between given and measured rotation rates

Figure 5 shows the linearity of the setup plotted from the data in fig.4 where the drift has been compensated.

4. Description of the Heterodyne Detection System

The homodyne system described above cannot realize a phase nulling detection which is useful for the gyro required to have a wide dynamic range and linearity. The pathlength difference in the system is also a problem. In the following sections, a novel optical heterodyne system is proposed, which has only a quite small pathlength difference to eliminate the output drift.

4.1 Optical Configuration

Figure 6 shows the conceptual diagram of the proposed system with an optical heterodyne technique. Two laser waves, divided by a grating and a beamsplitter, are launched into a fiber loop to propagate in both directions. Two waves from the fiber ends are recombined by AOM, which acts not only as a optical frequency shifter but also as a beamsplitter. For example, photo detector D_1 accepts the nondiffracted part (frequency f_0, equal to the laser oscillation frequency of the clockwise propagated wave and the diffracted part (f_0+f_1, f_1 is the AOM driving frequency) of the counterclockwise propagated wave, then outputs the intermediate frequency f_1. The Sagnac effect appears as the phase difference between the detected beat signal and the AOM driving signal, which can be measured electronically by phase nulling. The orthogonally polarized waves guided into a short reference path can monitor the phase fluctuation in the optical circuit except for the fiber, if required.

Two reflected waves or two transmitted waves at the beamsplitter can also be used as those to excite the fiber.

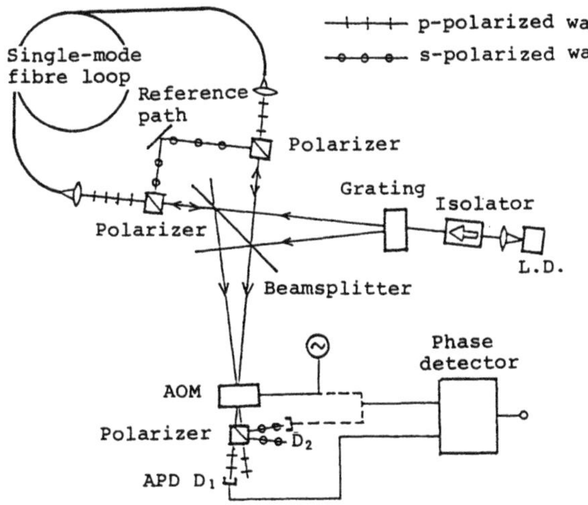

Fig.6 Conceptual diagram of proposed system with optical heterodyne technique

4.2 Evaluation of Pathlength Difference

There is no pathlength difference between the waves in the proposed system when optical components are correctly positioned. Furthermore a position error of components gives rise to only a quite small pathlength difference.

The pathlength difference between the waves $\Delta\ell$ that is induced by a displacement d of the AOM or the grating from its correct position along the optical path, is calculated as

$$\Delta\ell = d(1-\cos\phi) \simeq \frac{\phi^2}{2} d \qquad (4)$$

where ϕ denotes the diffraction angle of the AOM and the grating. The size of $\Delta\ell$ is very small because ϕ has a small value. The required accuracy for manufacture becomes less strict. For example, 1mm in position error causes only 50nm pathlength difference when $\phi = 10$mrad. Moreover fluctuation of the optical media hardly affects the output drift because two waves propagate very closely even in the separated region.

The homodyne detection can also be realized with a precisely controlled phase-difference bias as (4) replacing the AOM to a grating.

4.3 Effect of Frequency Difference Between Two Waves in Fiber

Temperature variation in the fiber dT is shown to induce an output phase drift $d\Theta$, when there is frequency difference $\Delta\omega$ between two waves in the fiber, as

$$\frac{d\Theta}{dT} = \frac{\Delta\omega\, N_1\, L}{c}\left(\frac{1}{N_1}\frac{dN_1}{dT} + \frac{1}{L}\frac{dL}{dT}\right) \qquad (5)$$

where L denotes the fiber length, N_1 is the group index, and c is the vacuum light velocity. For example, a temperature variation of 0.01 deg. induces an output drift of 0.1mrad with $\Delta\omega = 2\pi\cdot 40$MHz, $N_1 = 1.5$, $L = 1$km, and $(1/N_1\cdot dN_1/dT + 1/L\cdot dL/dT) = 10^{-5}$. This example shows that a system with a frequency difference in the fiber has considerable difficulty for practical use, but this point is not noticed in the previous works with frequency shifter(s) [2,6]. The proposed system does not suffer from the temperature variation, because no frequency difference between the waves exists in the fiber loop.

4.4 Electronics Circuits for Phase Nulling Detection

Electronic phase nulling can be realized, for example, by the circuits as shown in Fig.7. Two double balanced mixers (DBMs) convert two input signals into the second intermediate frequency f_2. The IF signal of the second is fed into a precisely fixed delay line such as a charge coupled device, which shifts the phase in proportion to f_2. The output phase of the delay line and that of the other DBM are compared by the third DBM, and a feedback loop controls f_2 to make the comparator operate in quadrature. The controlled frequency f_2 is expressed as

$$f_2 = \frac{\Delta\Theta + 2n\pi - \frac{\pi}{2}}{2\pi T_d} \qquad (6)$$

where $\Delta\Theta$, T_d, and n indicate the input phase difference, the delay time in CCD, and an arbitrary integer, respectively. $\pi/2$ is the comparator quadrature condition in (6). It shows that output frequency f_2 follows $\Delta\Theta$ linearly when selecting n so that $f_2 > 0$.

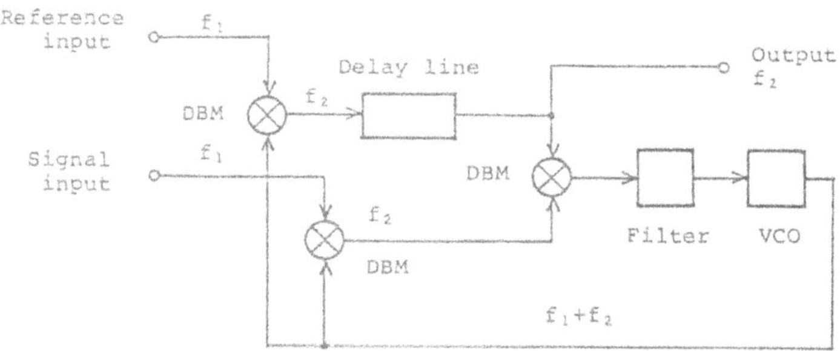

Fig.7 Diagram of phase nulling electronics

5. Preliminary experiment

A preliminary experiment has been carried out to show heterodyne detection. A test setup without reference path has been constructed with the same single-

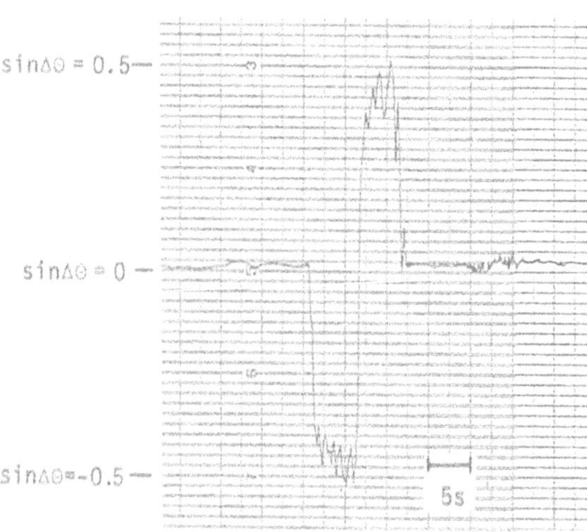

Fig.8 Preliminary experimental result of rotation detection

mode optical fiber and the 10mW single-mode GaAlAs laser diode that were used in the above experiment, and discrete optical components on a interferometer base. A DBM is used as a phase detector, with the AOM driving signal as a local oscillator instead of phase nulling electronics. Therefore, the output varies as $\cos(\Delta\Theta+\Theta_m)$ where $\Delta\Theta$ and Θ_m is due to the Sagnac effect and phase lag in the AOM, respectively. Θ_m is adjusted to realize the quadrature condition by trimming the AOM position perpendicular to the optical path.

Figure 8 shows the preliminary experimental result of rotation detection, rotating only the fiber drum at a rate of about ±90mrad/s by hand. The output drift shown in Fig.8 is caused mainly by mechanical fluctuation at laser mount or polarization instability in the fiber.

6. Conclusion

Two optical systems for the optical fiber laser gyro have been described. The one is a homodyne detection system, which has been considered theoretically and experimentally. The other system with an optical heterodyne technique has no difference in the pathlength nor the frequency in the fiber between the counterpropagating waves. An output drift appears when there exists a difference, which is not noticed in previous works. A preliminary experiment has been performed. The experimental setup with the phase nulling electronics for rotation detection is now under construction. The system is suitable for optical integrated circuits.

The authors would like to thank N.Inagaki and K.Okamoto of Ibaraki Electrical Communication Laboratory, NTT., for providing the fiber. Thanks are also due to I.Sagehashi, Y.Yoshida of our Institute for discussions.

References

1. K.Hotate, Y.Yoshida, M.Higashiguchi and N.Niwa: Electron. Lett. 16, 941 (1980)
2. D.E.Thompson, D.B.Anderson, S.K.Yao and B.R.Youmans: Appl. Phys. Lett. 33, 940, (1978)
3. T.Okoshi and K.Kikuchi: Electron. Lett. 16, 179 (1980)
4. F.Favre and D.Le Guen: Electron. Lett. 16, 709 (1980)
5. T.Horimatsu, M.Sasaki and K.Aoyama: Appl. Opt. 19, 1984 (1980)
6. R.F.Cahill and E.Udd: Appl. Opt. 19, 3054 (1980)

Part 5

Limiting Factors

5.1 Non-Reciprocal Error Sources

Intensity Dependent Nonreciprocal Phase Shift in a Fiberoptic Gyroscope

S. Ezekiel and J.L. Davis

Research Laboratory of Electronics, Massachusetts Institute of Technology
Cambridge, MA 02139, USA

R.W. Hellwarth

Departments of Electrical Engineering and Physics, University of
Southern California, Los Angeles, CA 90007, USA

We have predicted and observed an intensity-induced nonreciprocity in a fiberoptic Sagnac interferometer. This interferometer, which has been described previously [1] is being developed for absolute rotation sensing. We found that a nonreciprocal phase shift of 1.4×10^{-6} radians can be generated by a one microwatt power difference between the oppositely propagating light beams. Our fiber is 200 m long with a core diameter of 4.5 microns, and wound on a 19 cm diameter spool. The intensity dependent nonreciprocal phase shift, which is attributed to a four wave mixing process in the quartz medium [2-6] is equivalent to a rotation rate of 0.2 °/hr in the present geometry, and therefore stresses the need for a strict intensity control in precision fiber rotation sensors.

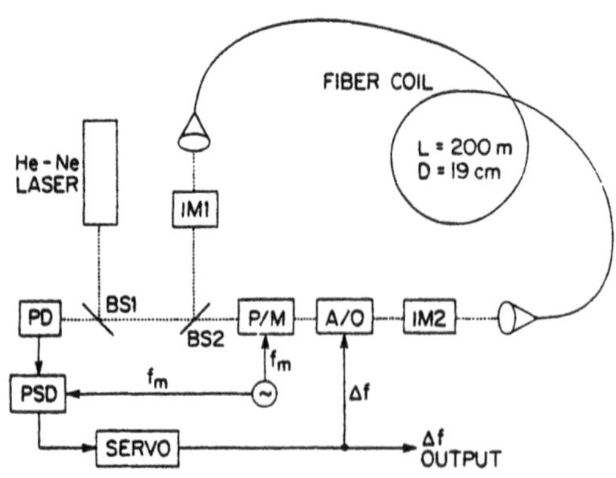

Fig.1 Experimental set-up

Our measurements were performed using a configuration that is essentially that which has been described in detail elsewhere [1]. The experimental arrangement is shown schematically in Fig. 1. Light from a 5 mW He-Ne laser is split by a 50/50 beam splitter (BS2) into clockwise and counterclockwise beams that propagate in the 200 meter long fiber. That portion of the light that returns back to the laser is picked off at beam splitter BS1 and directed onto a photodetector (PD). As described elsewhere [1], we use a closed loop nonreciprocal phase measurement technique which employs a nonreciprocal phase modulator and acousto-optic frequency shifters. The phase modulator (P/M) is placed near the beam splitter BS2 as shown in Fig. 1. In this way a discriminant is obtained when the output of the photodetector is synchronously demodulated in a phase sensitive demodulator (PSD). Closed loop operation is accomplished by feeding the output of the PSD into a pair of acousto-optic shifters (A/O) placed in the beam path of the interferometer [1]. In this way a frequency difference Δf is created between the clockwise and counterclockwise beams so as to null out any nonreciprocal phase shift ϕ_{NR} present in the interferometer so that

$$\Delta f = \frac{c}{2\pi n L_0} \phi_{NR}$$

where c is the speed of light in vacuum, n is the refractive index at the axis of the fiber core and L_0 is the total length of the interferometer which is approximately equal to the length L of the fiber.

Measurements of intensity dependent nonreciprocal phase shift were performed with the aid of two electro-optic intensity modulators IM1 and IM2 placed near each end of the fiber coil. Figure 2, lower trace, shows the variation in magnitude and sign of the power difference ΔP between cw and ccw beams propagating in the fiber obtained by driving IM1 with a sawtooth voltage. The upper trace in Fig. 2 is the corresponding nonreciprocal frequency difference Δf generated by ΔP. Figure 3 shows a plot of Δf vs. ΔP as obtained from data such as that presented in Fig. 2 using either IM1 or

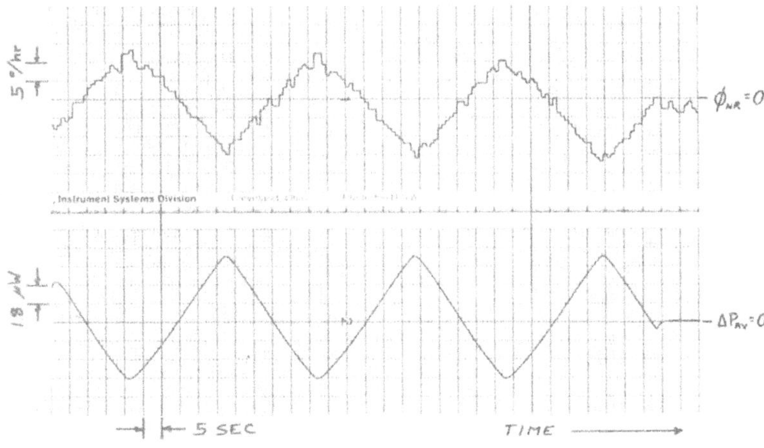

Fig.2 Nonreciprocal phase shift vs. power difference data

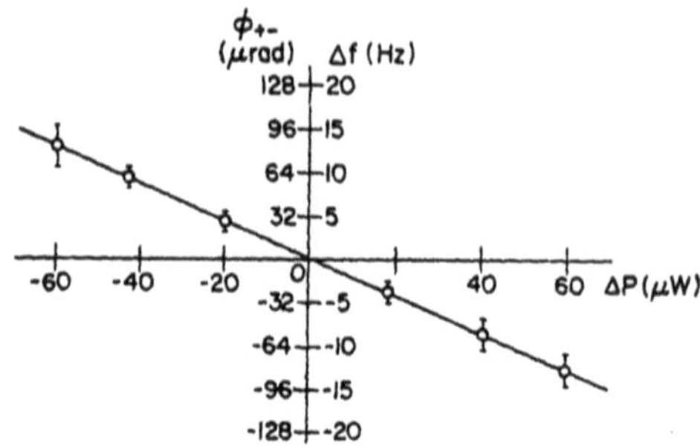

Fig.3 Plot of measured nonreciprocal phase shift vs. power difference in counter propagating beams

IM2. As can be seen, the relationship between Δf and ΔP is linear with a slope of about -0.2 Hz per microwatt or -1.4 microradians per microwatt. This corresponds to -0.2°/hr per microwatt power difference for our rotation sensor configuration. We also observed that the $\Delta f/\Delta P$ relationship was independent of the magnitude of the absolute power along either direction.

The magnitude and sense of the nonreciprocal phase shift ϕ_{+-} arising from the difference ΔP between the powers in the CW and CCW beams is $\int_0^L \Delta k dz$. Here Δk is the intensity-induced difference between the propagation constants k_+ and k_- of the counterpropagating beams at distance z along the length L of fiber. When this difference is assumed to arise primarily from the nonlinear polarization density that is third order in the optical field, as is the case in centro-symmetric materials, Δk has been found to be given by [6]

$$\Delta k = -\frac{1}{2}\beta\Delta P/S , \tag{1}$$

where β is a nonlinear susceptibility coefficient, and S is an effective beam area (of the order of the fiber core area) that are defined through the following relations among the propagating optical electric vector $E_i(\underset{\sim}{x},t)$, i = x,y,z, and the nonlinear electric polarization density $P_i^{NL}(\underset{\sim}{x},t)$. We consider first the case of perfectly monochromatic waves, so that $E_i = \mathrm{Re}E_i(x)e^{-i\omega t}$. We also assume that both beams propagate in the same one transverse guide mode so that

$$E_i = e_i(x,y)(E_+e^{ik_+z} + E_-e^{-ik_-z}), \qquad i = x,y,z \tag{2}$$

where the transverse field pattern is normalized according to $\int dxdy e_i^* e_i = 1$. Here, as elsewhere, repeated space indices are assumed to be summed. With these definitions, the third-order polarization density at ω is written conventionally (in esu) as [2-6]

$$P_{i\omega}^{NL}(\underset{\sim}{x},t) = \mathrm{Re}\ 3\chi_{ijk\ell}(x,y;\ -\omega,\omega,\omega,-\omega)\ E_jE_kE_\ell^*\ e^{-i\omega t} \ . \tag{3}$$

In terms of the nonlinear susceptibility $\chi_{ijk\ell}$ defined by (3) the nonlinear coefficient in (1) is [6]

$$\beta = \frac{96\pi^2\omega\chi^\circ_{eeee}}{n^2c^2} , \tag{4}$$

where

$$\chi^{\circ}_{eeee} \equiv \frac{\chi_{ijk\ell}(0)e_i^*(0)e_j(0)e_k(0)e_\ell^*(0)}{[e_n{}^*(0)e_n(0)]^2} \quad . \tag{5}$$

Here the argument x,y = 0 signifies the guide axis. When, as in (5), the frequency arguments of $\chi_{ijk\ell}$ are omitted, they are assumed to be as in (3). Finally, the effective guide area of (1) is defined by

$$S \equiv \chi^{\circ}_{eeee}/\int\!\!\int dxdy\ \chi_{ijk\ell}e_i^*e_je_ke_\ell^* \quad . \tag{6}$$

One sees that if the nonlinear susceptibility and the mode function were constant across the guide core, and zero outside it, then S would be the area of the core.

The physical origin of the intensity-induced nonreciprocal phase shift can be seen quickly by considering the effect on each other of counter-propagating plane waves. For this, let us assume that $e_i(x,y)$ equals δ_{ix}, a unit vector in the x direction, and that the nonlinear susceptibility is independent of position. Then, if one tries the form

$$P_{i\omega}{}^{NL} = \mathrm{Re}(P_+^{NL}\ e^{ik_+z} + P_-^{NL}\ e^{-ik_-z})\ e^{-i\omega t} \tag{7}$$

in (3), one sees that the nonlinear polarization density for the CW beam has complex amplitude

$$P_+^{NL} = 3\chi_{xxxx}(2|E_-|^2 + |E_+|^2)E_+ \tag{8}$$

and similarly for P_-^{NL} with exchange of subscripts + and -. The wavevector k_+ for the CW beam is, from Maxwell,

$$k_+{}^2 = \frac{\omega^2}{c^2}\,(n^2 + 4\pi P_+^{NL}/E_+) \quad , \tag{9}$$

and similarly for k_- with exchange of subscripts. Now (8) and (9) give (1) directly if one uses the standard expression for the intensity difference of the two beams:

$$\frac{\Delta P}{S} = \frac{nc}{8\pi}\,(|E_+|^2 - |E_-|^2) \ . \tag{10}$$

To estimate the intensity-induced nonreciprocal phase shift ϕ_{+-}, we assume linearly polarized beams. For fused quartz $\chi_{xxxx} = 5 \pm 1 \times 10^{-15}$ esu at 633 nm [5,8-10] giving $\beta \sim 7.4 \times 10^{-11}$ cm/W. Assuming negligible attenuation ($\alpha L << 1$) and an effective area S equal to the core area, we predict from (1) that

$$\phi_{+-}/\Delta P \sim -4.6 \text{ rads/W} \ . \tag{11}$$

From Fig. 1 our measurements showed a value -1.4 rW^{-1}, which is significantly lower than in (11). We feel this is most likely due to random deviations of the polarizations e_{i+} and e_{i-} of the CW and CCW waves both from being equal and from being linear as z varies from 0 to L. Analysis shows that β/S will be markedly less than from (4) and (6) when the waves are differently polarized. Also, the spilling of electric field into the evanescent wave, especially in a bent guide, could make the effective area larger than the core area.

The intensity-induced nonreciprocal phase shift ϕ_{+-} contributes to ϕ_{NR}, and hence to Δf, exactly as does the ϕ_r which arises from the rotation rate Ω of the apparatus according to $\phi_r = \omega LD/c^2$, where D is the fiber coil diameter [2]. In our set-up a $\phi_{+-} = 1.4 \times 10^{-6}$ rad generated by a 1 μW power difference corresponds to a rotation rate Ω of 0.2°/hr. Clearly ϕ_{+-} can be a major source of error in precision fiberoptic rotation sensors. The magnitude of ϕ_{+-} can be minimized by ensuring that the intensities in the counterpropagating directions are equal.

A number of possibilities exist for equalizing the two intensities which depend on the particular configuration of the sensor. For example, in our set-up where acousto-optic shifters are employed, equalization of the intensities may be accomplished by adjusting the r.f. power driving the Bragg cells. A simple feedback loop can be designed to maintain equal intensities. Another way of adjusting the intensities is by using a variable waveguide coupler instead of the fixed beam splitter (BS2).

There may also be some advantages to the intensity induced nonreciprocal phase shift. For example, it could be used as a means of calibrating the scale factor of a fiber rotation sensor during operation, or as a way of enhancing [10] the Sagnac effect in a passive resonator rotation sensor[11].

This research was supported by the Joint Services Electronics Program at the Massachusetts Institute of Technology. One of us (RWH) would like to thank the Air Force Office of Scientific Research for support under Grant No. 78-3479.

References

1. J.L. Davis and S. Ezekiel, Opt. Lett. 6, 505 (1981)
2. P.W. Maker and R.W. Terhune, Phys. Rev. 137, A801 (1965)
3. R.Y. Chiao, P.L. Kelley and E. Garmire, Phys. Rev. Lett. 17, 1158 (1966)
4. R.W. Hellwarth, Progress in Quantum Electronics 5, 1 (1977)
5. J.H. Marburger and R. Shockley, Appl. Phys. Lett. 30, 44 (1977)
6. R.W. Hellwarth, IEEE J. Quant. Electron. QE-15, 101 (1979)
7. A. Owyoung, R. Hellwarth and N. George, Phys. Rev. B5, 628 (1972)
8. R. Hellwarth, C. Cherlow and T.T. Yang, Phys. Rev. B11, 964 (1975)
9. E.S. Bliss, D.R. Speek and W.W. Simmons, Appl. Phys. Lett. 25, 728 (1974)
10. A.E. Kaplan and P. Meystre, Opt. Lett. 6, 590 (1981)
11. G.A. Sanders, M.G. Prentiss and S. Ezekiel, Opt. Lett. 6, 569 (1981)

Fibre Gyro Performance in the Presence of External Magnetic Fields

K. Böhm, K. Petermann, and E. Weidel

AEG-Telefunken, Forschungsinstitut, D-7900 Ulm, Fed. Rep. of Germany

Abstract

It is shown that a fibre optic gyro is influenced by external magnetic fields, yielding an increased bias instability with increasing field strength. In an experimental set-up a bias uncertainty of about 10 degrees/hour is obtained, if the earth magnetic field is acting upon the fibre coil. By protecting the fibre coil against external magnetic fields an improvement by at least one order of magnitude is obtained.

A fibre gyro is designed to detect the non-reciprocal Sagnac-phase shift between the co-rotating and counter-rotating beam through a fibre coil. By using a true single-mode filter before entering the fibre coil [1] the fibre gyro will be insensitive to any reciprocal phase shift as long as the variations are slow. However, besides the non-reciprocal Sagnac-effect there is the non-reciprocal Faraday-effect [2] and the question arises whether a fibre gyro is sensitive also to magnetic fields. It is obvious that a fibre gyro responds to a current flowing along the fibre coil axis [3] since then the line integral $\int \vec{H}\, d\vec{l}$ ($\vec{H}$ = magnetic field) along the fibre loop does not vanish. For external magnetic fields, however, we have a vanishing line integral $\int \vec{H}\, d\vec{l} = 0$ and the question arises how a fibre gyro behaves in such an external magnetic field.

The model to be considered is shown in Fig. 1 where the light is fed through a linear polarizer and the ports 1, 2 into both ends of a fibre loop which is exposed to a magnetic field $\vec{H}$. In this simple model the fibre is assumed to be ideal without any birefringence and two retarders $\underline{\underline{R}}_1$, $\underline{\underline{R}}_2$ are inserted.

If the state of polarisation would be constant along the fibre an external magnetic field would have no effect on the detected Sagnac phase shift. If, however, the state of polarization changes along the fibre the interaction between the magnetic field and the optical field changes as well, so that an external magnetic field may yield an erroneous output signal of the gyro.

In the model of Fig. 1 the change of the polarization within the fibre loop is represented by the retarders $\underline{\underline{R}}_1$, $\underline{\underline{R}}_2$, so that a linear state of polarization is obtained between port 2 and $\underline{\underline{R}}_2$ and an elliptical state of polarization is obtained between $\underline{\underline{R}}_1$ and $\underline{\underline{R}}_2$, the degree of ellipticity depending on the retardation angle α of the retarders $\underline{\underline{R}}_1$, $\underline{\underline{R}}_2$.

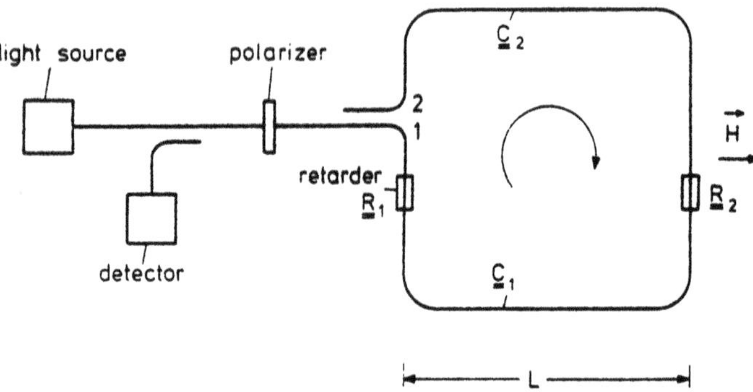

Fig. 1. Fibre gyro model

The resulting error for the Sagnac phase shift has been calculated in [4] and is given as

$$\Delta\varphi_S = 2\kappa \cdot \sin\alpha \qquad (1)$$

with κ the Faraday-rotation between the retarders $\underline{\underline{R}}_1$, $\underline{\underline{R}}_2$ and α the retardation angle of the retarders $\underline{\underline{R}}_1$, $\underline{\underline{R}}_2$. Without retarders ($\alpha = 0$) the error vanishes and the worst case is obtained for $\alpha = \pi/4$ where we have circular polarization between $\underline{\underline{R}}_1$, $\underline{\underline{R}}_2$.

It should be noted that the evolution of the state of polarization in an actual fibre is random, which may be expressed in eq. (1) by a random α, yielding also a random error for the detected Sagnac-phase shift. Even for a relatively short length $L = 5$ m between the retarders $\underline{\underline{R}}_1$, $\underline{\underline{R}}_2$ the earth magnetic field may yield an error for the Sagnac phase shift of up to 10^{-3} rad. which represents a considerable error.

Experiments have been done with a set-up which has been described in [5]. If the fibre gyro is operated without protection against magnetic fields a long term behaviour as shown in Fig. 2 [5] is obtained, where the gyro output is observed for 75 minutes with rotation rates of ± 0.02 degrees/second applied. This bias uncertainty is about ± 15 degrees/hour (= rms value of 10 degrees/hour), corresponding to a bias uncertainty of the

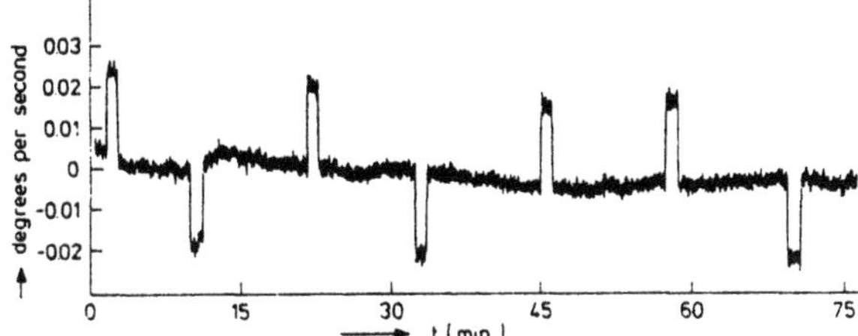

Fig.2. Output signal of the fibre gyro without protection against magnetic fields

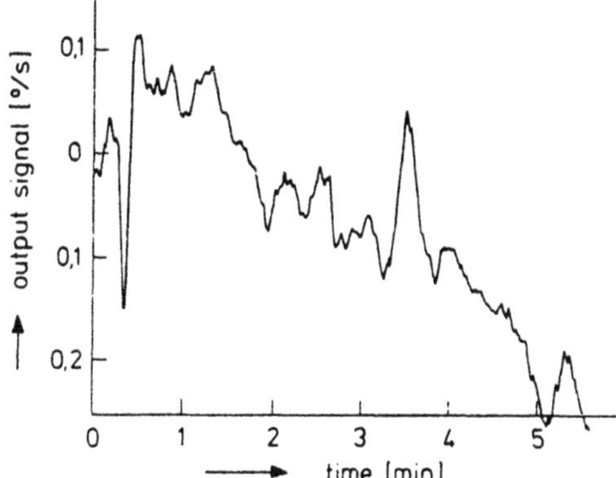

Fig.3. Output signal if a permanent magnet is placed close to the fibre coil

specific set-up of about $\pm$ $3 \cdot 10^{-4}$ rad. Because of the random change of polarization within the fibre loop the effect is much weaker than in the theoretical model.

By placing a permanent magnet, exhibiting a magnetic field which is by a factor of about 100 larger than the earth magnetic field, close to the fibre coil a gyro output as shown in Fig. 3 occurs. The variations of the output signal are at least $\pm$ 0.2 degrees/second and thus by about two orders of magnitude higher than the variations without permanent magnet.

In a third experiment, the fibre coil was placed in a μ-metal housing, in order to protect the fibre against the earth magnetic field. Fig. 4 shows the fibre gyro output signal of this set-up with the turntable being rotated with 14.4 degrees/hour opposite to the earth rotation yielding a net rotation at Ulm

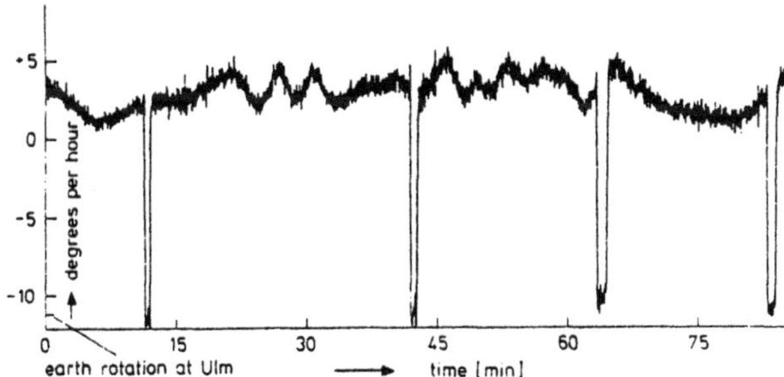

Fig.4. Output signal of the fibre gyro with the fibre coil being protected against the earth magnetic field

of about 3.2 degrees/hour. For periods of about 1 min the turntable was stopped so that the gyro is then detecting just the rotation of earth.

The bias uncertainty is about one order of magnitude smaller than in the set-up without protection against the magnetic field. It should be noted in this respect, that only the fibre coil was protected. The phase modulator and the main beam splitter were still within the earth magnetic field.

The authors wish to thank R. Ulrich for helpul and stimulating discussions.

References

1 R. Ulrich, "Fiber-optic rotation sensing with low drift", Opt. Lett. 5 (1980), pp. 173 - 175

2 H. M. Smith, "Polarization and magnetooptic properties of single-mode optical fiber", Appl. Opt. 17 (1978), pp. 52-56

3 H. J. Arditty, Y. Bourbin, M. Papuchon and C. Puech, "Current sensor using state-of-the-art fiber-optic interferometric techniques", IOOC, April 1981, San Francisco, paper WL 3

4 K. Böhm, K. Petermann, E. Weidel, in preparation

5 K. Böhm, P. Marten, K. Petermann, E. Weidel and R. Ulrich, "Low-drift fibre gyro using a superluminescent diode", Electron. Lett. 17 (1981), pp. 352 - 353

A Nonreciprocal Optical Effect in Optical Gyroscopes

Jiang Yanan
Department of Precision Instrument, Tsinghua University, Beijing, China

In a fiberoptic rotation sensor a nonreciprocal phase shift modulation is usually introduced to determine the difference of phase produced by the rotation of the sensor. It is also necessary that some frequency bias technique with nonreciprocal propagation properties be used to prevent lock-in in ring laser gyroscopes. Faraday effect, polar and transverse magneto-optical Kerr effects and optical activity of quartz rotators, which are used in the four frequency differential laser gyro and magnetic mirror laser gyro, can be generalized in the form of dielectric tensors containing off-diagonal and symmetrical complex conjugate elements.

This paper discusses the basic form of the wave-vector surface and normal mode of nonreciprocal gyrotropic dielectric tensors. The nonreciprocal effects mentioned are discussed and their applications in laser gyros are presented. Some errors, which are produced from using nonreciprocal elements, are also analyzed theoretically.

Introduction

When a ring laser rotates in an inertial space with a rotation rate perpendicular to the plane of the laser cavity, a frequency difference between two counterpropagating traveling waves CW and CCW due to Sagnac Effect is generated. This frequency difference can be expressed by

$$\Delta\nu = \nu_{CW} - \nu_{CCW} = \frac{4S}{\lambda L}\,\Omega$$

where S is the enclosed area, L is the optical length of the RLG, and λ is the optical wavelength. This is the principal formula when the ring laser is used as a rotation sensor. The gain process of the active medium in the laser cavity makes the bandwidth of the laser light more narrow than the bandwidth of the empty cavity. Thus the measuring error of the frequency difference

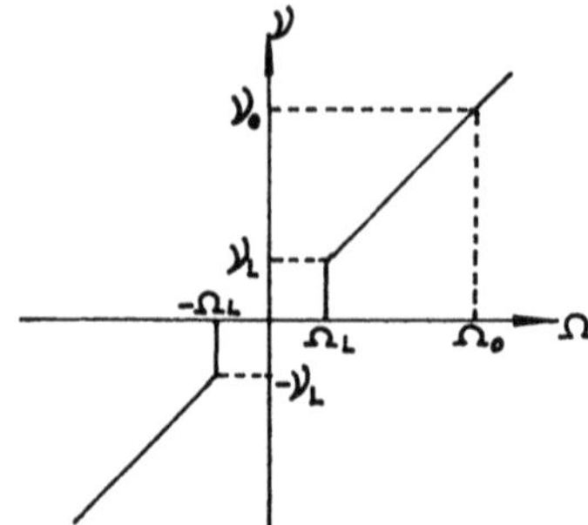

Fig. 1. Principle curve and lock-in region in laser gyros

$\Delta\nu$ has greatly been decreased. At present the random drift of the output signal of a laser gyro is about 10^{-3} deg/h (sample time $\tau = 1$ h), which is nearly equal to the accuracy limited by quantum noise from spontaneous radiation.

Because of the existence of backscattering and nonhomogeneous losses in the ring laser cavity, there will be coupling between two counterpropagating traveling waves, hence the output signal will vanish for low rotation rates as $|\Omega| < \Omega_L$, where Ω_L is the lock-in threshold (about 10^3 - 10^4 deg/h). It is well known as a bias technique in a laser gyro to prevent it from working in the lock-in region. There are mainly three means of biasing. These are mechanical dithering, a magnetic mirror (using the transverse magneto-optic Kerr effect) and the four frequency differential laser-gyro (using the Faraday Effect). All of them use effects based on one of the nonreciprocal effects for two counterpropagating traveling waves. The nonreciprocal effect provides a nonreciprocal phase shift $\Delta\phi$ between two traveling waves. As a result, there will be a frequency difference between them

$$\nu_0 = \frac{\Delta\phi C}{2\pi L} ,$$

where C is the light velocity. Under the condition that this nonreciprocal effect is strong enough, the bias frequency ν_0 may be much greater than the lock-in threshold ν_L [$\nu_L = (4S/\lambda L)\Omega_L$], see Fig. 1, so that bias technology is very important in a laser gyro.

In a four frequency laser gyro [1,2], two laser gyros, left and right circularly polarized, operate respectively in the same ring cavity, in which travel four different waves. A 90° quartz rotator (see Fig. 2) is used to provide a phase shift difference between left and right circularly polarized waves (LCP and RCP) due to their nonreciprocity. Hence, the frequencies of the LCP and RCP laser gyros split, each from other, in a half-mode separation (C/2L). In addition, a Faraday cell with a longitudinal magnetic field can

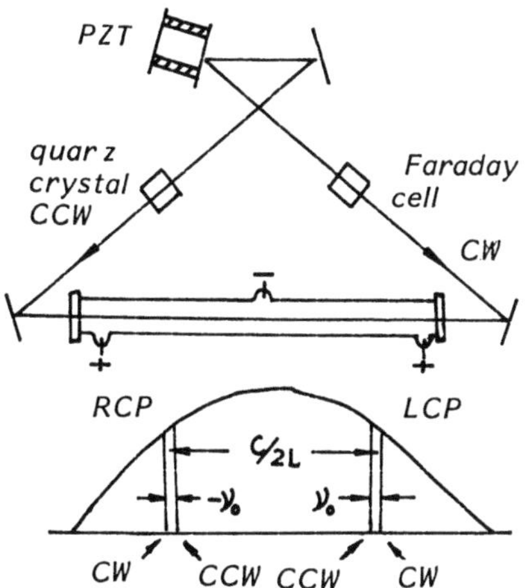

Fig. 2. Principle of four-mode laser gyro

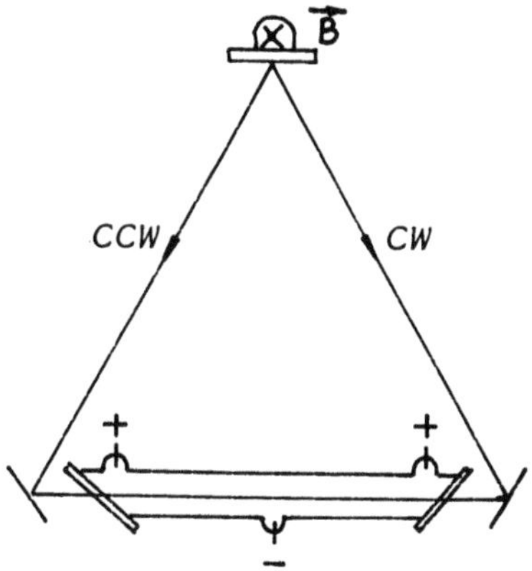

Fig. 3. Principle of magnetic mirror laser gyro

also be used, whose nonreciprocity provides the left and right circularly polarized gyros with biases $\pm\nu_0$, which are equal in magnitude and opposite in sign. Then the left and right polarized gyros shift off symmetrically from the lock-in region. At this time the frequencies of the output signals of the left and right circularly polarized gyros are given by

$$\nu_L = \nu_0 + \frac{4S}{\lambda L}\Omega \quad , \qquad \nu_R = \nu_0 - \frac{4S}{\lambda L}\Omega$$

respectively. As a result, the final differential frequency is

$$\Delta\nu = \nu_L - \nu_R = \frac{8S}{\lambda L}\Omega \quad ,$$

in which the Faraday bias term is cancelled out and only the information related to the rotation rate Ω remains. Because of the direct bias in the four frequency laser gyro, the measuring range of the gyro is limited by the magnitude of the bias ν_0. In practice, its magnitude is about 1 MHz.

With regard to a laser gyro with magnetic mirror bias, (see Fig. 3) a ferro-magnetic mirror (for example, a rare earth-iron garnet) coated by a stack of dielectric thin films has been used. It provides a nonreciprocal reflective coefficient for P linearly polarized light. Hence, a nonreciprocal phase difference $\Delta\phi = \phi_{CW} - \phi_{CCW}$ is introduced when a reflection takes place. An alternate bias is presented when the magnetic field is alternatively applied. The laser gyro with magnetic mirror usually works at an alternative bias. A bias 1-2 orders of magnitude greater than the lock-in threshold ν_L is sufficient, i.e., ν_0 is about 50 kHz.

All the nonreciprocal effects mentioned above can be characterized by a dielectrical tensor containing off-diagonal and symmetrical complex conjugate elements $\pm i\varepsilon_{12}$. It is called the non-reciprocal gyrotropic tensor $\underline{\underline{\varepsilon}}$,

$$\underline{\underline{\varepsilon}} = \varepsilon_0 \begin{bmatrix} \varepsilon_{11} & i\varepsilon_{12} & 0 \\ -i\varepsilon_{12} & \varepsilon_{22} & 0 \\ 0 & 0 & \varepsilon_{33} \end{bmatrix} \quad (1)$$

where ε_0 is the permittivity of the vacuum (MKS units), with the magnetic field or crystal axis being along the z-axis. As is well known, the energy loss of absorption in nonisotropic media is proportional to

$$\sum_{ij=1}^{3} (\varepsilon_{ij}^{*} - \varepsilon_{ji}) E_i E_j^{*} \quad .$$

Therefore, in a nonreciprocal medium in the absence of absorption this gyrotropic tensor $\underline{\underline{\varepsilon}}$ has to be of the Hermitian type, i.e., $\varepsilon_{ij} = \varepsilon_{ji}^{*}$. It means that all the diagonal elements must be real i.e., $\varepsilon_{ii} = \varepsilon_{jj}^{*}$, and that all the off-diagonal elements $\pm i\varepsilon_{12}$ must be purely imaginary, i.e., ε_{12} is a real number. The elements of the gyrotropic tensor $\underline{\underline{\varepsilon}}$ of the medium with absorption, ε_{ii} and ε_{12}, are complex. Their imaginary parts are related to the coefficient of absorption. In this case, the tensor $\underline{\underline{\varepsilon}}$ may be resolved as the sum of a Hermitian tensor and an anti-Hermitian tensor, the anti-Hermitian tensor satisfying $\varepsilon_{ij} = -\varepsilon_{ji}^{*}$.

The Faraday cell used in a laser gyro is made of a sort of glass with isotropic characteristics. But the rare earth-iron garnet crystal, which belongs to the cubic crystal system, has $\varepsilon_{11} = \varepsilon_{22} = \varepsilon_{33} = N_0^2$. The quartz rotator is a single axis crystal, hence $\varepsilon_{11} = \varepsilon_{22} = N_0^2$ and $\varepsilon_{33} = N_e^2$. In the following discussion we can treat it as a single axis crystal. For a cubic crystal, putting Ne = No will be enough.

Propagation of Light in a Crystal Characterized by a Gyrotropic Tensor-Normal Mode and Wave Vector

The dielectric tensor $\underline{\underline{\varepsilon}}$ is expressed as

$$\underline{\underline{\varepsilon}} = \varepsilon_0 \begin{bmatrix} N_0^2 & i\varepsilon_{12} & 0 \\ -i\varepsilon_{12} & N_0^2 & 0 \\ 0 & 0 & N_e^2 \end{bmatrix} \quad . \quad (2)$$

The general wave equation, which is satisfied by the electrical vector of the light wave, may be written as

$$\nabla \times (\nabla \times \underline{E}) + \frac{1}{c^2}\frac{\partial^2 \underline{E}}{\partial t^2} = -\frac{1}{c^2}\underline{\underline{\chi}}\frac{\partial^2 \underline{E}}{\partial t^2} \tag{3}$$

where $\underline{\underline{\chi}}$ is the susceptibility tensor. We introduce a displacement vector written as

$$\underline{D} = \varepsilon_0(1 + \underline{\underline{\chi}})\underline{E} = \underline{\underline{\varepsilon}}\underline{E} \quad .$$

Substituting this expression into (3), we obtain

$$\nabla \times (\nabla \times \underline{E}) + \frac{1}{c^2}\frac{\underline{\underline{\varepsilon}}}{\varepsilon_0}\frac{\partial^2 \underline{E}}{\partial t^2} = 0 \quad . \tag{4}$$

If a monochromatic plane wave $\underline{E}\exp[i(\omega t - \underline{k}\cdot\underline{r})]$, whose wave vector direction is $\underline{K} = \underline{k}/k = \{\alpha_1, \alpha_2, \alpha_3\}$, can maintain its polarization mode, when it travels through a nonreciprocal crystal characterized by a gyrotropic tensor, then we define this wave as the normal mode in the $\underline{K}$ direction of the crystal. In order to determine the polarization of the normal mode electrical vector $\underline{E}$ and the corresponding refraction index n or the wave-number $k = k_0 n$, on substituting the monochromatic plane wave into (4) we obtain (5), related to $\underline{E}$,

$$\underline{k}\cdot(\underline{k}\cdot\underline{E}) - k^2\underline{E} + k_0\frac{\underline{\underline{\varepsilon}}}{\varepsilon_0}\underline{E} = 0 \tag{5}$$

where $k_0 = \omega/c$ is the wave number in vacuum of this monochromatic wave. Equation (5) can be written in the form of a matrix as follows:

$$\begin{bmatrix} k_0^2N_0^2 - k^2(1-\alpha_1)^2 & k^2\alpha_1\alpha_2 + ik_0^2\varepsilon_{12} & k^2\alpha_1\alpha_3 \\ k^2\alpha_1\alpha_2 - ik_0^2\varepsilon_{12} & k_0^2N_0^2 - k^2(1-\alpha_2^2) & k^2\alpha_2\alpha_3 \\ k\alpha_1\alpha_3 & k^2\alpha_2\alpha_3 & k_0^2N_e^2 - k^2(1-\alpha_3^2) \end{bmatrix} \begin{bmatrix} E_x \\ E_y = 0 \\ E_z \end{bmatrix} . \tag{6}$$

These are linear homogeneous equation in Ex, Ey and Ez. If the determinant of the coefficients vanishes, a nontrivial solution exists. Then we obtain an equation in fourth order of the wavenumber k of the normal mode, i.e., a Fresnel wave-vector surface equation. There will be two values of k^2 in any direction of the wave-vector $\underline{K}$. Putting these values back into (6), we will obtain the polarization mode of the corresponding normal mode, and discuss it in detail in the following sections.

Quartz-Rotator Effect and Faraday Effect of Magneto-Rotator

The similarities in both cases mentioned above are that the wave-vector $\underline{K}$ is along the light axis or the direction of the magnetic field, i.e., along the z-axis. Hence, $\alpha_1 = \alpha_2 = 0$, $\alpha_3 = 1$. Substituting into (6) we obtain

$$\begin{bmatrix} k_0^2N_0^2 - k^2 & ik_0^2\varepsilon_{12} & 0 \\ -ik_0^3\varepsilon_{12} & k_0^2N_0^2 - k^2 & 0 \\ 0 & 0 & k_0^2N_e^2 \end{bmatrix} \begin{bmatrix} E_x \\ E_y \\ E_z \end{bmatrix} = 0 \quad . \tag{7}$$

Ez = 0 means that the electrical vector is perpendicular to $\underline{K}$, i.e., the normal modes are transverse waves. The condition for the existence of a non-trivial solution is $k_\pm^2 = k_0^2(N_0^2 \pm \varepsilon_{12})$, or $k_\pm = k_0(N_0^2 \pm \varepsilon_{12})^{\frac{1}{2}}$ if we take only the positive value. Putting it back into equation (7), we find that the corresponding polarization is $\{1, \pm i, 0\}$. This means that the normal modes are left and right circularly polarized light and are orthogonal. The difference of refraction index between them is

$$\Delta n = n_+ - n_- = (N_0^2 + \varepsilon_{12})^{\frac{1}{2}} - (N_0^2 - \varepsilon_{12})^{\frac{1}{2}} \doteq \varepsilon_{12}/N_0 \quad .$$

(1) *Rotary Power of Quartz-Crystal*

The dielectrical tensor of quartz crystal can be expressed as (2), where No and Ne are the refraction index of O light and E light in the quartz crystal respectively. ε_{12}^Q, No and Ne are all real because the absorption is negligible. In the case of $\lambda = 0.63\ \mu$, $N_0 = 1.544$ and $\varepsilon_{12}^Q = 1.0 \times 10^{-4}$ (the upper index Q represents left and right rotation quartz crystals, signs of ε_{12}^Q being opposite). When a beam of linearly polarized light travels through the crystal along its light axis, the light will be resolved in the left and the right circularly polarized normal modes, given by (Ne = 1.553)

$$E_0\begin{bmatrix} 1 \\ 0 \end{bmatrix} = \frac{E_0}{2}\begin{bmatrix} 1 \\ i \end{bmatrix} + \frac{E_0}{2}\begin{bmatrix} 1 \\ -i \end{bmatrix} \quad .$$

Both travel independently in the crystal with the different phase velocities showing as

$$\Delta n = n_R - n_L = \varepsilon_{12}^Q/N_0 \quad .$$

As they travel through a certain length in the crystal, a phase difference proportional to this length will arise. In other words, the polarization plane of the resultant linearly polarized light will rotate through an angle proportional to this length. We define the specific rotatory power δ as the angular shift of the polarization plane when a light beam travels through unit length. For the quartz crystal,

$$\delta_Q = \pi\varepsilon_{12}^Q/\lambda N_0[\mathrm{rad/mm^{-1}}]$$

or

$$\delta_Q = 180\varepsilon_{12}^Q/\lambda N_0[\mathrm{deg/mm^{-1}}] \quad .$$

For $\lambda = 0.63\ \mu$, $\varepsilon_{12}^Q = 1.0\times 10^{-4}$, we obtain $\delta_Q = 18.7$ deg/mm. We use a thickness of 4.81 mm cut normally to the light axis of the crystal as a 90° rotator in a four-frequency differential laser gyro to generate a frequency split of half-mode C/2L between the left and the right circularly polarized gyro (see Fig. 2).

(2) *Faraday Effect*

In the case of the Faraday Effect, ε_{12} is not a constant, and it behaves as an odd function of the magnetic induction intensity $\underline{B}$. In the linear region, it can be considered as $\varepsilon_{12} = \pm\beta B$. Because the wave-vector $\underline{K}$ is along the positive direction of the z-axis, so the direction of $\underline{B}$ along and opposite the z-axis corresponds to the + and - sign respectively in this expression. In practice, the Faraday cell used in a laser gyro is isotropic (as optical glasses) or is material of a cubic crystal system. Here No = Ne. In general practice, the Faraday cell has obvious absorption (as in the iron-garnet crystal), which means that No and ε_{12} are all complex. Define the complex specific rotatory power

$$\delta_F = \delta_F' + i\delta_F'' \frac{\pi}{\lambda}\frac{\varepsilon_{12}}{N_0} \quad .$$

Therefore, the normal modes of left and right circularly polarized light travel in the Faraday cell with different phase speed and absorption. As they travel through a certain length their phases and magnitudes will be different. In other words, when a beam of linearly polarized light travels in a Faraday cell with absorption, elliptical polarized light will be formed. In the general case, the absorption is weak. After the light has traveled through a unit length, the ratio of short to long axis of the ellipse is

$$a/b = \tanh \delta_F'' = \delta_F'' = \frac{\pi}{\lambda}\,\mathrm{Im}(\varepsilon_{12}/N_0)$$

(Im designates the imaginary part), and rotating the angle of the polarization plane designated by the long axis of the ellipse related to the polarization plane of the original light is

$$\delta_F' = \frac{\pi}{\lambda} \mathrm{Re}\left(\frac{\varepsilon_{12}}{N_0}\right)$$

(Re designates the real part). These expressions provide a general definition of the Faraday specific rotatory power. By means of the Faraday Effect measurement of δ_F' and δ_F'' to determine the saturate magnetic rotatory term ε_{12} is very effective. For example, in the laser gyro with a rare-earth iron garnet magnetic mirror, a measured value of the saturate rotatory power δ_F' [3] of the thin film of (Yb, Gd, Pr, Bi) $(FeAl)_5O_{12}$, that means ε_{12} may be evaluated

$$\lambda = 0.63\ \mu \quad \delta_F' = 1.3 \times 10^4 \text{ deg/cm} \qquad \varepsilon_{12} \doteq 1.0 \times 10^{-2}$$

$$\lambda = 1.15\ \mu \quad \delta_F' = 1.8 \times 10^3 \text{ deg/cm} \qquad \varepsilon_{12} \doteq 0.3 \times 10^{-2} \quad .$$

In a four frequency laser gyro, if we use K9 glass as the Faraday cell and neglect its absorption, then $\delta_F = VB$, where $V = 180 \cdot \beta / \lambda N_0$, is known as the Verdet constant (deg/Oe cm) in the general definition of the Faraday rotatory effect. The Verdet constant of K9 glass is

$$V_{K9} \doteq 2 \times 10^{-4} \frac{\text{deg}}{\text{Oe cm}} \quad .$$

Assuming, the cavity length of a ring laser L = 60 cm, then the space between two longitudinal modes is C/L = 500 MHz. Let thelength of the Faraday cell of K9 glass be d = 1 cm and the axial magnetic field B = 1000 Oe. In this case, $\varepsilon_{12} = \beta B = 1 \times 10^{-7}$ and the nonreciprocal phase shift $\Delta\phi \doteq 0.4$ deg. The bias frequency is then

$$\Delta\nu_b = \frac{VBd}{180} \cdot \frac{C}{L} \doteq 0.56 \text{ MHz} \quad .$$

From the descriptions in American literature in 1977 and other reports studying four frequency laser gyros in America we know that the quartz rotator is also used as a Faraday cell in their schemes. This scheme can minimize the number of optical elements in the cavity, but also provides a lot of errors. We try to analyze it in the following. When a longitudinal magnetic field is applied to the quartz crystal, then $\varepsilon_{12} = \varepsilon_{12}^{Q} \pm \beta B$. In this case, the refraction index of CW and CCW traveling waves of the left and the right circularly polarized gyro may be expressed for the RCP gyro:

$$n_R^{CW} = (N_0^2 + \varepsilon_{12}^Q + \beta B) \quad ,$$

$$n_R^{CCW} = (N_0^2 + \varepsilon_{12}^Q - \beta B) \quad ,$$

and

$$\Delta n_R = n_R^{CW} - n_R^{CCW} = \beta B(N_0^2 + \varepsilon_{12}^Q)^{-\frac{1}{2}} = \beta B/n_R \quad .$$

For the LCP gyro:

$$n_L^{CCW} = (N_0^2 - \varepsilon_{12}^Q + \beta B)^{\frac{1}{2}} \quad ,$$

$$n_L^{CW} = (N_0^2 - \varepsilon_{12}^Q - \beta B)^{\frac{1}{2}} \quad ,$$

and

$$\Delta n_L = n_L^{CW} - n_L^{CCW} \doteq - \beta B(N_0^2 - \varepsilon_{12}^Q)^{-\frac{1}{2}} = -\beta B/n_L \quad .$$

This shows that the bias frequency of the left and the right gyro are unequal, i.e., $\nu_0^L \neq \nu_0^R$ (see Fig. 2). The output differential frequency

$$\Delta\nu = (\nu_0^L - \nu_0^R) + \frac{8S}{\lambda L}\Omega \quad ,$$

includes a term $\Delta\nu(B) = \nu_0^L - \nu_0^R$ related to the applied magnetic field. This term will introduce a systematic error. This term of null shift may be evaluated by

$$\Delta\nu/\nu_0 = (n_R - n_L)/N_0 = 4 \times 10^{-5}$$

(taking $\nu_0 \doteq \nu_L \doteq \nu_R$). If $\nu_0 = 1 \times 10^6$ Hz, then $\Delta\nu_0 = 40$ Hz will result. This is a systematic null shift, which may be cancelled by its foreward measurement in the operation of the laser gyro. The random drift of applied magnetic field and laser beam will cause the harmful drift of bias frequency, because the bias frequencies ν_0^L and ν_0^R will not be cancelled completely and their random drifts will be added to the output of the laser gyro.

Transverse Kerr Effect

(a) Normal Mode

This is a nonreciprocal effect caused when light reflects on the surface of iron-magnetic material. As shown in Fig. 4, $\underline{B}$ is perpendicular to the plane of incidence and the coordinates are selected as before, i.e., $\underline{B}$ is along the z-axis. In this case the dielectrical tensor takes the form of (1) and

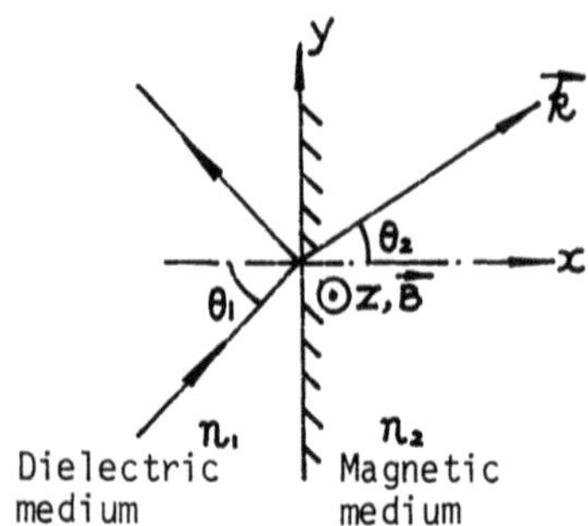

Fig. 4. Diagram showing transverse Kerr effect

the wave vector is in the x-y plane. We have $K = \{\alpha_1, \alpha_2, 0\}$, $\alpha_1 = \cos\theta_2$, $\alpha_2 = \sin\theta_2$, and No = Ne, because the iron garnet crystal is a cubic crystal. Thus the dielectrical tensor may be simplified as

$$\underline{\varepsilon} = \varepsilon_0 N_0^2 \begin{bmatrix} 1 & iQ & 0 \\ -iQ & 1 & 0 \\ 0 & 0 & 1 \end{bmatrix} \tag{8}$$

where $Q = \varepsilon_{12}/N_0^2$, and (6) may be simplified as

$$\begin{bmatrix} k_0^2N_0^2 - k^2(1-\alpha_1^2) & k^2\alpha_1\alpha_2 + ik_0^2N_0^2Q & 0 \\ k^2\alpha_1\alpha_2 - ik_0^2N_0^2Q & k_0^2N_0^2 - k^2(1-\alpha_2^2) & 0 \\ 0 & 0 & k_0^2N_0^2 - k^2 \end{bmatrix} \begin{bmatrix} E_x \\ E_y \\ E_z \end{bmatrix} = 0 \; . \tag{9}$$

For the purpose of clearly understanding the polarization mode of normal modes, we rotate the coordinates through an angle θ_2 around the z-axis, to make the x'-axis coincident with $\underline{K}$ as in Fig. 5. This is equivalent to a transform

$$\begin{bmatrix} x' \\ y' \\ z' \end{bmatrix} = A \begin{bmatrix} x \\ y \\ z \end{bmatrix} \quad \text{and} \quad A = \begin{bmatrix} \alpha_1 & \alpha_2 & 0 \\ -\alpha_2 & \alpha_1 & 0 \\ 0 & 0 & 1 \end{bmatrix} \; .$$

Therefore, in the new coordinates (x', y', z'), (9) transforms into

$$\begin{bmatrix} k_0^2N_0^2 & ik_0^2N_0^2Q & 0 \\ -ik_0^2N_0^2Q & k_0^2N_0^2 - k^2 & 0 \\ 0 & 0 & k_0^2N_0^2 - k^2 \end{bmatrix} \begin{bmatrix} E_{x'} \\ E_{y'} \\ E_{z'} \end{bmatrix} = 0 \; . \tag{10}$$

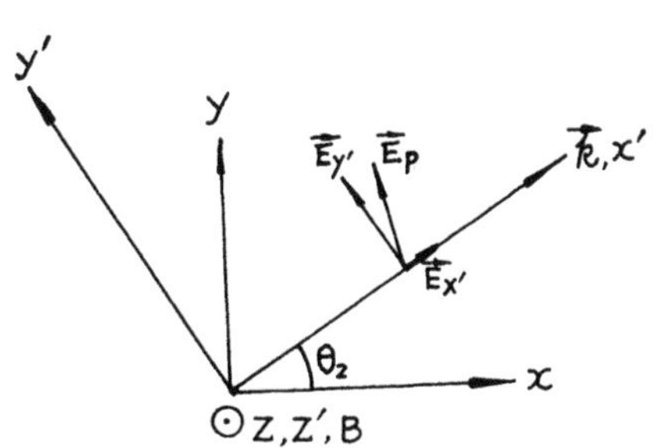

Fig. 5. The coordinate transformation and longitudinal and transverse components of P-wave

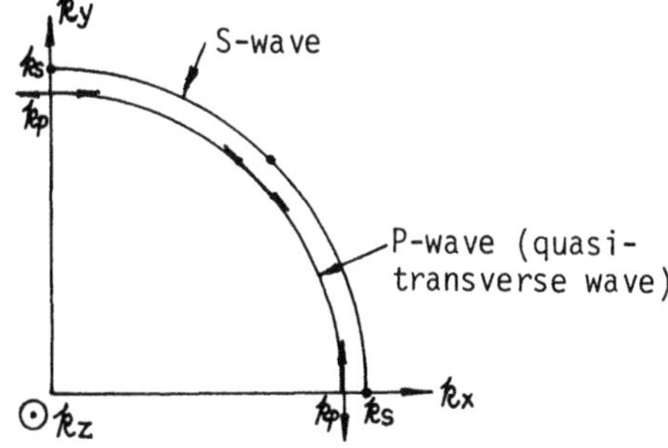

Fig. 6. The wave-vector in x-y plane

It is obvious, that the polarization mode of the normal mode corresponding to $k_s = k_0N_0$, is {0, 0, 1}, which represents linearly polarized light. Its electrical vector $\underline{E}$ is along the z-axis. It is an S-polarized wave related to the interface (see Fig. 4). Its refractive index Ne = No, which is not related to the magnetic-optic parameter Q. This shows that an S-polarized wave will not be modulated in reflection by the magnetic field. The value of k_s is not related to the direction angle θ_2, which means that the locus of wave-vector $\underline{k}_s$ is a circle with radius of k_0N_0, as shown in Fig. 6. From (10) we obtain another solution, whose normal mode polarization is in the (x', Y') plane, i.e., a P-polarized light. Also

$$k_p = k_0n_p = k_0N_0(1 - Q^2/2) \quad , \quad E_{x'}/E_{y'} = -iQ$$

which represents the polarization mode of the normal mode, {-iQ, 1, 0}. This means that the electrical vector of the P-wave is not perpendicular to the direction of the wave vector (see Fig. 5), but is a quasi-transverse wave. The ratio of the small longitudinal component $E_{x'}$ to the transverse component $E_{y'}$ equals -iQ ($Q \ll 1$). The complex Q is an odd function of $\underline{B}$, so that both magnitude and phase of the longitudinal component will be modulated by $\underline{B}$. The appearance of the longitudinal component in the P-wave is the cause of nonreciprocal reflection. Hence, the transverse Kerr effect used in magnetic mirror laser gyros is invariably related to the P-wave. In the (k_x, k_y) plane the locus of $\underline{k}_p$ is a circle with radius k_p, as shown in Fig. 6.

b) Transverse Kerr Effect

In the following, the reflection of P-polarized light at the interface of dielectric and iron-magnetic materials will be discussed. Let the wave-vectors of incident, reflected and refracted waves be represented by $\underline{k}^i$ $\underline{k}^r$ $\underline{k}^t$ respectively. Their directions are

$$\underline{K}^i = \{\cos\theta_1, \sin\theta_1, 0\} \quad ,$$

$$\underline{K}^r = \{-\cos\theta_1, \sin\theta_1, 0\} \quad ,$$

$$\underline{K}^t = \{\cos\theta_2, \sin\theta_2, 0\}$$

and $n_2 = n_0(1 - Q^2)^{\frac{1}{2}}$. Their electrical vectors are $\underline{E}^i$, $\underline{E}^r$ and $\underline{E}^t = \underline{E}_T^t + \underline{E}_L^t$, whose positive directions are defined as in Fig. 7. Here, $\underline{E}_T^t$ and $\underline{E}_L^t$ are the transverse and longitudinal components of the electrical vector $\underline{E}^t$ in the gyrotropic dielectric medium, and $E_L^t = -iQE_T^t$. From the continuity of the tangential components of the electrical vector at both sides of the interface,

$$\cos\theta_1(E^i + E^r) = (\cos\theta_2 - iQ\sin\theta_2)E_L^t \quad . \tag{11}$$

From $\underline{H} = n(\underline{K} \times \underline{E})$, the directions of the three wave-vectors are all parallel to the z-axis. From the continuity of the tangential components of the magnetic vector (notice $\underline{K}^t \times \underline{E}_L^t = 0$),

$$n_1E^i - n_1E^r = n_2E_T^t \quad . \tag{12}$$

From (11) and (12) we obtain

$$r_p = \frac{E^r}{E^i} = \frac{\eta_1(1 - iQ') - \eta_2}{\eta_1(1 - iQ') + \eta_2} \tag{13}$$

and

$$t_p = \frac{E_T^t}{E^i} = \frac{2(n_1/n_2)\cdot\eta_2}{\eta_1(1 - iQ') + \eta_2} = \frac{n_1}{n_2}(1 - r_p) \tag{14}$$

where

$$\eta_1 = \frac{n_1}{\cos\theta_1} \quad , \quad \eta_2 = \frac{n_2}{\cos\theta_2} \quad , \quad Q' = Q\tan\theta_2 \quad . \tag{15}$$

If $B = 0$, then $Q = 0$, and

$$r_p^0 = \frac{\eta_1 - \eta_2}{\eta_1 + \eta_2} \quad , \quad t_p^0 = \frac{n_1}{n_2}(1 - r_p^0) \quad .$$

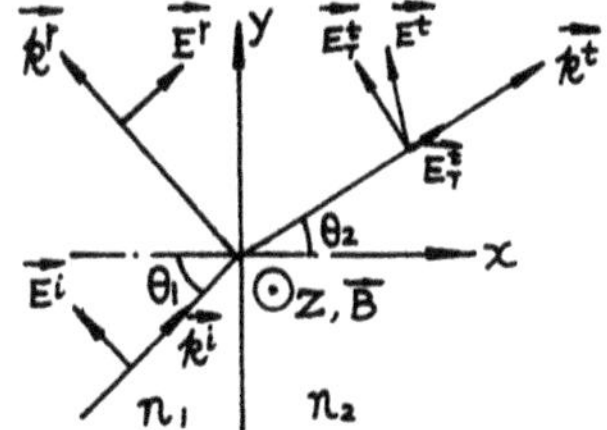

Fig. 7. The transverse magneto-optic scattering coefficients of P-wave

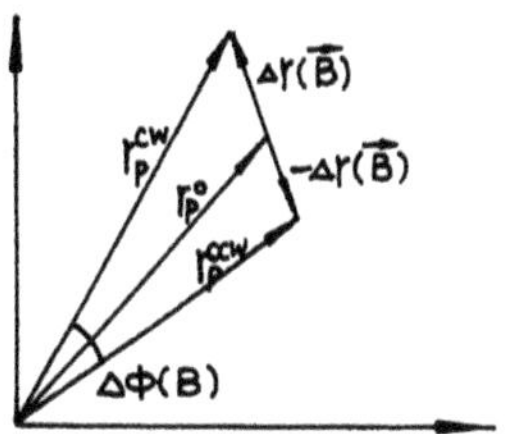

Fig. 8. Nonreciprocal reflection in transverse Kerr effect

Because Q' is a small quantity, expression (13) may be expanded in serie and taken to first order,

$$r_p = r_p^0 + \Delta r \quad , \quad \Delta r = -\frac{iQ}{2}(t_p^0)^2 \tan\theta_1 \quad . \tag{16}$$

As Q is an odd function of $\underline{B}$, the reflective coefficients of P-polarized CW and CCW traveling wave electrical vectors at the magnetic mirror, r_p^{CW} and r_p^{CCW}, are

$$r_p^{CW} = r_p^0 + \Delta r(B) \quad ,$$

$$r_p^{CCW} = r_p^0 + \Delta r(-B) = r_p^0 - \Delta r(B)$$

respectively. This causes a nonreciprocal reflectance as shown in Fig. 8. $\Delta\phi(B)$ is the phase difference introduced by this nonreciprocal reflection, and it is needed to get a bias in the laser gyro. In practice, a controlled layer of dielectric thin film is necessary to cause the nonreciprocal amplitude difference introduced by reflection to vanish. An inequality in the intensities of the CW and CCW traveling waves will occur, causing a drift in the differential losses in the operation of the gyro.

Normal Modes in Longitudinal and Polar Kerr Effect

In these cases, the wave-vector $\underline{k}$ is in the (x,z) or the (y,z) plane. Because the wave-vector surface presents a cylindrical symmetry around the z-axis in the cubic system and the magnetic field is applied along the z-axis, it is only necessary to discuss the case for any one section plane which includes the z-axis. Let $\underline{K} = \{0, \sin\varphi, \cos\varphi\}$, then the wave vector is in the (y,z) plane and has an intersection angle φ with the z-axis. (6) is simplified as follows, where $k = k_0 n$,

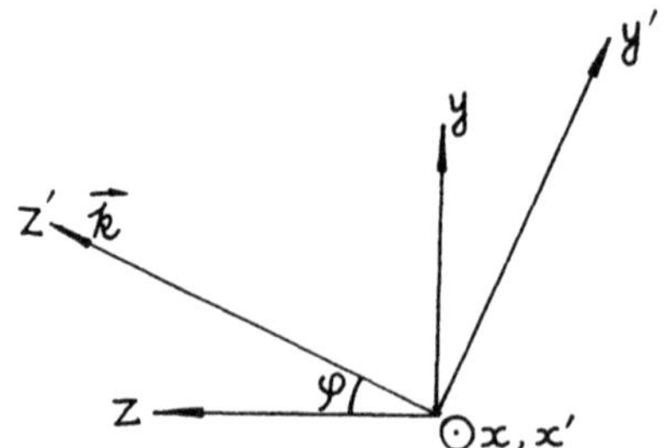

Fig. 9. The coordinate transformation

$$\begin{bmatrix} N_0^2 - n^2 & iN_0^2Q & 0 \\ -iN_0^2Q & N_0^2 - n^2\cos^2\varphi & n^2\sin\varphi\cos\varphi \\ 0 & n^2\sin\varphi\cos\varphi & N_0^2 - n^2\sin^2\varphi \end{bmatrix}\begin{bmatrix} E_x \\ E_y \\ E_z \end{bmatrix} = 0 \quad . \tag{17}$$

Similarly, we make a coordinate rotation with an angle φ around the x-axis and make the z-axis coincident with $\underline{k}$ as shown in Fig. 8. In the coordinate system (x', y', z') (17) transforms to

$$\begin{bmatrix} N_0^2 - n^2 & iN_0^2Q\cos\varphi & iN_0^2Q\sin\varphi \\ -iN_0^2Q\cos\varphi & N_0^2 - n^2 & 0 \\ -iN_0^2Q\sin\varphi & 0 & N^2{}_0 \end{bmatrix}\begin{bmatrix} E_{x'} \\ E_{y'} \\ E_{z'} \end{bmatrix} = 0 \quad . \tag{18}$$

The determinant of the matrix vanishes when a nontrivial solution does exist. An equation to the 4th order in refractive index n is obtained for normal mode in the $\underline{K}$ direction,

$$n^4 - n^2N_0^2(2 - Q^2\sin^2\varphi) + N_0^4(1 - Q^2) = 0 \quad .$$

Thus

$$n_\pm^2 \doteq N_0^2(1 \pm Q\cos\varphi) \quad ,$$

$$n_\pm \doteq N_0^2(1 \pm 1/2Q\cos\varphi) \quad ,$$

$$E_{y'}/E_{x'} = i\,\frac{N_0^2Q\cos\varphi}{N_0^2 - n^2} \doteq \mp i \quad . \tag{19}$$

The longitudinal component $E_L = E_Z'$, and the transverse component is E_T. Their ratio is

$$\frac{E_L}{E_T} = \frac{iQ}{\sqrt{2}} \sin\varphi \quad \text{and} \quad E_T = (|E_{x'}|^2 + |E_{y'}|^2)^{1/2} \quad .$$

It can be seen that in the (y,z) or (x,z) plane the normal modes are the left and right circular quasi-transverse waves with a small longitudinal component E_L.

Wave-Vector Surface and Normal Mode of Nonreciprocal Magneto-Optic Medium

For a nonreciprocal magneto-optical medium with dielectrical tensor as shown in (8), such as the isotropic Faraday rotatory media, (GM1), (GM2), (GM3) [3], YIG, $[(Bi, Tm)_3 (Fe,Ga)_5O_{12}]$ etc., of iron garnet type (recently developed magneto-optical media), a complete diagram of the wave-vector may be made in (k_x, k_y, k_z) space. When the applied magnetic field is equal to zero, it is a spherical surface. While a magnetic field is applied, it splits into two layers of rotatory symmetrical curve surface around the z-axis (not a rotatory ellipse). Because of its symmetry, in Fig. 10 only one quadrant of these curve surfaces and the corresponding polarization mode of the normal mode are shown. In this figure

$$n_{\pm}(\varphi) = N_0(1 \pm Q/2 \cdot \cos\varphi) \quad ,$$

$$n_{\pm}(\varphi = 90^\circ) = N_0(1 - Q^2/4 \pm Q^2/4) \quad ,$$

$$k_s = k_0N_0 \quad , \quad k_p = k_0N_0(1 - 1/2 \cdot Q^2) \quad .$$

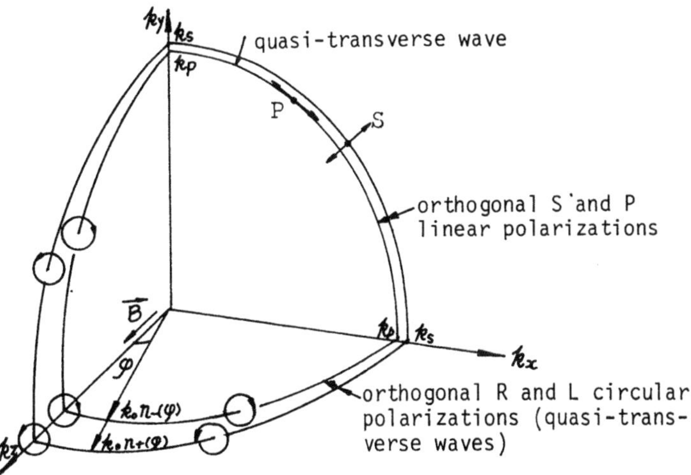

Fig. 10. The wave-vector surface and normal mode in magnetic (gyrotropic) medium

The dielectric tensor of single axis crystal quartz is expressed in (2) for light traveling along the z-axis. It differs from (8) only by $N_e = N_0$. Hence, its wave-vector surfaces appear to be similar to Fig. 10. The difference between them is only in $k_s = k_0 N_e$, and in the (y,z), (x,z) planes the polarization modes of the normal mode are orthogonal and elliptically polarized.

Conclusion

The wave-vector for the transverse Kerr Effect used in the magnetic mirror bias laser gyro is in the (k_x, k_y) plane. The nonreciprocity of reflection of the P-polarized normal mode is used. In the four-frequency differential laser gyro, the wave-vectors of the four traveling waves are all along $\underline{k}_z$. Because of Faraday effect, the nonreciprocity of orthogonal left and right polarized normal mode has been used, and it is noticed that in the four-frequency differential gyro, orthogonal quasi-circularly polarized normal modes in the (k_x, k_z) or (k_y, k_z) planes may also be used, i.e., we may use the nonreciprocal polar Kerr Effect to form a magnetic mirror instead of a Faraday cell. It can thus reduce the number of optical components in the cavity, and decrease the loss and scattering in the cavity. The polar Kerr Effect has strong non-reciprocity. We may select the MnBi magnetic mirror with a stack of dielectric thin films to construct a four-frequency orthogonal circular polarized laser gyro cavity [5,6].
4110045 (Aug. 1978)

Acknowledgements. The author would like to offer his sincere thanks to Feng Tiesun, Liu Wei-min and Bao Cheng-zu for their help.

Reference

1 Jiang Yanan: Laser Journal *8* (4), 1-8 (1981), in Chinese
2 J. Bresman, H. Cook, D. Lysobey: Navigation: Journal of the Institute of Navigation *24* (2), Summer (1977), USA
3 J.J. Krebs, W.G. Maisch, G.A. Pinz, D.W. Forester: IEEE Trans. Magn. *16* (5), 1179 (1980)
4 R.E. McClure: IEEE Trans. Magn. *16* (5), 1185 (1980)
5 T.A. Dorschner, T.W. Smith, H. Statz: Proc. 1978 IEEE Nat. Aerospace Electronics Conf., p.569
6 T.W. Smith, Jr., T.A. Dorschner: U.S. Patent No. 4110045 (Aug. 1978)

5.2 Detection Noise

Analysis of Noise in Phase Detection in a Fiber Optic Rate Sensor

R.L. Phillips

Department of Electrical Engineering and Communication Sciences
University of Central Florida, Orlando, FL 32816, USA

L.C. Andrews

Department of Mathematics and Statistics, University of Central Florida
Orlando, FL 32816, USA

1. Introduction

The mathematical analysis of the effect of noise on the detection of optical phase shift in a fiber optic rate sensor must be based upon appropriate mathematical modeling of the noise. Noise may be categorized as being (1) additive to the optical field, (2) additive to the phase of the optical field, (3) multiplicative to the detector signal current, or (4) additive to the signal current. Because the rate information is contained within the phase of the detector current, it is the level crossings of the detector current that contain the desired information. We consider the effects of two types of noise on the level crossings of the signal current -- that due to the optical source which is additive to the optical field, and that due to the polarization fluctuations of the two counter-propagating optical fields in the fiber gyro which is multiplicative to the signal current.

2. Additive Field Noise

In the analysis of the additive field noise it is assumed that the total complex representation of the optical signal field at the detector is the sum of clockwise and counterclockwise fields, i.e.,

$$E(t) = \left(A_1 + A_2 e^{i\Delta\theta}\right) e^{i\omega_0 t} , \tag{1}$$

where A_1 and A_2 are constant amplitudes and $\Delta\theta = \Delta\omega t + \phi$. Here ϕ is the rotation phase and $\Delta\omega$ is the frequency difference between the two frequency modulated counter-propagating fields. By defining

$$A^2 = A_1^2 + A_2^2 + 2A_1A_2 \cos(\Delta\theta) \tag{2}$$

and

$$\alpha = \tan^{-1}\left(\frac{A_2 \sin(\Delta\theta)}{A_1 + A_2\cos(\Delta\theta)}\right) , \tag{3}$$

the signal field can then be expressed as

$$E(t) = Ae^{i\omega_0 t + i\alpha} . \tag{4}$$

Narrowband noise added to the signal field yields the total field

$$U(t) = E(t) + n(t), \tag{5}$$

where the narrow band noise has the standard representation [1]

$$n(t) = x(t)\cos \omega_0 t - y(t)\sin \omega_0 t. \tag{6}$$

The signal current i is proportional to the intensity $I = |U(t)|^2$, and thus we seek to find the average number of crossings $< N(T) >$ of the intensity through a specified threshold T, which can be computed from [2]

$$< N(T) > = \int_{-\infty}^{\infty} |\dot{I}| \; p(T,\dot{I}) \; d\dot{I} . \tag{7}$$

Here $\dot{I}$ is the time derivative of intensity and $p(T,\dot{I})$ is the joint probability density function of I and $\dot{I}$ evaluated at $I = T$. Under the assumption that $x, y, \dot{x}$, and $\dot{y}$ are jointly Gaussian independent variables with zero means and variances $\sigma_x^2 = \sigma_y^2 = b$ and $\sigma_{\dot{x}}^2 = \sigma_{\dot{y}}^2 = c$, it can be shown that the above integral leads to

$$< N(T) > = \frac{1}{b}(2cT/\pi)^{\frac{1}{2}} \exp\left[- \frac{1}{2b}(T + A^2) - \frac{\dot{A}^2}{2c}\right]$$

$$\times \sum_{k=0}^{\infty} \frac{1}{(k!)^2}\left(\frac{A\,T^{\frac{1}{2}}}{2b}\right)^{2k} {}_1F_1(\tfrac{1}{2};k+1;\tfrac{\dot{A}^2}{2c})\,{}_3F_1(-k,\tfrac{1}{2}-k,1;\tfrac{1}{2};\tfrac{2\dot{A}^2b^2}{cTA^2}) , \tag{8}$$

where ${}_1F_1$ and ${}_3F_1$ are hypergeometric-type functions. If we retain only the first term of the above series, we find

$$< N(T) > \cong \frac{1}{b}(2cT/\pi)^{\frac{1}{2}} \exp\left[- \frac{1}{2b}(T + A^2) - \frac{\dot{A}^2}{4c}\right] I_0(\dot{A}^2/4c) , \tag{9}$$

where $I_0(.)$ is the modified Bessel function of order zero.

Next, we compute the average fade period

$$\bar{t} = \frac{P(I < T)}{\frac{1}{2} < N(T) >} , \tag{10}$$

which gives the average time that the intensity will remain below threshold once it falls below threshold. Here $P(I < T)$ is the probability that the intensity is less than the threshold T. The probability density function of the intensity can be derived from the Ricean density for the envelope of a Gaussian random process by a simple change of variables. Thus, we find

$$p(I) = \frac{1}{b} e^{-(I + A^2)/b} I_o \left(\frac{2A}{b}\sqrt{I} \right) , \qquad (11)$$

and therefore

$$P(I < T) = 1 - \int_T^{\infty} p(I) dI$$
$$= 1 - Q(A\sqrt{\frac{2}{b}} , \sqrt{\frac{2T}{b}}) , \qquad (12)$$

where

$$Q(a,b) = \int_b^{\infty} x \exp\left(- \frac{a^2 + x^2}{2} \right) I_o(ax)\, dx \qquad (13)$$

is the Marcum Q function [3].

The intensity of the reference waveform is given by (see Fig. 1 a)

$$I_R = 2B^2 [1 + \cos(\Delta\omega t)]. \qquad (14)$$

Here we have set $A_1 = A_2 = B$ (i.e., clockwise and counterclockwise signals have the same amplitude B). The additive noise leads to a distorted waveform as shown in Fig. 1 b . If t_1 denotes the time of the first level crossing of the reference waveform and t_2 the time of the first level crossing of the distorted waveform, then the system interprets this difference as an apparent phase shift given by

$$\theta = \Delta\omega t_1 - \Delta\omega t_2 . \qquad (15)$$

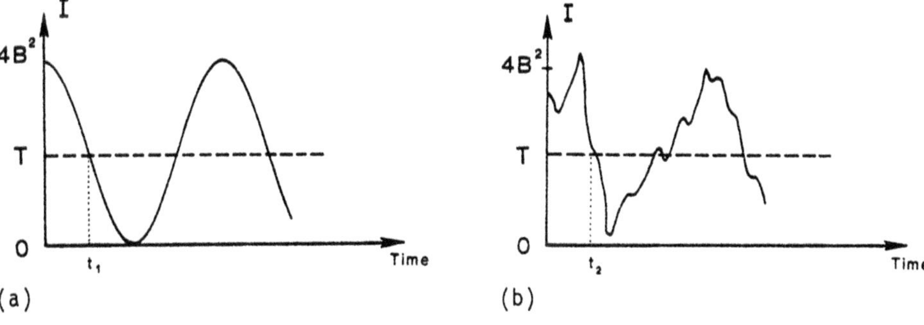

Fig. 1 (a) Level crossings of the intensity of the reference waveform (b) Level crossings of the intensity of the reference waveform with additive Gaussian field noise

On the average, $t_2 = \frac{1}{2}\bar{t}$ whereas $t_1 = \pi/2\Delta\omega$, and so we get the average phase shift at the first level crossing given by

$$< \theta > = \frac{\pi}{2} - \frac{\Delta\omega\bar{t}}{2} . \tag{16}$$

In actuality, of course, the intensity of the distorted waveform has a frequency different from that of the reference waveform so that the apparent average phase shift is really a linear function of time rather than a constant. If we assume that the reference and distorted waveforms are in phase at time $t = 0$, then the average phase shift of the distorted waveform for any time t is simply

$$< \theta > = (1 - \frac{\Delta\omega\bar{t}}{\pi})\Delta\omega t. \tag{17}$$

Remarks: The above analysis is based upon the threshold being set at $T = 2B^2$.

3. Multiplicative Current Noise

The relative polarization fluctuation manifests itself as a multiplicative effect, leading to the intensity (see Fig. 2)

$$I = 2B^2[1 + \cos\psi\cos(\Delta\omega t + \phi)], \tag{18}$$

where ψ is the polarization angle between the two counter-propagating fields. If the angle ψ fluctuates in such a way that causes $\cos\psi$ to become zero, the system will interpret the resulting amplitude fluctuation as a zero crossing of the function $\cos(\Delta\omega t + \phi)$. This means there is an apparent rotation phase angle ϕ_a such that

$$\cos\psi\cos(\Delta\omega t + \phi) = \cos(\Delta\omega t + \phi_a). \tag{19}$$

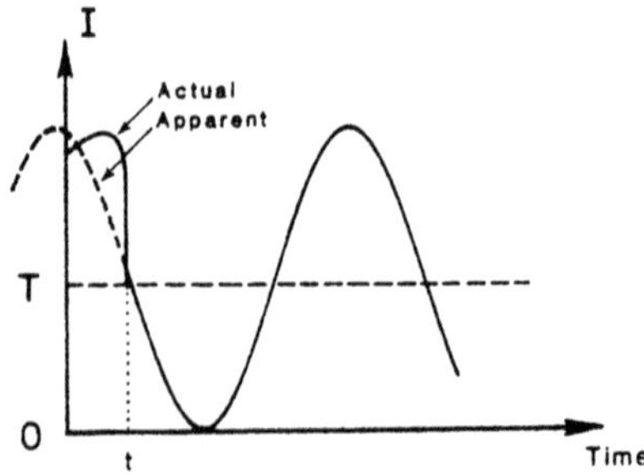

Fig. 2 Actual waveform with rotation phase angle ϕ and apparent waveform due to multiplicative noise

Solving this expression for the angle ϕ_a yields the result

$$\phi_a = \cos^{-1}[\cos(\Delta\omega t + \phi)\cos\psi] - \Delta\omega t. \tag{20}$$

If we assume that the polarization angle ψ is uniformly distributed, then it can be shown that the average value of the apparent rotation phase angle is

$$< \phi_a > = \frac{\pi}{2} - \Delta\omega t. \tag{21}$$

Similarly, it can be shown that the variance is given by

$$\sigma^2_{\phi_a} \cong \frac{1}{2} \cos^2(\Delta\omega t + \phi), \tag{22}$$

where we have retained only the first term of an infinite series.

Acknowledgement

This work was supported by the United States Army Ballistic Missile Defense Advanced Technology Center, Huntsville, Alabama. Contract No. DASG60-80-C-0050.

References

1. S. O. Rice, "Mathematical analysis of random noise", Bell Sys. Tech J. Vol. 23, pp. 282-332 (July 1944); Vol. 24, pp. 46-156 (Jan. 1945)

2. D. Middleton, An Introduction to Statistical Communication Theory, New York: McGraw-Hill (1960)

3. J. I. Marcum and P. Swerling, "Studies of target detection by pulsed radar", IRE Trans. Infor. Theory Vol. IT - 6, pp. 59-308 (April 1960)

Part 6

Advanced Concepts

Re-Entrant Fiber Optic Rotation Sensors

G.A. Pavlath and H.J. Shaw

Edward L. Ginzton Laboratory, Stanford University
Stanford, CA 94305, USA

A recirculating fiber optic rotation sensor is described which employs multiple transits of pulsed light around a closed coil of optical fiber. An all-fiber implementation of this re-entrant rotation sensor is presented. The experimental results verify the basic operation of the device.

Present fiber optic rotation sensors propagate cw light once around an N-turn optical fiber coil. In re-entrant rotation sensors [1],the N-turn optical fiber coil is closed so that pulsed light can propagate many times around the coil. The Sagnac phase shift is increased proportionally to the number of recirculations of the light around the optical fiber coil. Thus, in effect, the length of the fiber is increased by the factor N.

Fig.1 shows the schematic of the present re-entrant rotation sensor. It is constructed in all-fiber form, without splices, using single-mode fiber components.

An optical pulse obtained from a Nd:YAG laser by an acousto-optic beam deflector is coupled in to the fiber by a coupling lens. This pulse propagates to the splitter/combiner coupler, which is a fiber-to-fiber evanescent field coupler [2,3],where it is split into two equal parts. These pulses propagate to the loop I/O coupler where portions of the pulses are injected into the loop in opposite directions. This coupler is adjusted for a coupling coefficient near unity. The pulses propagate around the loop many times. After each transit around the loop, a portion of each pulse is coupled out. These sampled pulses are recombined by the splitter/combiner coupler and detected by a germanium photodiode. The polarization controllers [3,4] compensate for birefringence in the fiber, and allow the interferometer to be easily balanced at zero rotation rate.

We have found in practice that the requirements of this interferometer, namely efficient and balanced injection of pulses into the sensing coil together with efficient and balanced reinjection of pulses after each circulation, are difficult to satisfy with bulk optics components external to the fiber requiring a large number of alignments and adjustments which are

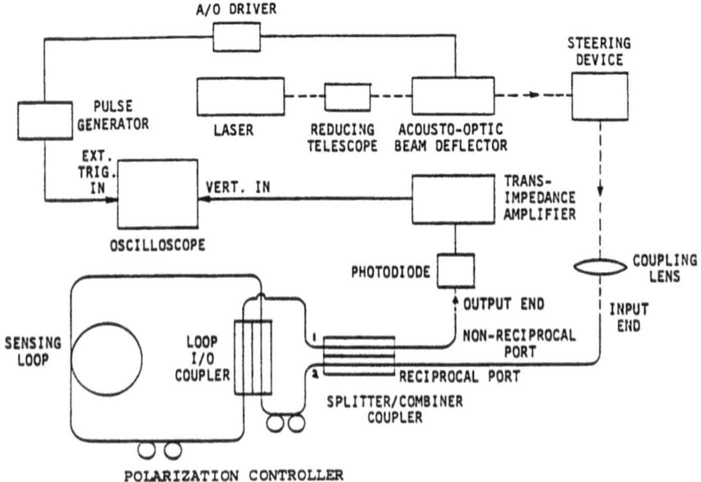

Fig.1 All fiber optic, passive, reentrant rotation sensor schematic

not independent. The present all-fiber version involves no alignments and contains only four adjustments (two coupler and two polarization controllers) which are non-interacting and very easily made and maintained, leading to excellent performance of the interferometer.

Each input pulse gives rise to a train of pulses at the photodiode with pulse spacing equal to the loop transit time (Fig.2). The envelope of the pulse train has a exponentially decaying component due to loss in the sensing loop and a sinusoidal component whose frequency is proportional to the rotation rate. The latter occurs because the Sagnac phase shift increases linearly with the number of transits around the loop while the pulse height is proportional to the cosine of the Sagnac phase shift.

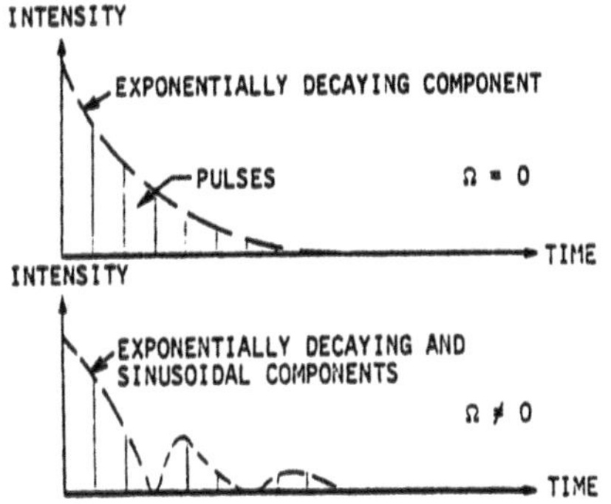

Fig.2 Detected pulse trains at rest and during rotation

For a circular sensing loop of diameter D, the frequency, F, of the sinusoidal component of the pulse train envelope is

$$F = \frac{D}{\lambda}\Omega \tag{1}$$

where λ is the wavelength and Ω is the rotation rate. The accumulated phase $\phi(t)$ of the sinusoidal envelope component in a time T is proportional to the angle turned through $\theta(T)$ as

$$\phi(T) = \int_0^T F(g)dt = \frac{D}{\lambda}\theta(T) \quad . \tag{2}$$

Thus with a re-entrant rotation sensor the integration which yields the angular position change is accomplished optically.

Fig.3 shows the measured pulse trains at various rotation rates for the re-entrant rotation sensor depicted in Fig.1. The sensing loop was an 830 meter length of Corning single-mode fiber wound on a circular form of 15 cm diameter. The attenuation of the fiber was measured at 1.8 dB/km at 1.06 μm. At rest a 180 degree phase shift, which results from the splitting/recombining process, exists between the counterpropagating pulses. This results in a straight base line as the output if the rotation sensor is properly balanced. The thickness of the line is due to the thermal noise of the detector and corresponds to 50 nW of optical power.

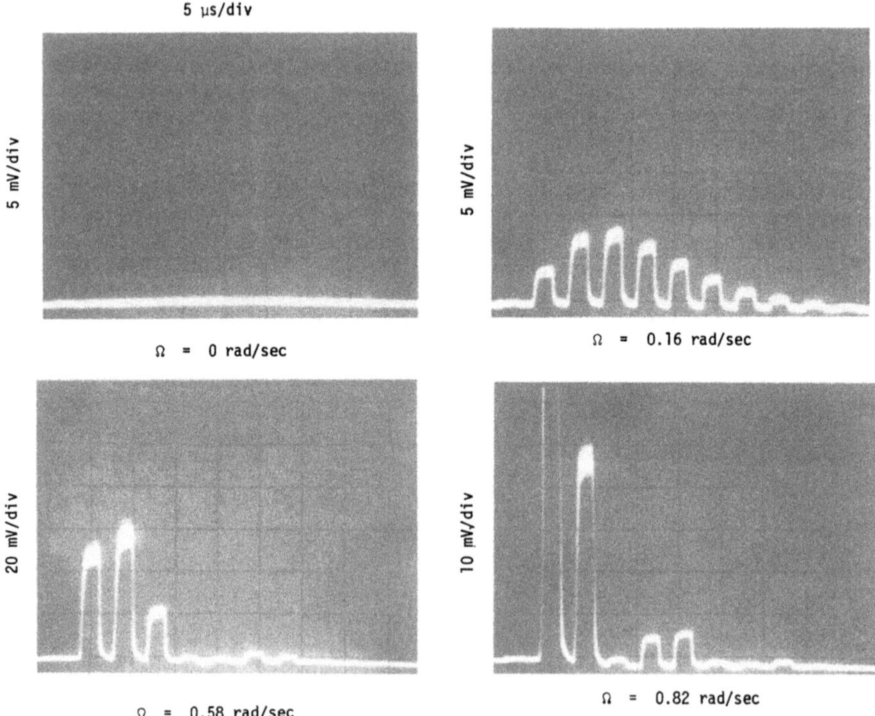

Fig.3 Experimental output pulse trains

When the sensor is rotated, the envelope of the pulse train is a decaying sinusoid. The rotation rates in Fig.3 were chosen to give approximately 1, 2, and 3 cycles of the sinusoidal component. The measured pulse trains are in agreement with theory.

In conclusion, an all-fiber re-entrant rotation sensor has been constructed. It has shown that pulses can be recirculated many times (10) and that the Sagnac phase shift accumulates for multiple transits of the sensing. An optical amplifier will be placed in the sensing loop to compensate for loss in order to extend the number of recirculations by a very large factor.

The potential of reentrant systems for sensing very small rotation rates will be evaluated, including the use of large diameter sensing loops where the use of such loops is applicable. This reentrant system is also a good vehicle for studying the effects of Rayleigh backscattering in fiber optic rotation sensors.

References

1 H. J. Arditty, H. J. Shaw, M. Chodorow and R. Kompfner, "Re-entrant Fiber Optic Approach to Rotation Sensing", Proc. Soc. Photo-Opt. Instrum. Eng. 157, 138-148 (1978)

2 R. Bergh, G. Kotler and H. J. Shaw, "Single-mode Fiber-optic Directional Coupler", Electronics Letters 16, 260-261 (1980)

3 M. Digonnet and H. J. Shaw, "Analysis of a Tunable Single Mode Optical Fiber Coupler", to be published in the April 1982 issue of IEEE Journal of Quantum Electronics

4 H. Lefevre, "Single-mode Fiber Fractional Wave Devices and Polarization Controllers", Electronics Letters 16, 778-780 (1980)

This work was supported by the Air Force Office of Scientific Research under Contract F49620-80-C-0040.

Forced Reciprocity Using Phase Conjugation

Ph. Graindorge, H.J. Arditty, M. Papuchon, and J.P. Huignard

Thomson-CSF, Laboratoire Central de Recherches, BP10, Domaine de Corbeville F-91401 Orsay, France

Ch. Bordé

Laboratoire de Physique des Lasers, Université de Paris-Nord F-93430 Villetaneuse, France

The various physical effects that are detected with interferometers can be classified into two types : the reciprocal effects and the non-reciprocal effects. The corresponding interferometers are the Michelson interferometer and the Sagnac interferometer. A particular interferometer is chosen according to the phenomenon to be detected. On the other hand, this interferometer is sensitive to all the phenomena of the corresponding class, and all the non-wanted effects add noise to the interesting signal. The signal is usually extracted by electronic filtering, which is difficult and expensive. We propose a new scheme of interferometer, sensitive either to reciprocal or non-reciprocal effects, depending on the conditions. In particular, low frequency reciprocal effects are not detected, so that there is no problem of long-time drift, often due to slowly varying reciprocal perturbations, whereas medium frequency reciprocal signals are detected. This interferometer has the stability properties of a non-reciprocal interferometer (i.e. Sagnac), but detects reciprocal events. Those properties are realized by replacing the mirrors of a Michelson interferometer by "conjugate mirrors", made out of a photorefractive material.

I - CONJUGATE MIRRORS

Photorefractive materials are both photoconductive and electrooptic, so that they can be used as sensitive materials for holographic recording. To build a phase volume hologram within such a crystal, a light interference pattern is created by directing two coherent laser beams (the reference and the signal beams) onto the crystal (fig. 1). Photocarriers are then generated proportional to the local intensity of the light, i.e. the intensity interference pattern. Those carriers move inside the crystal through different mechanisms (diffusion or drift under an electric field) and are trapped, so that the charge varies proportional to the illumi-

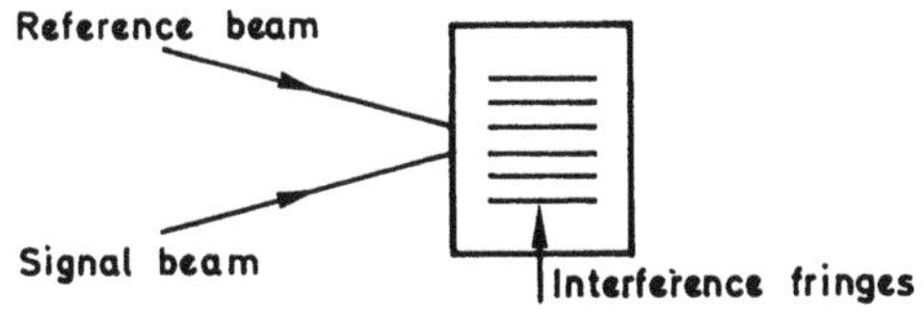

Fig. 1 Hologram in a photorefractive crystal

nation. This means that a space charge field is created, and varies as the local illumination, but with a phase displacement [3] (fig. 2). Finally, through electrooptic modulation, a phase hologram is created which bears the information of the signal beam. The materials used must be photoconductive, but this is true for most materials, because of impurities often trapped in the lattice. The other important feature is the electrooptic coefficient of the material. This is why the crystals commonly used are $Bi_{12}SiO_{20}$ [1] or $BaTiO_3$. The coefficient of the baryum titanate, for example, is equal to $820 \cdot 10^{-12}$ e.s.U. which is one of the highest known electrooptic coefficients and it is the reason for recent work done on this material in connection with phase conjugation [2,4].

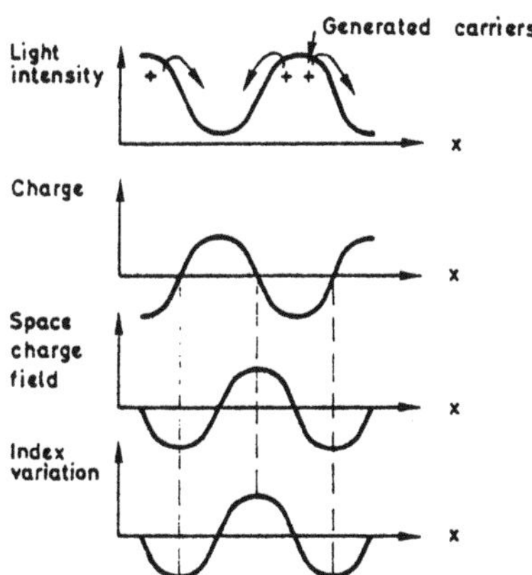

Fig. 2 Index modulation mechanism

A "conjugate mirror" is composed of a classical mirror and such a crystal, where the reference is reflected back to read out the hologram and generate the conjugate beam of the signal beam (fig. 3). This conjugate beam's amplitude is proportional to the complex conjugate of the amplitude of the signal beam. In particular, if the signal beam's phase is ϕ with respect to the reference beam, then the conjugate beam's phase is $-\phi$. This is due to the term

$$E^*_{signal} \cdot E_{reference}$$

of the amplitude of the hologram recorded ; the amplitude of the image beam after readout is that of the hologram multiplied by the amplitude of the readout beam

$$E_{conj.} = E^*_{signal} \cdot E_{reference} \cdot E_{readout}$$

This gives the expression of the phase of the conjugate beam.

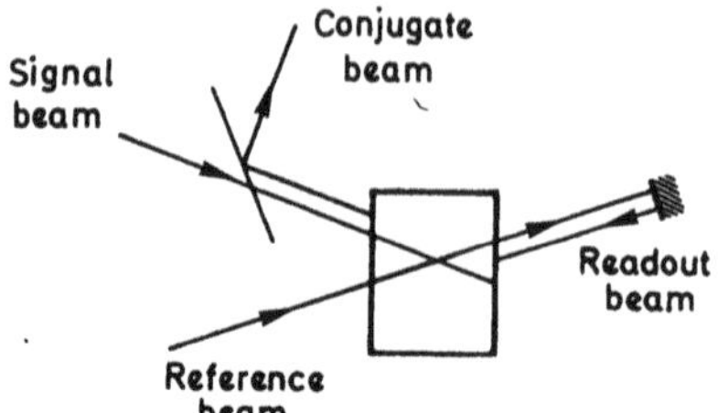

Fig. 3 A conjugate mirror

In the case where the readout beam is the reflected reference beam, and has the same phase, this gives

$$E_{conj.} = E^*_{signal} \cdot |E_{reference}|^2 \quad .$$

We will now show that both terms are equally important for the conjugate mirror interferometer.

II - MICHELSON INTERFEROMETER WITH CONJUGATE MIRROR

A Sagnac interferometer is sensitive to non-reciprocal phase perturbations, whereas a Michelson is mainly sensitive to reciprocal phase perturbations on one of the arms (fig. 4). If we replace the mirrors of a Michelson interferometer by conjugate mirrors, it becomes insensitive to reciprocal phase perturbations (fig. 5). But if there is a non-reciprocal phase shift on one of the arms, it is now detected (fig. 6) [7]. We will show that only one conjugate mirror is needed, so that the device becomes much simpler (fig. 7). In this device, the pump beam is actually one of arms of the interferometer. If there is a reciprocal phase shift on one of the arms, there is no change of the interference fringes at the output, whereas a non-reciprocal phase shift on either arm causes a fringe displacement.

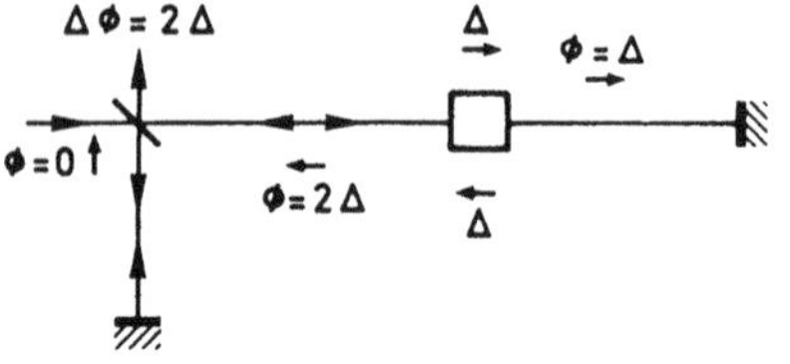

Fig. 4 Michelson interferometer

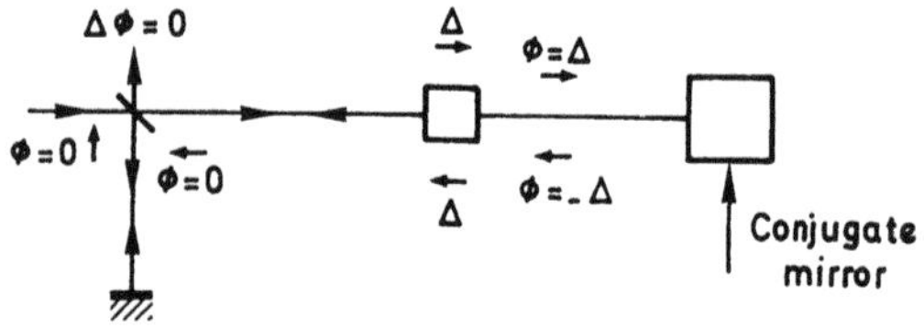

Fig. 5 Reciprocal perturbation on a conjugate mirror interferometer

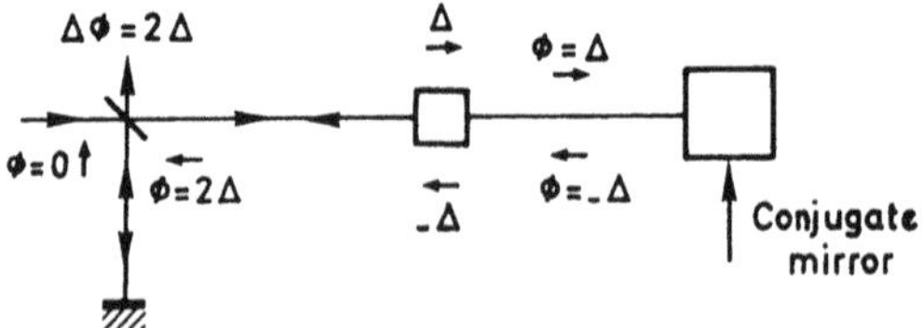

Fig. 6 Non-reciprocal perturbation on a conjugate mirror interferometer

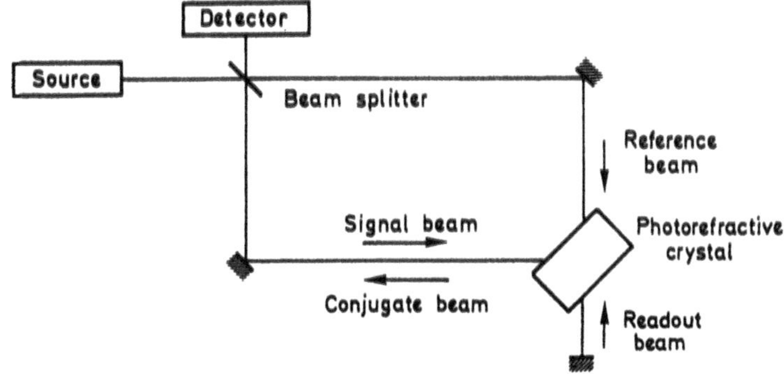

Fig. 7 Conjugate mirror Michelson interferometer

This is described in the following figures (figs. 8 to 11) (all phases are given with respect to the unperturbed configuration). If a reciprocal perturbation is placed on the signal arm, it is straightforward to see that the phase of the conjugate beam cancels exactly the phase variation due to the crossing of the perturbation on the return path, so that at the output, the phase of the conjugate beam is still equal to zero. On the other hand, in the case of a reciprocal phase perturbation placed on the reference path, the mechanism is the following.

After crossing the perturbation, the phase of the reference is Δ. Thus near the crystal, the various phases are

$$\phi_{ref.} = \Delta \;,$$
$$\phi_{readout} = \Delta \;,$$
$$\phi_{signal} = 0 \;,$$
$$\phi_{conj.} = 2.\Delta \;.$$

After going through the perturbation, the readout's phase increases by Δ, and its value is 2Δ, so that at the output, the conjugate's and the readout's phases are equal and there is no change in the interference pattern. On the other hand, if the perturbation is non-reciprocal, it will cause a fringe shift at the output.

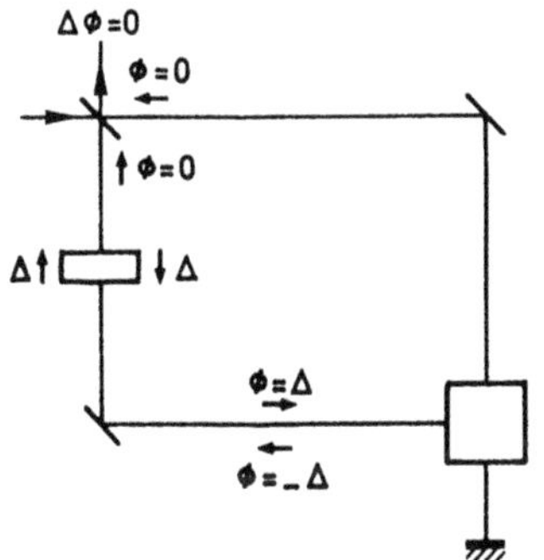

Fig. 8 Reciprocal perturbation on the signal arm

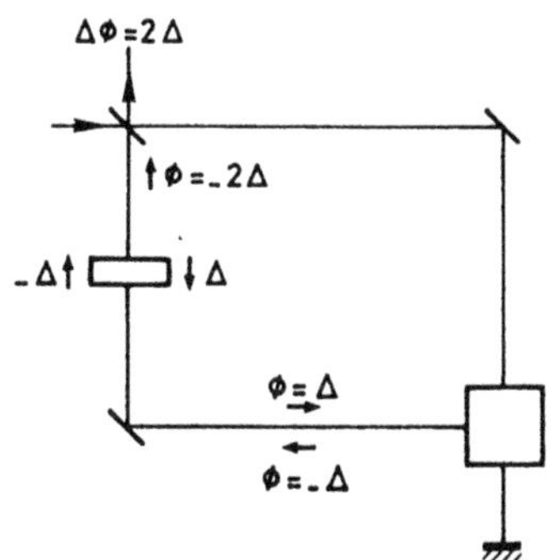

Fig. 9 Non reciprocal perturbation on the signal arm

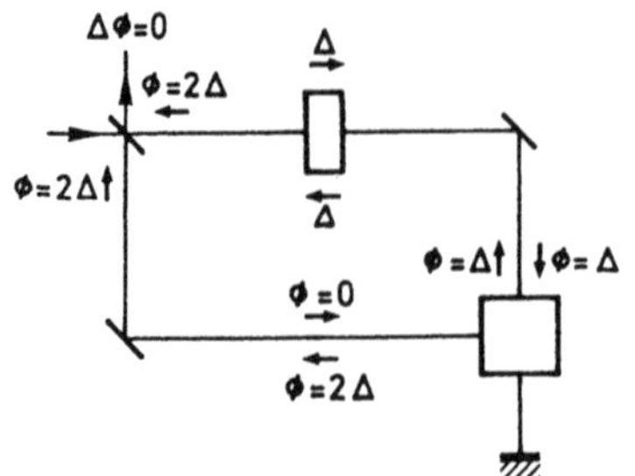

Fig. 10 Reciprocal perturbation on the reference arm

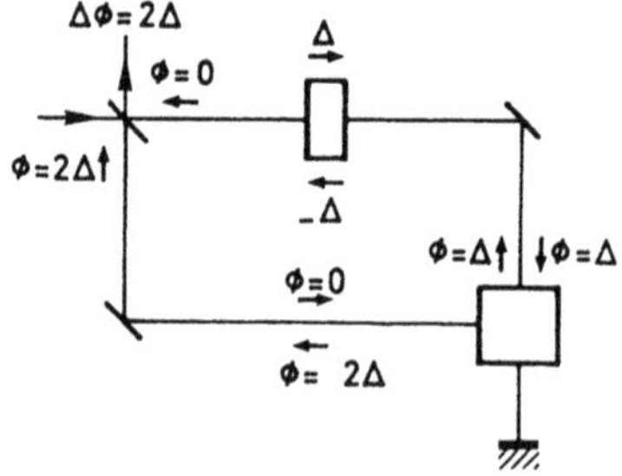

Fig. 11 Non reciprocal perturbation on the signal arm

III - PROPERTIES OF THE CONJUGATE MIRROR INTERFEROMETER

We have shown that the conjugate mirror interferometer is insensitive to reciprocal phase perturbations, and detects non-reciprocal perturbations. But this is in fact limited to phenomena which vary slowly with respect to the time constant of the crystal. The mechanism of the construction of the hologram involves the displacement of carriers in the crystal, and as these displacements are not instantaneous, the build-up takes some time, which can range from a few nanosèconds to a few minutes, depending on the material and the incident power [4,5]. This implies that the mechanisms we have just described are only valid for slowly varying phase perturbations. If the frequency of a perturbation is higher than the inverse of the response time of the crystal, the crystal only records the average of this signal, so that the conjugate beam's phase is the opposite of the average of the variable phase perturbation. Consequently, the output of the interferometer will detect the high frequency phase shift, while the D.C. component will be filtered out. This interferometer is then sensitive to varying phase signals or non-reciprocal phase signals, and ensures an automatic long term zero stability.

As a conclusion to this section we propose a scheme for a fiber interferometer using a conjugate mirror (fig. 12). The path of each arm is an optical fiber, which is monomode for the reference arm, and may be multi-

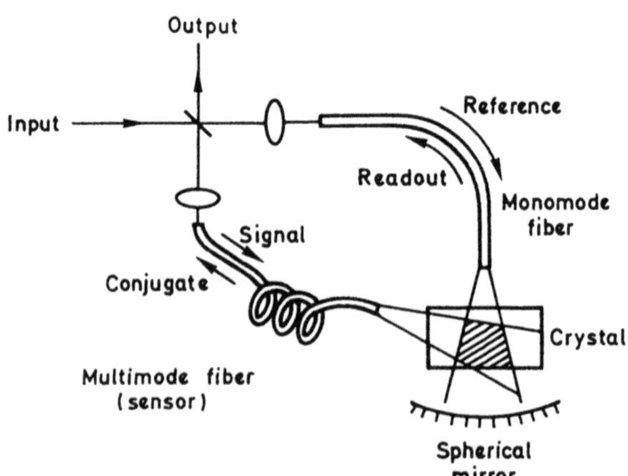

Fig. 12 Optical fiber Michelson interferometer

mode for the signal arm. As the reference beam is reflected back into the fiber with a spherical mirror, the reference fiber must be monomode to ensure reciprocity. But the signal beam is reflected by the conjugate mirror, and the wavefront of the return beam is exactly the same as that of the incident beam, so that even with a multimode fiber, the return path will be identical to the direct path, thus ensuring reciprocity. This interferometer has the same properties as the free wave conjugate mirror interferometer ; moreover, the multimode fiber on the signal arm can be used as a long interaction, low cost sensor for this interferometer.

IV - EXPERIMENTAL RESULTS

To demonstrate the conjugate mirror interferometer, we realized the setup represented in fig. 13. On one of the arms, we placed a cell in which we could change the air pressure from 1 to 5 atmospheres. This caused a change of the path length of about 10 λ. Because of the very low diffraction efficiency of the four wave mixing in the $Bi_{12}\ Si\ O_{20}$ crystal we used, we had to attenuate the readout beam by placing a $\lambda/4$ plate between the crystal and the mirror, and polarizers on the path of the readout beam and at the output of the interferometer.

This ensured maximum contrast of the fringes at the output. We used an argon laser at a wavelength of 514 nm, and with an output power of 100 mW.

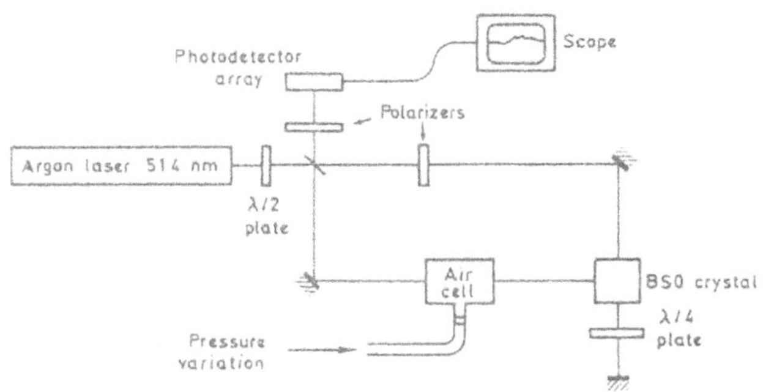

Fig. 13 Experimental setup

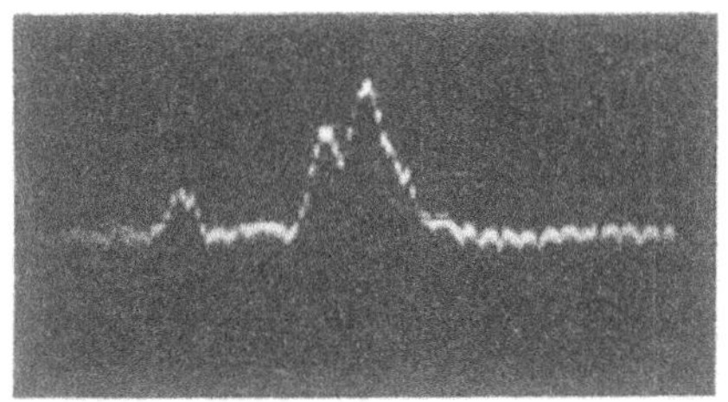

Fig.14. Output at one atmosphere

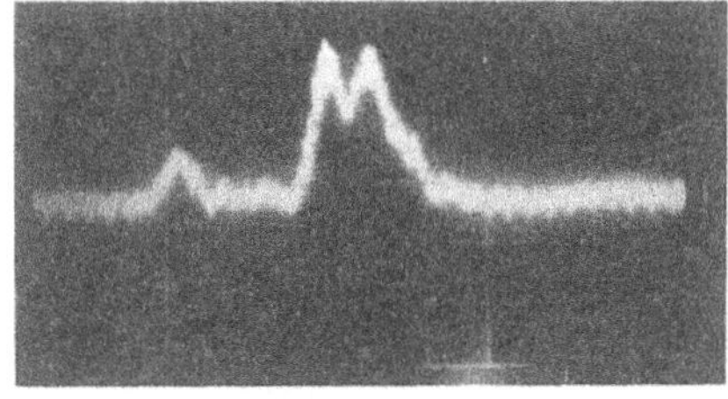

Fig.15. Superimposed outputs at 1 and 5 atmospheres

When we changed the pressure in the cell, we observed no change in the interference pattern, except a fast transient due to the settling of the modified hologram.

Fig. 14 shows the recording of the interference pattern at the output of the interferometer ; in fig. 15 there is a double exposition, the first one with a pressure of one atmosphere in the cell, and the second with a pressure of 5 atmospheres. The interference patterns are perfectly identical thus proving the feasability of the conjugate mirror Michelson interferometer.

We have demonstrated that the conjugate mirror Michelson interferometer is insensitive to slowly varying reciprocal phase variations on either arm, but detects both non-reciprocal or high frequency (with respect to the time response of the crystal) phase variations. The arms of the interferometer can either be free wave, or guided by optical fibers, one of which can be multimode.

References

[1] J.P. Huignard, J.P. Herriau, Appl. Opt. Vol 76 (77) p 1807

[2] J. Feinberg, R.W. Hellwarth, Opt. Lett. Vol 5 (80) p 512

[3] D.L. Staebler, J.J. Amodei, J. Appl. Phys. Vol 43 (72) p 1042

[4] J.O. White, M. Cronin Golomb, B. Fischer, A. Yariv, Appl. Phys. Lett. Vol 40 (82) p 450

[5] J.P. Herman, J.P. Huignard, J.P. Herriau, Opt. Lett. Vol 20 (81) p 2173

[6] T.J. Hall, M.A. Fiddy, H.S. Nar, Opt. Lett. Vol 5, 485 (1980)

[7] Ch. Bordê, Nato Advanced Study Institute on Quantum Optics and Experimental General Relativity - Bad Windsheim (1981)

Large Enhancement of the Sagnac Effect in a Nonlinear Ring Resonator and Related Effects

A.E. Kaplan

Francis Bitter National Magnet Laboratory*, Massachusetts Institute of Technology, Cambridge, MA 02139, USA

P. Meystre

Max-Planck-Institut für Quantenoptik
D-8046 Garching, Fed. Rep. of Germany

Abstract

A fundamentally novel way is discussed to enhance the Sagnac effect and to obtain directionally asymmetric bistability by using a nonlinear ring resonator. These effects are based on nonlinear nonreciprocity induced by counterpropagating waves due to formation of nonlinear index grating. The enhancement of the Sagnac effect could be as large as 10^4 - 10^5 which allows one to accordingly increase the sensibility of the passive resonant rotational sensors.

1. The Main Principles

It is well known that the Sagnac [1] effect provides a powerful way of measuring rotation rates with extreme accuracy — better than 10^{-4} x the earth's rotation rate [2]. It can be used for geophysical applications (e.g., for measurement of the earth's rotation, its precession, etc. [3]), and in general relativity (e.g., for measurement of the curvature of space [4]). Most important is its application as a rotation sensor, i.e., "optical gyroscopes" in intertial navigation systems [1,5] (e.g., in aircrafts and missiles).

We discuss here a fundamentally novel way to dramatically enhance the Sagnac effect in the passive ring resonator, which is based on nonreciprocity, induced in a nonlinear sample set in the resonator. The basic idea

*Supported by the National Science Foundation.

for this effect was first proposed by Kaplan and Meystre [6,7]. The proposed idea takes advantage of the different speed of propagation of counterpropagating waves of different intensities $|E_1|^2$ and $|E_2|^2$ in a nonlinear medium, which could be considered as nonlinearly-induced nonreciprocity. It is well known [6-10] that due to formation of a nonlinear index grating, the effective nonlinear susceptibilities ($\Delta\varepsilon_1^{NL}$ and $\Delta\varepsilon_2^{NL}$) of each of these waves are different provided that intensities of these waves are different. For instance, if a nonlinear material has a Kerr-like component of susceptibility,

$$\Delta\varepsilon^{NL} = \varepsilon_2|E(x)|^2 \tag{1}$$

where E(x) is the total field at location x at the axis of propagation, then, in the presence of two counterpropagating waves, the effective nonlinear components $\Delta\varepsilon_j^{NL}$ $(j=1,2)$ for both of these waves are different:

$$\Delta\varepsilon_1^{NL} = \varepsilon_2\left(|E_1|^2 + 2|E_2|^2\right) \; ; \quad \Delta\varepsilon_2^{NL} = \varepsilon_2\left(|E_2|^2 + 2|E_1|^2\right) \tag{2}$$

provided that $|E_1|^2 \neq |E_2|^2$. Here, E_1 and E_2 label the right and left (or clockwise and counterclockwise) propagating field, such that total field E is

$$E(x) = E_1(x)e^{ikx} + E_2(x)e^{-ikx} \quad . \tag{3}$$

The nonreciprocal factors in Eq. (2) result from the formation of a nonlinear index grating which leads to the cross-interaction of E_1 and E_2, as can be shown from the following argument. The nonlinear part of the polarization is given by $P^{NL} \propto E\Delta\varepsilon^{NL}$. Using Eq. (3), in the case of Kerr-nonlinearity (1) this gives [6]

$$P^{NL} \propto E|E|^2 = \left(E_1e^{ikx} + E_2e^{-ikx}\right)\left[|E_1|^2 + |E_2|^2 + \left(E_1E_2{}^*e^{2ikx} + c.c.\right)\right] =$$
$$E_1\left(|E_1|^2+2|E_2|^2\right) + E_2\left(E_2{}^2 + 2\,E_1{}^2\right) + \text{(rapidly varying terms)} \; , \tag{4}$$

so that the relevant polarizations for the clockwise and counterclockwise fields are as given in Eq. (2). We would like to stress two important facts with regard to this result. First, nonlinearly-induced nonreciprocity (2) is solely due to formation of a nonlinear index grating which is caused by the periodic term $[E_1E_2{}^*\exp(2ikx) + c.c.] \propto \cos 2kx$ in $|E|^2$. If this periodic grating is washed out by some process (e.g., diffusion), then nonreciprocity in Eq. (2) should decrease. Second, cross interaction (2) caused by the Kerr-nonlinear index grating, does not lead to the mutual back reflection of both of the counterpropagating waves at the grating, as has been explicitly shown in Ref. [9]. Hence, the effect discussed here is caused solely by the nonlinear change of optical paths, not by the induced reflection. The linear Sagnac effect is caused by the linear nonreciprocity (i.e., different optical paths of two counterpropagating waves), which is due to rotation of the waveguide system, e.g., a ring interferometer. In a resonator filled with a nonlinear medium, the optical paths are further varied as functions of light intensity. If this nonlinear change is the same for both waves, it has no influence on the Sagnac effect. However, if the nonlinear changes of optical paths are different for the left and right propagating waves, Eq. (2), it can lead to considerable increase in the total nonreciprocal effect, provided that intensities of these waves are different.

Suppose now that the ring resonator is pumped in both directions by two incident beams of the same intensity $|E_{in}|^2$ (and frequency ν) (see Fig. 1).

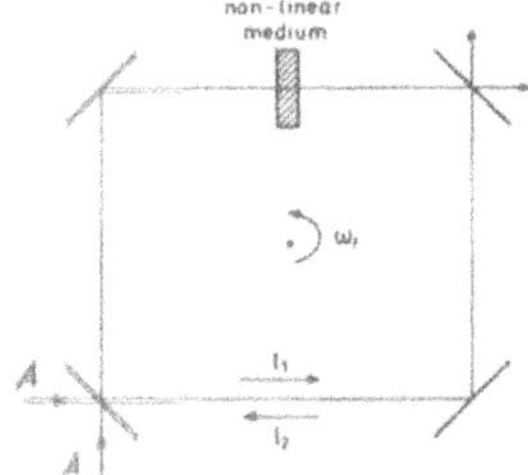

Fig.1 Rotating ring resonator with nonlinear medium and two pumping beams of equal intensity.

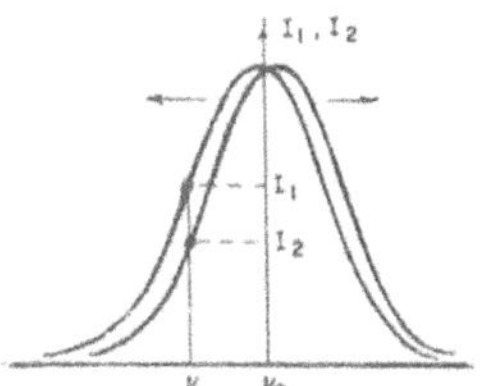

Fig.2 The Sagnac splitting of resonance for clock- and counterclockwise propagating waves.

In the absence of rotation, the intensities of the waves excited inside the resonator should be the same, $|E_1|^2 = |E_2|^2 \equiv |E_o|^2$. Suppose also that the laser frequency is set on the side of the resonant curve (Fig.2) in the vicinity of one of the eigenfrequencies of the resonator ν_o. Then, in the presence of rotation, the Sagnac effect causes a different change of the intensities of the waves. This is because of the sharp dependence of each of the intensities on Sagnac detuning at the slope of the resonant curve: one of the intensities decreases while the other increases. Therefore, the initial small difference of intensities provided by the usual Sagnac effect can be used to generate a large additional difference via the nonlinearly-induced nonreciprocity of the optical paths of the counterpropagating waves.

The initial phase shift between both of these waves $\Delta\phi$ due to the linear Sagnac effect is $\Delta\phi_o = 4S\omega_r Q/cL$, where S is the area of the ring resonator, Q and L are its finesse and total optical path, ω_r is the rotation rate. By using a resonator filled by a nonlinear material and by tuning the frequency of the light to the proper slope of the resonant curve, one can obtain a large increase of the measured Sagnac phase shift, $\Delta\phi = \eta\Delta\phi_o$, where η is the factor η of nonlinear enhancement (see Section 2 below). The factor of enhancement η could be as large as $10^4 - 10^5$, which allows one to accordingly increase the sensitivity of the passive resonant rotational sensors based on the Sagnac effect.

Under some conditions, the factor of enhancement becomes infinite which results in some other new effects due to nonlinear nonreciprocity in the ring resonators (besides enhancement of the Sagnac effect), in particular, directionally asymmetrical optical bistability [7]. Inside a domain whose boundary corresponds to $\eta = \infty$, the symmetrical waves generated in the ring resonator become unstable even without any rotation. This instability leads to damping of one of the propagating waves and to an increase of intensity of another counterpropagating wave. This asymmetrical excitation (under symmetrical pumping) results in a novel kind of optical bistability which provides another opportunity to obtain spatial (directional) switching of radiation, in addition to some previously proposed and experimentally observed spatial nonlinear switching and bistability (at nonlinear interfaces [11,12], and in mutually self-acting beams [13]).

2. Enhancement of the Sagnac Effect

We consider a rotating passive ring resonator filled with a nonlinear medium. The system is pumped in both directions by two incident beams of the same frequency ν and amplitude E_{in}. Taking the frequency ν of the driving field to be close to one of the eigenfrequencies ν_o of the resonator, the steady-state amplitude E_j of the field j $(j = 1,2)$ is

$$E_j = \frac{(\gamma_c/\gamma)E_{in}/\sqrt{T}}{1 + i\gamma^{-1}\left[\nu-\nu_o + (-1)^j\omega_s + \nu_o\Delta\varepsilon_j^{NL}L_N/L\right]} \quad . \tag{5}$$

Here, T is the mirror transmittivity, L the total optical length of the resonator and L_N the optical length of the nonlinear sample. $\gamma_c = cT/L$ is the empty cavity bandwidth, and γ_s gives the linear losses of the medium inside the cavity, so that $\gamma = \gamma_c + \gamma_s$ gives the total bandwidth of the system, $k_o = \omega_o/c$, and

$$\omega_s = 4S\omega_r k_o/L \tag{6}$$

is the Sagnac frequency shift, where S is the area of the ring resonator and ω_r its rotation rate. Note that Eq. (5) is an approximate form valid for the good finesse case, that is, $\gamma << c/L$. We introduce the dimensionless frequencies $\Delta = (\nu - \nu_o)/\gamma$ and $\Omega = \omega_s/\gamma$, and the dimensionless intensities

$$I_j = \varepsilon_2|E_j|^2L_N\nu_o/L\gamma \qquad (j = 1,2) \quad ,$$

$$A = \varepsilon_2|E_{in}|^2L_N\nu_o\gamma_c^2/L\gamma^3T \quad , \tag{7}$$

so that A corresponds to the driving field, and I_j to the generated waves. (Note that I_j, A, and Δ change sign for $\varepsilon_2<0$.) The equation for the intensity I_j is readily obtained by multiplying Eq. (5) by its complex conjugate:

$$A = I_j\left\{1 + \left[\Delta + (-1)^j\Omega + I_j + 2I_{3-j}\right]^2\right\} \quad . \tag{8}$$

We assume that in the absence of rotation the intensities I_1 and I_2 are equal (symmetrical regime). That is $I_1 = I_2 = I_o$, where I_o satisfies the equation

$$A = I_o \left[1 + (\Delta + 3I_o)^2 \right] . \tag{9}$$

We also restrict ourselves to the most important case of small rotations ($\Omega << 1$), so that we can assume that the perturbations $\xi_j = I_j - I_o$ of both waves are small ($\xi_j << I_o$). Under these conditions, we immediately find the solution for ξ_j:

$$\xi_1 = -\xi_2 = (\eta - 1)\Omega ,$$

where

$$\eta = \frac{1 + (\Delta + 3I_o)^2}{1 + (\Delta + 3I_o)(\Delta + I_o)} . \tag{10}$$

Therefore, the total phase difference between the two counterpropagating waves is given by $\eta\Omega$ instead of Ω. Hence, the result of the nonlinear medium is to scale the Sagnac effect by a factor of η. Eq. (10) is an implicit form for η, and I_o must still be determined from Eq. (9). One expects η to be much larger than unity for values of the driving field and detuning such that the system is near the onset of an instability.

Calculations based on Eqs. (9) and (10) show that the boundary of the unstable regime ($\eta \to \infty$) is given by

$$A = (2/3) \left[-\Delta(3\Delta^2 - 5) \pm (3\Delta^2 - 1)\sqrt{\Delta^2 - 3} \right] . \tag{11}$$

This domain of instability in the plane of parameters A and Δ is shown in Fig. 3 as curve A_1. As shown in Section 3 below, this instability leads to a new steady-state regime such that $I_1 \neq I_2$ (non-symmetrical regime) in the absence of rotation of the ring interferometer. The detuning threshold of this instability occurs for $\Delta_{th} = -\sqrt{3}$, which corresponds to $A_{th} = 8/\sqrt{3}$.

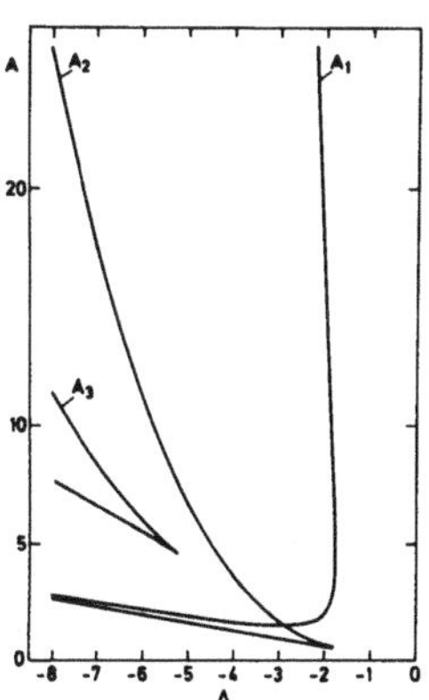

Fig.3 Domains of operation in the plane of parameters A and Δ. Curve A_1 defines the domain of instability of the symmetrical solution; A_2 the domain of bistability of the symmetrical solution; and A_3 the domain of multistability of the asymmetrical solution

We find that the enhancement factor η increases dramatically as one approaches the instability boundary from the symmetrical regime of operation, i.e., as one decreases Δ. Note that for $\Delta > \Delta_{th}$, the regime of operation for $\Omega = 0$ is symmetrical and stable for any value of the driving field.

In Fig. 4 we show the enhancement factor η as a function of the driving field A and for various detunings $\Delta \geq \Delta_{th}$. These curves were obtained by solving Eqs. (9) and (10) consistently. As expected, η becomes infinite for $\Delta = \Delta_{th}$ and $A = A_{th}$. It follows from Eq. (10) that for a fixed detuning, the highest enhancement is given by

$$\eta_{max} = 3/\left(2 - \sqrt{\Delta^2 + 1}\right) \quad . \tag{12}$$

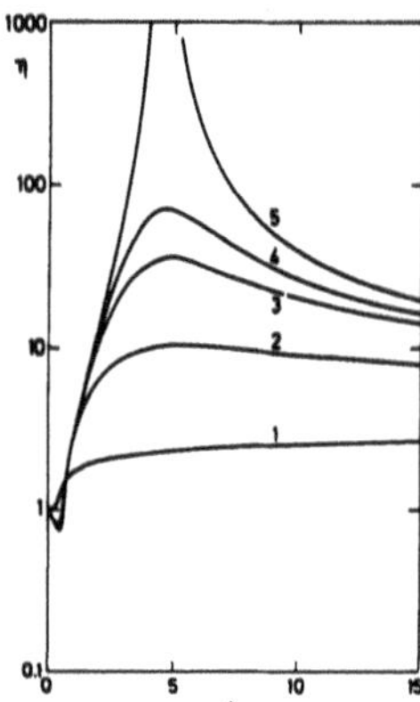

Fig.4 Enhancement factor η as a function of the dimensionless pump intensity A for various values of the dimensionless frequeny detuning Δ: 1) $\Delta=0$, 2) $\Delta = -\sqrt{3}(1-0.2)$, 3) $\Delta = -\sqrt{3}(1-2/35)$, 4) $\Delta = -\sqrt{3}(1-1/35)$, 5) $\Delta = -\sqrt{3}$

In the vicinity of the threshold ($\Delta - \Delta_{th} \ll |\Delta_{th}|$), Eq. (12) reduces to $\eta_{max} \cong 2\sqrt{3}/(\Delta - \Delta_{th})$. The corresponding amplitude of the driving field is

$$A_{max} = -2(\Delta^2 + 1)\left(\sqrt{\Delta^2 + 1} + 1\right)^2 / 3\Delta^3 \quad . \tag{13}$$

It is interesting to note that even far from the instability threshold, the enhancement remains very large. In the limit $A \to \infty, \eta \to 3$, independently of Δ. We also remark that the enhancement is only weakly sensitive to changes of the driving field, which indicates the possibility of using pump lasers with relatively high intensity fluctuations.

Under reasonable requirements for stability of parameters of the resonator and laser radiation, the proposed system has the potential to achieve enhancement as large as $10^3 - 10^4$, which allows one to accordingly increase the sensitivity of the passive rotation sensor. For example, provided that the frequency stability of the resonator and laser radiation is about 10^{-9}, $Q \sim 10^3$, and the amplitude stability of the laser [14] is about 10^{-3}, Eqs. (10,12) show th feasibility of obtaining $\eta \sim 10^4$ with stability $\sim 10^{-2}$.

One of the most important facts about nonlinear enhancement of the Sagnac effect is that due to the relatively slow time scale of its conventional applications (usually, it is more than a few msec), one can use huge but slow nonlinearities, recently discovered and studied in the optical and IR ranges, for example, in InSb.[15] We illustrate this with the following example. We assume small mirror transmittivity ($T \cong 0.1$), neglect the internal losses ($\gamma_s << \gamma_c$), and take advantage [15] of the large $\chi^{(3)}$ in InSb ($\chi^{(3)} = 10^{-2}$esu). Considering $L_s \cong 100$ µm and $k_0 \cong 1.2 \times 10^4$ cm^{-1}, we obtain, with the threshold condition $A_{th} = 8/\sqrt{3}$, the required incident field, $|E_{in}|_{th} \cong 45$ V/cm. This corresponds to an incident power of about 50 mW/mm^2, which can readily be achieved in the cw regime.

3. Directionally Asymmetrical Bistability

As was already mentioned above, the cross-interaction of two counterpropagating waves via the nonlinear index grating that they generate could result in instability of the symmetric regime of propagation, even without any rotation of the ring resonator. This instability leads to the non-symmetric regime, in which the amplitudes of the two counterpropagating waves become different, while the ring resonator is symmetrically pumped by two beams of the same intensity and opposite directions. We immediately note that this effect could not occur in the case of single beam pumping (which is characterized by symmetrically optical bistability [16]). Direction of the dominantly generated wave could be clockwise as well as counterclockwise under the same conditions. This corresponds to the situation which could be described as directional optical bistability. Optical bistability has been the object of intense activity over the last few years [17]. Up to now, the bulk of the research has been concerned with intensity switches in a nonlinear Fabry-Perot resonator (except some nonresonator devices [11-13]), or in ring cavities pumped by a single beam. The existence of the directionally asymmetric regime is of great interest for applications where directional switching of light is required.

We consider again the same nonlinear ring resonator (Fig. 1), now without any rotation ($\omega_r = 0$; $\Omega = 0$). In such a case, the equation (8) for the waves, intensities I_j ($j = 1,2$) can be written in the form

$$A = I_j \left\{ 1 + [\Delta + I_j + 2I_{3-j}]^2 \right\} \quad . \tag{14}$$

We consider two different solutions, the symmetrical and asymmetrical ones. In the symmetrical case, ($I_1 = I_2 = I_0$), the intensities satisfy Eq. (9). The intensity I_0 as a function of the incident intensity A is shown in Fig.5 (symmetrical branch) for various values of the detuning Δ. As can be seen from Figs.5b, c, d, the symmetrical regime exhibits hysteresis for sufficiently large values of Δ and A. By requiring that $dI_0/dA = \infty$ at the points of hysteretic jump, we find readily from Eq. (9) that the domain of usual symmetrical bistability is given in the plane of the parameters A and Δ by

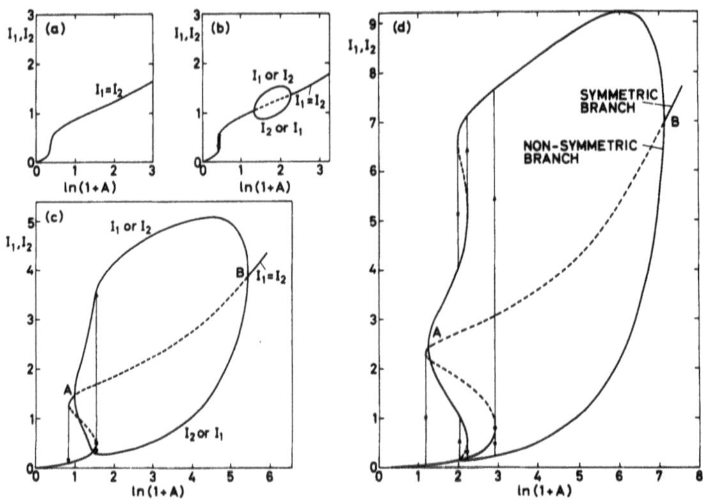

Fig.5 Normalized intensities I_1 and I_2 as a function of the dimensionless pump intensity A. The dashed parts of the curves indicate the regions of instability. The curves labeled by $I_1 = I_2$ give the symmetrical regime, and the points A' and B' the crossings between symmetrical and asymmetrical regimes. The four figures are for various values of the dimensionless detuning Δ. (a) $\Delta = -1.65$ (b) $\Delta = -1.8$ (c) $\Delta = -4$ (d) $\Delta = -7$

the relation

$$A = (2/81)\left[-\Delta(\Delta^2 + 9) \pm (\Delta^2 - 3)^{3/2}\right] . \tag{15}$$

This domain in shown in Fig.3 as curve A_2. The detuning threshold for the onset of symmetrical bistability is given by $\Delta_2 = -\sqrt{3}$, which corresponds to a driving intensity $A = 8/9\ \sqrt{3}$. However, it can easily be shown by a linear stability analysis of Eq. (14) that under appropriate conditions, the symmetrical regime determined by Eq. (9) becomes unstable. The domain of instability of the symmetrical solution is given by Eq. (11), which is shown as curve A_1 in Fig.3. The detuning threshold for this regime is $\Delta_1 = -\sqrt{3}$ which is equal to Δ_2, but corresponds to the larger pump intensity $A = 8/\sqrt{3}$. Inside this domain, the stable behavior of the system is characterized by the fact that I_1 and I_2 are unequal. For $I_1 \neq I_2$, the two coupled third-order Eq. (14) can be reduced to a single third-order equation for $S = I_1 + I_2$,

$$A = \left[1 + (S + \Delta)^2\right]\left[2\Delta + 3S\right] , \tag{16}$$

which can be solved readily for any values of A and Δ. The separate values of I_1 and I_2 can then be obtained via the equation

$$A = P(2\Delta + 3S), \tag{17}$$

where $P = I_1 I_2$.

The asymmetric solution is shown in Figs.5b,c,d. Note that I_1 and I_2 play completely symmetrical roles in Eqs. (16,17), so that which of them will become larger is determined by noise or slight initial asymmetries in pumping.

It is interesting to point out that the asymmetrical regime can itself become multivalued for large enough Δ and A, as seen in Fig.5d. Applying the same criteria as in the symmetrical case ($dI_j/dA = \infty$, $j = 1,2$) we find that the domain of multistability of the asymmetrical branch is given by

$$A = (2/3^5)[-\Delta(\Delta^2 + 81) \pm (\Delta^2 - 27)^{3/2}] \ . \tag{18}$$

which is shown in Fig.3 as curve A_3.

4. Conclusion

The nonlinearly induced nonreciprocity of counterpropagating beams of light in the nonlinear ring resonator can be used to dramatically enhance the Sagnac effect and therefore to increase sensitivity of the passive optical rotational sensors. This enhancement of the Sagnac effect could obviously find applications in cases where high sensitivities are required (e.g., optical tests of general relativity). On the other hand, this effect provides a direct way of studying experimentally non-local interactions of light with matter, since it is due solely to the formation of a nonlinear index grating, which, in turn, can be formed if the nonlinearity depends only on the light intenstiy at a given point (local interaction). If the index grating is washed out by some fast process (for instance, diffusion in gases), the non-reciprocal factor 2 in Eq. (2) is replaced by $1 + \alpha$, where in the general case $0 \leq \alpha \leq 1$. This leads to a reduced enhancement factor $\eta_\alpha = 1 + \alpha(\eta - 1)$. Thus, the measurement of η_α provides a novel spectroscopic method of analyzing non-local interactions (non-reciprocal Sagnac spectroscopy).

The effect of directionally asymmetrical bistability is based as well on nonlinear nonreciprocity, and could be applied for directional switching, e.g., in optical signal processing. In addition, this effect should be very useful in connnection with enhancement of the Sagnac effect. Observation of asymetrical bistability will immediately help to locate the zone of the large enhancement which should occur in the vicinity of the onset of this bistability.

We acknowledge several fruitful discussions with P.L. Kelley, S. Ezekiel, M. Sargent III, M.O. Scully, and H. Walther. One of us (A.E.K.) is grateful to H. Walther, S. Witkowsky, and M.O. Scully for their hospitality during his stay at Max-Planck-Institut. The work of A.E.K. was supported by the U.S. Air Force Office of Scientific Research and by Max-Planck-Institut für Quantenoptik, Garching, Germany.

References

1. For a comprehensive review, see E. Post, Rev. Mod. Phys. 39, 475 (1967); J. Anandan, Phys. Rev. D. 24, 338 (1981)
2. F. Aronowitz, in Proc. Soc. Photo-Opt. Instrum. Eng. 157, 2 (1978); T.A. Dorschner, H. Haus, M. Holz, I.W. Smith and H. Statz, IEEE J. Quant. Elect. QE-16, 1376 (1980); G.A. Sanders, M.G. Prentiss and S. Ezekiel, Opt. Lett. 6, 569 (1981)
3. D.M. Eckhardt, Proc. Soc. Photo-Opt. Instrum. Eng. 157, 172 (1978)
4. M.O. Scully, in Proc. Fifth Int. Conf. Laser Spectr., H. Walther and K. Rothe, eds. (Springer-Verlag, Berlin, 1979), p.21; M.P. Haugan, M.O. Scully and K. Just, Phys. Lett. 77A, 88 (1980)
5. F. Aronowitz, in Laser Applications, (Academic Press, 1971), vol. 1, p.133; J. Killpatrick, IEEE Spectrum 4, 44 (1967)
6. A.E. Kaplan and P. Meystre, Opt. Lett. 6, 590, 1981
7. A.E. Kaplan and P. Meystre, to appear in Opt. Commun.
8. R.Y. Chiao, P.L. Kelley and E. Garmire, Phys. Rev. Lett. 17, 1158 (1966); R.L. Carman, R.Y. Chiao and P.L. Kelley, Phys. Rev. Lett. 17, 1281 (1966)
9. B.Ya. Zeldovich, Brief Communications in Physics of the Lebedev's Institute, No. 5, 20 (1970) (in Russian)
10. Nonlinear nonreciprocity was recently confirmed in experiments with nonlinear glass fibers by S. Ezekiel, see these Proceedings, p.
11. A.E. Kaplan, JETP Lett. 24, 114 (1976); Sov. Phys. JETP 45, 896 (1977); Sov. J. Quan. Elect. 8, 95 (1978); Radiophysics and Quant. Elect. 22, 229 (1979); Appl. Phys. Lett. 38, 67 (1981); IEEE J. Quant. Elect. QE-17, 336 (1981)

12. P.W. Smith, J.-P. Hermann, W.J. Tomlinson and P.J. Maloney, Appl. Phys. Lett. 35, 846 (1979); P.W. Smith, W.J. Tomlinson, J.P. Hermann and P.J. Maloney, IEEE J. Quan. Elect. QE-17, 340 (1981)
13. J.E. Bjorkholm, P.W. Smith, W.J. Tomlinson and A.E. Kaplan, Opt. Lett. 6, 345 (1981); A.E. Kaplan, Opt. Lett. 6, 360 (1981); A.E. Kaplan, JETP Lett. 9, 33 (1966)
14. These data are far from the best obtained, e.g., for the stabilized argon laser, in which case frequency stability is $\sim 10^{-12}$ and amplitude stability is $\sim 10^{-4}$, see L.A. Hackel, R.P. Hackel and S. Ezekiel, Metrologia 13, 141 (1977)
15. D.A.B. Miller, S.D. Smith and B.S. Wherrett, Optics Commun. 35, 221 (1980)
16. See, for instance, R. Bonifacio, R. Gronchi and L. Lugiato, "Optical Bistability" [17], p.31; J.D. Cresser and P. Meystre, "Optical Bistability" [17], op. cit. p. 265, and references therein
17. For recent comprehensive reviews, see "Optical Bistability," (Ed. by C.M. Bowden, M. Ciftan and H.R. Robl, Plenum, 1981), and special issue of IEEE J. Quantum Electronics on Optical Bistability, QE-17 (March 1981). For an introduction to the subject, see P.W. Smith and W.J. Tomlinson, IEEE Spectrum 18, 26 (1981)

Synchronous Fiber-Optic Gravitational Telescopes

S.A. Kingsley

Optical Sciences Group, Battelle Columbus Laboratories, 505 King Avenue
Columbus, OH 43201, USA

Abstract

Optical fibers may be used to sense a number of parameters including rotation. A new form of fiber-optic device is proposed, which is sensitive to gravitational radiation, and is based on the BRAGINSKY-MENSKII synchronous electromagnetic gravitational-wave detector. The fiber-optic gravitational telescope employs a very large Sagnac interferometer to detect the gravitationally-induced phase modulation.

1. Introduction

It is well-known that optical fibers conveying coherent light may be used as very sensitive strain, pressure and temperature sensors and the resulting induced phase modulation detected by means of a Mach-Zehnder interferometer [1]. A Sagnac interferometer can be employed to detect fiber-optic phase modulation if the phase modulation is non-reciprocal. Two such non-reciprocal phenomena are inertial rotation effects [2] and magnetic field effects (discussed elsewhere in these proceedings). This paper proposes that there is a third non-reciprocal effect which will allow the use of optical fibers as gravitational-wave sensors.

Figure 1 illustrates the basic Weber-type bar gravitational-wave detector [3] and a three-dimensional representation of a gravitational-wave, the latter being reproduced from MTW [4], which is an excellent source of information on gravitational-wave physics. The positive results of Joseph Weber's experiments of the late 1960's and early 1970's are now thought to have been erroneous. However, his pioneering work has inspired many other research groups to construct their own gravitational-wave antenna. Weber's original detector was capable of measuring strains of about 10^{-16} while the state-of-the-art detectors are aiming for about 10^{-21} [3-7].

While it is possible to conceive of a mechanical Weber-type bar gravitational-wave detector employing a fiber-optic strain-gauge [1] instead of more conventional transducers, e.g. piezoelectric, laser probing and

SQUIDS (Superconducting Quantum Interference Devices) we shall not be describing such a technique here. Where the sensitivity of the detector, even when cooled, is limited by mechanical kT noise, there is no advantage in employing a fiber-optic strain-gauge.

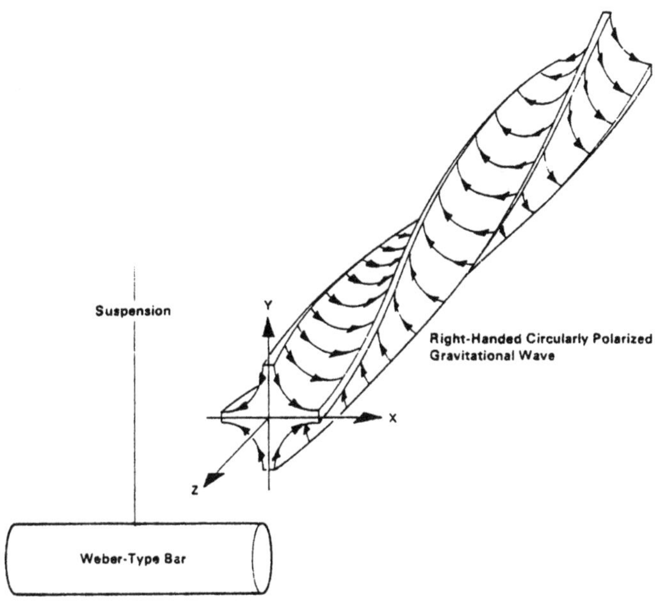

Fig.1 Basic Weber-type bar gravitational wave detector and a three-dimensional representation of a right-handed circularly-polarized gravitational-wave

In 1971, BRAGINSKY and MENSKII [4, 7-9] proposed a non-mechanical form of gravitational-wave detector in which an electromagnetic wave at microwave frequencies is circulated around a small superconducting Niobium ring [10] to interact with high frequency gravitational waves. Unfortunately, astrophysicists predict that the most powerful sources of gravitational waves should radiate in the frequency range below 10kHz [3-7]; there is unlikely to be much gravitational radiation at microwave frequencies. More recently, LYAMOV and RUDENKO [11] have suggested a low-frequency acoustic-surface-wave version of this system, where the diameter of the detector can be made small, even at low gravitational frequencies, because of the low velocity of acoustic surface waves. This approach appears promising but does require a single crystal quartz bar of about 1 meter diameter for a 1kHz resonance.

With the advent of low-loss fiber-optic waveguides, it now becomes possible to conceive of a fiber-optic version of the BRAGINSKY-MENSKII synchronous gravitational-wave detector and this forms the subject of this paper [12].

2. Principle of Operation

Figure 2 shows the basic form of the proposed fiber-optic gravitational-wave telescope, which is used with some form of optical interferometer [1, 13] for detecting the frequency (or phase) changes induced on the guided optical carrier by the gravitational wave. An unmodulated optical carrier $E_C \sin\omega_C t$

from a laser enters the loop of length L, and a phase-modulated attenuated optical carrier $\sqrt{10^{-\alpha L}}\, E_c \sin[\omega_c t + \Delta\phi \sin\omega_g t + \psi]$ leaves the loop, where ω_g is the angular frequency of the gravitational wave, $\Delta\phi$ is the peak phase deviation produced, ψ is an arbitrary phase angle and α is the attenuation factor.

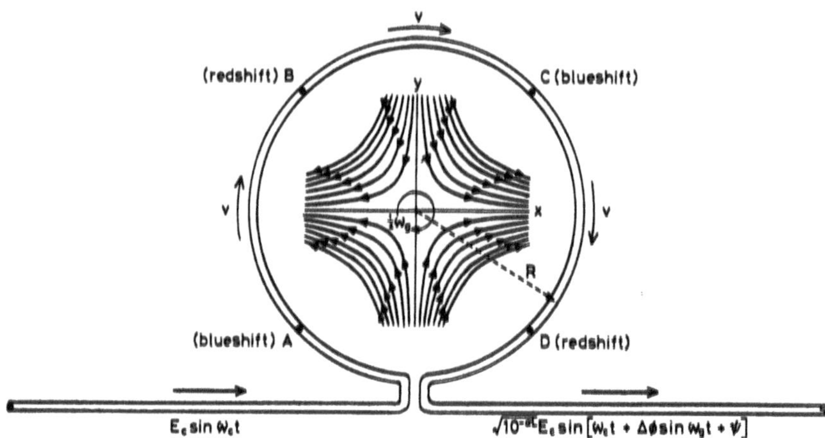

Fig.2 Basic form of the fiber-optic gravitational telescope

The gravitational-wave sensing region of the single-mode fiber consists of one or more turns arranged into a circular loop or coil. For clarity the loop has been shown to be slightly less than one turn. The transverse nature of the gravitational wave is represented by the lines of force (or acceleration) shown within the loop of fiber. If we imagine that the wave is polarized in the x-y plane, then the wave propagates in the z direction which is out of the plane of the page (see Fig.1).

Unlike plane-polarized electromagnetic waves which have orthogonal E and H components, a plane-polarized gravitational wave has two orthogonal force (or acceleration) components [3, 4]. A plane-polarized gravitational wave, like its electromagnetic counterpart, can be resolved into two counter-rotating circularly polarized components. If the diameter of the loop is arranged so that the angular velocity of the optical carrier around the loop is the same as the angular velocity of one of the circularly polarized components, then cumulative (traveling-wave modulation) interaction can occur. Thus it is possible to obtain a resonance between the electromagnetic wave in the waveguide and the passing gravitational wave [4, 7-9].

A wavetrain or photon in positions A or C in the loop is in an accelerating part of the gravitational field, so that it becomes continually blueshifted as it rotates around the loop. Similarly, a photon in positions B or D is in a decelerating part of the field, so that it becomes continually more and more redshifted. This phenomenon is basically the same as the Gravitational Redshift effect [4]. The net effect of the other circularly-polarized component will be zero.

If the loop contains two counter-rotating optical wavetrains, as would occur in a Sagnac interferometer, the interaction is more complicated. It is not clear to the author, because of the uncertainty as to whether gravitational radiation will be linearly or circularly polarized, or a combination of both, that there will always be a non-reciprocal phase shift in a Sagnac interferometer. However, it is obvious that a circularly polarized gravitational-wave will induce a non-reciprocal phase shift.

The phase-sensitivity of the fiber-optic gravitational-wave antenna will now be derived in a somewhat unconventional manner without recourse to Riemannian tensor calculus [4].

2.1 The Gravitational Redshift

When a photon leaves a static gravitational field it loses energy corresponding to the work done against the field and hence its frequency is reduced [4]. This fractional change in frequency is given by the classic redshift relationship

$$\frac{\Delta f}{f_c} = \frac{-gH}{c^2} \quad , \tag{1}$$

where Δf is the frequency change, f_c is the optical frequency, g is the acceleration due to gravity, H is the distance through which the photon rises and c is the velocity of light (3×10^8 m/s). For He-Ne laser light at 0.6328 µm ($f_c = 4.74 \times 10^{14}$ Hz) the redshift at the earth's surface amounts to 52Hz/km.

2.2 Synchronous Interaction

Now the angular frequency of the gravitational wave ω_g is twice the angular velocity with which the pattern of 'lines of force' rotates [4, 7, 8]. Hence, for synchronous interaction of the gravitational field with that of the electromagnetic field, the radius of the coil must be

$$R = \frac{2c}{\omega_g n} \quad , \tag{2}$$

where n is the refractive index of the optical fiber (1.447). Substituting the given values of c and n into (2), we find that a gravitational-wave telescope at 10kHz has a diameter D = 13.2km and a total fiber length for one turn of 41.5km.

2.3 Relative Acceleration

The wave-induced accelerations relative to the center of mass of the detector are

$$\frac{d^2x}{dt^2} = \frac{1}{2}(\ddot{A}_+ x + \ddot{A}_\times y) \quad , \tag{3a}$$

$$\frac{d^2y}{dt^2} = \frac{1}{2}(-\ddot{A}_+ y + \ddot{A}_\times x) \quad , \tag{3b}$$

$$\frac{d^2z}{dt^2} = 0 \quad , \tag{3c}$$

where A_+ and $A_\times$ are the dimensionless amplitudes of the two independent states of polarization, and $\cdot$ denotes differentiation with respect to time. For polarization X, the force lines are rotated by 45° from that shown in Fig.2., which indicates a plane-polarized wave $\ddot{A}_\times = 0$, $\ddot{A}_+ > 0$. The closer are the force lines, the greater is the force and acceleration [4].

Letting $\ddot{A}_\times = 0$ and considering the acceleration along the x-axis relative to the center of mass,

$$\frac{d^2x}{dt^2} = \frac{1}{2}\ddot{A}_+ x \quad . \tag{4}$$

If this plane-polarized wave is decomposed into two counter-rotating circularly-polarized components, then the acceleration due to one of the components at radius R is given by

$$a = \frac{1}{2\sqrt{2}}\ddot{A}_+ R \; . \tag{5}$$

2.4 The Gravitational Poynting Vector

The intensity of a gravitational wave averaged over many wavelengths is derived by consideration of the gravitational Poynting Vector [4]. It may be shown that the intensity is given by

$$I = \frac{c^3}{16\pi G}\left\langle \dot{A}_+^2 + \dot{A}_\times^2 \right\rangle, \tag{6}$$

where G = Universal gravitational constant ($6{\cdot}673 \times 10^{-11}$ $m^3/kg.s^2$).

Note that in a conventional mechanical detector, e.g. the Weber-bar, the (non-resonant) strain produced by a plane-polarized wave, when the detector is small compared to the gravitational wavelength, is given by

$$\frac{\Delta \ell}{\ell} = \frac{A_+}{\sqrt{2}} = \frac{1}{\omega_g}\sqrt{\frac{8\pi G I}{c^3}} \quad , \tag{7}$$

where ℓ is the length of the bar and $\Delta\ell$ is its change in length. For $I = 1 W/m^2$ and $f_g = 10 kHz$, we find that $\Delta\ell/\ell \simeq 10^{-22}$.

2.5 Phase Sensitivity

We are now in a position to derive the relationship linking the peak induced phase deviation $\Delta\phi$ to the gravitational-wave intensity I.

Using (6) to eliminate the dimensionless wave amplitude A_+ from (5), noting that $A_\times = 0$, we may write that the relative acceleration

$$a = \omega g R\sqrt{\frac{2\pi G I}{c^3}} \quad . \tag{8}$$

Replacing -g in the redshift relationship (1) by the acceleration a of (8) and using the synchronous relationship (2), we can now write

$$\Delta f = \frac{\omega_c R}{cn}\sqrt{\frac{8\pi G I}{c^3}} \quad . \tag{9}$$

The phase change $\Delta\phi$ is found from the product of the optical path length (in radians) and the mean frequency change around the loop. Since the frequency change is a linear function of interaction length and time, we may thus write

$$\Delta\phi = \frac{2\pi Ln}{c}\left(\frac{\Delta f}{2}\right) \quad . \tag{10}$$

Finally, using (9) to eliminate Δf from (10), and knowing that the interaction time $\tau = Ln/c$ we obtain

$$\Delta\phi = \frac{\omega_c \tau^2}{n^2}\sqrt{\frac{2\pi GI}{c^3}} \quad . \tag{11}$$

This expression is very similar to that given by BRAGINSKY and MENSKII for their microwave resonator, except for the presence of the n^2 factor in the denominator and a small numerical factor [8].

Since τ is proportional to n, the phase deviation is not dependent on the refractive index of the fiber, just as is the case for the fiber-optic rotation sensor [2].

If we now insert some typical values into (11) for a one-turn 10kHz antenna, at an optical wavelength $\lambda = 1.3\ \mu m$ ($f_c = 2.31 \times 10^{14}$Hz) for which $L = 41.5$ km and $\tau = 4.82\ \mu s/km$, the calculated value of peak phase deviation $\Delta\phi$ is 1.1×10^{-10} radians for an intensity $I = 1 W/m^2$.

Note that in the above derivation we have ignored any simultaneous mechanical effects, such as changes in fiber length and photoelastic effects [1]. Fig.3 illustrates the same gravitational-wave detector concept but with the

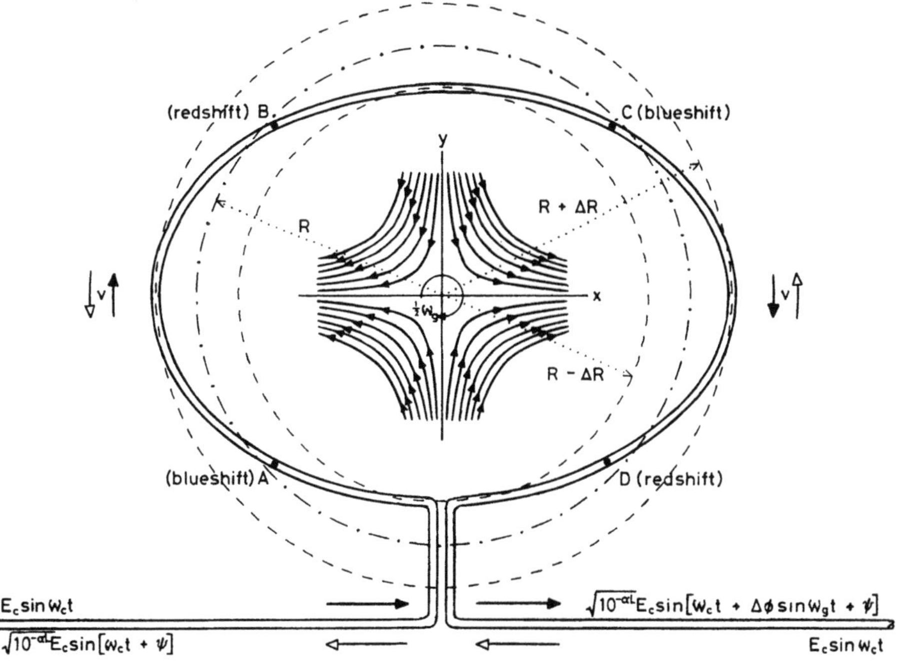

Fig.3 Fiber-optic gravitational telescope employing a Sagnac interferometer. Frequency-shift and possible mechanical effects are indicated

mechanical effect included. Here, we assume that the strain $\Delta\ell/\ell$ of (7) may be approximately replaced by the ratio $\Delta R/R$, even though the diameter of the detector is comparable to the gravitational wavelength.

If we assume the case of a circularly-polarized gravitational wave instead of a plane-polarized wave, then the major axis of the loop, which has been greatly exaggerated for clarity, will trace out a longer path length than the minor axis. This too is a non-reciprocal phenomenon whose effect on the optical phase appears to be comparable to that due to the frequency effect. For the purposes of this present paper and because of uncertainties in the polarization state of the gravitational radiation, we will ignore the mechanical effects. This should not alter our order-of-magnitude approximation for the sensitivity of the detector.

3. Gravitational Detectability

3.1 Signal-to-Noise Ratio

Although we have been implying the use of a Sagnac interferometer for detecting this gravitational-wave induced phase deviation, as indicated by Figs.3 and 5, we shall assume for this ball-park calculation of gravitational detectability that a shot-noise-limited heterodyne receiver is employed with a powerful local oscillator laser acting as a reference. For obvious reasons, i.e. no common-mode rejection of environmental (mechanical) noise, we would not build the device using such detection techniques [2]. The recovered baseband signal-to-noise ratio (SNR) for such a coherent detection system is given by:

$$SNR = \frac{\eta Po 10^{-\alpha L} \Delta\phi^2}{2hf_c B} \quad , \qquad (12)$$

where η = quantum efficiency of the photo-detector (0.5), P_o = optical power launched into the fiber (10mW), α = fiber-optic attenuation constant (0.2 dB/km), h = Planck's constant (6.63×10^{-34} J.s) and B = baseband bandwidth (1Hz) [13].

By substituting the above bracketed values into (12) along with the above calculated value of $\Delta\phi$, we find the SNR = -45.4 dB. Clearly, we cannot detect a monochromatic gravitational flux of $1W/m^2$ with this gravitational sensor. If we define the threshold of detectability as an SNR = 0dB, then this antenna is 45.4dB below detectability for a 1Hz detection bandwidth.

3.2 Gravitational Wave Intensity

The intensity of a gravitational wave in the vicinity of the earth is given by:

$$I = \frac{Mc^2}{4\pi d^2 \tau_{GW}} \quad W/m^2 \quad , \qquad (13)$$

where Mc^2 is the energy liberated by the conversion of mass M into gravitational radiation, d is the distance of the gravitational 'event' from the earth, and τ_{GW} is the duration of the pulse of gravitational radiation [4].

It is more useful to define our detectability in terms of the gravitational-wave spectral density, viz:

$$I(\upsilon) \approx \frac{Mc^2}{4\pi d^2 \tau_{GW} \Delta\upsilon} \quad W/m^2.Hz \quad , \qquad (14)$$

where $\Delta\upsilon$ is the bandwidth of the gravitational-wave pulse, which may often be approximated to the inverse of τ_{GW}.

3.3 Gravitational Spectrum Levels

We can now define a Threshold Gravitational Spectrum Level (TGSL) which gives the gravitational flux level just detectable in a 1Hz bandwidth, with respect to a Gravitational Spectrum Level (GSL) of 1 W/m^2.Hz. Thus, the above 10kHz gravitational-wave detector has a TGSL = +45.4 dB. Fig. 4 shows the TGSL for 1-turn gravitational telescopes using optical fibers of differing attenuations over a frequency range of three decades. It is assumed that $\tau_{GW} \geq \tau$.

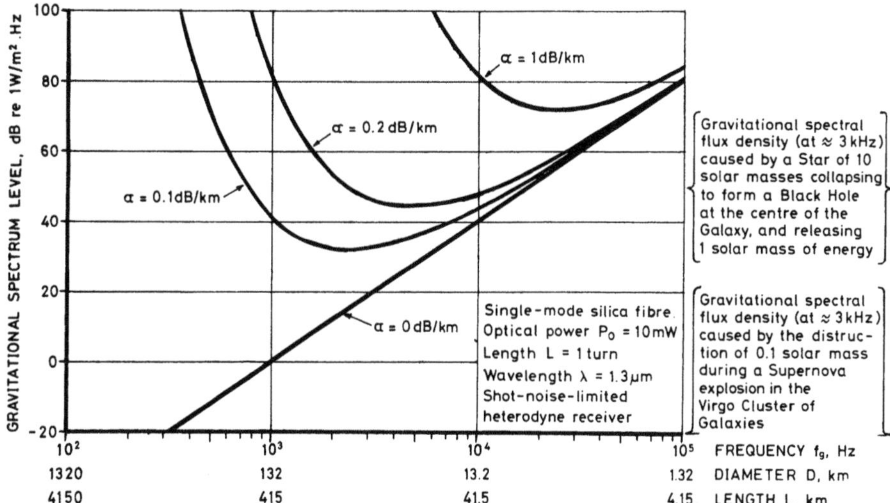

Fig.4 Detectability of a synchronous fiber-optic gravitational telescope and gravitational flux spectrum levels. This graph of ball-park sensitivity is based on ignoring all noise sources other than quantum noise

A typical, but rather rare gravitational 'event' (1 or more per 30 years) would be the conversion of 1 solar mass (1.99×10^{30}kg) into gravitational energy by the formation of a Black Hole at the center of our Galaxy, for which d = 3.1×10^{20} m, $\tau_{GW} \approx 10^{-2}$s and $\Delta\upsilon \approx 100$ Hz [4]. This could produce an isotropic pulse of gravitational radiation, at a center-frequency of a few kHz, and with a GSL ≈ +50dB re 1W/m^2.Hz (=10^5 W/m^2.Hz). The space-time strain produced by this intensity may be obtained from (7) and is about 10^{-19}.

A much more common'event' (10 or more per year) would be the conversion of 0.1 solar mass (1.99×10^{29} kg) into gravitational energy by Supernovae in the Virgo Cluster of Galaxies for which d = 3.4×10^{23}m, $\tau_{GW} \approx 10^{-2}$s and $\Delta\upsilon \approx 100$ Hz [4]. This could produce an isotropic pulse of gravitational radiation, again at a center frequency of a few kHz, and with a GSL ≈ -20dB re 1W/m^2.Hz (=$10^{-2}W/m^2$.Hz). The space-time strain produced by this intensity may also be obtained from (7) and is about 4×10^{-23}. The GSLs corresponding to these 'events' have been indicated on the right-hand ordinate of Fig.4.

Even though these flux intensities are very high in relation to electromagnetic intensities, we have difficulty in detecting them because our detectors are very inefficient. Gravitational waves pass right through the earth with negligible attenuation [4].

The lowest optical fiber attenuation reported is 0.2dB/km at $\lambda = 1.55\ \mu m$ [14]. Substantial further reductions in α over this wavelength region are unlikely, as the limit set by Rayleigh scattering has been reached. Non-linear and stimulated scattering effects, as discussed elsewhere in these proceedings, will restrict the launched optical power to below 10mW at these wavelengths [2, 15].

It may become possible to construct in-line low-noise quantum amplifiers so that the overall optical attenuation around the loop becomes very low, thus allowing the TGSL to approach the fictitious 0dB/km line shown on the graph of Fig.4. Note that the rising TGSL values to the right of the minima indicate that more than one turn of fiber is required to optimize the gravitational detectability. To the left of the minima (sensitivity maxima) the sensitivity is limited by optical attenuation, i.e. the telescope contains too much fiber. With very low attenuation fibers and/or quantum amplifiers, recirculating techniques could be employed for long interaction times [2, 16].

3.4 Thermal Phase Noise

The 'thermal phase noise' caused by the thermal jitter of the optical fiber molecules could limit the gravitational detectability. This quantum-mechanical noise source has not been experimentally confirmed, but one theory (considering only longitudinal thermal phonons and ignoring transverse effects) indicates that the mean-squared level of this noise is given by

$$\Delta\phi_N^2(f) \simeq \left(\frac{n}{\lambda f_g}\right)^2 \frac{8kT}{\rho AL} \quad \text{rad.}^2/\text{Hz} \tag{15}$$

where $\Delta\phi_N^2(f)$ = phase noise spectral density, k = Boltzmann's constant (1.38×10^{-23} J/°K), T = environmental temperature (300 °K), ρ = density (2.2×10^3 kg/m³) and A = cross-sectional area of the fiber (7.85×10^{-9} m² for a 100 μm diameter fiber). Since the recovered signal power is proportional to $\Delta\phi^2$, i.e. $\propto L^4$, the signal-to-phase-noise ratio is proportional to L^5.

Again, substituting the above values into (15) for a single turn 10kHz antenna, we find that $\Delta\phi_N^2(f) \simeq 5.7 \times 10^{-16}$ rad.²/Hz. The Gravitational Spectrum Level equivalent to this thermal phase noise may be found by taking the ratio between $\Delta\phi_N^2(f)$ and $\Delta\phi^2$. At 10kHz the room temperature GSL equivalent is at +47dB re 1W/m².Hz, and has a positive slope of 50dB/decade.

Clearly, the thermal phase noise may limit the gravitational detectability at 10kHz. The thermal phase noise can of course be reduced by operating the fiber-optic gravitational telescope at cryogenic temperatures, though the physical and optical performance of optical fibers at or near absolute zero temperatures is unknown.

Other expressions have been developed for predicting the level of this noise source [17, 18] but there remains considerable uncertainty as to the magnitude of this effect, particularly in relation to the Sagnac interferometer configuration.

4. Discussion and Conclusions

There is little doubt that a gravitational telescope mounted in a terrestrial environment would require the fiber to be enclosed in a cooled evacuated toroidal tube, since optical fibers are both very microphonic and temperature sensitive. Ideally, the telescope should be mounted in a space environment

One particular advantage of the fiber-optic gravitational-wave detector over many of its mechanical counterparts is its relatively low Q and high bandwidth. A single turn telescope should have a bandwidth of about ±20%, although this would decrease as more fiber turns are employed. Fig.5 shows a fiber-optic gravitational-wave antenna using the basic Sagnac interferometer configuration and a toroidal winding technique. Helical toroidal windings of differing pitch would allow for center-frequency retuning by the simple expedient of switching different fibers into the optical interferometer circuit. The penalty we pay for using this slow-wave construction technique is that for every factor of ten by which we shrink the resonant size of the telescope, the TGSL will increase by 20dB (assuming that the total length of fiber remains the same).

A fiber-optic gravitational-wave detector several kilometers in diameter would not be impossible to construct, though it would of course be expensive. Such a structure would have a diameter comparable to the 2km diameter 400 GeV Proton Synchrotron at CERN, Geneva. Indeed, the possibility exists that the latter could be used to house the fiber-optic antenna. There are even plans to build a new ring at CERN which will be 11 km in diameter.

The above first-order analysis indicates that a non-mechanical gravitational-wave detector could be constructed from a fiber-optic waveguide, with a sensitivity that might allow for the detection of gravitational 'events' at the Galactic-center. Future technological developments, e.g. very low loss far

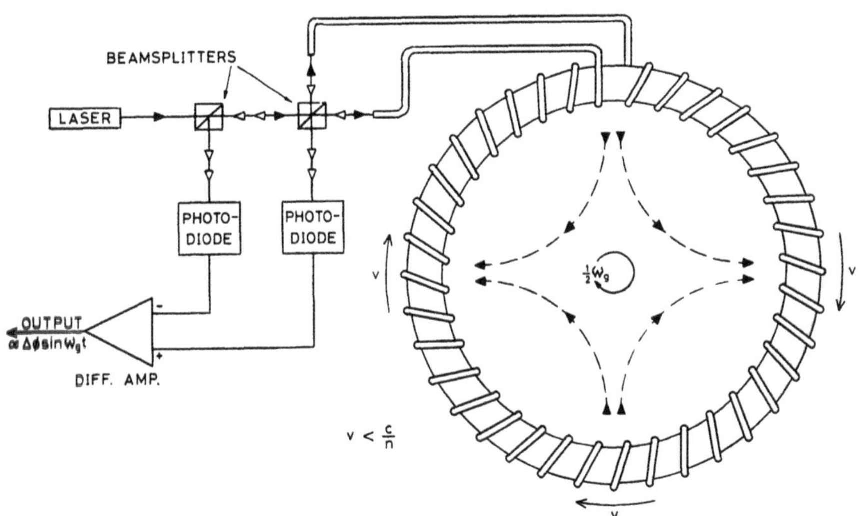

Fig.5 Fiber-optic gravitational telescope with 'slow-wave' toroidal construction and Sagnac interferometer

infra-red fibers [19] and low-noise quantum amplifiers may allow the detection threshold to be substantially reduced, with the resulting possibility that Supernova explosions in the Virgo group of Galaxies might then become detectable.

Note that in this first-order determination of sensitivity we have completely neglected many of the problems (discussed elsewhere in these proceedings) that severely limit the realisable sensitivity of fiber-optic gyroscopes at this time, e.g. non-reciprocal noise phenomena caused by the polarization degeneracy of the HE_{11} mode. The 10kHz fiber-optic gravitational telescope discussed here, is of course a very sensitive gyroscope with a sensitivity to earth rotation rate of about 1300 radians.

At the present time the most promising approaches to wide-band gravitational-wave detection are based on either laser ranging between drag-free satellites [7] in space and folded-cavity (FABRY-PEROT) Michelson interferometers [3], such as those being constructed at Caltec, MIT and Glasgow University. Whereas present-day optical fibers have severe power restrictions, limiting launched optical powers to the milliwatt level, these interferometers will employ laser beam powers of several watts. Presently, the Michelson interferometers under construction have arm lengths of about 10 meters, but there are plans for arms up to 10km in length giving sensitivities approaching 10^{-22} strain and thus allowing low-frequency flux levels below $10mW/m^2$ to be detectable. One way to overcome the power limitations of the fiber-optic approach to gravitational-electromagnetic resonance is to replace the fiber by a ring system of heliocentric satellites [7].

One thing that we can conclude from the above analysis, is that there is no danger of gravitational-waves interfering with the workings of a fiber-optic gyroscope.

5. Acknowledgements

This work was done while the author was with the Department of Electronic and Electrical Engineering, University College London, and was supported by the UK Science & Engineering Research Council. The author wishes to thank Professor D. E. N. Davies for his encouragement, and Dr. M. F. Fiddy for his helpful comments during the conceptual stages of this idea.

6. References

1. B. Culshaw, D. E. N. Davies and S. A. Kingsley: 'Fiber-optic strain, pressure and temperature sensors', Proceedings of the 4th European Conference on Optical Communication, 12-15 Sept. 1978, Genoa, Italy, pp. 115-126

2. 'Laser Inertial Rotation Sensors', Proceedings of SPIE, 157, 1978

3. P. C. W. Davies: "The search for gravity waves', (Cambridge University Press, Cambridge, 1980)

4. C. W. Misner, K. S. Thorne and J. A. Wheeler: 'Gravitation', (W. H. Freeman and Company, San Francisco, 1973)

5. J. A. Tyson and R. P. Giffard: 'Gravitational-wave astronomy', Ann. Rev. Astron. Astrophys. 16, 1978, pp. 521-554

6. J. Hough and R. Drever: 'Gravitational waves-a tough challenge', New Scientist, 17 August 1978, pp. 464-467

7. V. B. Braginsky and A. B. Manukin: 'Measurement of weak forces in physics experiments', (University of Chicago Press, Chicago, 1977)

8. V. B. Braginsky and M. B. Menskii: 'High-frequency detection of gravitational waves', Sov. Phys. JETP Lett. 13, 11, 5 June 1971, pp. 417-419

9. V. B. Braginsky, L. P. Grishchuk, A. G. Doroshkevich, Ya. B. Zel-Dovich, I. D. Novikov and M. V. Sazhin: 'Electromagnetic detectors of gravitational waves', Sov. Phys. JETP 38, 5 May 1974, pp. 865-869

10. J. P. Turneaure and I. Weissman: 'Microwave surface resistance of superconducting Niobium', Jour. App. Phys. 39, 9 August 1968, pp. 4417-4427

11. V. E. Lyamov and V. N. Rudenko: 'An acoustic heterodyne detector of gravitational radiation', Sov. Phys. JETP 40, 5, 1975, pp. 787-791

12. S. A. Kingsley: 'Fiber-optic gravitational telescopes', Internal Report, Department of Electronic and Electrical Engineering, University College London, March 14th 1979

13. W. K. Pratt: 'Laser communication systems', (John Wiley & Sons, New York, 1969)

14. T. Miya, Y. Terunuma, T. Hosaka and T. Miyashita: "Ultimate low-loss single-mode fiber at 1.55 μm', Electron. Lett. 15, 4, 15 February 1979, pp. 106-108

15. S. C. Lin and T. G. Giallorenzi: 'Sensitivity analysis of the Sagnac-effect optical fiber ring interferometer', Applied Optics 18, 6, 15 March 1979, pp 915-931

16. H. Arditty, H. J. Shaw, M. Chodorow and R. Kompfner: 'Re-entrant fiberoptic approach to rotation sensing', Proceedings of SPIE 157, Laser Inertial Rotation Sensors, August 1978, pp. 138-148

17. B. Culshaw and S. A. Kingsley: 'Thermal phase noise in coherent optical fiber systems', Electron. Lett. 16, 3, 31 January 1980, pp. 97-99

18. M. R. Seiler and E. R. Leach: 'Analysis of thermal phase noise in fiber-optic sensors', Battelle Columbus Labs Report No. R-6109, DARPA, August 1981, Unpublished

19. S. Mitachi, T. Miyashita and T. Kanamori: 'Fluoride-glass-cladded optical fibers for mid-infra-red ray transmission', Electron. Lett. 17, No. 17, 20 August 1981, pp. 591-592

Part 7

Related Fiber-Optic Sensors

Geometrical Fiber Configuration for Isolators and Magnetometers

R.A. Bergh, H.C. Lefèvre, and H.J. Shaw

Edward L. Ginzton Laboratory, Stanford University
Stanford, CA 94305, USA

1. Introduction

Due to the Faraday effect, the presence of a longitudinal magnetic field $\vec{B}$ modifies the phase of a circularly polarized light wave, by an amount determined by the Verdet coefficient of the medium. The sign of this phase shift depends on the left or right handed character of the polarization, and also on the relative senses of the field and light propagation vectors. The phase shift may be detected in a ring interferometer where identical circularly polarized waves counterpropagate around the ring, or it may manifest itself as a change in the orientation of linearly polarized light resulting from the opposite phase shifts of its co-propagating left and right hand circularly polarized components.

The phase shift per unit length in presently available optical fibers is small due to the small Verdet coefficient of silica, but the overall phase shift may be enhanced through efficient use of the length of the optical path. If the fiber is free of linear birefringence, the total Faraday effect along a given path is proportional to the vectorial integration of $\vec{B}$ along the path. For a closed path, the result is different from zero by Ampere's law only if the path encloses an electrical current. A toroidal closed path configuration has been used to form a current driven optical isolator [1] and an electric current sensor [2].

2. Basic Structure

We present a new fiber configuration using bend induced linear birefringence to make a Faraday cell which has high sensitivity to a uniform magnetic field. It allows one to use available permanent magnets or electromagnets to form optical isolators, or to detect environmental magnetic fields with high sensitivity using a ring interferometer.

A uniform magnetic field $\vec{B}$ applied to a single mode fiber rotates the direction of a linearly polarized mode through an angle $\theta = V \cdot \vec{B} \cdot \vec{L}$

where $\vec{L}$ is the length vector associated with the optical path and V is the Verdet coefficient. Since V is very small for quartz ($V = 3$ rad $\cdot$ $m^{-1} \cdot T^{-1}$ for $\lambda = 633$ nm) long lengths of fiber are required to produce appreciable angles θ, and this in turn presents two problems, one concerned with how to coil up the fiber into a small space and the other associated with birefringence in the fiber.

Consider first a case in which there is no birefringence in any portions of the fiber. If we wrap the fiber into any closed path, Ampere's law specifies that the line integral of $\vec{B}$ around that path will be zero, and thus there will be no net rotation θ produced unless the path encloses a current (Fig.1). A current driven fiber isolator has been described [1] in which a multiturn toroidal solenoid is wound around a multiturn fiber loop. If we wish to avoid threading conductors through the fiber loop, we must alter the polarization of the light as a function of distance along the path in order to produce a net angle θ.

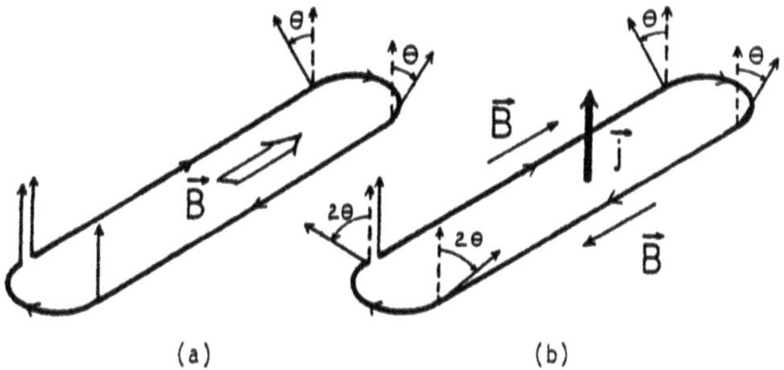

Fig.1 (a) Closed fiber loop in a uniform field $\vec{B}$
(b) Closed fiber loop which encloses electrical current

Consider light propagating around the path of Fig.1 a consisting of parallel equal-length straight sections of fiber joined at their ends by curved sections, all immersed in a uniform $\vec{B}$ field, and again assume for the moment that there is no birefringence in any portion of the fiber. The rotation angle θ produced in one straight section will be cancelled in the next, because on traversing the curved section, the polarization vector is symmetrized (mirror symmetry) with respect to the saggital plane (Fig.1 a). To avoid this cancellation we now design the curved sections to produce a half-wave differential delay due to stress-induced birefringence resulting from the bending of the fiber. This results in a second symmetrization of the polarization vector, which just compensates for the previously described symmetrization and allows the θ values produced in successive straight sections to add (Fig.2).

Wrapping many straight lengths interconnected with such half wave loops can give a sizeable Faraday rotation even without too large a field. However, such a design requires a very precise half wave value for the birefringence of the loops. A better approach is to orient loops in orthogonal planes (Fig.3). In this way the fast and slow axes of the successive coils are interchanged and the net birefringence after an even number m of loops is near zero rather than $m\lambda/2$. This reduces the systematic error in the precision of the half wave loops and, in addition, increases the wavelength tolerance of the device.

3. Experimental Isolator

An experimental model was constructed with 32 straight segments each 12 cm in length. The half wave birefringence was obtained with one and one-half turn coils rather than the half turn loops indicated in Fig.3. A simple half turn would require a radius too small to avoid curvature loss. The proper radius of curvature is given by [3]

$$R = \frac{2\pi a\, r^2}{\lambda} \cdot N$$

where $a = 0.133$ for $\lambda = 633$ nm and N is the number of half turns. Applying a 0.09 T uniform field gives a rotation of 45 degrees of a linearly polarized light wave which is needed for an optical isolator. This result is consistent with the value of the Verdet constant. Parasitic effects produced a small ellipticity in the polarization (1/10 in power). A principal feature of this device is that it can be inserted into the field of convenient permanent magnets or electromagnets. To simplify the winding, it could be possible to have only half a turn with the same radius and to increase the birefringence by applying tension on the fiber [4].

To avoid parasitic effects due to the intrinsic birefringence of the fiber we used a low birefringence spun fiber [5].

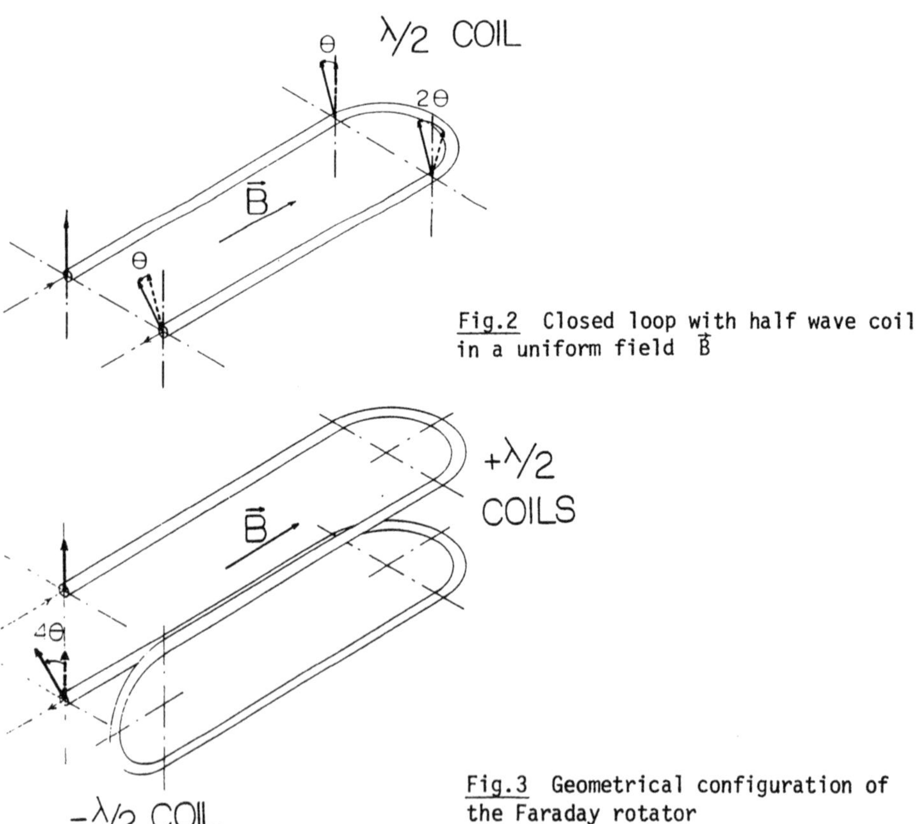

Fig.2 Closed loop with half wave coil in a uniform field $\vec{B}$

Fig.3 Geometrical configuration of the Faraday rotator

4. Experimental Magnetometer

Connecting this Faraday probe into a fiber ring interferometer (Fig.4) allows one to use the technology developed for fiber gyroscopes to measure very small non-reciprocal phase shifts. The important notion of reciprocity still applies and a polarizer is required at the common input-output port of the system. The use of all fiber components [6] improves the stability and the compactness of the system. A coupler splits and recombines the interfering counter-propagating waves, and two $\lambda/4$ coils oriented at 45 degrees transform the input linear polarization into circular polarization in the straight sections of the probe. In this case a longitudinal field does not give a polarization change but produces a phase shift $\Delta\phi_F$ between the counterpropagating waves. The value of $\Delta\phi_F$ is twice the angle of linear polarization rotation created by the same field. It is again important to orient successive end loops in mutually orthogonal planes.

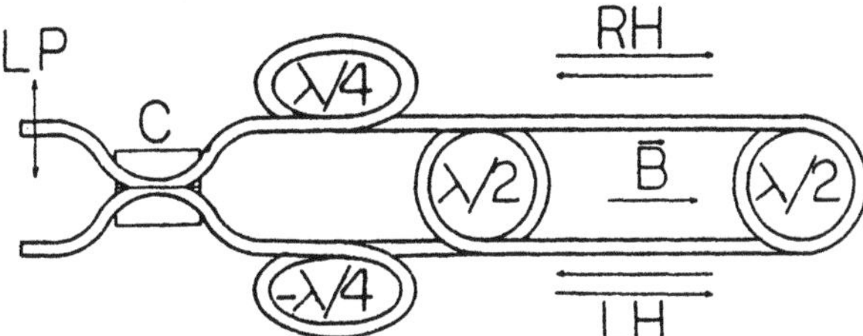

Fig.4 Magnetometer configuration

To test the feasibility of such a system, we connected the Faraday rotator described in the previous section in series with the rotation sensing loop of an all-fiber gyroscope [7]. Coupling was accomplished by butting fiber ends together (Fig.5). The source is a solid state GaAs laser. Two directional couplers DC_1 and DC_2 and a polarizer P are used to assure reci-

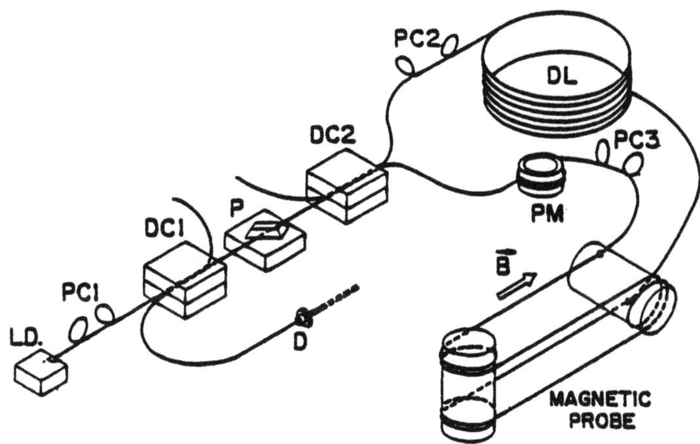

Fig.5 Magnetometer experimental set-up

procity. Polarization controllers PC_1, PC_2, and PC_3 allow light to pass efficiently through the polarizer and provide circularly polarized light in the magnetic probe. The rotation sensing loop serves as a delay line, allowing the use of a phase modulation technique [7] to bias the signal. The optimum modulation frequency is $1/2\tau$, where τ is the group transit time through the fiber delay line and probe. A short fiber would require a frequency too high for efficient use of the piezoelectric ceramic phase modulator PM.

The rms value of the noise is in this system equivalent to 8×10^{-7} T with 3 sec integration time (Fig.6). The use of a longer fiber would improve the result. With 500 meters the scale factor would be $\Delta\phi_F$ (rad) = $2 \times 10^3 \times$ B (Tesla). A phase sensitivity of 10^{-7} rad would allow one to detect 5×10^{-11} T (i.e., 5×10^{-7} G). Sophisticated cryogenic magnetometers have sensitivities of this order of magnitude. The use of a magnetostrictive jacket on the fiber of a Mach Zender interferometer was recently demonstrated and predictions of sensitivities of 10^{-11} G have been made. However, such systems respond only to time varying magnetic fields, while the present device operates on time-independent fields as well.

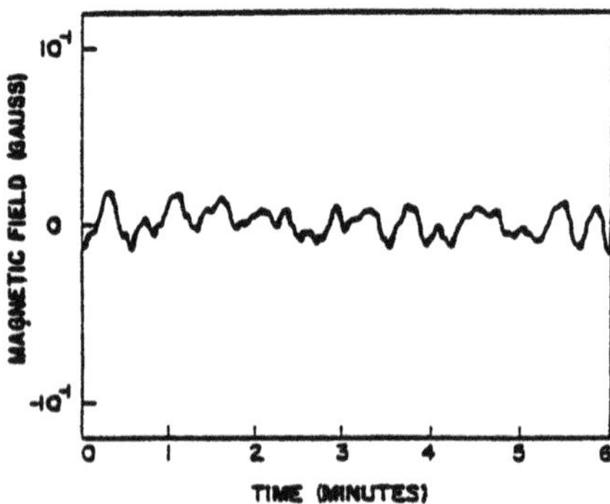

Fig.6 Output signal of the magnetometer

This all-fiber DC magnetometer appears to be an interesting device to investigate. It would perturb the field to be measured only slightly since the magnetic permeability of silica is small. The sensitivity might be improved with a dopant which could increase the value of the Verdet coefficient without introducing excessive loss.

References

1 T. K. Findakly, "Single Mode-Fiber Optic Isolator", Paper TuF5, Integrated Optics and Optical Fiber Communication (1981)

2 H. J. Arditty, Y. Bourbin, M. Papuchon and C. Puech, "Current Sensor using State-of-the-Art Fiber-Optic Interferometric Techniques", Paper WL3, Integrated Optics and Optical Fiber Communication (1981)

3 H. C. Lefevre, Electronics Letters 16, 778-779 (1980)

4 S. C. Rasleigh and R. Ulrich, Optics Letters 5, 354-356 (1980)

5 A. J. Barlow, J. J. Ramshov-Hansen and D. N. Payne, Applied Optics 20, 2962-2968 (1981)

6 R. A. Bergh, M. F. Digonnet, H. C. Lefevre, S. A. Newton and H. J. Shaw, "Single Mode Fiber Optic Components", same Proceedings

7 R. A. Bergh, H. C. Lefevre and H. J. Shaw, "All Single Mode Fiber Optic Gyroscope", same Proceedings

This research was supported by the Atlantic Richfield Co. We would also like to acknowledge D. N. Payne (University of Southampton, UK) who kindly provided the low birefringence spun fiber.

Fiber Optic Strain Sensors*

E. Snitzer, J.R. Dunphy, and G. Meltz

United Technologies Research Center
East Hartford, CT 06108, USA

Some theoretical and experimental results are presented for the temperature, wavelength, and strain dependence for strain sensors based on modifying the cross talk between closely spaced, identical single mode cores in a common cladding. The strains considered are axially symmetric. It is shown that by proper selection of geometric and materials properties, strain gauges can be made which are temperature independent.

The first analytical treatment of cross talk in coupled parallel waveguides came from the microwave literature [1] in which an exact solution was obtained for dielectric sheets and an approximate one for circular cylindrical cores. The asymmetric three layer slab and the effect of strain on the beat length in this structure were later treated exactly [2]. Since then, many studies have been made of sheets and cylinders using coupled mode theory [3-5], of lossy coupled guides [6], and of dissimilar cylinders based on exact formulations [7-9]. These analytical treatments have been concerned with the field distribution in multicore fibers and the beat length or rate at which energy is exchanged between sheets or cores that guide the light in these structures. The first experimental observations of cross talk in multimode cores were made in the early work on visible light propagation in fused fiber bundles [10]. Later it was qualitatively demonstrated that cross talk could be temperature sensitive [11].

The theoretical expressions derived for cross talk sensors use a scalar coupled mode formulation to evaluate the effects of various perturbations on the light distribution in the cores. Recent exact calculations [9] based on an integral representation technique show that coupled-mode theory provides good accuracy in most cases. The coupled-mode results provide simple

*This work has been supported in part by the Naval Avionics Center, Indianapolis, Indiana.

formulas that can be used to select material and geometrical parameters for the design of fiber optic longitudinal strain sensors. Using these design relationships, it is possible to evaluate the effect of core size, position and shape factors on measurement sensitivity. The theory has also been applied to cross talk between inhomogeneous cores, because in some of the early twin-core fibers core-cladding diffusion had changed the core index profile from a uniform step to a Gaussian-like distribution.

The exchange of energy between cores can be analyzed in terms of modal interference. To a very good approximation, the twin-core normal modes are linear combinations of the lowest order HE_{11} single core excitations. There are two orthogonally polarized, symmetric and anti-symmetric pairs of HE_{11} modes. Illumination of a single core is equivalent to the excitation of a pair of normal modes, namely, a symmetric and anti-symmetric combination with the same polarization. Complete energy exchange from the illuminated core and back takes place in a beat length λ_b. The net effect is a change in beat phase for the exit light from the fibers, which changes the ratio of light intensities emerging from the two cores, with beat phase $\phi \equiv \pi L/\lambda_b$.

Coupled mode theory [5] can be used to derive an expression for the beat length in terms of a field overlap integral, which is a measure of the interaction between the individual single core modes. The geometry for the twin cores is shown in Fig.1.

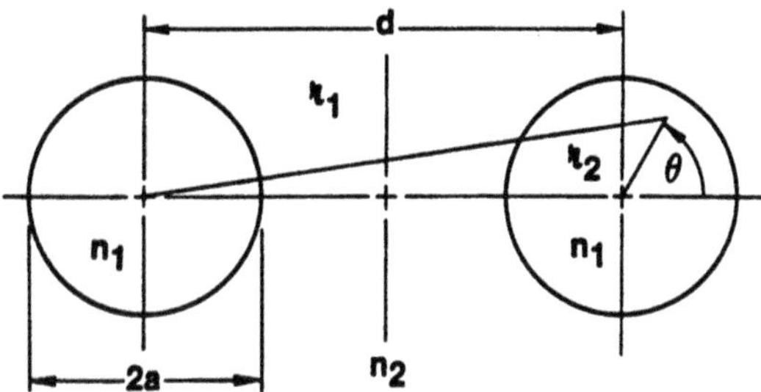

Fig.1 Twin core geometry for calculation of the coupling factor

$$\pi/\lambda_b = (NA^2/\lambda n_1)(1/a^2)\int_{A_2} \delta(r_2^2/a^2)\ \psi_1\ (r_1)\ \psi_2\ (r_2)\ r_2 dr_2 d\Theta \qquad (1)$$

where the radial variation of the refraction index is described by the profile function

$$\delta(r_2^2/a^2) = [n^2(r^2/a^2)-n_2^2]/NA^2 \qquad (2)$$

with $NA^2 \equiv n^2(0) - n_2^2$. The functions ψ_1 and ψ_2 are proportional to the principal components of the transverse electric field of a single core mode, centered on core 1 or core 2, respectively. It is convenient to introduce a coupling factor F(V; d/a) defined by

$$F(V;d/a) = V/2\pi \int_{A_2} \delta\ \psi_1\psi_2(r_1/a)d(r_2/a)d\theta. \qquad (3)$$

Eq. (3) can be evaluated in closed form for both step index and Gaussian profile cores. The coupling factor for these two cores is shown in Table 1.

Table 1 Coupling factor

Core profile	$\delta(r_2^2/a^2)$	$F(V; d/a)$, where $V = 2\pi/\lambda \cdot a \cdot NA$
Step	$1, r/a \leq 1$	$(U^2/V^3)\, K_0\, (Wd/a)/K_1^2\, (W)$
	$0, r/a > 1$	$W = (V^2 - U^2)^{1/2}$
		$U = (1 + \sqrt{2})V/[1+(4+V^4)^{1/4}]$
Gaussian	$\exp(-r^2/a^2)$	$(V-1)V^3/(V+1)^2 \exp[(V-1)^2/V+1]$
		$\cdot K_0 \left\{(V-1)d/a\right\}$

The influence of one core on another can be shown in another way by plotting the phase refractive indices (Fig.2) of the HE_{11} mode, and the symmetric and anti-symmetric twin-core modes as a function of V. The presence of an adjacent core splits the HE_{11} mode into two other modes with slightly larger and smaller propagation constants. The splitting is greater for smaller V values and closer spaced cores. As V is increased, for example by decreasing the wavelength λ, the fields become more tightly confined to the cores and thus are influenced less by an adjacent core. As a result, for large V, the phase velocities of the normal modes are nearly the same as the HE_{11} mode. To obtain cross talk the V value must be greater than the cut-off for the anti-symmetric combination of HE_{11} modes, but V should be less than 2.405 to avoid propagation of the next higher order modes (TE_{01}, TM_{01}, and HE_{21}).

Eqs. (1) and (3) can be combined to give the more useful form

$$\lambda_b/\lambda = 1/2 \cdot n_1/NA^2 \cdot V/F \quad . \tag{4}$$

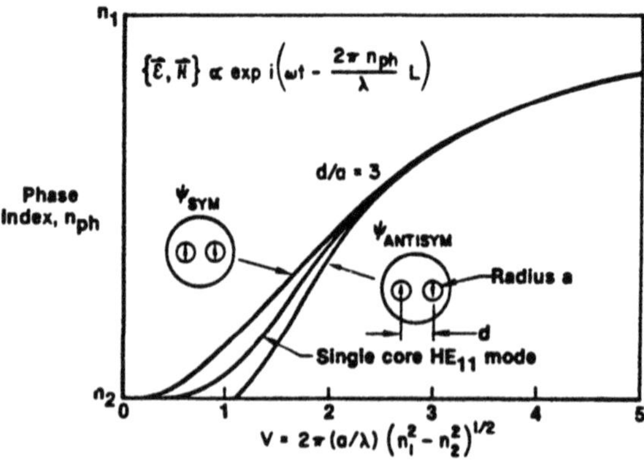

Fig.2 Phase refractive vs. V parameter

Typical values of λ_b/λ for various core separations can be read from Fig.3 for both step index and Gaussian core profiles. The beat length can be reduced to a fraction of a millimeter by an appropriate selection of glasses, core size and spacing.

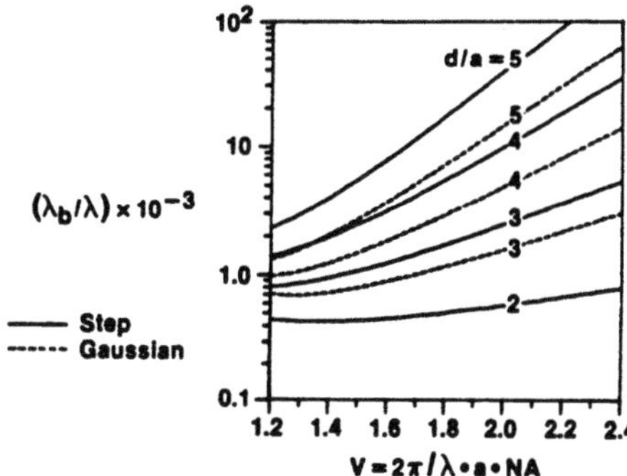

Fig.3 Beat length vs. V for various core separations $n_1=1.5$ NA=0.15

The fractional change in beat phase $\Delta\phi/\phi$ due to a cylindrically symmetric elastic deformation is given by:

$$\frac{\Delta\phi}{\phi} = \varepsilon_z - \varepsilon_r + \frac{n_2^2}{n_1^2-n_2^2}\left(\frac{\Delta n_1}{n_1} - \frac{\Delta n_2}{n_2}\right) + \frac{V}{F}\frac{dF}{dV}\left[\varepsilon_r + \frac{\Delta n_1}{n_1} + \frac{n_2^2}{n_1^2-n_2^2}\left(\frac{\Delta n_1}{n_1} - \frac{\Delta n_2}{n_2}\right)\right] \quad (5)$$

where ε_z and ε_r are the longitudinal and radial strains, and $\Delta n_1/n_1$ and $\Delta n_2/n_2$ are the changes in the indices of refraction of the cores and cladding, respectively, due to the elastic deformation. This equation neglects any localized stresses in the vicinity of the cores due to a mismatch in elastic properties of the materials. For simplicity in later calculations, assume the cores and cladding have the same thermal expansion coefficient α, Young's modulus E, and Poisson's ratio ν.

The index of refraction for polarization in the x-direction and propagation in the z-direction is a linear function of the three principal strains; it is dependent on polarization but not the direction of propagation [12], i.e.,

$$\Delta n_x/n = -(n^2/2)\left[p_{11}\,\varepsilon_x + p_{12}\,(\varepsilon_y + \varepsilon_z)\right] \quad . \quad (6)$$

For a force f applied in the z-direction on a fiber of cross sectional area A, the strains are $\varepsilon_z = f/EA$ and $\varepsilon_r = \varepsilon_x = \varepsilon_y = -\nu f/EA$. It is convenient to define an effective photoelastic constant

$$p_e = (n^2/2)\left[p_{12} - \nu(p_{11} + p_{12})\right] \quad .$$

For a fiber with $\Delta n_1/n_1 = \Delta n_2/n_2$ and using (6) and (7), (5) reduces to

$$\Delta\phi = (\pi L/\lambda\hat{b})\left[(1+\nu) - (\nu + p_e)(V/F)\,dF/dV\right] f/EA \quad . \quad (8)$$

Typical values for alkali, alkaline earth silicate glasses are $\nu = 0.2$, $E = 10^7$ psi and $p_e = 0.2$.

For uniform hydrostatic pressure P applied to a long fiber $\varepsilon_z = \varepsilon_r = -(1-2\nu)\ P/E$. For equal photoelastic responses of the cores and cladding,

$$\Delta n_1/n_1 = \Delta n_2/n_2 = (n_1^2/2)\ (p_{11} + 2p_{12})\ (1-2\nu)\ P/E \tag{9}$$

and the change in beat phase is

$$\Delta\phi/\phi = -(V/F)\ dF/dV \left[1 - (n_1^2/2)(p_{11} + p_{12})\right] (1-2\nu)\ P/E \quad . \tag{10}$$

The temperature dependence of the beat phase is

$$\Delta\phi/\phi = \left\{ \frac{n_2^2}{n_1^2-n_2^2} (\zeta_1-\zeta_2) + V/F\ dF/dV\ [\alpha+\zeta_1 + \frac{n_2^2}{n_1^2-n_2^2} (\zeta_1-\zeta_2)] \right\} \Delta T \tag{11}$$

where ζ = thermo-optic coefficient = $1/n\ dn/dT$. For $\zeta_1 = \zeta_2$, the sensitivity to temperature variation simplifies to

$$\frac{d\phi}{dT} = (L/a)(NA/n_1)(\alpha+\zeta)\, V\, \frac{dF}{dV} \quad . \tag{12}$$

This is plotted for one special core-cladding combination for various V and d/a ratios in Fig.4. These curves are interesting because they are proportional to $V\,\frac{dF}{dV}$.

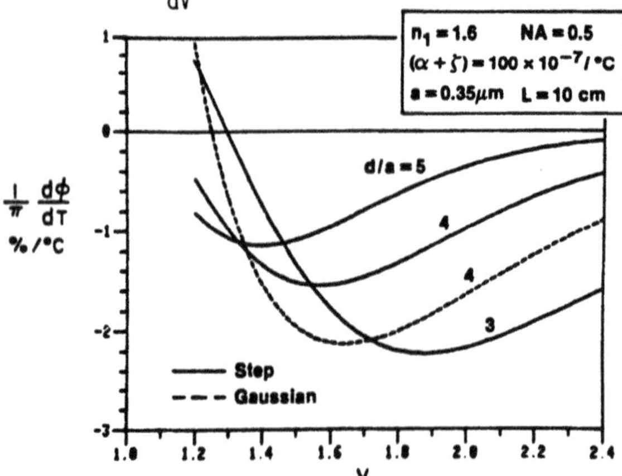

Fig.4 Temperature dependence of beat phase

The wavelength dependence of the beat phase can be derived from (4).

$$\Delta\phi/\Delta\lambda = -\pi L\ \lambda_b^{-2}\ \partial\lambda_b/\partial\lambda = -2\pi\ \ \lambda^{-2}\ NA^2\ n_1^{-1}\ dF/dV \quad . \tag{13}$$

With the approximations made above for identical elastic properties for the core and cladding, temperature independent beat phase results when

$$dF/dV = 0. \tag{14}$$

It is interesting that this is also the condition for the beat phase to be independent of wavelength or uniform hydrostatic pressure, but not of axial tensile loading. To obtain different conditions at which the various dependencies of the beat phase go to zero, the materials of the cores and cladding can have different elastic properties or fibers can be fabricated with multiple cladding which have different expansion coefficients.

Figure 5 gives the experimentally observed change in beat phase due to axial stretching of the fiber. The irregularities in the curve are due primarily to friction in the spring-loaded slider on which a microscope objective was mounted so as to observe the change in light intensities existing from the cores while tension was applied to the fiber. The 25 cm length of fiber can be used for an unambiguous measurement of an axial force up to 90 grams or approximately 0.2 lb. For the 250 μm fiber O.D. this corresponds to an axial stress of 3,000 psi, and with Young's modulus of 10^7 psi a strain of 300 microstrains. Photon shot noise limited measurement of the exit light intensity could in principle permit a measurement of 10^{-8} of the maximum change in beat phase. Since the sensitivity is inversely proportional to fiber length, for a fiber length of 1 mm, the limiting strain that could be detected for the fiber shown in Fig.5 would be 0.75 X 10^{-3} microstrain.

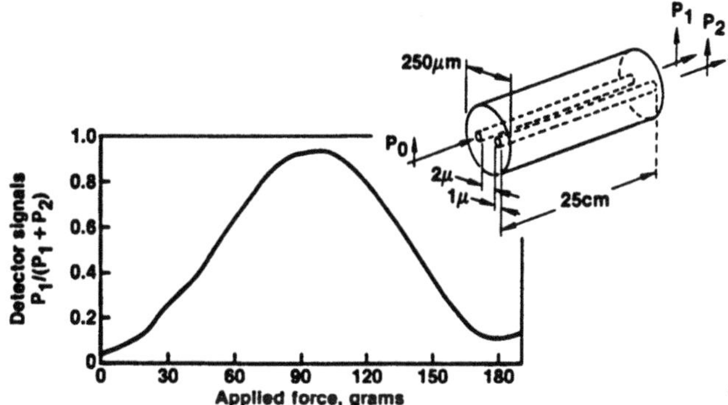

Fig.5 Fiber optic strain sensor $n_1=1.62$ $n_2=1.52$

Work is now in progress on the wavelength and temperature dependence. Preliminary results do indicate that they have a minimum influence on the beat phase at the same V value.

The authors wish to acknowledge the help of W. W. Morey, A. L. Wilson, and L. J. Muldrew in preparing and characterizing the fibers and J. Ryan for technical assistance in making the cross talk measurements.

References

1 M. F. Bracey, A. L. Cullen, B. F. F. Gillespie and J. A. Staniforth, IRE Transactions on Antennas and Propagation Vol. 7, p. 219 (1959)
2 J. Kane, IEEE Transactions on Geoscience Electronics GE-4, p. 1 (1966)
3 A. L. Jones, J. Opt. Soc. Am. Vol. 55, p. 261 (1965)
4 R. Van Clooster and P. Pariseau Physica Vol. 47, p. 498 (1970)
5 A. W. Snyder, J. Opt. Soc. Am. Vol. 62, p. 1267 (1972)

6 D. Marcuse, Bell Sys. Tech. J. Vol. 50, p. 1791 (1971)
7 W. Wijngaard, J. Opt. Soc. Am. Vol. 63, p. 944 (1973)
8 L. Eyges, "Integral Representation Techniques for Calculating Modes of Coupled Electro-Magnetic Dielectric Guides", Acoustic, Electromagnetic and Elastic Wave Scattering-Forces on T-Matrix Approach, Pergamon Press, 1979
9 L. Eyges and P. Wintersteiner, "Modes of an Array of Dielectric Waveguides", to be published
10 W. Hicks and W. L. Hyde, private conversation to E. S. (1959)
11 N. S. Kapany and J. J. Burke, "Optical Waveguides", p. 254, Academic Press, New York City (1972)
12 D. A. Pinnow, p. 478 in Handbook of Lasers, Edited by R. J. Pressley, Chemical Rubber Company, Cleveland, Ohio (1971)

Fiber Optic Sonar Sensor

M. Rudd

Bolt Beranek and Newman Inc., Cambridge, MA 02138, USA

Abstract

BBN has been working with the Naval Research Laboratory to develop a Fiber Optic Sonar System. The acoustic sensing element is 10-100 meters of single mode optical fiber. The acoustic pressure modulates the length and refractive index of the fiber, and hence the optical path length. The fiber is made one arm of a Mach-Zehnder Interferometer and is illuminated by a gas or solid-state laser. Phase modulations as small as 10^{-6} rads/$Hz^{.5}$ can be detected. Sensitivity can be enhanced by mounting the fiber on a compliant base; the fiber then senses the strain in its support. Strains as small as 10^{-14} are detected. This corresponds to a sound pressure of 10^{-4} Newtons/m^2. This technique will not operate at very low frequencies (less than 20 Hz) or DC because of ambient temperature variations which, at low frequencies, will swamp acoustic pressures.

Fiber Optic Sonar System

1. *Introduction*

It was demonstrated several years ago (BUCARO and HICKMAN 1979), that optical fibers can be used to sense sound pressure directly. The pressure squeezes the fiber and causes both its density to increase and the length to decrease. If the fiber is made one arm of an interferometer, then the change in phase of light traveling through the fiber can be measured. The total phase, ϕ, of the light is

$$\phi = nk_0\ell \tag{1}$$

where n = refractive index of the fiber, k_0 = wavenumber in free space and ℓ = length of the fiber. Now

$$\frac{d\phi}{dp} = k_0 \left(\frac{\ell dn}{dp} + \frac{n d\ell}{dp}\right) \tag{2}$$

where p = acoustic pressure.

$$\text{Now } \frac{1}{\ell}\frac{d\ell}{dp} = -\frac{1}{3\rho}\frac{d\rho}{dp} = -\frac{K}{3\rho} \tag{3}$$

where ρ = density of fiber and K = compressibility. Further, from the Clausius-Mossotti relation,

$$\frac{dn}{dp} = \frac{(n^2-1)(n^2+2)}{6n}\frac{K}{\rho} \quad . \tag{4}$$

Hence

$$\frac{d\phi}{dp} = \frac{k_0 \ell K}{3\rho}\left(\frac{(n^2-1)(n^2+2)}{2n} - n\right) \tag{5}$$

$$\frac{d\phi}{dp} = \frac{k_0 \ell K}{6n\rho}(n^2+1)(n^2-2) \tag{6}$$

$$= k_0 \ell \times 4.5 \times 10^{-14} \text{ per dyne/cm}^2$$

for fused silica.

If we have 100 meters of fiber, $k_0 \ell = 10^9$ and

$$\frac{d\phi}{dp} = 4.5 \times 10^{-5} \text{ radian cm}^2\text{/dyne.}$$

Note that if $n^2 = 2$, there is zero sensitivity for the density changes and the length changes cancel. For silica $n^2 = 2.12$.

If we wind the silical fiber on a rubber mandrel which is much softer than glass, then this will be compressed by the sound pressure and drag the fiber with it. The fiber is then sensitive to the strain induced in its support rather than the pressure itself. Thus, if the mandrel (or for that matter the jacket of the fiber) is much more compliant than the fiber, then the second term in (2), $d\ell/dp$ is greatly increased and is not offset by the first term.

2. *Strain Sensitivity of Fiber*

The practical hydrophone operates by measuring the strain induced in its compliant support. This strain produces a fractional phase change $\Delta\phi/\phi$ given by

$$\frac{\Delta\phi}{\phi} = (\Delta\ell/\ell - \Delta v/v) = \varepsilon_2 + \Delta n/n \tag{7}$$

where ℓ = length of fiber, v = speed of light, ε_2 = strain along fiber, and n = refractive index. Now, changes in refractive index are related to the strain by the photoelastic tensor (YARIV, 1975),

$$\Delta\left(\frac{1}{n^2}\right)_{ij} = p_{ijk\ell}\varepsilon_{k\ell} \quad . \tag{8}$$

For a light wave traveling along the z = 3 axis, we are only interested in $\Delta(1/n^2)_{11}$ and $(1/n^2)_{22}$. Further, the transverse components of strain for the fiber are equal, $\varepsilon_{11} = \varepsilon_{22}$ and

$$\Delta(1/n^2)_{11} = \Delta(1/n^2)_{22} = -2\Delta n/n^3 . \tag{9}$$

Further, since unstressed silica is an isotropic material, there are only two independent non-zero photoelastic constants which we shall take as p_{11} and p_{44}, and a third, p_{12}, is related to them by $p_{12} = p_{11} - 2p_{44}$.

For silica, $p_{11} = 0.13$, $p_{44} = -0.075$, and $n = 1.46$ (BUDIANSKY, DRUCKER, KINO and RICE, 1979). These values yield,

$$\frac{\Delta\phi}{\phi} = \varepsilon_{33} - \frac{n^2}{2} [\varepsilon_{11}(p_{11}+p_{12}) + \varepsilon_{33}p_{12}) \ , \tag{10}$$

$$\frac{\Delta\phi}{\phi} = \varepsilon_{33} - \frac{n^2}{n} [2\varepsilon_{11}(p_{11}-p_{44}) + \varepsilon_{33}(p_{11}-2p_{44})] \tag{11}$$

$$= \varepsilon_{33} - 0.44\ \varepsilon_{11} - 0.30$$

$$= 0.70\varepsilon_{33} - 0.44\varepsilon_{11} \tag{12}$$

for silica. For simple axial strain, $\varepsilon_{11} = -\nu\varepsilon_{33}$, where ν = Poisson's ratio = 0.17 for silica. Thus, $\Delta\phi/\phi = 0.775\varepsilon_{33}$. (13)

In conclusion, the fractional phase change of the light in a silica fiber subject to axial strain is .775 times that strain. In turn that strain is Kp/3 where K = bulk modulus of the mandrel and p = acoustic pressure. For a mandrel with a compressibility of 10^{-10} cm^2/dyne, and $\phi = 10^9$ radians

$$\frac{d\phi}{dp} = 2.58 \times 10^{-2} \text{ radians/cm}^2\text{/dyne}$$

or 570 times the free fiber sensitivity.

3. Optical System

The optical system which BBN has used for the Fiber Optic Sonar System is shown in Fig. 1. It is a heterodyne interferometer.

The reason for this is to overcome problems due to differential thermal drift in the interferometer arms. The output photocurrent is given by

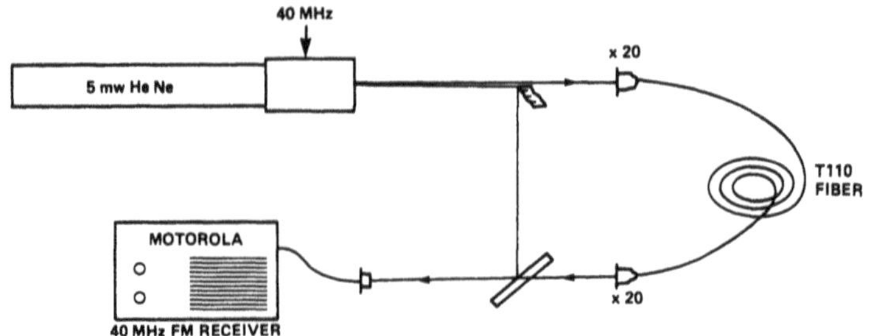

Fig.1 Heterodyne interferometer for fiber optic sonar system

$$I = \overline{[R\cos\omega t + S\cos((\omega+\omega_s)t + \phi)]^2} \tag{14}$$

where R = reference beam amplitude, S = signal beam amplitude, ω = optical frequency, ω_s = modulation frequency, ϕ = phase difference between interferometer arms and the overbar denotes an average over a time long compared with the optical period.

$$I = \frac{1}{2}(R^2+S^2) + 2RS\cos(\omega_s t + \phi) \; , \tag{15}$$

$$\frac{dI}{d\phi} = -\,2RS\sin(\omega_s t + \phi) \; . \tag{16}$$

In a homodyne interferometer, there is no frequency shift and $\omega_s = 0$. The sensitivity, $dI/d\phi$, is then proportional to $\sin\phi$. Now if we are concerned with small changes only, the signal from these changes will depend on the mean value of ϕ. If the mean value is zero, then there is no signal and if the mean value is 90°, then the sensitivity is a maximum. The mean value of ϕ continually drifts due to differential temperatures.

One solution to this is to incorporate an active phase shifting element to compensate for temperature changes. This would produce an optical phase locked loop which would hold the mean value of ϕ at 90°. Another approach is to shift the frequency of the signal beam by 40 MHz relative to the reference beam. This frequency shift is achieved by a Bragg cell, or Acousto-Optic Modulator. The output of the photo-diode is now a 40 MHz signal which is phase modulated.

The phase modulated signal is demodulated either by a phase locked loop or a frequency discriminator. Phase drift, due to differential temperature changes, are then tracked electrically instead of mechanically. A Motorola communications receiver was used for demodulation.

The system was set up in the BBN anechoic chamber to measure the acoustic sensitivity and directivity (Fig.2). Since the hydrophone senses pressure, it works equally well in water or air. The hydrophone can be seen mounted in the top left corner and the interferometer is mounted on the floor. The hydrophone consists of 16 meters of single mode optical fiber wound on a rubber mandrel. The sensitivity was compared with a condenser microphone mounted just above the hydrophone.

The sensitivity was very close to that predicted in the previous section. The directivity showed a 10 dB increase at endfire over a certain frequency range due to coincidence between the speed of sound in air and the bar wavespeed in the rubber.

Fig.2 Experimental arrangement: Note hydrophone at upper left

4. Vibration Sensitivity

One major problem to be overcome was the vibration sensitivity of the hydrophone. Since the optical fiber senses stress due to pressure so too can it sense stress due to vibration. If the fiber is subjected to an axial strain ε_z then from (13)

$$\frac{\Delta\varepsilon}{\phi} = .775\ \varepsilon_z\ . \tag{17}$$

If the mandrel is a cylinder accelerated at a rate a from one end, the mandrel is compressed with a strain ε_y, given by

$$\varepsilon_y = a\rho\ell/Eg \tag{18}$$

where ρ = density, E = Young's modulus of mandrel, ℓ = distance from free end and g = acceleration due to gravity. The circumferential strain, which the fiber sees is ε_z given by

$$\varepsilon_z = \nu_m \varepsilon_y \tag{19}$$

where ν_m = Poisson's ratio for the mandrel. The ratio of the strains induced by vibration and sound is

$$\frac{\nu_m \rho L}{2Eg} \frac{3}{K} \tag{20}$$

where L = total length of mandrel.
But K = $(1-2\nu_m)/E$.

Therefore the ratio is

$$\frac{3\nu_m \rho}{2(1-2\nu_m)g}$$

which is only a function of Poisson's ratio and the density of the mandrel. For rubbers, ν_m is close to 0.5 and this ratio is large (high vibration sensitivity). For metals ν is about .33 giving a ratio of 1.5 ρ/g or 4 μbar/g for an aluminum mandrel. This is comparable to the vibration sensitivity of piezo-ceramic hydrophones (SQUIER 1971).

A solution to this problem is the vibration compensation of the hydrophone. If the hydrophone is centrally supported, then when it is accelerated, one end will be under tension and the other under compression. The two strains will be opposite and approximately equal leading to cancellation. A reduction of 15 dB has been obtained in the vibration sensitivity in this manner.

Another solution, applicable to rubber mandrels, is to stiffen the mandrel axially with wire, so as to increase the effective Young's modulus without reducing the compressibility.

In practice the vibration sensitivity of fiber-optic hydrophones can be made as low or lower than piezo-ceramic hydrophones.

Conclusions

Fiber Optic Hydrophones operate by measuring the strain induced in a fiber mounted on a compliant mandrel or with a compliant jacket. The fiber is made one arm of an interferometer and phase changes of the light in the arm are monitored. Phase changes of 10^{-6} radians on a 1 Hz bandwidth can be observed. However, compensation for phase changes due to temperature variations must be made. This is done either by a phase locked loop or by heterodyne interferometer. With care the vibration sensitivity can be made less than for piezo-ceramic hydrophones.

References

J.A. Bucaro and T.R. Hickman, "Measurement of Sensitivity of Optical Fibers for Acoustic Detection", *Applied Optics* 18, pp. 938-940 (1979)

Amnon Yariv, Quantum Electronics (John Wiley & Sons, New York, 1975)

B. Budiansky, D.C. Drucker, G.S. Kino and J.R. Rice, "Pressure Sensitivity of a Lead Optical Fiber", *Applied Optics* 18, pp. 4085-4088 (1979)

Earl D. Squier, "Survey of Hydrophone Acceleration Response", Report 71-19, NTIS Reference AD735931, 1971

Fiber Optic Accelerometer

A. Dandridge

Naval Research Laboratory, Washington, DC 20375, USA

At NRL the primary interest in optical fiber sensors has been the acoustic [1] and magnetic sensor [2]. However, the optical fiber sensor can be configured to act as a sensitive accelerometer with sensitivities of 1 μg or better for sinusoidal accelerations [3]. Although many different configurations of accelerometer are possible, the configurations tested so far have been in the form of a simple harmonic oscillator (sho) consisting of a mass suspended between two fibers as shown in Fig. 1a or a mass suspended from a single fiber as in Fig. 1b. These two configurations are attractive in the sense that they employ the standard fiber Mach Zehnder configuration used in the acoustic and magnetic sensors. Another possible configuration is that of a Michelson interferometer in which the mirrored ends of the fiber (replacing the two mirrors in the bulk device) are attached to the mass as shown in Fig. 1c. The success of the accelerometer is contingent on the following: a) an interferometer of rugged construction with μrad sensitivity, b) a fiber configuration to maximize sensitivity in one preferred direction with a crosstalk (to the other orthogonal directions) of better than 10^{-5}, c) a large dynamic range ($10^5 \rightarrow 10^6$) over which the accelerometer will respond linearly.

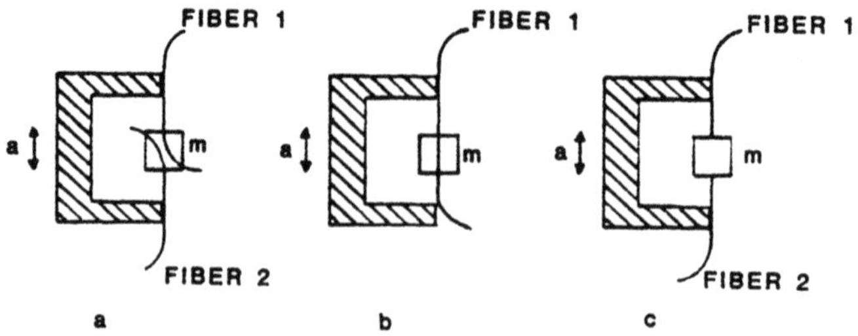

Fig. 1 Different configuration of the fiber optic accelerometer: a) double fiber Mach-Zehnder b) single fiber Mach-Zehnder c) double fiber Michelson

In the laboratory set-ups investigated to date, the acceleration has been only applied to the sensing region of the accelerometer, not to the whole interferometer; however, many applications of this sensor would require the complete device to be subjected to the acceleration. Thus, ruggedization of the interferometer is necessary to insure that a) no spurious signals are induced by the components of the interferometer or b) in extreme cases no damage is caused by large accelerations. Recently there have been advances in ruggedization of the laser fiber pigtail [4], fiber beam splitter [5] and the active homodyne compensation scheme [6]. A typical configuration is shown in Fig. 2.

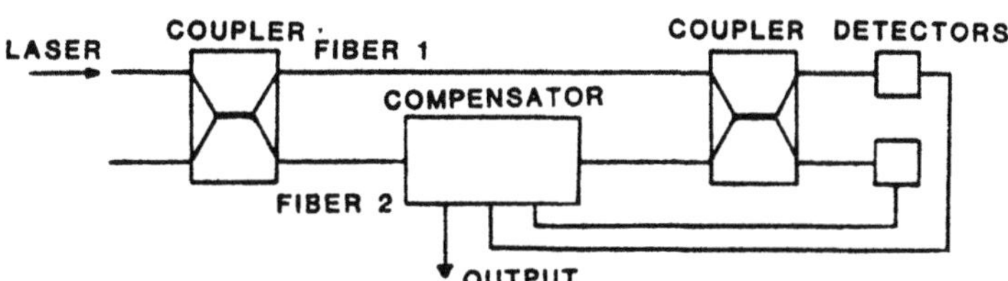

Fig. 2 Typical Mach-Zehnder arrangement used in the configurations shown in Fig. 1a and 1b

Although configurations to date have not been specifically designed to be insensitive to acceleration, it appears that the recent advantages in packaging [6] may allow the interferometer to operate with a minimum detectable phase shift in the μrad range over a broad range of frequencies, in the presence of an acceleration ($\sim$ 1 g). However, interferometer ruggedization to this degree has not yet been completely tested.

Although fiber interferometers (powered by diode lasers) have μrad sensitivities above 100 Hz, it should be noted that many noise sources in these systems have a $(\text{frequency})^{-1/2}$ dependence [7]. Thus at low frequencies the minimum detectable phase shift in the interferometer tends to increase with decreasing frequency. An added problem at still lower frequencies ($\sim$ 1 Hz) is the spurious signals produced by differential thermal drifts between the two arms of the interferometer. While this effect may be reduced by careful packaging, it is expected to present a formidable problem for near d.c. operation of the interferometer.

The use of diode lasers in Mach-Zehnder and Michelson configurations requires accurate matching of the interferometer's paths for good sensitivity. This is not suitable for laboratory set-ups where a number of different configurations of sensor are spliced into the interferometer, tested and then removed. Consequently, all the work presented here was performed using a single mode HeNe laser as the source, which allowed the interferometer to operate with many meters of path difference.

The two configurations shown in Fig. 1a and 1b have been constructed and typical frequency response of the sensitivity (for Fig. 1b) is shown in Fig. 3. The solid horizontal line is the low frequency theoretical sensitivity calculated from

$$\frac{\delta\psi}{a} = 8Lmn/\lambda Yd^2 \qquad (1)$$

where $\delta\psi$ is the induced optical phase shift for the light (wavelength λ), a is the acceleration, d is the fiber diameter, m is the mass, L the fiber length, n and Y are the fiber's refractive index and Young's modulus. The sensitivity of the accelerometer may be increased by adjusting a number of parameters but at the expense of lowering the fundamental mechanical resonance frequency, thus lowering the useful frequency range of the device. In general, the longitudinal vibration of such a device limits its useful frequency range. The longitudinal fundamental frequency for the configuration in Fig. 1b is

$$f_L = \frac{1}{4}\sqrt{\frac{Yd^2}{\pi L m}} \tag{2}$$

or in terms of the accelerometer sensitivity

$$f_L = \sqrt{\left(\frac{n}{2\pi\lambda}\right)} / \sqrt{\left(\frac{\delta\psi}{a}\right)} \quad . \tag{3}$$

The f_L for the configuration in Fig. 1a is a factor of $\sqrt{2}$ larger than for that shown in Fig. 1b. Eq. (3) shows how intimately the fundamental frequency and sensitivity are related. However, in the configuration used to produce the curve shown in Fig. 3, f_L corresponds to a $\sim$ 1200 Hz, a factor of three greater than the resonance observed in Fig. 3. The fundamental resonant frequency of the transverse modes of vibration would, however, produce a signal output in the interferometer of 2 f_T where

$$f_T = \frac{1}{2L}\sqrt{\frac{T}{\mu}} \tag{4}$$

where T is the tension in the fiber and μ the mass per unit length of fiber. The values of f_T ($\sim$ 550 Hz for Fig. 1b) may be raised by using the configuration shown in Fig. 1a; the configuration of 1c would also allow f_T to be raised.

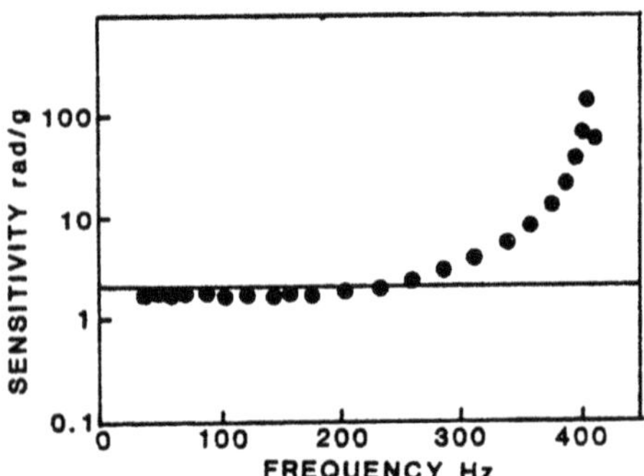

Fig.3. The sensitivity of the accelerometer (Fig.1b) in radians/g (g = 9.81 m/Ω) as a function of frequency. The low frequency theoretical sensitivity is shown by the horizontal line

Each of the three configurations shown in Fig. 1 has advantages and disadvantages. Figure 1a allows f_T to be increased, but the lack of fiber symmetry and the two "loose" fiber ends present a problem. Fig. 1b, although simple, has a low value of f_T and one "loose" end. Although it may be possible to design the housing to effectively remove contributions from the "loose" ends, they are certainly undesirable. The configuration of Fig. 1c has three distinct advantages, namely, elevation of f_T by tension in the two fibers, fiber symmetry and no "loose" ends. Two problems with this configuration are severe; firstly, the Michelson configuration attempts to feed back into the laser half the output of the interferometer. It has been shown previously that feedback above 10^{-7} into the laser cavity can produce line narrowing [8], mode hopping [9], satellite mode generation, multimode behavior (cavity and external cavity modes) and line broadening [10]. Consequently, it will be necessary to isolate the laser from the interferometer to reduce the feedback into the laser by at least four orders of magnitude. The second problem concerns the fabrication of the mirrored ends of the fiber and their mounting on the mass m. All these configurations are being investigated at NRL, but it is clear that best configuration is not self evident.

The cross axis sensitivity is defined as the error in axial accelerations per g, caused by a cross axis acceleration of 1 g. For the single fiber case, the device cannot operate in the presence of large lateral accelerations. However, the value of the cross axis sensitivity for the two fiber case can be calculated (assuming no cross axis constraint is applied). It can be seen that if the tension in both fibers is equal, no phase shift is observed as both arms are lengthened equally. However, if a 1 g longitudinal acceleration occurs simultaneously with the lateral acceleration, an error is introduced. Calculations indicate the error is $\simeq 2 \times 10^{-4}$ (approximately 200 $\mu g/g^2$). To reduce this error lateral displacement of the mass must be reduced to < 20 μm for one g lateral acceleration. This constraint may be provided by a diaphragm, although the initial alignment of the fiber must be to this accuracy. Care must be taken in the design of the diaphragm to insure that the longitudinal motion of the mass is relatively unimpaired by the diaphragm. Care will also be necessary in the tensioning of the fibers either side of the diaphragm to insure linearity of the device.

At NRL we are at the stage of testing the performance of the diaphragms interferometrically to determine their sensitivity. Typically, sensitivities of $\sim$ 0.5 rad/g have been obtained. However, the frequency response contained a number of resonances. It appears that further work is necessary in the mechanical design of the diaphragm so as to allow sub μg sensitivities to be achieved.

The author would like to thank his coworkers A. B. Tveten, C. M. Davis and T. G. Giallorenzi for their help.

References

1 J. A. Bucaro, H. D. Dardy and E. Carome, J. Acoust. Soc. Am. 62, 1302, (1977)

2 A. Dandridge, A. B. Tveten, G. H. Sigel, Jr., E. J. West and T. G. Giallorenzi, Electron. Lett. 16, 408, (1980)

3 A. B. Tveten, A. Dandridge, C. M. Davis and T. G. Giallorenzi, Electron. Lett. 16, 854,(1980)

4 Developed by Laser Diode Labs under contract #N0014-81-C-2259 supported by NAVAIR

5 K. P. Koo, D. C. Tran, S. K. Sheem, C. A. Villarruel and R. P. Moeller, Technical Digest of 3rd Int. Conf. on Integrated Optics and Optical Communication, IOOC '81, San Francisco, CA, April 1981

6 D. A. Jackson, R. Priest, A. Dandridge and A. B. Tveten, Appl. Opt. 19, 2926, (1980)

7 A. Dandridge, A. B. Tveten, R. O. Miles and T. G. Giallorenzi, Appl. Phys. Lett. 37, 526, (1980); A. Dandridge, A. B. Tveten, Appl. Phys. Lett. 38, 530, (1981)

8 S. Saito, Y. Yamamoto, Electron. Lett. 17, 325,(1981); L. Goldberg, A. Dandridge, R. O. Miles, T. G. Giallorenzi and J. F. Weller, Elect. Lett. 17, 677, (1981)

9 R. O. Miles, A. Dandridge, A. B. Tveten, T. G. Giallorenzi and H. F. Taylor, Appl. Phys. Lett. 38, 848, (1981)

10 R. O. Miles, A. Dandridge, A. B. Tveten, H. F. Taylor and T. G. Giallorenzi, Appl. Phys. Lett. 37, 990, (1980)

Part 8

Market Considerations

An Affair to Remember

L.D. Copeland

Northrop Corporation, Precision Products Devision

A distinguished German scientist once said to me - when I inquired as to his progress with a new gyro technology - "ah, but that was last year's affair - now we have switched to a newer technology".

In fact, he scolded me for not being aware of his new interest!

For my part, I could not help remarking that it was time he abandoned these casual flirtations and settled down!

Of course he was indignant. This new technology was the real thing! The others, as the French say, were just "passing affairs".

In my many years in the gyro business I have seen brilliant ideas come and go. My question to you is: "Are you really serious this time, or is this "just another affair?"

Let me explain that my years in the business have been spent doing one thing: trying to convince someone that he needs what you have invented. I am just a humble marketing man.

I have always understood that my job was to find out what it was that the customer wanted and to tell the "creators" - you engineers and scientists - so that you could evolve something to fill the need.

Of course, you and I know that this is not the way things happen in real life. In practice, I am called into the Chief's office, informed about some marvelous new device whispered in the Chief's ear at a cocktail party, and ordered to "get out there and sell it!"

I have often wondered why this was so. How is it that my brilliant but down-to-earth analyses of the market get short shrift when stacked up against the equations and formulas of the engineers and scientists?

Can it be that, like the Indian chiefs of yore, Senior Executives feel the need for the magic and incantations of the Soothsayer and Medicine Man? Is it that they feel reassured in the face of the terrible unknown "Gods of the marketplace", by having a few spells cast in some unintelligible language?

Nor am I at all certain that this process of having hard-boiled businessmen awed by their technical staffs is all that wrong. For without this process we may not have had the micro-processor or the laser and countless other innovations which have been adapted to so many, originally unforeseen, applications.

On the other hand, perhaps we might not have had the "Spruce Goose" either!

Nevertheless, the process which I have described continues. This distinguished gathering is proof that someone has put up a great deal of money to see what the Medicine Men can come up with next!

But perhaps we warriors in the field (ancient warriors I might add) and you Medicine Men can get together on this matter. To begin, I have now heard from you all about your magic (although I understand it not one whit more than the Tribal Chiefs understood the meaning of two eagle feathers and a panther howl); now may I tell you what I know of the trails ahead; of the grasslands and the deserts; of the mountains and the rivers?

In the words of another kind of Medicine Man: let me show you the promised land!

The promised land is money.

I am sorry to appear so crass, but there is really no other way to get at the heart of the matter unless we are willing to talk about Margins, and Return on Investment, and Pricing Structures and Overhead, and all those distasteful things!

(Sadly, it turns out that we battle-wise Warriors and you distinguished Medicine Men are finally at the mercy of those little grey scribblers who tell the Chief how many buffalo we have caught and how many arrows we shot to do it.)

First, the promised land:

If the fiber-optic gyro can meet certain specifications by a certain date, there is a potential market of one billion dollars beginning about 1987.

An inertial guidance system for tactical missile type applications, that is to say performance of about 1°/hour, employing fiber-optic or laser gyro sensors, can compete in this market. Pay close attention to what I'm saying - it will compete in a market of more than $1.0 billion, it does not have that market yet nor is it necessarily assured of any of the market if it fails to meet the specifications! These are not technical specifications. Rather they are defined by a marketplace far more concerned with profit than exotic technology.

This market can be segmented any number of ways. It is 50% air-to-ground missiles, 25% ground-to-air and ground-to-ground, 15% tracked vehicles, 10% air-to-air. Mind you, these are only those programs not currently committed to other technology. The potential and percentages would increase and change as the laser-based system gains credibility and attractiveness. This will cause currently "committed" programs to reevaluate their commitment and consider the laser as a better solution to their requirement.

The market is an international one. I estimate it to be 40% North America (mostly U.S. of course), 25% Europe, 20% East Asia and 15% Middle East. Again, arms exports are overwhelmingly from the "committed technology" programs. As commitments change, so will percentages.

These are "tactical" requirements, which differ from strategic applications such as ICBMs principally in terms of performance. There are over one hundred individual programs contributing to the potential. These numbers are, furthermore, conservative if anything. Product inertia and market synergies will probably yield even greater potential.

Are there other markets than these military applications? Of course there are, just as digital watches and pocket calculators were not even dreamed of by inventors and developers of the transistor. What are they? They are the "UNK-UNKS", the unknown-unknowns which are truly the business of Medicine Men and Lucky Geniuses. I am definitely not a lucky genius.

Now what is it the fiber-optic gyro must do in order to even be considered for these tactical markets? The following is my projection of specifications for a three axis inertial guidance system:

Size:	less than 5" diameter
Weight:	less than 4#
Performance:	about 1 degree/hour, 3 sigma day-to-day
Price:	less than $10,000

Now, if you do what I have cited as requirements, and if you do it on time, what are the chances of the fiber-optic gyro succeeding?

First there is competition. Much as we are here today a band of brothers, joining together in the search for truth, tomorrow we will be back at our respective companies scheming to do each other in.

In other words, if the fiber-optic gyro succeeds beyond all imagining -- if it wipes out the competing technologies -- there will always be at least two fiber-optic gyros -- the marketplace will see to that!

Therefore, for starters, you can divide the promised land by 2.

Secondly, what about the competing technologies? Are they going to meekly lie down and die?

Well, if we go back to the specifications which I presented earlier I can tell you that there is a technology which exists now; a proven technology, flight tested in the field, which meets every single one of these specifications except price, and it is not far off in this parameter.

In other words, in my simple shorthand, the only thing you have to offer the marketplace is price, and the gap in price between you and the existing, proven technology is not that large!

So, if you do all that you dream of, you had better divide the remaining portion of the promised land by another 2!

I am reminded of the old story about a parlor game in which a lovely young actress was asked if she would sleep with a total stranger for a million dollars. When she replied in the affirmative, she was then asked if she would do so for five dollars? "Of course not," she replied indignantly, "What do you think I am?"

"We know what you are," came the rejoinder, "we're only trying to determine the price!"

And the question of price brings up nasty little items like: Return on Investment. For example, if you achieve the price I have indicated is what the market will pay, what is the profit margin? What is the margin if we amortize the past investment? For that matter, have you calculated what the investment is already? What it will be?

Someone will have to do it.

If, and when, you do all these sums, you may conclude that we would be better off to stick with Money Market Funds!

Personally, I don't think this will happen -- that your project (no matter what the figures show) will be abandoned for some mercenary pursuit like making money! Chief Executives are just too enamored of spells and incantations.

Finally, I would like to offer you a practical suggestion. It has something to do with the old advertising adage about "selling the sizzle, not the steak," and even though you are selling a concept, it still applies.

What you need is a catchy name.

The name "fiber-optic gyro" simply doesn't have it, and the abbreviation "FOG" is worse.

I would like to suggest that henceforward it be called the "FLOOGIE" (which stands for Fiber Laser Optical Orthogonal Gyro Interferometer Equipment).

I thank you for patiently listening to me. I do not know what will become of the "FLOOGIE." I only hope that like this pleasant evening, it will be:

"AN AFFAIR TO REMEMBER."

Inertial System Market Potential

R.G. Brown

R.G. Brown Associates, Inc., P.O. Box 74
Topsfield, MA 01983, USA

Growing Role Predicted for Inertial Technology

In recent years the DoD has placed increased demands on the inertial industry to provide more reliable, higher performance, smaller and less expensive inertial components and systems. The inertial industry of the free world responded to this challenge along two basic paths, low risk, innovative spinning wheel gyro development and the more high risk solid state approach to sensor development. The active and passive ring laser gyroscopes are typical of this latter category.

A recent comprehensive study by R.G. Brown Associates, Inc., forecasts an increased application rate of inertial equipment to all types of military weapons and expanded utilization in the commercial world. Figure 1 depicts the large U.S. inertial system market currently available to industry suppliers. The projected annual procurement of inertial systems is conservatively predicted to exceed $650 million annual sales.

Competing Gyro Technologies

The forecast for the 1980-1995 time period includes eight major market segments. The military segments are: aircraft, tactical missiles, strategic missiles, land vehicles and weapons, naval weapons and space systems. The commercial portion includes aircraft and marine applications of inertial devices and systems.

Table 1 identifies the various gyro technologies expected to compete within each market segment.

Impact of Mature Gyro Technologies

This category covers those gyro approaches regarded as conventional or traditional inertial sensors. These include single degree of freedom (1DOF) floated or unfloated gyros, two degree of freedom (2DOF) unfloated

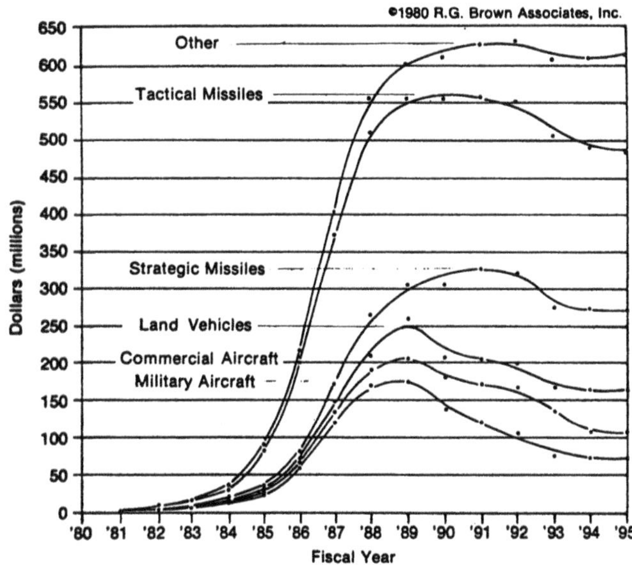

Figure 1: Available Inertial System market penetration (uncommitted market).

Table 1: Competing gyro technologies (1980-1995).

Segment	Candidate Technology
Military Aircraft	1DOF, 2DOF, ESG, RLG
Tactical Missiles	1DOF, 2DOF, FG, RLG, Passive Optical Rate Sensor (PORS), Nuclear Magnetic Resonance Gyro (NMR), SONIC, SCAG
Strategic Missiles	1DOF, 2DOF, RLG
Naval, Marine, Commercial and Space	1DOF, 2DOF, RLG, PORS, NMR, ESG
Commercial Aircraft	1DOF, 2DOF, RLG, PORS, NMR
Land Vehicles/Weapons	1DOF, 2DOF, RLG, PORS, NMR, SCAG, SONIC

or dry-tuned gyros, the electrostatic (ESG) gyro, the free gyro (FG) and the spin-coupled accelerometer gyro (SCAG) and other such multisensor approaches.

The 1DOF gyro continues to mature in development. Conventional 1DOF gyros, probably employing new materials and manufactured by automated means, will continue to satisfy many needs where small size, weight and low cost are dominant considerations. This will be particularly true where performance requirements allow the utilization of 1DOF gyros in strapdown configurations rather than the more familiar gimballed systems.

Probably the most exciting new development in the traditional gyro category are the two degree of freedom (2DOF) dry-tuned gyro. Many U.S. and foreign companies now include 2DOF gyro in their product lines. The gyro's application is ideal where size, weight, fast reaction time and medium cost are important. The future application of new materials and automated manufacturing techniques can improve the 2DOF gyros' ability to compete in lower cost markets such as tactical missiles, land vehicles and weapons, commercial aircraft and marine application.

Impact of Advanced Concept Gyro Technologies

The advanced concept gyro category includes the ring laser gyro (RLG), the passive optical-rate sensor (PORS), the nuclear magnetic resonance gyro (NMR), the cryogenic gyro (CG) and the sonic gyro. Of these, the ring

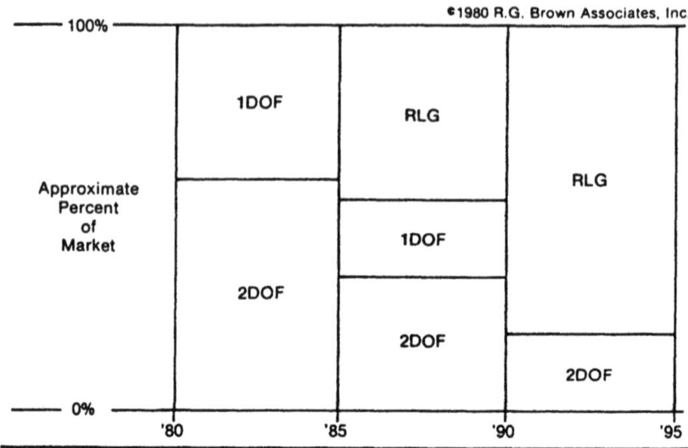

Figure 2: Probable gyro technology market share - military aircraft segment - 1mi/h.

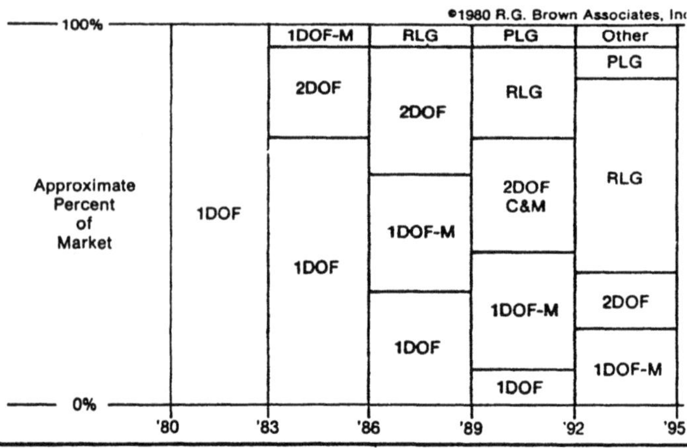

Figure 3: Probable gyro technology-tactical missile-market sharing.

laser gyro appears to offer the greatest potential for satisfying the widest range of future market requirements. Recent rapid progress in laser gyro technology has attracted increased DoD interest and resulted in rekindling of the entire inertial industry activities in this field.

Figures 2 and 3 show the predicted impact of the advanced concept and improved traditional gyroscopes for two of the major markets, military aircraft navigation systems and tactical missile guidance and control.

In the near-term tactical missile market, conventional single degree of freedom rate and 2DOF gyros will tend to prevail because of the ever-present demand for low-cost and small-size interial sensors. Figure 3 shows the expected penetration in this weapon area by the two degree of freedom gyro and molded plastic versions of 1DOF and 2DOF instruments. It isn't until the mid '80s that laser gyro is predicted to significantly contribute to the tactical missile market. However, breakthroughs which lead to early cost and size reductions in the laser gyro could greatly accelerate the entry of these solid state devices into this large and demanding market-place.

Figure 3 reflects the predicted gradual increase in laser gyro applications.

At this time the laser gyroscope appears to be the most promising new gyro concept candidate for replacing its spinning wheel counterpart. The maximum expected penetration of the available market is shown in Figure 4. More than 50% of future inertial needs are expected to be satisfied by the normal evolution of generations of laser gyroscopes. Initially, navigation, guidance and control needs in the U.S. will be met by what are now conventional, dithered HeNe ring laser gyroscopes of Honeywell, Sperry and Litton. Eventually, more advanced laser gyro techniques, such as the multi-oscillator and passive optical rate sensors, can be expected to evolve to meet a broad spectrum of systems requirements.

The increased emphasis and concentration of the free worlds' technical capabilities for developing passive (or fiber optics) approaches to angular rotation sensors should accelerate the application of these devices. On

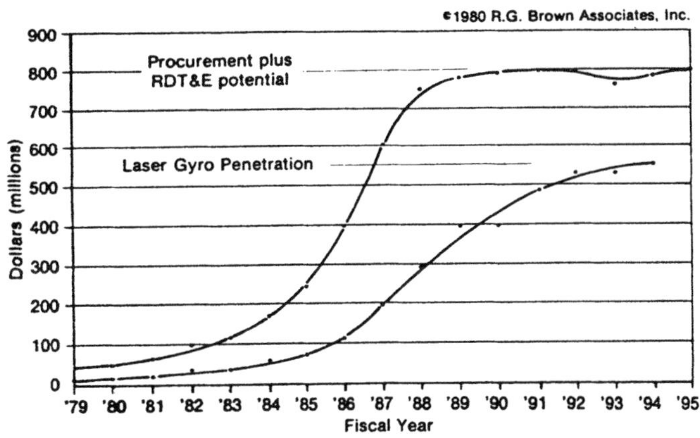

Figure 4: Total INS and laser gyro penetration

the other hand, the potential user industry, which range from land navigators and aircraft flight control systems to man portable anti-tank weapons will require that the inertial sensors be competitive in cost. (In many perceived future applications the PORS are expected to be so low in cost as to be essentially throw-away items.) Systems and weapons which employ passive optical rotation sensors will demand low power operation, long shelf life, and full mil-spec environment capability.

Index of Contributors

Subject Index

B. Saleh
Photoelectron Statistics
With Applications to Spectroscopy and Optical Communication
1978. 85 figures, 8 tables. XV, 441 pages
(Springer Series in Optical Sciences, Volume 6). ISBN 3-540-08295-6

"... The material described in the book represents the work of many authors which have been widely scattered throughout the literature in various technical journals and books. Dr. Saleh has made a major contribution to the field by gathering most of the pertinent results in a single book and presenting the material concisely, correctly, and with a single consistent notation ..."
Applied Optics

Vertebrate Photoreceptor Optics
Editors: **J. M. Enoch, F. L. Tobey, jr.**
With contributions by numerous experts
With a Foreword by W. S. Stiles
1981. 164 figures. XV, 483 pages
(Springer Series in Optical Sciences, Volume 23). ISBN 3-540-10515-8

Contents: Introduction. - The Retinal Receptor: A Description. - The Stiles-Crawford Effects. - Retinal Receptor Orientation and Photoreceptor Optics. - Waveguide Properties of Retinal Receptors: Techniques and Observations. - Theoretical Considerations of the Retinal Receptor as a Waveguide. - Theoretical Consideration of Optical Interactions in an Array of Retinal Receptors. - The Visual Receptor as a Light Collector. - Microspectrophotometry and Optical Phenomena: Birefringence, Dichroism, and Anomalous Dispersion. - Tapeta Lucida of Vertebrates. - A Comparison of Vertebrate and Invertebrate Photoreceptors. - Additional References with Titles. - Subject Index.

Springer-Verlag
Berlin
Heidelberg
New York

A. B. Sharma, S. J. Halme, M. M. Butusov
Optical Fiber Systems and Their Components
An Introduction
1981. 125 figures. VIII, 246 pages
(Springer Series in Optical Sciences, Volumes 24). ISBN 3-540-10437-2

... "A single slim book alone could not possibly solve this problem, but this one, by providing lucid discussion of essential matters, a unified viewpoint over a wide range and a uniform mathematical treatment, certainly provides indispensible help which will be appreciated by hardpressed lecturers in this important area. ...
Physics Bulletin

Lasers and Applications
Proceedings of the Sergio Porto Memorial Symposium, Rio de Janeiro, Brasil, June 29 - July 3, 1980
Editors: **W. O. N. Guimaraes, C.-T. Lin, A. Mooradian**
1981. 200 figures. IX, 337 pages
(Springer Series in Optical Sciences, Volume 26). ISBN 3-540-10647-2

"... It is an excellent general information book for the scientist and professional using lasers, as well as for the graduate or advanced undergraduate student who is seriously interested in the field of optics and lasers."
Applied Optics

R. G. Hunsperger
Integrated Optics Theory and Technology
1982. 167 figures. XIV, 299 pages
(Springer Series in Optical Sciences, Volume 33). ISBN 3-540-11667-2

Contents: Introduction. - Optical Waveguide Modes. - Theory of Optical Waveguides. - Waveguide Fabrication Techniques. - Losses in Optical Waveguides. - Waveguide Input and Output Couplers. - Coupling Between Waveguides. - Electro-Optic Modulators. - Acousto-Optic Modulators. - Basic Principles of Light Emission in Semiconductors. - Semiconductor Lasers. - Heterostructure, Confined-Field Lasers. - Distributed Feedback Lasers. - Direct Modulation of Semiconductor Lasers. - Integrated Optical Detectors. - Applications of Integrated Optics and Current Trends. - References. - Subject Index.

www.ingramcontent.com/pod-product-compliance
Ingram Content Group UK Ltd.
Pitfield, Milton Keynes, MK11 3LW, UK
UKHW021859190726
13853UKWH00003B/1342

* 9 7 8 3 6 6 2 1 3 5 2 6 6 *